# Restless Genius

# RESTLESS GENIUS

ROBERT HOOKE AND HIS EARTHLY THOUGHTS

**Ellen Tan Drake**

New York   Oxford
OXFORD UNIVERSITY PRESS
1996

Oxford University Press

Oxford   New York
Athens   Auckland   Bangkok   Bogota   Bombay
Buenos Aires   Calcutta   Cape Town   Dar es Salaam
Delhi   Florence   Hong Kong   Istanbul   Karachi
Kuala Lumpur   Madras   Madrid   Melbourne
Mexico City   Nairobi   Paris   Singapore
Taipei   Tokyo   Toronto

and associated companies in
Berlin   Ibadan

Copyright © 1996 by Ellen Tan Drake

Published by Oxford University Press, Inc.
198 Madison Avenue, New York, New York 10016

Oxford is a registered trademark of Oxford University Press

All rights reserved. No part of this publication may be reproduced,
stored in a retrieval system, or transmitted, in any form or by any means,
electronic, mechanical, photocopying, recording, or otherwise,
without the prior permission of Oxford University Press.

Library of Congress Cataloging-in-Publication Data
Drake, Ellen T.
Restless genius : Robert Hooke and his earthly thoughts / Ellen Tan Drake.
p.   cm.
Includes bibliographical references and index.
ISBN 0-19-506695-2
1. Hooke, Robert, 1635–1703.   2. Geologists—England—Biography.
3. Hooke, Robert, 1635–1703—Contributions in geology.
4. Hooke, Robert, 1635–1703—Influence.   5. Earthquakes.   6. Volcanism.
7. Hooke, Robert, 1635–1703.   Lectures and discourses of earthquakes and
subterraneous eruptions.   I. Hooke, Robert, 1635–1703.
Lectures and discourses of earthquakes and subterraneous eruptions.   II. Title.
QE22.H76D74   1996
550'.92—dc20
[B]   95-42703

1 3 5 7 9 8 6 4 2

Printed in the United States of America
on acid-free paper

To my husband Charles
and all our children and grandchildren

# PREFACE

Many historians of science understandably have an innate distrust of histories written by scientists who have not been trained in the methods and principles of history. On the other hand, many scientists, especially those nearing retirement, just love to reflect on the history and development of the field in which they have spent their entire professional life. Some histories written by such scientists are very amateurish in the eyes of historians, while others are rather good. In geology, I find that some very insightful histories have been written by geologists. Geology, after all, is a science that requires a historical outlook—searching out what must have occurred in the past from evidence written in the rocks themselves and establishing a sequence of events and a chronology. Organizations such as the History of Geology Division of the Geological Society of America (GSA) have promoted the works of both geologists and historians of geology. The same is true of the History of Earth Sciences Society (HESS), started by geologist Gerald Friedman, which is publishing a journal of international stature, with members from both history and geology. This is the kind of interaction between scientists and historians I feel is needed. Too often, however, I hear the opinion that historians and scientists intrinsically cannot communicate. I have even heard the analogy that as a biologist does not interact with the bacteria he or she sees at the end of a microscope, so a historian should not interact with scientists! This attitude surely is counterproductive, for we are all humans together endeavoring to seek truth each in our own way.

Robert Dott, in the September 1990 issue of the GSA History of Geology Division Newsletter, characterized the internalist-externalist tension that is creating two cultures in the history of geology. Internalists are scientists writing about the history of their science, primarily interested in the origins and development of the salient ideas in their subject and their authors, while externalists are historians, philosophers, or sociologists of science, concerned with placing the science and the scientist in the context or framework of a particular period in history. While each group may have interest in the research results of the other, a suspicion exists that somehow the work of the other group lacks legitimacy. The internalist, whether justifiably or not, feels that in order to write the history of a science, one needs to have understanding of not only where that science was at some period in time but also its development through time. This researcher wishes to know whether an idea is significant in light of later development to assess its influence and therefore importance, on the progression of that science to the present day. In order to do this, one needs to have knowledge and understanding of the science itself, not just the state of that science during some snapshot in time. The externalist, on the other hand, feels that the domain of the history of science belongs to historians or other social scientists interested in the historical-sociopolitical and intellectual milieu and ramifications of a science. Dott, therefore, emphasized the importance of "bilingualism"—in this case, being versed in both fields, history *and* science. Although this state is desirable, usually it can be achieved only by scholars who have the time to devote to the learning and practice of a second field. Cooperative research is one way to solve the dilemma. A GSA Penrose Conference to discuss the insider/outsider dilemma in the discipline of the history of earth science was held March 1994 in San Diego, California. At least a clarification and some comprehension on both sides, if not a resolution, came out of this conference.

Through a series of lucky accidents, I actually find myself in an integrated internalist-externalist microcosm of my own and feel equipped to embark upon writing this volume. As a geologist, on the one hand, I was curious about the geology of the one place in the world that Robert Hooke knew best, his birthplace, the Isle of Wight. I spent a delightful few days on that beautiful island with my physicist husband and our daughter Linda (also a geologist), traipsing up and down its cliffs and ferreting out its geological history. I saw firsthand what Hooke himself, both as a child and later as an adult when he revisited his birthplace, saw and marveled at in the cliffs of his island. Being able to interpret the geologic events that took place, I could almost sense what must have impressed the highly imaginative Robert Hooke: How came these marine shells there sixty feet above the high-water mark? What forces tilted up these layers of sediment laid down horizontally by the sea, so that they are now vertical? And lo, and behold, here are some layers that seem to have been even *over-turned*! I realized then that, without a doubt, the wheels of Hooke's mind turned on earth science because of his intimate knowledge of his birthplace. My own knowledge of geology and my geologically-trained inclination to work *in the field*, therefore, provided me with this added dimension in studying Hooke. Few historians, in my opinion, would have this urge to participate as a scientist in their research, nor would they think that such activity would be particularly desirable.

By the same token, I am grateful for the training I received among the historians when I was a doctoral student in the history of science, having embarked on research in that field before I switched, for personal reasons, to the field of oceanography (marine geology). Applying the principles of history of science to my study of Hooke's earthly thoughts has enriched my outlook. They showed me that while Hooke was undoubtedly an extraordinarily remarkable man, one who accomplished more than most ordinary people can in several lifetimes, a true founder of the science of geology, he was both product and victim of his time. And while he showed great originality in breaking away from much of the medievalism that still prevailed in the 17th century, he was at the same time frustratedly locked in by the tenets of his age. The great stumbling block in his thinking was the element of *time*; while he knew instinctively that the biblical chronology could not have served all that he witnessed—a giant step in itself— he could not fathom the billions of years of geologic time, or at least there was no explicit evidence that he did. Modern geology was possible only when James Hutton, almost a century later, essentially ruled out time altogether in his famous maxim, "No vestige of a beginning; no prospect of an end." To understand why Hooke could not have made this important break, one must understand the socio-politico-religious milieu of the 17th century in which he lived—that is, the domain of history.

I am hopeful, therefore, that my approach to writing this volume, focusing on Hooke's earthly thoughts, is a marriage of both traditions of epistemology, that of history and that of science, and will be of interest to professionals in both, as well as their students.

## ACKNOWLEDGMENTS

Many people have helped me in this effort. Within the College of Oceanic and Atmospheric Sciences at Oregon State University (OSU), I shall always be grateful to my former thesis adviser and now colleague, Paul D. Komar, who excited the interest in me to research Hooke. In an atmosphere of reductionist science I find it most gratifying to be able to converse with him about our seventeenth-century savant. I thank the support of G. Ross Heath, Jack Dymond, Larry Small and Nick Pisias who believed that my maverick research interests deserve a place under the sun in a highly technical university besieged with budgetary problems. I also acknowledge with thanks OSU professors Robert Duncan, Shaul Levi, and Oberlin emeritus professor Robert Weinstock, all of whom read parts of my manuscript and offered helpful suggestions and constructive criticism. OSU professor of history of science Paul Farber offered encouragement. Craig Wilson, the then assistant director of the OSU Kerr Library, provided me with letters of introduction to libraries in England. Also, the reference librarians at the Kerr Library, notably Robert Lawrence, were always willing to help in locating a source in spite of great turmoil in the library itself while it converted to an on-line system. Barbara McVicar, word-processing specialist, gave me kind advice and assistance throughout the preparation of the manuscript, as has Sue Pullen, secretary in Marine Geology. David Reinert's assistance in photographing illustrations and in design was very helpful. June Wilson, faculty research assistant in oceanography, at my request, went out of her way to take pictures of the "Worthies" window at St. Helen's Church, Bishopsgate, London, while she was vacationing in Britain. I thank my contacts at Oxford University Press in New York: Joyce Berry, acquisition editor, Helen McInnis, vice president, and especially Cynthia Garver, production editor, for bringing the project to a successful finish. I thank the editors of the *American Journal of Science, Journal of Geological Education,* and *History of Earth Sciences Journal* for allowing me to draw freely from my own articles published in these journals.

Most of the research assistance, of course, I had to receive from wonderfully cooperative people in England itself. Such friendly hospitality to a foreign researcher cannot go without acknowledgment with grateful thanks and appreciation; I mention Mr. William G. Hodges, superintendent of the reading rooms, Duke Humphrey Library at the Bodleian Library in Oxford University, who brought me rare and valuable original manuscripts, even on a Saturday when such manuscripts are ordinarily not to be seen; Mr. Ron Cleeveley of the department of paleontology, British Museum (Natural History), London; Mrs. H. E. Weir, librarian, Palæontology/Mineralogy Library, the Natural History Museum; Mr. John Fisher, deputy keeper of prints, and other helpful staff at the manuscripts division of the Guildhall Library, London, where I read Hooke's *Diary* in his own inimitable hand; Miss Norma Aubertin-Potter, librarian, All Souls College, Oxford University, who brought out for me the original design of Christopher Wren for the London Monument even though she was officially on leave that week to work on her own research; Mr. John Cooper, department of palæontology, British Museum (Natural History); Mr. Bruce Wilcock, geology editor, Oxford University Press, Oxford, who provided me with entrées into various libraries and museums; Mr. John Wing, librarian, Christ Church College, Oxford University, Oxford; Mr. Clifford Stephen Marcus (M.A. in classics, Oxford University) who helped to translate and transliterate some of the Greek passages in the *Discourse of Earthquakes*; Mr. Reginald A. Mills, whom I met at the Prince of Wales tavern in Oxford, for helpful conversation concerning some of the Latin.

Finally, I wish to devote a special paragraph to the following special people: Mr. Alan J. Clark, deputy librarian of the Royal Society of London, was extraordinarily generous in giving me of his time, expertise and scholarship. Being allowed to browse through all the unpublished Hooke manuscripts I wished that were reposited in the Royal Society was an experience I shall treasure always; Miss Sandra Cumming, information officer of the Royal Society Library, adopted the same generous spirit shown by Mr. Clark. My work at the Royal Society was most productive and enjoyable. Through Mr. Clark, also, I was introduced to the distinguished engineer Dr. Edmund C. Hambly, now deceased, who had a special interest in Hooke's engineering contributions, and Mr. Trevor Clarke, retired engineer and deputy archivist for the Church of All Saints, Freshwater, Isle of Wight, the church where Hooke's father was curate. I spent pleasant afternoons with both of these gentlemen. Mr. Clarke guided us on what he called the "micrographia trail" in "Hooke country." Thanks are due our daughter, Linda Drake Neshyba, who helped me assess the geological history of the western cliffs of the Isle of Wight, and to her chemist husband, Steven, who advised me on Hooke's chemical experiments. Our daughter Judy, technical writer, read the manuscript and, with our lawyer son Robert, gave me general advice and moral support. Last, I wish to express my thanks to and appreciation of my physicist husband of over forty-three years, Professor Charles Whitney Drake Jr., who, in spite of a traditional male-chauvinist upbringing characteristic of his time, always feels that my intellectual pursuits are also important; he supports me in all my endeavors, assisted me during our many weeks of research in England, critically read drafts of my chapters, and patiently and helpfully listened to my prepared talks prior to professional meetings.

Corvallis, Oregon  E. T. D.
December 18, 1995

# CONTENTS

Illustrations and Credits, *xii*

Introduction, *3*

### I. Robert Hooke's Life and Work

1. The Life of Robert Hooke, *9*

2. The Isle of Wight and Its Influence on Hooke's Earthly Thoughts, *60*

3. Other Theories of the Earth, *69*

4. Hooke's System of the Earth, *77*

5. Hooke's Concept of Polar Wandering on an Oblate Spheroid Earth, *87*

6. Hooke's Theory of Evolution and Attitude toward God and Time, *96*

7. Plagiarism or Paranoia?, *104*

8. Final Assessment, *112*

Notes, *129*

Bibliography, *137*

### Part II. Hooke's *Discourse of Earthquakes* and *Subterraneous Eruptions* (1667–1694)

Introduction by E. T. Drake, *155*

Hooke's *Discourse of Earthquakes*
Transcribed and Annotated by E. T. Drake, *159*

Appendix A, *367*

Appendix B, *372*

Index, *375*

# ILLUSTRATIONS AND CREDITS

### PART I. ROBERT HOOKE'S LIFE AND WORK

1-1. A view of Hook Hill and All Saints Church. (Reproduced with the kind permission of the Reverend Brian Banics, Rector of Freshwater, Isle of Wight.)

1-2. Willen Church. (Reproduced with the kind permission of the Reverend Ian Jagger, Team Vicar of Milton Keynes.)

1-3. Plaque on the wall of University College in High Street, Oxford. (Photograph by C. W. Drake.)

1-4. Hooke's pneumatic engine or air pump. (From Gunther, 1930, vol. 6.)

1-5. Hooke's gas pressure/volume apparatus. (From *Micrographia* 1665.)

1-6. Gresham College. (From Gunther, 1921, vol. 1.)

1-7. Facsimile reproduction of Hooke's sketch of a sounding gear without a line and a way to sample sea water at any depth. (Reproduced with the kind permission of the President and Council of the Royal Society, London.)

1-8. Title-page of *Micrographia* (1665.)

1-9. Hooke's dedication in *Micrographia* (1665.)

1-10. Hooke's sketch of his microscope. (From *Micrographia* 1665.)

1-11. Hooke's depiction of microfungi in *Micrographia* (1665).

1-12. Hooke's drawing of the structure of cork in reference to which he coined the term "cells." (From *Micrographia*, 1665.)

1-13. Hooke's drawing of a fly. (*Micrographia*, 1665.)

1-14. Hooke's illustration of clusters of the compound eyes of flies. (From *Micrographia*, 1665.)

1-15. Hooke's depiction of a louse holding a human hair. (From *Micrographia*, 1665.)

1-16. Hooke's illustration of interference. (From *Micrographia*, 1665.)

1-17. Hooke's drawing of part of the lunar surface. (From *Micrographia*, 1665.)

1-18. The Monument, London, designed and built by Hooke commemorating the London Fire of 1666. (Photograph by P. D. Komar.)

1-19. One of Wren's designs for the London fire monument. (Reproduced with the kind permission of the Warden and Fellows of All Souls College, Oxford.)

1-20. Hooke's drawing of the gates and dome of the Royal College of Physicians. (From Gunther, 1930, vol. 7.)

1-21. Bethlehem Hospital designed and built by Hooke 1675. (From Gunther, 1930, vol. 6, frontispiece.)

1-22 Montague House designed and built by Hooke 1676. (From Gunther, 1930, vol. 7, frontispiece.)

1-23. Ragley Hall designed and built by Hooke 1680. (Photograph by E. T. Drake.)

1-24. Some of Hooke's meteorological instruments: barometer, hygrometer, and wind gauge. (From Gunther, 1930, vol. 6.)

1-25. Telescope set-up through two floors of Hooke's lodgings in Gresham College. (From Gunther, 1930, vol. 7.)

1-26. Hooke's universal joint. (From Gunther, 1930, vol. 7.)

1-27. Hooke's diagrams showing the principles of reflection and refraction. (From Cutlerian Lecture *Lampas,* 1677.)

1-28. Hooke's drawing of balance wheels for watches. (From *Lampas,* 1677.)

1-29. Hooke's sketch of felt-making. (From Andrade, 1950.)

1-30. Hooke's diagram of the Comet of 1677. (From Cutlerian Lecture *Cometa,* 1678.)

1-31. Hooke's Law. (From lecture *Of Spring*, 1678.)

1-32. Hooke's optical telegraph. (From Gunther, 1930 vol. 7.)

1-33. Illustration of Hooke's lecture on the Chinese abacus and the Chinese language. (From *Philosophical Transactions*, no. 180, 1686.)

1-34. Robert Boyle. (From an engraving by R. Woodman in *Gallery of Portraits*, 1833.)

1-35. Representation of Hooke in the "Worthies" window in St. Helen's Church, Bishopsgate, London. (Photograph by June Wilson.)

1-36. Facsimile reproduction of Hooke's list of things accomplished. Undated. (Reproduced with the kind permission of the President and Council of the Royal Society, London.)

1-37. Commemorative block on Hook Hill, Isle of Wight. (Photograph by C. W. Drake.)

2-1. Outline map of Isle of Wight. (Drawing by E. T. Drake.)

2-2. Western end of the Isle of Wight showing the Needles. (Photograph by C. W. Drake.)

2-3. Vertical Eocene beds in Alum Bay. (Photograph by C. W. Drake.)

2-4. Hooke's drawing of a foraminifer. (From *Micrographia*, 1665.)

2-5. Part of the cliff on the southwest coast of Isle of Wight. (Photograph by E. T. Drake.)

2-6. A typical fossiliferous rock along the southwestern coast of Isle of Wight. (Photograph by E. T. Drake.)

2-7. Massive white chalk showing flint nodules, Isle of Wight. (Photograph by E. T. Drake.)

2-8. The effect of dissolution on the chalk, Isle of Wight. (Photograph by E. T. Drake.)

2-9. Roman road shows extent of erosion in Freshwater Bay, Isle of Wight. (Photograph by E. T. Drake.)

2-10. Huge blocks that broke off the main island, Isle of Wight. (Photograph by E. T. Drake.)

3-1. Frontispiece of Thomas Burnet's book *The Sacred Theory of the Earth*, 1691.

3-2. Burnet's figures illustrating break-up and collapse of Earth's crust. (From *The Sacred Theory of the Earth*, 1691.)

3-3. William Whiston's illustration of the cause of the Deluge by a collision of a comet with Earth. (From *A New Theory of the Earth*, 1696.)

5-1. Diagram showing that on the surface of an oblate spheroid, the ground distance subtended by one degree between two radii of curvature is greater near the poles than near the Equator. (Drawing by E. T. Drake.)

5-2. Wegener's equipment used in demonstrating his "pole-flight" force. (Reproduced with the kind permission of Friedr. Vieweg und Sohn, from the 4th revised edition of *Die Entstehung der Kontinente und Ozeane*, 1929, Braunschweig.)

5-3. Diagrams showing the relationship among continental drift, polar wandering, and biogeography. (Drawing by E. T. Drake.)

7-1. John Ray. (From an engraving by H. Meyer in *Gallery of Portraits*, 1833.)

8-1. Illustration of a shark's head and teeth used by Steno in his report on dissecting a shark. (From *Canis Carcharia Dissectum Caput*, 1667.)

8-2. Steno's six aspects of Tuscany. (From *Prodromus*, 1669.)

8-3. Steno's diagrams of crystals. (From *Prodromus,* 1669.)
8-4. Hooke's diagrams of crystals. (From *Micrographia,* 1665.)
8-5. An eighteenth century caricature of Hutton (From Wright, 1865.)
8-6. Title page of *The History and Philosophy of Earthquakes* (1757.)
9-1. Hooke's four effects of earthquakes. (Chart by E. T. Drake.)

**PART II. HOOKE'S *DISCOURSE OF EARTHQUAKES***

Hooke's plates:

Tables I, II, III, IV, V

Appendix A. Waller's plates:

Tables VI, VII

# RESTLESS GENIUS

# Introduction

*As he is of prodigious inventive head,
so is a person of great virtue and goodnes.*

—John Aubrey (1626–1697)

The few geologists who have read Robert Hooke's *Discourse of Earthquakes* have been astonished by his almost clairvoyant postulations concerning the formation of geomorphological features, the origin and usefulness of fossils, biological evolution, and all the dynamic changes that constantly take place on this planet. Hooke's ideas have been shown to reach beyond those of his contemporary, the Dane Nicholas Steno (1638–1686), who is widely hailed as a founder of geology. There is also evidence that James Hutton (1726–1797), who had some knowledge of Hooke's hypothesis of the terraqueous globe, incorporated much of what Hooke pronounced into his own theory, a theory that laid the foundation of modern geology. Hooke was therefore important in the history of Earth science, as he was in the development of many other fields of scientific and technological endeavour. The purpose of this volume is not to engage in a search for Hutton's precursors but to present the ideas of Robert Hooke concerning the Earth and to analyze them in light of the prevailing knowledge in the 17th century, placing the man in historical context and revealing his vital role in the development of the science of geology.

The story of Hooke's life as recounted in Chapter 1 shows him to be a restless genius whose significant influence in the development of many fields of science is still felt today. That such a giant figure in history should be so obscure in reputation today underscores the peculiar irony of his circumstance. He undoubtedly suffered in the shadow of Isaac Newton. In spite of much research over the last several decades detailing Hooke's contributions that counter the handed-down history, our received opinion has not changed very much. This situation is especially true with respect to the universal law of gravitation to which Hooke contributed in a profound way. But legends regarding Newton and his apple persist in both children's books and text books for higher learning, so that Hooke's importance is often summarily dismissed. Sadly, such seeming avoidance of the truth gives some credence to the postmodernists' viewpoint that most histories are fatally flawed because of the subjectivity and selectivity of the authors, and that our very methodology and the inadequacy of historical documents result in the publication of a great deal of fiction, particularly when individuals like Newton are adept in manipulating their own historical images, often to the detriment of others. Historians and other writers, in propagating the received opinion, have tended to withhold credit to Hooke for many of his insightful ideas and conclusions. They criticize him for claiming more credit than he deserved, then intimate that he sometimes did not realize the significance of his own inventions and complain that he left things unfinished. They even find fault with his character and physical appearance.

No portrait or likeness of Hooke exists today, although the evidence shows that in line with custom there should have been a portrait in the Royal Society, where he was curator of experiments and secretary ('Espinasse, 1956; Diary). The description of Hooke most commonly referred to is by Richard Waller, Hooke's biographer and the editor of Hooke's *Posthumous Works* (1705, p. xxvi–xxvii):

> As to his Person he was but despicable, being very crooked, tho' I have heard from himself, and others, that he was strait till about 16 Years of Age when he first grew awry, by frequent practicing, turning with a Turn-Lath, and the like incurvating Exercises, being but of a thin weak habit of Body, which increas'd as he grew older, so as to be very remarkable at last: This made him but low of Stature, tho' by his Limbs he shou'd have been moderately tall. He was always very pale and lean, and lately nothing but Skin and Bone, with a meagre Aspect, his Eyes grey and full, with a sharp ingenious Look whilst younger; his Nose but thin, of a moderate height and length; his Mouth meanly wide, and upper Lip thin; his Chin sharp, and Forehead large; his Head of a middle size. He wore his own Hair of a dark Brown colour, very long and hanging neglected over his Face uncut and lank, which about three Years before his Death he cut off, and wore a Periwig. He went stooping and very fast (till his weakness a few Years before his Death hindred him) having but a light Body to carry, and a great deal of Spirits and Activity, especially in his Youth.
>
> He was of an active, restless, indefatigable Genius even almost to the last, and always slept little to his Death, seldom going to Sleep till two, three, or four a Clock in the Morning, and seldomer to Bed, oftener continuing his Studies all Night, and taking a short Nap in the Day. His Temper was Melancholy, Mistrustful and Jealous, which more increas'd upon him with his Years. He was in the beginning of his being made known to the Learned, very communicative of his Philosophical Discoveries and Inventions, till some Accidents made him to a Crime close and reserv'd. He laid the cause upon some Persons, challenging his Discoveries for their own, taking occasion from his Hints to perfect what he had not; which made him say he would suggest nothing till he had time to perfect it himself, which has been the Reason that many things are lost, which he affirm'd he knew. He had a piercing Judgment into the Dispositions of others, and would sometimes give shrewd Guesses and smart Characters.
>
> From his Youth he had been us'd to a Collegiate, or rather Monastick Life, which might be some reason of his continuing to live so like an Hermit or Cynick too perniriously, when his Circumstances, as to Estate, were very considerable, scarcely affording himself Necessaries.

The image of Hooke as the "restless, indefatigable genius," sleeping little and becoming mistrustful in his later life, can be supported by his own *Diary*. But the description of his physical appearance is probably drawn from Waller's remembrance of him as an old man.

John Aubrey, who was a good friend of Hooke's for over 30 years, described him in the following way:

> He is but of midling stature, something crooked, pale faced, and his face but little belowe, but his head is lardge; his eie full and popping, and not quick; a grey eie. He haz a delicate head of haire, browne, and

of an excellent moist curle. He is and ever was very temperate, and moderate in dyet, etc.

As he is of prodigious inventive head, so is a person of great virtue and goodnes. Now when I have sayd his Inventive faculty is so great, you cannot imagine his Memory to be excellent, for they are like two Bucketts, as one goes up, the other goes downe. He is certainly the greatest Mechanick this day in the World.[1]

The depiction of Hooke used by most writers, however, is that presented by Waller. Aubrey's picture of Hooke as having "midling stature, something crooked," became "despicable" and "very crooked" under Waller's pen. Several other items in Waller's description of Hooke are simply not true. For example, he claims that Hooke cut off his unkempt hair (which to Aubrey was "delicate" and "of an excellent moist curle") and wore a periwig only three years before his death. This statement can be disproved just by reading Hooke's *Diary*; a number of entries exist about cutting his hair or paying someone to curl his periwig as far back as August 26, 1672 (he was 37 then), when he bought a periwig for 28 shillings and then cut off his hair two days later. His *Diary* also shows him to enjoy buying expensive fabric, having new clothes made and wearing new suits and silk stockings,—e.g., July 10, 1673: "Put on new coloured suit"—which belies Waller's statement that Hooke denied himself "Necessaries." There are also innumerable entries recording buying necklaces and other pretty things for his beloved niece Grace.

Aubrey's description of Hooke as a person of "great virtue and goodnes," has never been repeated by modern writers, but Waller's words such as "Melancholy, Mistrustful and Jealous" are often quoted. These adjectives, however, are not apropos for most of Hooke's life. He might have become that way in later life; mostly he was generous and open, and as Waller wrote, he was "very communicative of his Philosophical Discoveries and Inventions." Only after finding many of his ideas adopted by others who claimed them as their own without even an acknowledgment of simultaneity did he begin to become "mistrustful."

Hooke's *Diary* clearly shows him to be a gregarious, friendly, and open man whose favorite pastime was good conversation over a beer or a hot cup of chocolate. He mentions at least 150 taverns and coffee-houses frequented by him and his friends. Some 2,000 names of friends and acquaintances, mostly contemporaries, are mentioned in his diary. Both Andrade (1935) and 'Espinasse (1956) describe Hooke's active social life. He dined with his old friend Robert Boyle at least once a week at Boyle's home. He moved easily through all the social levels in the 17th century:—discoursed with scientists and technicians alike; made love with his maid, Nell, who visited him and made clothes for him even after she left his service; and was acquainted with an extraordinary number of people in London, including Charles II.

Shapin (1989) goes to great lengths to contrast Robert Boyle, an undisputed "Christian gentleman," with Robert Hooke, who he considered had an uncertain and "problematic" social status. Shapin relegates Hooke to servant status, at best on the level of a tradesman, a "mechanic," perhaps, but would withhold the labels of "experimental philosopher," "Christian gentleman," and even plain "gentleman" from him. His thesis is that Hooke's uncertain social status produced some enormous difficulties in his life. It is possible that it could have underlain some of the controversies he faced with those contemporaries who felt free to take his ideas as their own. It could also be, however, that Hooke's ideas were so freely and openly expressed that they were there for the taking and his successes were a source of others' jealousy. As Hooke grew older he realized his own naiveté, becoming withdrawn and mistrustful.

It is to Hooke's credit that his ability to function effectively was not hampered by his so-called lack of "gentlemanly" status. Clearly, he never gave it a thought, nor is there evidence that any of his friends did. His own intellectual achievements afforded his place as a respected member of the intellectual community to which he belonged. The king himself was friendly with him and had several audiences with him. Hooke recorded in his Diary for Friday, May 23, 1673, (by then he was famous both for his *Micrographia* and his work in rebuilding London), that while he was walking in the park with his friend and colleague Christopher Wren, the "King seeing me cald me told me he was glad to see me recoverd asked for [measurement of] degre by water." It was Hooke who devised a thermometer that used the freezing point of water as zero, and King Charles was obviously familiar with Hooke's activities and was curious about his progress. Also, the king was congnizant of Hooke's state of health. No mere servant or tradesman would have evoked such royal attention in a public park.

Although well known and well respected in his time for his ideas and inventions, somehow over the centuries of adulation of his contemporary, Isaac Newton, Hooke has slipped into obscurity. We have been too overawed by the spectre of Newton for too long; it is time that we allow a more balanced view of the Age of Enlightenment.

It is gratifying to know that Hooke's insights on Earth science have been hailed by the distinguished Newton scholar, Richard Westfall, as Hooke's greatest scientific achievement. In geology, in Westfall's words (Westfall, 1969, p. xxiv), Hooke "yielded preeminence to no man in the seventeenth century." Indeed, stripped of platitudes that crept into his later lectures, but essential to the time, regarding God, Genesis, and the Flood, and such favorite pastimes of the seventeenth century intelligentsia as divining the verisimilitude of myths and legends, Hooke's methods and ideas reveal a clear and logical mind, one that should be respected and admired in any age. His observations were perceptive, and many of his conclusions were far ahead of his time. In many ways Hooke was freer of medievalism than all of the other virtuosi of his age. His achievements and contributions speak for themselves.

The following chapters in Part I present a brief account of Hooke's remarkable life, attempting to analyze his place among the intelligentsia of the 17th century by presenting his ideas on the Earth along with other contemporary theories and showing how he broke from the medievalism that still prevailed in his time. They show how the geology of his birthplace, the Isle of Wight, influenced his thinking in formulating his Hypothesis of the Terraqueous Globe. They detail the various aspects of his hypothesis, including his gravitational theory (essential to his concept of the Earth in space), and show just how original and creative many of his ideas were and how useful they still are today. In particular, Hooke's geological contributions will be assessed vis-à-vis those of Steno and Hutton. Finally, in Part II, Hooke's ideas will be presented by setting his *Discourse of Earthquakes* in modern type, arranged in chronological order (after Rappaport, 1986)[2] with marginal annotations to guide the reader.

# I

Robert Hooke's

# LIFE AND WORK

# 1

# *The Life of Robert Hooke*

*He was of an active, restless, indefatigable Genius even almost to the last.*

—Richard Waller, 1705

The intellectual biography of Robert Hooke should be written by a committee of experts—a large committee. Hooke has done so much and contributed to so many fields of human endeavor that no single author could possibly do him justice. Besides Richard Waller's 1705 summary of Hooke's life in *The Posthumous Works*, the best biographies remain those of Margaret 'Espinasse and E. N. da C. Andrade, dating to the 1950s. Gunther's multivolume chronicle of Hooke's life and work is an invaluable source. Yet even these excellent accounts can only touch upon certain aspects in his life while emphasizing others. And of course, the publication of Hooke's diary in 1935 has illuminated both his personality and facts in his life.[1] Many articles dealing with individual points in Hooke's life have been written over the last 40 years. When a team of scholars led by Michael Hunter and Simon Schaffer (1989) recently convened for a much-needed symposium on Hooke, what came forth, albeit of distinct value, was a collection of articles on some specific but disjointed aspects of Hooke's life and work. Although it was not the goal of this symposium to represent Hooke in one conference and volume, their effort, with its broad range of topics, demonstrates the difficulty or impossibility of such a task.

By the same token, the present book, because it focuses on Hooke's contributions to geology, should be but one volume in a series of volumes. Other volumes should cover Hooke's work in astronomy, meteorology, physics, chemistry, biology, architecture, oceanography, instrumentation, microscopy and telescopy, horology, music, and cartography. They should discuss his calculating and flying machines, his design of vehicles of transportation, and many other projects he undertook. Can such a team be found and coordinated to work on the life of this extraordinary man? Yet a brief account of his life is needed here to give the reader a glimpse of his genius and enormous productivity and to place his earthly thoughts in the context of his life's work. Without an overview, the understanding of Hooke and his earthly thoughts is bound to be incomplete—much like the blind men trying to discover the nature of an elephant.

## THE EARLY YEARS

It might be said that Hooke was born unlucky. His family was not rich; he was neither handsome nor endowed with robust health. Life was never easy for him, partly because he drove himself to such extreme lengths and partly because, having an open and friendly personality, he was vulnerable to small and large hurts. These eventually taught him to be wary and on guard, which, in turn, prompted critics to call him suspicious and paranoid. But he was gifted with an extraordinary mind that was resourceful, creative, imaginative, and even visionary. He was born at noon on July 18, 1635, in the village of Freshwater at the western end of the Isle of Wight and was christened eight days later by his father, John, the curate of the All Saints Church in Freshwater (Fig. 1–1);[2] Hooke's mother, Cicely Giles, was John's second wife. He had an older sister, Katherine, and an older brother, John, who later became mayor of Newport, Isle of Wight. John Aubrey (1626–1697) in his *Brief Lives* (1949) claims that the curate John Hooke was from the ancient and respected family of Hooke, variously spelled with or without the *e*, of Hampshire. Others place him in the "yeoman" class, who were freeholders but below the gentry. The young Hooke was a sickly child; no food agreed with him except milk, and for several years his parents believed he would not survive his childhood. He suffered from bad headaches, and all through his life he was troubled by indigestion, colds, dizziness, and insomnia. His health as an adult was hardly improved by constant self-doctoring. His diary records frequent ingestion of mercury, which could not have been very good for him.[3]

No portrait of Hooke exists, although the Royal Society probably should have had one (see below concerning Hooke's brief career as an apprentice artist and footnote no. 5 of this chapter). Newton's succeeding to the presidency of the society shortly after Hooke's death might have had something to do with the absence of a portrait. As evidenced by his behavior toward his rivals (e.g. Leibnitz, co-founder of calculus), Newton was quite capable of intense hatred. Because Hooke had criticized some of his ideas and claims of priority, Newton extended this hatred to Hooke. It seems probable, then, that he did not wish to have any reminders of the man around the Royal Society. The society itself, soon after Hooke's death, also moved out of Gresham College, where Hooke had been professor of geometry and as such had lodgings. Blame must also be placed partially on Newton that none of the instruments Hooke made for the society over a period of some forty years had been preserved. Is it any wonder that Hooke's portrait also disappeared?

*Figure 1-1. A view of Hook Hill and All Saints Church, where Hooke was christened by his father, the curate John Hooke. (Reproduced with the kind permission of the Reverend Brian Banics, Rector of Freshwater, Isle of Wight.)*

The descriptions of Hooke's appearance are generally unflattering (see Introduction); he was thin and apparently of average height for that time, but he appeared shorter because at a very young age he became somewhat hunched from working at a lathe, making such things as models of sailing ships and wooden models of machinery. (He even took apart a brass clock and built one out of wood that worked. These devices he referred to as "toys" in his autobiography, which he started to write at age 62 on Saturday April 10, 1697. It is a pity he could not finish his life story because of the many projects with which he was still involved six years before his death [Waller, 1705]). His bent frame grew worse as he grew older (see Waller's description of Hooke quoted in the Introduction). His unimpressive appearance in contrast to his most impressive mind even warranted a much-quoted entry in Pepys' diary in which the diarist described Hooke as being "the most, and promises the least, of any man in the world that ever I saw" (Pepys, Feb. 15, 1665).

Hooke lost his father when he was only 13 years old. When the artist John Hoskyns came to Freshwater to paint, young Hooke decided he could do as well, whereupon he "getts him chalke, and ruddle, and coale, and grinds them, and putts them on a trencher, gott a pencill, and to worke he went, and made a picture" (Aubrey's *Lives*, p. 242). Thereafter he copied several pictures that hung on the parlor walls, writing in his incipient autobiography that "Mr *Hoskins* (Son to the famous *Hoskins Cowpers* Master) much admired one not instructed could so well imitate them." After the death of his father, because of this demonstrated talent in art, Hooke was sent to London with £100, £40 of which we know were left him by his father, to be apprenticed to the well-known portrait painter Sir Peter Lely.[4]

Typically, however, the odor of oil paints aggravated his propensity to get headaches. Another student of Lely's was Mary Beale, who later painted portraits of several of Hooke's friends and associates, some of whom were members of the Royal Society. It seems likely, therefore, that a portrait, or at least a sketch, of Hooke existed, as there is evidence that he had posed for Mary Beale.[5]

Because of his aversion to oil paints, Hooke was forced to abandon his career as portrait-painter. He was then sent to school at Westminster where the headmaster at that time was Dr. Busby, much-respected but, according to 'Espinasse, a "redoubtable flogger" (1956, p. 3). Busby immediately recognized the genius in the boy, who on his own had learned the first six books of Euclid in a week. Hooke also taught himself to play the organ, became quite competent in Latin and Greek, and even gained "some insight into the *Hebrew* and some other Oriental Languages" (Waller, 1705, p. iii). He had a life-long interest in music, partly because of its association of numbers and proportions. Kassler and Oldroyd (1983) deduced Hooke's music theory from fragmentary writings, concluding that his ideas on harmonics dovetail with his perception of world harmony. Mathematics per se, however, was Hooke's favorite subject. He lived as a scholar in Dr. Busby's own house, and remained a lifetime friend to his old schoolmaster. When in 1680 the latter wanted a church built in the village of Willen, Hooke designed it for him. This charming little church is still used (Fig. 1–2).

## OXFORD DAYS

In 1653 Hooke went to Oxford University. Waller writes, "but as 'tis often the Fate of Persons great in Learning to be small in other Circumstances, his were but mean" (1705, p. iii). He was a chorister-student at Christ Church, and then as today, choirs were important entities within the colleges. Schools attached to Oxford colleges still often offer scholarships to boys who can sing.[6] He also

*Figure 1-2. Willen Church, designed and built by Robert Hooke for his schoolmaster, Dr. Busby. (Reproduced with the kind permission of the Reverend Ian Jagger, Team Vicar of Milton Keynes.)*

worked as a "servitor" to a Mr. Goodman. Hooke's status as chorister and servitor at age 17 and his position later as an "employee" of the Royal Society as Curator of Experiments prompted S. Shapin (1989) to suggest that Hooke's reputation was forever thereafter branded. According to Shapin, because no "gentleman" would find himself in such a position, Hooke was not a gentleman; therefore, he did not behave as a gentleman and was treated as a servant for much of his life. This is an interesting thesis but much too simplistic and can be refuted. Although Hooke started at Oxford with impecunious means, a situation not uncommon at university, his birth, as a son of a cleric, was not considered so low as to exclude him from the higher intellectual and economic circles. The social standing of "yeoman" class, although not aristocratic, was well respected. Isaac Newton whose father was a small farmer was also of yeoman origin, and Newton was a "subsizar" at Cambridge, the equivalent of "servitor" at Oxford. No one today would deny that Newton was a "gentleman" because of his larger than life reputation. As with Newton, Hooke's contemporaries respected him for his abilities and talents.

This was a period of puritanism, and Oxford, which had been under Royalist control until 1646, was being closely scrutinized by a parliamentary committee seeking to control the university in every aspect of its existence. Talented young scholars, therefore, needed to be protected from such a restrictive atmosphere. Fortunately, the man who most appreciated Hooke's abilities was John Wilkins (1614–1672) who became the warden of Wadham College in 1648 and Bishop of Chester in 1668. Hooke's praise of Wilkins in the preface of *Micrographia* (printed here as in the original) expressed his true feelings for the man, as well as, incidentally, for organized religion at that time:

> *He* [Wilkins] *is indeed a man born for the good of mankind, and for the honour of his Country. In the sweetness of whose behaviour, in the calmness of his mind, in the unbounded goodness of his heart, we have an evident Instance, what the true and the primitive unpassionate Religion was, before it was sowred by particular Factions.*

Wilkins was an important figure in many intellectual movements, and Hooke became one of his protégés. At this time Oxford was accused by the "Visitors" from parliament of being "hostile to science," a charge with little foundation. Wilkins held a philosophy of moderation in both religious and political views, and he had gathered around him a group of extraordinarily talented men, regardless of their political or religious persuasions, for the express purpose of delving into the mysteries of the natural world through experimentation and observation. The curiosity of the group was boundless. Far from being hostile to science, in pursuing its scientific interests, it acquired a reputation of having a complete lack of religious fervor. The Oxford group at first met weekly at Dr. William Petty's lodgings in an apothecary's house so they could be close at hand to inspect the drugs there. After Petty moved to Ireland in 1651, they met in Wadham College which became a haven for Royalists and Anglicans in a society dominated by Puritans.

Earlier, when Wilkins lived in London, he had been a member of the "1645 group" which met weekly to discuss all kinds of scientific matters.[7] Some confusion exists in the literature as to whether the Royal Society was formed from this 1645 group which met in Gresham College in the rooms of the professor of astronomy or from Wilkins' group in Wadham College at Oxford, but it is hardly important, as the connection between the two was extremely strong. Charter members of both moved from one to the other. If a member from Oxford happened to be in London, he would attend meetings of the 1645 group and vice versa. Also, several of Wilkins' Oxford group, including Hooke, later became Gresham professors. Both groups had the same philosophy and barred discussions of religion and politics. In the 1650's the two parts, as indeed they can be looked upon as "parts" of the same general community of intelligentsia, merged and became the Royal Society.

Astronomer Seth Ward, a known Royalist and Anglican, came to live at Wadham, as did Christopher Wren, also a Royalist, in 1649. Mathematician Lawrence Rooke left Cambridge to be near Wilkins and Ward, as did William Neile, grandson of the Archbishop of York. John Wallis, a Presbyterian who had been a member of the 1645 group, also came to Wadham. Even such parliamentarians as the sons of Sir Francis Russell and General John Disbrowe (both allied to the Cromwell family by marriage) were part of the group (Shapiro, 1969, p. 118–119).

Hooke was indeed fortunate to have been under the influence and protection of Wilkins during this period. In 1656 Wilkins married Robina French, the widow of an Oxford colleague and the youngest sister of Oliver Cromwell, the Protector. Wilkins thus soon found himself in a key position in the administration of Oxford, especially after Cromwell's son, Richard, became chancellor of Oxford, and Wilkins was his Uncle John. Wilkins's philosophy of moderation allowed the peaceful coexistence of various and changing religious and political views at the university (Shapiro, 1969, p. 111), and it was probably through his interest

*Figure 1-3. Plaque on the wall of University College, Oxford, commemorating the site of Robert Boyle's laboratory. (Photograph by C. W. Drake.)*

and influence that Hooke obtained a position in 1655 as assistant to Dr. Thomas Willis to make chemical preparations. Willis was a Royalist and an Anglican who allowed his house to be used as a place for covert Anglican services. The fact that a large part of the university community knew about these services demonstrates the degree of tolerance shown by the academic population.

Wilkins took pride in discovering bright young men like Hooke and Wren. Hooke began to attend the meetings of the Oxford group in 1655 and became acquainted with its intellectually curious members. Willis introduced him to Robert Boyle, who had moved to Oxford two years earlier, later setting up his chemistry laboratory (Fig. 1–3). Boyle also joined the group at Wadham in 1655. Hooke wrote about his early experience with this group and the air pump he built for Boyle in his unfinished autobiography (Waller, 1705, p. iii):

*Figure 1-4. Hooke's pneumatic engine or air pump. (From Gunther, 1930, vol. 6.)*

> At these Meetings, which were about the Year 1655 (before which time I knew little of them) divers Experiments were suggested, discours'd and try'd with various successes, tho' no other account was taken of them but what particular Persons perhaps did for the help of their own Memories; so that many excellent things have been lost, some few only by the kindness of the Authors have been since made publick; among these may be reckon'd the Honourable Mr. *Boyle's Pneumatick Engine* and Experiments, first Printed in the Year 1660. for in 1658, or 9, I contriv'd and perfected the Air-pump for Mr *Boyle*, having first seen a Contrivance for that purpose made for the same honourable Person by Mr. *Gratorix*, which was too gross to perform any great matter.

As Gunther has pointed out, this "Pneumatic Engine" Hooke designed and built in Boyle's lab "prepared the way for the successful construction of the Atmosphere Engine of Newcomen, of the Steam Engine of James Watt, of the myriads of air, gas, steam, petrol motor-engines of today" (1930, 6:72) (Fig. 1–4).[8]

As for Boyle's Law—the experimental proof of the relationship of pressure and volume of a gas hypothesized by Henry Power and Richard Towneley and announced in the second edition of Boyle's *Spring of the Air* (1662) treatise—I. B. Cohen (1964) gives convincing evidence that it was Hooke who first performed the experiment. On August 2, 1661, Hooke repeated some experiments he had performed a year earlier with a bent tube and recorded the results in a table which he published in *Micrographia* (1665). He wrote in *Micrographia* (Observ. no. 58, p. 224–225) (Fig. 1–5):

Having lately heard of Mr. *Townly's Hypothesis,* I shap'd my course in such sort, as would be most convenient for the examination of that *Hypothesis*; the event of which you have in the latter part of the last Table.

The other Experiment was, to find what degrees of force were requisite to compress, or condense, the Air into such or such a bulk.

The manner of proceeding therein was this: I took a Tube about five foot long, one of whose ends was sealed up, and bended in the form of a *siphon*, much like that represented in the fourth Figure of the 37.*Scheme,* one side whereof AD, that was open at A, was about fifty inches long, the other side BC, shut at B, was not much above seven inches long; then placing it exactly perpendicular, I pour'd in a little Quicksilver, and found that the Air BC was 6inches, or very near to seven; then pouring in Quicksilver at the longer Tube, I continued filling of it till the Air in the shorter part of it was contracted into half the former dimensions, and found the height exactly nine and twenty inches; and by making several other tryals, in several other degrees of condensation of the Air, I found them exactly answer the former *Hypothesis.*

*Figure 1-5. Hooke's gas pressure/volume apparatus. (From* Micrographia *1665.)*

Hooke had thus proved the Power and Towneley hypothesis now known as Boyle's Law by 1661, if not 1660, when he first made the measurements and had not yet heard of the hypothesis. Hooke never disputed Boyle's priority of publication, however, and remained a loyal friend to Boyle throughout his life. Andrade (1950, p. 459) points out that Hooke would never have published his description of the experiment in *Micrographia* "if it was likely to give pain to, or be disputed by Boyle." Boyle's Law, therefore, might well have been the Boyle-Hooke Law had Hooke wished to make a claim. It was, after all, a sort of counterpart for gases of Hooke's only eponymous law.[9]

In 1655, Hooke also designed contraptions for human flight, a project in which both he and Wilkins were interested. One of his models of "flying chariots" supposedly sustained itself in the air for a short time. A true scientist, Hooke accepted his own evidence that human muscles, however, have limitations. His autobiography records the following (Waller, 1705, p. iv):

I contriv'd and made many trials about the Art of flying in the Air, and moving very swift on the Land and Water, of which I shew'd several Designs to Dr. *Wilkins* then *Warden* of *Wadham College,* and at the same time made a Module, which, by the help of Springs and Wings, rais'd and sustain'd itself in the Air; but finding by my own trials, and afterwards by Calculation, that the Muscles of a Mans Body were not sufficient to do any thing considerable of that kind, I apply'd my Mind to contrive a way to make artificial Muscles; divers designs whereof I shew'd also at the same time to Dr. *Wilkins,* but was in many of my Trials frustrated of my expectations.

By the age of twenty, Hooke must have felt that he had found his place in life among the brightest and most forward-looking men of the experimental-philosophical movement. According to Waller, he conducted many "curious Experiments, Observations and Inquires," and he constructed instruments for these various purposes (for example, the barometer). Waller (1705, p. vii–viii) records, "At this time I have heard Mr. *Hooke* say, it was first observ'd, that the height of the *Mercury* in the *Barometer* did not conform itself to the Moon's motion, but to that of the different Gravitation of the Air," i.e. pressure.

For his studies in astronomy, Hooke had the privilege of being guided by Seth Ward, Savilian Professor of Astronomy. Hooke relates (Waller, 1705, p. iv):

> About this time having an opportunity of acquainting my self with Astronomy by the kindness of Dr. *Ward*, I apply'd my self to the improving of the *Pendulum* for such Observations, and in the Year 1656, or 57, I contriv'd a way to continue the motion of the *Pendulum*, so much commended by *Ricciolus* in his *Almagestum*, which Dr. *Ward* had recommended to me to peruse; I made some trials for this end, which I found to succed [sic] to my wish.
>
> The success of these made me farther think of improving it for finding the Longitude, and *the Method I had made for my self for Mechanick Inventions,* quickly led me to the use of Springs instead of Gravity for the making a Body vibrate in any Posture, whereupon I did first in great, and afterwards in smaller Modules, satisfy my self of the Practicableness of such an Invention, and hoping to have made great advantage thereby, I acquainted divers of my Friends, and particularly Mr. *Boyle,* that I was possest of such an Invention, and crav'd their Assistance for improving the use of it to my advantage.

In his book on horology, J. E. Haswell (1951) cites 1656 as the year that Hooke invented the anchor escapement to replace the verge or crown wheel in clocks for better accuracy, just after the introduction of the pendulum. Then, in 1658, Hooke invented the balance wheel for pocket watches by using "Springs instead of Gravity for the making a Body vibrate in any Posture," as just described. The spiral balance-spring minimizes the effect of gravity that had plagued the pendulum. In 1675 after Huygens invented a similar watch with a balance wheel also controlled by a spiral spring, Hooke had the famous instrument-maker Thomas Tompion (1638–1713) make a spring-controlled watch, which he then presented to Charles II. It was inscribed *Rob. Hooke invenit 1658 T. Tompion fecit 1675.* A controversy then ensued as to priority. Undoubtedly, Hooke had invented such a watch in 1658; he had asked several witnesses for help in "improving the use of it to my advantage," as he recorded. Huygens is also always credited with the invention of the pendulum clock in 1657; the passage just quoted should also place this fact into question. Certainly John Aubrey had always claimed that Hooke invented the pendulum clock. Simultaneity, however, could very well be the answer.

## THE ROYAL SOCIETY

Charles II was restored to the throne in 1660, and soon after, the London Royal Society received its Charter. Wilkins helped draw up the initial charter, redrafting a new one in 1663.

In 1661, Hooke published his first treatise, an article on his observations on capillary action and surface tension, entitled *An Attempt for the Explication of the Phenomena Observable in an Experiment Published by the Honorable Robert Boyle Esq; in the xxxv Experiment of his Epistolical* Discourse touching the Aire, *In confirmation of a former conjecture made by R. H.* 8vo. In the same year, he published another tract, *A discourse of a New Instrument to make more accurate observations in Astronomy, than ever were yet made.* 4to. Gunther (1930, 6:45) feels that the experiments on capillary attraction must have been done in Boyle's laboratory in the High Street, Oxford. Hooke himself in *Micrographia* (p. 205), under Observation 50 (having to do with mites) records that in September and October of 1661, he was still in Oxford making these observations with his microscope.

By April 10, 1661, members of the newly formed Royal Society had read Hooke's publication on capillary action and had decided to debate the issue at its next meeting. Only 25 years old, Hooke was already well known among this elite group. Waller (1705, p. ix) records that this tract, "together with his former Performances, made him much respected by the *R. Society*." On November 5, 1662 the Journal Book of the Royal Society records the following entry (Gunther, 1930, 6:76–78): "Sir *Robert Moray* propos'd a Person willing to be employed as a *Curator* by the *Society*, and offering to furnish them every day on which they met, with three or four considerable Experiments, and expecting no recompense till the *Society* should get a stock enabling them to give it." Once again Wilkins recommended his protégé Hooke for the curatorship, later nominating him for the council. On November 12, Hooke was unanimously accepted as curator of experiments and "it was ordered, that Mr. Boyle should have the Thanks of the *Society* for dispensing with him for their use; and that Mr. *Hooke* should come and sit amongst them, and both bring in every Day three or four Experiments of his own, and take care of such others, as should be mentioned to him by the *Society*." It was Wilkins again who in 1665 helped Hooke get elected Gresham Professor of Geometry; lodgings in Gresham College went with the position (Fig. 1–6). Thereafter the society's meetings took place in Hooke's rooms at Gresham.

At this time the position of curator carried no recompense, although the society had agreed to pay Hooke a salary of £80 per annum, to be raised through subscriptions or a stock investment—when they could afford it. From the start, however, the society had financial difficulties. They even had trouble collecting the small amount of members' fees. It appears that in the vernacular of the age the word "ordered" was used freely, denoting less an unquestioned command than the sense of "recorded" or "noted"—that is, it is noted in the records that such and such should happen. The duties of the curator were constantly being

*Figure 1-6. Gresham College, where Hooke lived from 1665 until his death in 1703. His lodgings are located at the upper right-hand corner of the quadrangle. (From Gunther, 1921, vol. 1.)*

delineated, even when the demands far exceeded what any one person could accomplish. This situation was more symptomatic of the uncertainties faced by a young organization than anyone's attempt to assert social rank over a hired hand. Clearly, Hooke's position was not that of a "servant." Gunther (1930, 6:130) writes, "In view of the fact that [Hooke] was appointed Curator of that distinguished Society at an early date, it is of interest to note that before his appointment, indeed before the new Society had obtained a Charter and could thereby claim the title 'Royal,' the relationship of the Curator to an Operator had already been clearly defined." And he quotes from Birch's *History of the Royal Society*: "*Dec.* 12, 1660. At a Preliminary Meeting it was agreed that the salary of the Operator be four pounds a year; and for any other service, as the Curators, who employ him, shall judge reasonable." This implies that a curator would be a regular member of the Royal Society, entitled to "come and sit amongst" the members as such. On May 20, 1663, Hooke's name was included in a list of others "to be received, admitted, and ordered to be registered Fellows of the Royal Society" (Gunther, 1930, 6:130).

Hooke took upon himself the responsibility of the "repository of rarities" until Richard Shortgrave, the operator, took over in 1676. Hooke was also librarian for many years, keeping a catalogue of acquisitions. In 1664 a miserly, self-aggrandizing "philanthropist" merchant named Sir John Cutler offered £50 a year to Hooke to deliver a series of lectures to be named after himself—whereupon the society immediately reduced Hooke's ephemeral salary by that amount. Cutler, however, never paid him. In fact, Hooke still did not regularly receive the reduced amount (£30 per annum) from the society. It was only after Cutler's death some thirty years later that Hooke was paid what was owed him—and then only because he sued the Cutler estate. By then a rich man as a result of his efforts in rebuilding London after the Great Fire of 1666, Hooke still felt the need to take the Cutler matter to court, on principle's sake. All this points to the complexity of Hooke's personality—going to court to retrieve a sum of money, which was small relative to his wealth; putting away most of his money in a trunk untouched, which was found at his death; and being invariably generous to his friends and relatives. It is too facile, therefore, to attribute Hooke's difficulties with a few unscrupulous individuals who stole his ideas to his "ungentlemanly" birth or rank, especially because he was highly regarded by most of his contemporaries.

From the moment Hooke became associated with the Royal Society, there was little rest for him. During the next thirty years, he faithfully discharged his duties and delivered his Cutlerian and Gresham lectures. He devised so many experiments, invented so many instruments, discoursed on so many ideas and hypotheses that one marvels at how a single person could have accomplished so much in one lifetime. Robinson and Adams (1935) claim "There can be no doubt that Hooke was the one man who did most to shape the form of the new Society and maintain its active existence. Without his weekly experiments and prolific work the Society could scarcely have survived. It is scarcely an exaggeration to say that he was, historically, the creator of the Royal Society." Indeed, Hooke himself records in his *Diary* that he wrote the "lawes for R.S." The wonder is not that he left many things inconclusive and unfinished (often the criticism of present-day writers) but that he accomplished as much and as successfully as he did, given the circumstances. The evidence is clearly more in favor of 'Espinasse's assessment of Hooke as "the chief professional scientist" of the Royal Society than of Shapin's calling him servant or laborant. In spite of the difficulties that Hooke might have encountered in an organization that included rich members who attended meetings largely for amusement, he was able to pursue many of his own scientific and experimental interests to fruition. One simply wishes that he had had more time to pursue more of his original and creative ideas whose dénouement provided avenues of fame and fortune for others.

To Hooke the business of the Royal Society was "to improve the knowledge of naturall things, and all useful Arts, Manufactures, Mechanick practises, Engynes and Inventions by Experiments—(not meddling with Divinity, Metaphysics, Moralls, Politicks, Grammar, Rhetorick, or Logick)." These words are preserved in Hooke's handwriting in the archives of the Royal Society as part of his "Proposals for the good of the Royal Society." The fledgling society, although armed with a motto, hardly had a thematic goal. Many inquiries and experiments were done almost by whim of the members.

These members of the society were curious about all bodies and phenomena and eager to discover truth through experimentation. Their motto, *nullius in verba* (nothing in words) is taken from the Epistles of Horace, *Nullius addictus jurare in verba magistri* or "I am not bound to swear in the words of the master"—that is, I am not bound to accept received opinion or to adhere to any school of thought. Implicit in this motto is the distrust of authority and the conviction that knowledge should be sought by observation and experimentation rather than obtained through the arm-waving of people who came before. As a result, practically any investigation was fair game.

Because of the Hooke-Boyle pump, the society in its early days was able to perform many experiments having to do with the properties of air—its weight, its expansion and rarefaction, its pressure and temperature and combustion. Other air experiments were done on respiration of organisms (e.g. the length of time a fish could live in open air or in a closed glass, first filled with air, then removed). Not only did Hooke subject animals to respiratory experiments, he also enclosed himself in a space from which air was slowly being removed to note the effect until he almost lost consciousness.

Other concerns of Hooke included glass-making, refraction in hot and cold water, crystallization of water, weighing bodies in water, decrease of gravity as a body moves upward from the Earth's surface (he carried out such an experiment up the tower of Westminster Abbey to see if he could detect a decrease in gravity measurement), velocity of falling bodies, pressure at the bottom of the sea, eclipses of Jupiter's satellites, eclipses in general, the formation and growing of stones, and many others. In a letter to Boyle reporting on a meeting of the Society, Hooke relates the discussion of the formation of stone in animals or humans (letter dated June 5, 1663, Gunther, 6:133): "Monsieur Monconis related a story of a woman in France, who for a long time together every month voided the perfect bones of children." Such discussion reveals both the sophistication and the naiveté of the early Royal Society and all the hodgepodge that apparently interested the membership.

The members of the society followed Baconian tenets and methods of seeking knowledge by observation, collection of data, and open exchange of ideas. They sought to reform the English language in order to achieve a simple, direct, and natural way of expression and to bring out clarity of meaning by eliminating convoluted and florid constructions, referred to by Bishop Thomas Sprat (1667, p. 112–113) as "this vicious abundance of phrase, this trick of metaphors, this volubility of tongue, which makes so great a noise in the world." In an unprecedented declaration, Sprat added that members should bring "all things as near the mathematical plainness as they can; and preferring the language of artisans, countrymen, and merchants, before that of wits or scholars." True to its purpose, the society opened its membership to all classes of people who were curious about the new experimental philosophy.

Yet its goals and activities were also open to mockery and criticism. Samuel Butler (1612–1680) lampooned the pompous, ceremonious meetings of the Society in "The Elephant in the Moon" (1676).[10] Hooke himself was satirically depicted in *The Virtuoso,* a play by Thomas Shadwell. In it Hooke is represented by a Sir Nicholas Gimcrack, F.R.S., who is described, in reference to Hooke's *Micrographia*, as "a Sot, that has spent 2000 l. in Microscopes, to find out the nature of Eels in Vinegar, Mites in Cheese, and the Blue of Plums."[11]

More serious criticism came from certain members of the Church, who were fearful the new philosophy would lead to atheism, and from physicians, who felt the society was trampling on their territory when it investigated diseases and their cure—even though the membership included prominent churchmen, physicians, and surgeons. The most virulent attack came from Dr. Henry Stubbe, who had been a deputy librarian of the Bodleian Library at Oxford. In 1670, Stubbe published two articles attacking Joseph Glanvill, a cleric who came to the defense of the society. Stubbe was a friend of Boyle and declared strongly to the latter that his close association with the society was detrimental to his reputation and integrity (More, 1944, p. 112–113).

Thus in spite of its royal charter and the whole-hearted support of Charles II, the new society traveled a rough road. Still, the active members, personified by Hooke, were so engrossed in their work and their discourses that they weathered the criticism or mostly ignored it. Hooke did take pains in one of his discourses to defend the society, using the opportunity to expound on the methods of science in his response to the critics' question, "What *hath* the *Royal Society done for so many years?*" and the answer by others, "*Just nothing.*" [See discourse no. 3 in Part II of this volume.]

Hooke carried out his duties faithfully and diligently. He always approached a problem by exhausting all possible ways of tackling it—a habit that is reminiscent of his youth when he claimed that he had invented thirty ways of flying. He called his approaches "algebras," which are all the different possible methods listed and classified in relation to a particular problem. An example of an "algebra," entitled *Proposals for the Good of the Royal Society,* also reveals his own proprietary feelings about the society (Gunther, 1930, 6:xii–xiii):

> Allurements to Members present are
> >Desireable Acquaintance
> >Delightful Discourse
> >Pleasant Entertainment by Experiments
> >Instructive Observation by Tracts
> >Considerable Intelligence by Letters
> >New Discoveries by Inventors
> >Solution of Doubts and Problems
> >An easy way to know what is already known
> >Liberty to Peruse ye Repository
> >Liberty to Peruse ye Letters and Registers
> >Liberty to Peruse ye Library
> >Liberty to Peruse ye Modules and Instruments
> >Liberty to be present at Mechanick, Optick, Astronomick, Chymick, Physicall and Anatomick Tryalls.
>
> Allurement to members absent:
> >An Account of these once a month.

Since Hooke himself was responsible for most of these items, he was justifiably proud of a society in which he was the central figure. One of his early accomplishments, of interest to oceanographers, was the invention of an apparatus for

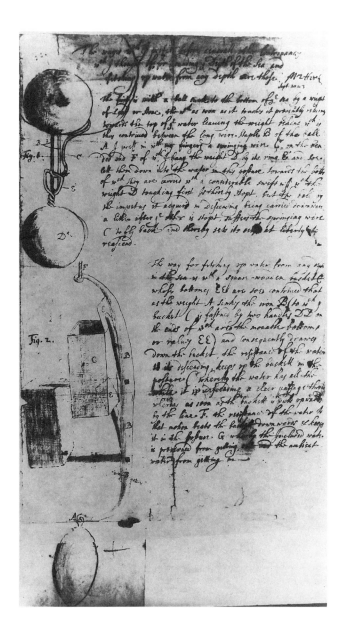

*Figure 1-7. Facsimile reproduction of Hooke's design for a sounding gear without a line and a way to sample seawater at any depth. (Reproduced with the kind permission of the President and Council of the Royal Society, London.)*

sounding the depth of the sea without a line and another for collecting deep-sea water samples from any depth (Fig. 1–7). He showed both designs to Royal Society members at their meeting of September 30, 1663, who were so impressed that they requested the sounding gear be made so it could be shown to King Charles II. He was also instrumental in designing and building a diving apparatus which was tried out in May 1664. The operator was able to breathe under water through a pipe connected to an air-box for four minutes.

Some contemporary reaction to the activities of the Royal Society and to Hooke as a scientist/experimentalist can be found in Samuel Pepys' diary. For example, on the day that Mr. Povy proposed Pepys to membership in the Royal Society, Hooke and Boyle had demonstrated experiments on combustion. Pepys's entry (February 15, 1665) reads: "But it is a most acceptable thing to hear their discourses and see their experiments, which was this day upon the nature of fire and how it goes out in a place where the ayre is not free, and sooner out where the ayre is exhausted; which they showed by an engine on purpose."

And on March 1, 1665, Pepys wrote in his diary:

> And thence to Gresham College, where Mr. Hooke read a very curious lecture about the late comett among other things, proving very probably that this is the very same comett that appeared before in the year 1618, and that in such a time probably it will appear again—which is a very new opinion—but all will be in print.

Here is one of many examples in his life of an important Hooke discovery that has been credited to someone else. Edmund Halley (1656–1742) is usually given credit for discovering that some comets are periodic. In 1665, when Hooke gave this lecture, Halley was only nine years old. Indeed, during this period Hooke was very much involved in astronomy, trying out different telescopes and making improvements, such as designing a new quadrant which, though small in size, enhanced the precision of celestial observations. On May 9, 1664, Hooke had discovered Jupiter's spot with a twelve-foot telescope observing that within two hours the spot had moved from east to west about half the planet's diameter, thus demonstrating as well the planet's rotation. He published his findings on March 6, 1665. The Italian astronomer Giovanni Cassini also studied this Jovian feature; at that time, it was known as "Hooke's Spot" (Schartzenburg, 1980).

On February 21, 1666, Pepys's diary reads: "Thence with my Lord Brouncker [who was president of the Society] to Gresham College, the first time after the sickness that I was there, and the second time any met. And hear a good lecture of Mr. Hookes about the trade of felt-making, very pretty." This lecture on felt-making is apparently lost, as no other record of it remains, testifying to the fact that unpublished Hooke manuscripts might yet turn up. "The sickness" Pepys referred to was, of course, the plague. The Royal Society recessed on June 28, 1665, and the members retired to places away from London where the plague was most prevalent. Wilkins, Sir William Petty, and Hooke went to Durdans, near Epsom, seat of Lord Berkeley. The three accomplished much during their sojourn there. John Evelyn, a frequent visitor, found them "contriving Chariotts, new riggs for shipps, a Wheele for one to run races in, & other mechanical inventions." It was Evelyn's opinion that "perhaps three such persons together were not to be found elsewhere in Europe, for parts and ingenuity" (Shapiro, 1969, p. 199–200). They also made plans for experiments on temperature, gravity, pressure, motions of the pendulum, sound, respiration, combustion, condensation, and others. Other members of the society kept in constant touch with the three at Durdans. Hooke and Boyle, especially, corresponded regularly. Hooke's first letter to Boyle written from Gresham College after his return to London was dated February 3, 1666. The plague abated by then and meetings of the Royal Society resumed later that month. Most certainly Hooke always remained at the center of intellectual activity among his peers both on an academic and on a social level. On August 8, 1666, Pepys wrote in his diary:

> Discoursed with Mr. Hooke a little, about the nature of musicall sounds made by strings, mighty prettily; and told me that having come to a certain number of vibracious proper to make any time, he is able to tell how many strokes a fly makes with her wings (those flies that hum in their flying) by the note that it answers to in musique during their flying. That, I suppose, is a little too much raffined; but his discourse in general of sound was mighty fine.

In 1666, Hooke invented the sextant and reported to the society that he had a method for observing positions and distances of fixed stars from the moon by reflection; he illustrated his explanation by diagrams. Much later, in 1691, Halley designed an instrument so similar that he was forced to withdraw his claim of invention when he was shown that Hooke had the idea a quarter of a century earlier (Chapin, 1953).

Another entry of Pepys's diary, November 14, 1666, relates the success of a blood-transfusion experiment in which Hooke participated along with some "doctors of physic" who were assigned to observe the procedure:

> Dr. Croone told me that at the meeting at Gresham College tonight (which it seems they now have every Wednesday again) there was a pretty experiment, of the blood of one dogg let out (till he died) into the body of another on one side, while all his own run out on the other side. The first died upon the place, and the other very well, and likely to do well. This did give occasion to many pretty wishes, as of the blood of a Quaker to be let into an Archbishop, and such like. But, as Dr.Croone says, may if it takes be of mighty use to man's health, for the amending of bad blood by borrowing from a better body.

Regarding a similar experiment, Pepys records on 8 August 1666: "This noon I met with Mr. Hooke, and he tells me the dog which was filled with another dog's blood, at the College the other day, is very well, and like to be so as ever, and doubts not its being found of great use to men; and so do Dr. Whistler, who dined with us at the taverne." These records of Pepys certainly underscores the variety of activities in which Hooke participated.

## MICROGRAPHIA (1665)

Hooke's microscopical investigations resulted in one of the most famous books in the history of science (Fig. 1–8), and the dedication to Charles II demonstrates Hooke's beautiful prose (Fig. 1–9). The book was much more than pictures of objects and organisms enlarged under the microscope (Fig. 1–10). Hooke essentially discovered microorganisms; his description of a fossil Foraminifera (see Fig. 2–4) shows his comprehension of its nature, and his depiction of the two living species of microfungi was the earliest recorded (Fig. 1–11; Bardell, 1988). In Observ. no. 18, Hooke examines thin slices of cork under the microscope and, coining the term "cells" to describe the structure, was the first to discern the cellular structure of plants (Fig. 1–12). Suddenly a whole world was revealed, the existence of which no one had ever suspected. His own drawings of what he saw under the microscope were much admired by all who perused the book. Some of his depictions must have been rather disconcerting; for example, the common house fly (Fig. 1–13), the clusters of the composite eyes of a fly (Fig. 1–14) or a louse holding a human hair, which may have prompted some to scurry to wash their hair (Fig. 1–15). In the book he also expounds on all sorts of phenomena, offering plausible hypotheses in explanation. Pepys records in his diary that he stayed up till two o'clock in the morning reading *Micrographia*, and we know that Newton read the book carefully taking copious notes ('Espinasse, 1956, p. 58).

*Micrographia* contains 57 microscopic and three telescopic observations; the latter are placed at the end of the book, almost as an afterthought, but some points he made in relation to the telescopic observations are of significance in earth science and will be discussed later in relation to crater formation (Observ. no. 60).

*Figure 1-8. Title-page of* Micrographia *(1665).*

*Figure 1-9. Hooke's dedication in* Micrographia *(1665).*

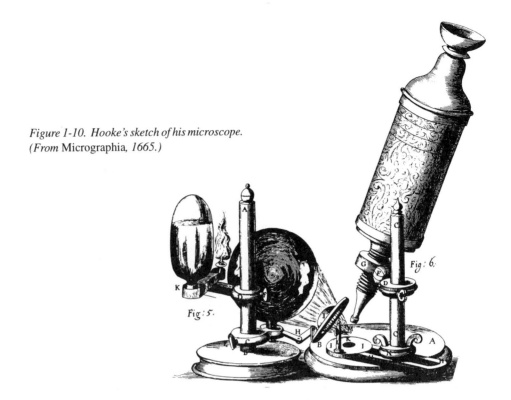

*Figure 1-10. Hooke's sketch of his microscope. (From* Micrographia, *1665.)*

*Figure 1-11. Hooke's depiction of microfungi in* Micrographia *(1665).*

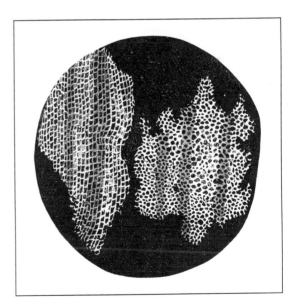

*Figure 1-12. Hooke's drawing of the structure of cork in reference to which he coined the term "cells." (From* Micrographia, *1665.)*

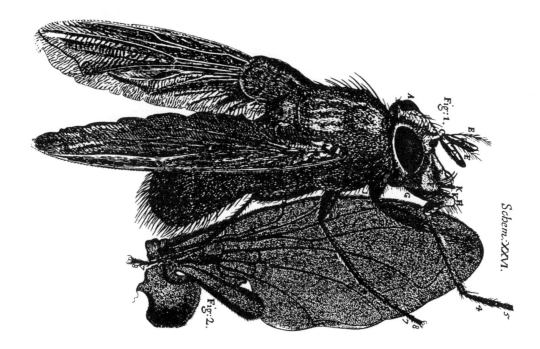

*Figure 1-13. Hooke's drawing of a fly. (From* Micrographia, *1665.)*

*Figure 1-14. Hooke's illustration of clusters of the compound eyes of flies. (From* Micrographia, *1665.)*

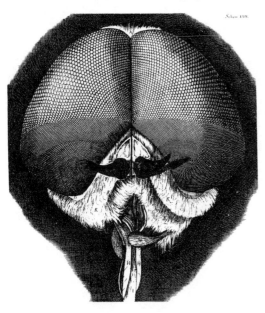

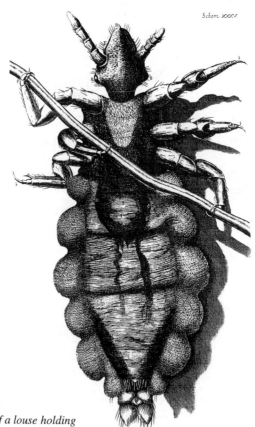

*Figure 1-15. Hooke's depiction of a louse holding a human hair. (From* Micrographia, *1665.)*

The book also contains descriptions of some of his inventions and innovations, such as his improvements to the thermometer, described in Observ. no. 7. Hooke was the first to use the freezing point of water as zero. He wrote, "A *Thermometer*, thus marked and prepared, will be the fittest Instrument to make a Standard of heat and cold that can be imagined. For being sealed up, it is not at all subject to variation or wasting, nor is it liable to be changed by the varying pressure of the Air, which all other kind of *Thermometers* that are open to the Air are liable to." He was keenly aware that without accuracy in quantitative measurements and universal standards by which all scientists can communicate, science could not develop. He was essentially the first meteorologist, having invented the wheel barometer, the wind gauge, and the hygrometer for measuring the moisture in air. With the aid of these inventions, he devised a way of keeping track of weather and long-range climate records.

Hooke made observations and conclusions of far-reaching importance in *Micrographia*—on optics, theories of combustion, respiration, the cellular structure of plants and, notably, geology. He described heat as a vibratory motion or agitation and distinguishes it from combustion for which air is required. His theory of combustion (Observ. no. 16) is another achievement that has been largely ignored by history. The theory, published nine years before that of John Mayow, asserts that burning cannot take place without air because some substance in the air is necessary, and, in fact, after combustion "the air [is] diminished one twentieth part." And in Observ. no. 22 he described the importance to respiration of this substance inherent in air, which is like "that which is fixed in Saltpetre." He maintained that air "loses" something in the lungs, so that renewing "that property in the Air which it loses in the Lungs, by being breath'd, that one square foot of Air might last a man for respiration much longer, perhaps, then ten will now serve him of common Air." He was moving toward the discovery of oxygen. While Hooke's theory was expressed simply and clearly, Mayow's was indirect and obscure, in John Robison's (1803) words, "complicated, and wavering, mixed with much mechanical nonsenses of wedges, and darts, and motions, &c." Hooke, however, never disputed Mayow about priority of the discovery, even proposing Mayow as Fellow of the Royal Society on November 30, 1676. Robison considered Hooke "one of the greatest geniuses and most ardent inquirers into the operations of nature," and, he wrote, "I do not know a more unaccountable thing in the history of science, than the total oblivion of this theory of Dr. Hooke, so clearly expressed, and so likely to catch attention."[12] Interestingly, after Hooke's death in 1703, Newton published his ideas on the same subject in 1704 appended to his book on optics. As de Milt (1939) notes, Newton's ideas on combustion are "very, very much like those of Hooke."—de Milt's use of the double adverb is, I believe, a deliberate redundancy.

In Observ. no. 9, Hooke described the iridescent interference colors seen when light falls on a layer of air between two thin glass plates, explaining that the colors are produced by the combination of reflected light from the upper and the lower surfaces (Fig. 1–16). Yet posterity attributes the discovery of this phenomenon to Newton and names it "Newton's rings." In practically all encounters with Newton, Hooke's reputation comes off second best, not because what he had to say was inferior in quality or later in priority, quite the contrary, but because chroniclers seem always to give Newton the advantage. Those who recognize the importance of Hooke's work are often not heeded. For example, Whittaker (1951, p. 14) regards Hooke's theory of light as of great importance in the history of science, as it was transitional "from the Cartesian system to the fully developed theory of waves." In the Cartesian view, light is only a tendency to motion, and Hooke attacked this view by asserting that light is an actual motion, a rapid vibratory motion of small amplitude. Hooke also discovered the

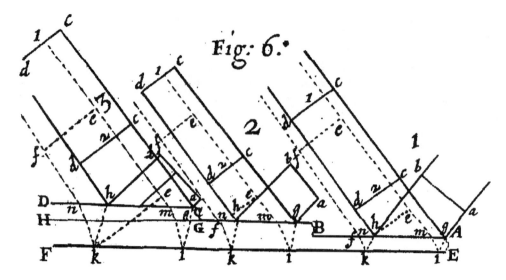

*Figure 1-16. Hooke's illustration of interference. (From* Micrographia, *1665.)*

phenomenon of diffraction (Observ. no. 58), which he refers to as "*inflection, differing both from reflection, and refraction,*" saying that light in air is not propagated exactly in straight lines "but that there is some illumination within the geometrical shadow of an opaque body." He also introduced the idea of the wave-front, even trying to determine what happens to the wave-front at the interface of two media. He concluded correctly that the side of the front arriving at the interface first will continue in the new medium at the velocity appropriate to that medium while the other side of the front is still moving with the old velocity in the first medium. In this way, the wave-front is deflected at the interface between two media. Hooke then supposed that this deflection causes prismatic colors, because in the process of refraction, white light, being a simple and uniform pulse orthogonal to the direction of propagation, is disturbed, and color is generated by the distortion that results.

Hooke, therefore, had failed to appreciate the significance of Newton's experiments with a prism. Even so, his own ideas on light are not insignificant, and his criticism of Newton's corpuscular theory certainly did not deserve the ire and lifelong hatred harbored by Newton. It is in another fundamental aspect of light, however, that Hooke proves to be representative of the thinking of his day, despite his many breakthroughs in other areas—the finite velocity of light, discussed later in this chapter.

But the greatest interest to geology in *Micrographia* lies in Hooke's hypotheses about the "figur'd stones," or fossils. Here his investigations into petrified wood (Observ. no. 17) led him to discussions about the origin of fossils. He alone, of all the 17th-century intellectuals, claimed an organic origin for fossils that is not related specifically to Noah's flood. He wrote (p. 111):

> all these and most other kinds of stony bodies which are formed thus strangely figured, do owe their formation and figuration, not to any kind of *Plastick virtue* inherent in the earth, but to the Shells of certain Shelfishes, which, either by some Deluge, Inundation, Earthquake, or some such other means, came to be thrown to that place, and there to be fill'd with some kind of Mudd or Clay, or *petrifying* Water, or some other substance, which in tract of time has been settled together and hardned in those shelly moulds into those shaped substances we now find them.

Hooke also depicted and described the first Foraminifera ever recognized as a tiny shell in Schem. V, between p. 44 and 45 of *Micrographia* (see Fig. 2–4 of this volume).

Observ. no. 13, on "small Diamants, or sparks in Flints," reveals Hooke's remarkable intuition about the nature of matter. He carefully examined under his microscope several different kinds of *Crystaline*, or *Adamantine*, bodies. Then, using round bullets, sometimes referred to by him as globules, he demonstrated that they could be arranged to represent the various regular geometric shapes of these bodies, thus concluding that the outward appearance of crystals is a reflection of the internal arrangements of particles. Crystallography probably had its birth right here in *Micrographia* (Schneer, 1960 and Drake and Komar, 1981). He illustrated this concept in his 7th Scheme, depicted in Chapter 8.

Hooke found bullets extremely useful in demonstrating other ideas in geology as well. He used them to bombard a soft clay surface to simulate the craters of the moon. He added Observ. no. 60 partly because he had some empty space on his last plate, Scheme 38, drawing as Fig. 2 a part of the surface of the moon with the aid of a 30-foot-long telescope. The plate (Fig. 1–17) also shows the same area as seen by the German astronomer Johannes Hevelius, who called the area Mons Olympus, (Figure X shown in Hooke's plate), and by Ricciolus, who called it Hipparchus (Figure Y shown in Hooke's plate). Hooke's drawing demonstrates how far short both Hevelius and Ricciolus came to depicting the actual landscape.

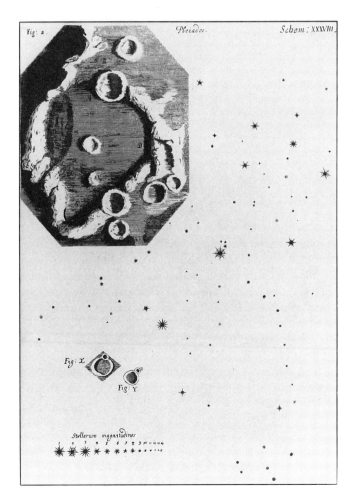

*Figure 1-17. Hooke's drawing of part of the lunar surface, showing craters, as seen through his telescope. (From* Micrographia, *1665.)*

A mild dispute between Hooke and Hevelius took place around this time. Hooke maintained logically that the use of telescopic rather than plain sights on astronomical instruments was advantageous, allowing for more accurate observations as well as being easier on the eyes. He wrote (Gunther, 1931, 8: 9): "I resolved to assist my eyes with a very large and good Telescope, instead of the Common sights, whereby I can with ease distinguish the parts of an object to Seconds." By the end of 1664, he had applied a telescope to a small quadrant for measuring seconds in contrast to the extremely large cumbersome instruments Hevelius and others were using. He showed his new quadrant to the Royal Society in February 1665. Gunther (1930, 6: 221, 237, 256) also records for March 21 of the same year that Hooke brought in yet another "small new quadrant which was to serve for accurately dividing degrees into minutes and seconds, and to perform the effect of a great one. It had an arm moving on it by the means of a screw, that lay on the circumference."[13] Hooke had been using this application of telescopic sight for his own observations, using crosshairs of spider's web or silk thread at the focus of the objective and eyepiece of the telescope. The same year he wrote to Hevelius, urging him to adopt telescopic sights. But Hevelius, a peevish conservative gentleman, felt he was doing just fine without such new-fangled improvements and would have none of it—and that was the end of the dispute (Cutlerian Lectures in Gunther, 1931, 8: 41ff). Typical of some historians' and other writers' neglect of Hooke is J. W. Olmstead's (1949) "definitive" exposition of the history of application of telescopic sights for astronomical measuring instruments, which fails to mention Hooke's use of telescopic sights and crosshairs (at least by 1665) and precedes what Olmstead considers the earliest application, 1667–69.[14]

What is significant for geology in Observ. no. 60 is Hooke's discussion of the two processes that could have produced the cratered lunar terrain. One raging controversy of today is whether it was a meteorite strike or some volcanic activity that caused the extinction of many living things at various periods during geologic history, such as the dinosaurs' demise at the end of the Cretaceous. Here was Hooke contemplating these two processes more than 200 years before G. K. Gilbert's 1893 classic publication on the origin of the lunar surface (see Drake and Komar, 1984).

When Galileo first looked at the moon through his telescope in 1609, he discovered that it was not the smooth Aristotelian sphere as was thought. From shadows cast by the lunar features Galileo realized that lunar mountains necessarily exceeded in height the mountains in his native Italy. He also observed and sketched circular depressions he called "spots," but did not speculate on their origin. The idea of craters on the moon was still so foreign to the latter part of the 17th century that Hooke apparently felt called upon to describe them in *Micrographia*. He depicted them (p. 243) as shaped "almost like a dish, some bigger, some less, some shallower, some deeper, that is, they seem to be a hollow *Hemisphere*, incompassed with a round rising bank, as if the substance in the middle had been digg'd up, and thrown on either side."

Hooke then devised two experiments to reproduce these "pits" and offered two hypotheses to explain their origin. His first experiment was to bombard a mixture of water and tobacco-pipe clay with a "heavy body, as a Bullet." This "would throw up the mixture round the place, which for a while would make a representation, not unlike these of the Moon." The second experiment involved boiling a pot of alabaster where "the eruption of vapours" reduced the powder "to a kind of fluid consistence." Hooke noted in Observ. 60 (p. 243–244) that if he gently removed the pot from the fire while it was boiling:

the Alabaster presently ceasing to boyl, the whole surface, especially that where some of the last Bubbles have risen, will appear all over covered with small pits, exactly shap'd like these of the Moon, and by holding a lighted Candle in a large dark Room, in divers positions to this surface, you may exactly represent all the *Phænomena* of these pits in the Moon, according as they are more or less inlightned by the Sun.

Picture the energetic Hooke thus in his bent frame shining a light this way or that over the hot alabaster, or perhaps slamming bullets onto a tray of soft clay late into the night!

Of the two explanations for cratering, eruption or impact, Hooke favored the eruptive one, therefore the volcanic theory, because in the impact idea he could not imagine "whence those [bombarding] bodies should come," as meteorites were not known as such in his day. He also realized that a pipe-clay mixture might not be analogous to the lunar surface, questioning that the moon should be made of such a soft substance. Hooke's difficulty in envisioning a crater-forming body could also have been a reaction to the prevalence of superstition and mysticism connected with such phenomena as shooting stars and other bright objects in the sky. Such superstition persisted well through the 18th century. In Paris during the Age of Reason, l'Académie des Sciences, for example, in an effort to put an end to irrational behavior on the part of most people, simply decreed that objects could not fall from the sky, and museums all over Europe, then, discarded their meteorite specimens that had been brought in by eyewitnesses of such events. The eminently rational 17th century Hooke probably instinctively separated himself from what he would have considered irrational explanations and thus could not conceive of objects falling from the sky to produce the pits (Drake and Komar, 1984).

An important corollary to Hooke's volcanic theory for the origin of lunar craters were the conclusions he made in *Micrographia* concerning the Earth itself. His interest in earth science began very early in his career and spanned the entire 30 years or so of his professional life. He wrote (p. 243–244):

And that there may have been in the Moon some such motion as this, which may have made these pits, will seem the more probable, if we suppose it like our Earth, for the Earthquakes here with us seem to proceed from such cause, as the boyling of the pot of Alabaster, there seeming to be generated in the Earth from some subterraneous fires, or heat, great quantities of vapours, that is, of expanded aerial substances, which not presently finding a passage through the ambient parts of the Earth, do, as they are increased by the supplying and generating principles, and thereby (having not sufficient room to expand themselves) extreamly condens'd, at last overpower, with their *elastick* properties, the resistence of the incompassing Earth, and lifting it up, or cleaving it, and so shattering of the parts of the Earth above it, do at length, where they find the parts of the Earth above them more loose, make their way upwards, and carrying a great part of the Earth before them, not only raise a small brim round about the place, out of which they break, but for the most part considerable high Hills and Mountains, and when they break from under the Sea, divers times, mountainous Islands; this seems confirm'd by the *Vulcans* in several places of the Earth, the mouths of which, for the most part, are incompassed with a Hill of a considerable height, and the tops of those Hills, or Mountains, are usually shap'd very much like these pits, or dishes, of the Moon.

Hooke's description of this eruptive process is essentially that of a steam explosion, a method for crater formation adhered to by many ever since. While there is no evidence that he believed the moon was created from a primeval soup similar to his pot of boiling alabaster, this idea has persisted through the centuries in such authors as Rozet (1846), St. Clair Humphreys (1891), and Mills (1969), who all performed experiments involving crater-formation on a boiling surface. Even the tidal theory for the formation of lunar craters popularized by George Darwin in the 1880s implies a hot, liquid moon. Steam explosion was also the idea most accepted for the origin of Meteor Crater in Arizona after the distinguished geologist G. K. Gilbert favored it over the impact hypothesis in spite of much evidence supporting impact.[15] Because Hooke's *Micrographia* was widely read in the 18th and 19th centuries it is reasonable to assume that his ideas had a profound influence far beyond his time.

## A MOST PRODUCTIVE IDEA

On March 21, 1666, Hooke lectured on gravity and discussed his attempts at measuring gravity with the pendulum. He also used a conical pendulum to describe the dynamics of orbital motion. The year 1666 was historically momentous for Hooke not only because of the Great Fire of London the rebuilding of which he became very much involved, but also because this was the year that he read his most significant paper to the Royal Society (on May 23) concerning the "Inflexion of a direct Motion into a Curve by a Supervening Attractive Principle." In this paper he first described his celestial mechanics, central to which is the expression of his idea that an attractive force at the center of the planetary system causes an object moving tangentially to be deflected from a straight line into a curve. This concept of centripetal force, necessary to the universal law of gravitation and later (in 1679) communicated to Newton, allowed the latter to think correctly about gravitation. Up to that moment, Newton wrote only in terms of the exact opposite effect, that of the *vis centrifuga* (centrifugal force). In the course of the correspondence between Newton and Hooke in the year 1679–1680, Hooke presented Newton with a complete statement of the gravitational problem including the following points:

(1) the force of attraction between two bodies is inversely proportional to the square of the distance between them;
(2) this force differs within the body of the Earth;
(3) the attraction decreases with the increasing centrifugal effect in low latitudes;
(4) the gravitational force of attraction provides the centripetal force necessary to keep planets in orbit—in Hooke's own words to Newton, "compounding the celestial motions of the planets of a direct motion by the tangent and an attractive motion towards the central body";
(5) and calculations should be made from the centers of the Sun and planets.

Hooke published his ideas on celestial mechanics in his *Attempt to prove the motion of the earth by observation*, London, 1674, 13 years before the publication of *Principia*. Newton, however, claimed to have arrived at his universal law of gravitation at his country home in Woolsthorpe during the plague years 1665 or 1666 (it is not clear which), during his *annus mirabilis* (his "marvelous year" when the legendary apple fell). This date, of course, would clearly predate

Hooke's expression of the law except that there is clear proof that as late as 1675, Newton still thought that the planets and Sun were kept apart by "some secret principle of unsociableness in the ethers of their vortices," and that gravity was due to a circulating ether that had to be replenished in the center of the Earth by a process like fermentation or coagulation" (letter to Oldenburg December 7, 1675, Turnbull, 1959, vol. 1: 368; Patterson, 1950, p. 32–33). Hooke, in contrast, possessed a highly sophisticated understanding of the gravitational theory at least by 1679, and most likely for at least a decade before. He definitely formulated the physical hypothesis and stated the mathematical problem, although there is no evidence extant that he followed this with mathematical analysis. Many of Hooke's papers, however, are lost, some perhaps have even been put into unsympathetic hands and therefore destroyed (Patterson, 1949, p. 339–341). Patterson considers that "the Leibniz-Hooke letters would be of particular interest could they be found, since they were from the period of Hooke's correspondence with Newton on the subject of gravitation, and touched upon the 'universal algebra' upon which Leibniz was then at work" (Patterson, 1949, p. 340–341). Leibniz, of course, was another enemy of Newton and a co-founder of the calculus. A diagram constructed by Hooke was found recently among the Wren papers in the Trinity College Library, Cambridge, and photocopied by Pugliese (1989). Physicist Michael Nauenberg (1994) analyzed the diagram and concluded that Hooke was indeed very close to at least a geometric solution of the elliptical orbit. By the end of 1679, Hooke had claimed to his friends that he had worked out the whole theory; he noted in his diary for January 4th, 1680, "— perfect Theory of Heavens." The only proof that Newton had developed his theory by the plague years is some unsubstantiated statements made by Newton himself to friends, one of them, William Whiston, almost 30 years after the supposed event. An example of Newton's untruthfulness is cited by Patterson (1950): Newton wrote a letter to Oldenburg on June 23, 1673, to be forwarded to Huygens, in which he claimed to have expressed his gravitational theory and then referred to this letter in his later correspondence with Halley as evidence of his priority over Hooke. Later authors have supported Newton's claims by quoting this passage from a copy of his June 23rd letter, reposited at the Royal Society. But Huygens' editors have proved with photostatic copies that the passage concerning gravitation does not appear in Huygens' copy of the letter. The Royal Society copy, therefore, must have been doctored to produce the desired effect. Newton gained his vast reputation, therefore, partly by planting evidence to establish priority, if not by himself, at least by his followers. He became less than civil to Hooke and always refused to give him credit for this most productive idea. Hooke was deeply saddened and hurt by this neglect of one of his greatest achievements. Hooke's discussion of his ideas on the Earth in space and the Earth's system is in Chapter 4.

## THE GREAT FIRE AND REBUILDING OF LONDON

On September 2, 1666, London was subjected to an even greater devastation than the Plague. Fire broke out in Pudding Lane not far from the wharves along the Thames near London Bridge, and within four days it spread over almost the entire city within the walls. As the fire raged across Cheapside, it engulfed the street, destroying houses, Goldsmiths' Row, Bow Church, and many famous old taverns such as Mermaid Tavern and The Mitre, frequented by Shakespeare and Ben Jonson. In its wake the fire also destroyed the Royal Exchange, the Guildhall, and St.Paul's Cathedral.Of the 109 churches in the city, 84 were totally destroyed;

373 acres within the walls and 63 acres outside the walls were burnt, and 13,200 houses in over 400 streets were demolished (Bell, 1923).

By chance, because the fire was swept westward by the wind, in the northeastern part of the city it stopped almost at the mouth of Bishopsgate where it meets with Cornhill and Grace Church Streets. Both Gresham College (where Hooke lived) and St. Helen's Church (where he was buried) were in Bishopsgate and so were spared. The college opened its doors to the city administrators to conduct their affairs in place of the burnt Guildhall and also invited the city merchants to use it as the locus for commercial exchange because of the destruction of the Royal Exchange. Through the kindness of Henry Howard, later duke of Norfolk, the Royal Society was moved to Arundel House in the Strand for their meetings.

The fire stopped spreading on September 12, and while the ashes were still smoldering, Hooke, talented in drafting and artistic skills, drew up plans and designed a model for rebuilding London. He exhibited his model first to the Royal Society on September 19 and then on September 21 to the Common Council, the City Corporation. Christopher Wren and John Evelyn also submitted plans. The lord mayor and the aldermen preferred Hooke's plan, while the king preferred Wren's. Hooke's model, unfortunately, has not survived, or at least it has not yet surfaced from its forgotten resting place. Bell (1923) describes the model as "the prototype of the modern American city." Somewhat similar to New York City, the main streets were laid out east to west, forming a grid with north-south cross streets. Bell remarks (p. 239), "Let us be thankful that this utilitarian scheme was not adopted, for it would have cleared from the ground as completely as marks are sponged off a slate every link with London's great and historic past."

As it turned out, all three models were considered too radical and none of them was adopted, as the city fathers quickly realized that the transfer of property among the citizenry necessary under any of the plans would have involved great confusion, suspicion, and boundary disputes. Rebuilding, therefore, could only take place on the old foundations. As a result of the City Council's initial approbation of his model, however, Hooke was appointed one of three city surveyors. The king appointed Wren surveyor general for the rebuilding while the city selected the city surveyors. Wren, along with Hugh May and Roger Pratt, formed the king's commissioners, with Wren being the principal figure, while Hooke, Peter Mills, and Edward Jerman were the city surveyors. Jerman died within two years and May died in 1673. Thus Hooke essentially became chief city surveyor and, because of his friendship with Wren, also Wren's close associate as architect.

In spite of the ungenerous statement of Wren's son in *Parentalia* (1750) in which Hooke's part in the rebuilding was relegated chiefly to "the Business of measuring, adjusting, and setting out the ground of the private Street-houses to the several Proprietors" while all the churches and "publick Works" were reserved to Wren's "own peculiar Care and Direction," Wren and Hooke worked together closely, almost as partners. The two had been good friends at Oxford, and now Hooke was Gresham Professor of Geometry and Wren had been Gresham Professor of Astronomy. Wren also may have been related to Hooke. One of Wren's sisters married a John Hooke, who appears to have belonged to the Hampshire Hookes of which Robert Hooke is reported to have been a member. Wren often addressed Hooke as "Cousin." Indeed, Hooke's diary entries indicate close working and social relationships between the two men. Generally, Wren was in charge of rebuilding most of the churches while Hooke directed the rebuilding of many of the public buildings and other structures, such as bridges, canals, and quays. But Wren often consulted with Hooke, even in regard to the churches.

A prime example is the rebuilding of St. Paul's Cathedral. The central and most impressive structure of the cathedral is its dome, and Hooke used his scientific knowledge as well as his ingenuity to advise Wren in its design. As has been pointed out by Margaret 'Espinasse in her biography of Hooke, between August 1672 and December 1680, there are more than a hundred entries in Hooke's diary about St. Paul's, and each one indicates a visit to the site, sometimes accompanied by Wren. Hooke's advice to Wren on the design of the dome came in the form of an ingenious model of the structure made from chain mail. He hung it upside down so that the links were tensionally stretched by gravity, then added weights and additional links at key places until he achieved the shape he desired. He was quite aware, therefore, that masonry could transmit only compressional forces, not tensional ones. This method produced exactly the right shape for the shell dome, and if one imagines that the perfected chain mail model dipped in plaster and allowed to set, then, when turned right side up, as the distinguished London engineer, E. C. Hambly (1987, p. 5–10) pointed out, "all the tension forces reverse to become pure compressions with a distribution ideal for masonry." This highly original method of design was adopted by Wren.

During the period of rebuilding, therefore, Hooke devoted many hours thinking about the physical principles involved in architecture. He formulated a general theory on arches, noting in his diary (June 5, 1675), "He [Wren] was making up of my principle about arches and altered his module by it." He announced his principle of the arch at the end of his Cutlerian lecture on Helioscopes as follows: "The true Mathematical and Mechanical form of all manner of Arches for Building, with the true butment necessary to each of them. A Problem which no Architectonick Writer hath ever yet attempted, much less performed." He then put the secret in the following anagram: abcccddeeeeefggiiiiiiiillmmmmnnnnn ooprrssstttttuuuuuuuux.[16]

Hooke was also involved with the famous London landmark erected in 1677 to commemorate the fire, the Monument (Fig. 1-18), another structure generally credited to Wren. It is now accepted by many that Hooke was the designer and chief architect of this structure. It is a 202-foot column—the distance from the king's baker's house in Pudding Lane, where the fire was suspected to have started, to the site of the monument on Fish Street Hill. Wren's designs for this monument were, in fact, not only not completely approved by the king, but Wren himself was not totally satisfied with them. One of his rejected designs, for example, shows a phoenix at the top of the column (Fig. 1-19).[17] Hooke's diary between 1673 and 1679 contains many references to work with "the Piller," including perfecting, on October 19, 1673, the "module of Piller." Today, the brochure one buys at the base of the Monument still credits Wren as the architect but notes "in collaboration with Robert Hooke." It would be more accurate to say that the Monument was designed by Robert Hooke, in collaboration with Christopher Wren.

Wren had so much to do that he must have placed Hooke in charge of many projects. With little or no sleep, Hooke was madly dashing around the city on foot (he apparently did not ride in any conveyance), seeing to the building of this or that structure or the laying out the lines of boundaries of public and private property. He was, of course, still giving his Cutlerian and Gresham lectures and carrying out many of his scientific experiments and making inventions. His architectural work in the streets of London made him quite rich, which was a good thing, because his income from the Royal Society was often in arrears and his grant from Sir John Cutler was nonexistent. His diary indicates that over and above what the City paid him as City Surveyor, Wren was also paying him for his assistance in the rebuilding of almost all the City churches (Batten, 1936–37).

*Figure 1-18. The Monument, London, designed and built by Hooke to commemorate the Great Fire of 1666. (Photograph by P. D. Komar.)*

*Figure 1-19. One of Wren's designs for the London fire monument. (Reproduced with the kind permission of the Warden and Fellows of All Souls College, Oxford.)*

It is now clear that more than a few structures constructed during the rebuilding were designed and executed by Hooke but were attributed to Wren. There are several reasons for this: (1) Hooke and Wren worked so closely together that it was difficult to untangle the responsibility of each; (2) Wren had a greater reputation as an architect, as he had designed the famous Sheldonian Theatre at Oxford and further he *was* the surveyor-general, so that in the absence of diligence in research, one is apt to credit him with the entire rebuilding; (3) Hooke was appointed city surveyor and therefore, again, if lacking a curiosity for the truth, one is apt to conclude that he must have simply surveyed the boundary lines; (4) a less tangible reason is probably simply Hooke's bad luck. He was in charge of many public works such as bridges, ditches, and the quay on the north bank of the Thames. The architects of such works are usually unnamed and tend to be replaced in time, while Wren's churches, with the help of Hooke, were built for eternity. Also, several of Hooke's buildings were destroyed by fire or demolished to make room for other construction.

Although architecture was not a primary occupation with Hooke, his work as an architect was significant, and his services were sought by well-known members of the aristocracy. He designed and built a number of important structures, some of them quite beautiful. His reputation, therefore, should not be as obscure as it is in this respect. Without Hooke's assistance surely Wren could not have accomplished all that have been credited to him. Contemporary praise and recognition of Hooke were not lacking, but somehow, through inadequate writing of history, Hooke's name is obscure today. Some historians, perhaps awed by the historical stature of household names like Wren and Newton, tend to find Hooke second-class and relegate a lesser place to him than he deserves, even when such assessment has been shown to be based on inadequate or erroneous evidence. It is a curious phenomenon involving academic personality, not having much to do with the true role in history of Hooke either as a scientist or as an architect. Some of Hooke's better known structures (and actually attributed to him) are listed in Appendix B. A few are pictured here (Figs. 1–20, 1–21, 1–22, 1–23).

*Figure 1-20. Hooke's drawing of the gates and dome of the Royal College of Physicians designed and built by him. (From Gunther, 1930, vol. 7.)*

*Figure 1-21. Bethlehem Hospital for the mentally ill, better known as Bedlam, 1675. (From Gunther, 1930, vol. 6, frontispiece.)*

*Figure 1-22 Montague House, 1676, Bloomsbury, London. (From Gunther, 1930, vol. 7, frontispiece.)*

*Figure 1-23. Ragley Hall, Warwickshire, designed by Hooke in 1680. (Photograph by E. T. Drake.)*

During the years 1666 and 1667, although deeply involved in the rebuilding effort, Hooke kept up with his scientific work and inventions. He produced a 2-foot contracted reflecting telescope that compared favorably with one of 6-foot length. He recognized that a reflecting telescope is most useful as a helioscope because it was safer for one's eyes. He devised a way of making climatic observations and keeping weather records using all the meteorological instruments he himself invented—a sealed thermometer scaled to the freezing point of water as zero, a hygroscope with a single beard of a wild oat, a barometer, and a wind guage that measures both direction and strength (Fig. 1–24). On November 28, 1666, he showed the Royal Society a bubble level he invented, useful for many purposes even today. He made and experimented with new types of pendulums, made a clock with a circular pendulum, constructed a sea bucket for sampling or "fetching things" from the bottom of the sea. He also proposed the measurement of the circumference of the Earth "with a 12-foot telescope and three stakes to be

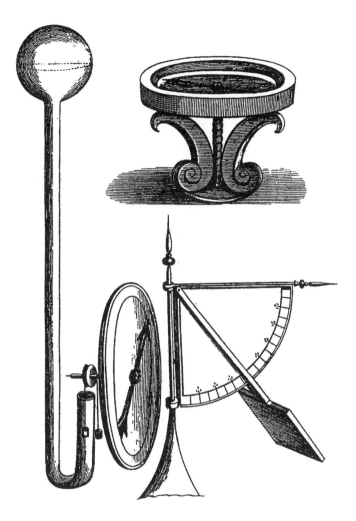

*Figure 1-24. Some of Hooke's meteorological instruments: barometer, hygrometer, and wind gauge. (From Gunther, 1930, vol. 6)*

practised in St. James's Park on a calm day." Between 1667 and 1670, Hooke conducted many other experiments involving the circulation of blood, transfusion of blood from one dog into another, respiration, and some experiments involving vivisection. He had a dog's thorax surgically opened and kept the animal alive by blowing into his lungs with a pair of bellows. Hooke's other activities included experiments with dyes and cloth-making, cataloguing the Royal Society library, measuring a degree of latitude on the Earth, and many others.

In spite of Hooke's diligence in his duties, John Cutler still failed to pay Hooke for his lectures, and his annual meager £30 salary from the Royal Society continued to be ephemeral and materialized only in a haphazard way. In November of 1670 Hooke was even censured by the council for "the neglect of his office" (Gunther, 1930, 6: 370). Without doubt the council recognized how important Hooke was to the survival of the society and was desperate to hold him to his work there. All too often the society's regular meetings apparently had little to discuss if Hooke was absent, as for example, the entry of November 24, 1670: "Mr. Hooke being absent from this meeting, no experiments were provided."

## THE MATTER OF THE WATCH

Hooke persisted in his horological work and the development of the chronometer throughout the decade following the fire. Earlier he had devised the anchor escapement and the spiral balance-spring—the former replacing the verge escapement for better accuracy and the latter minimizing the efffect of gravity that plagued the pendulum. His own account of his invention of the spring-controlled balance wheel follows (Waller, 1705, p. v):

> Immediately after his *Majesty's* Restoration [which was in 1660], Mr. *Boyle* was pleased to acquaint the Lord *Bouncher* [who was the president of the fledgling Royal Society] and Sir *Robert Moray* with it, who advis'd me to get a Patent for the Invention, and propounded very probable ways of making considerable advantage by it. To induce them to a belief of my performance, I shew'd a Pocket-watch, accommodated with a Spring, apply'd to the Arbor of the Ballance to regulate the motion thereof; concealing the way I had for finding the Longitude; this was so well approv'd of, that Sir *Robert Moray* drew me up the form of a Patent, the principal part whereof, *viz.* the description of the Watch, so regulated, is his own hand Writing, which I have yet by me, the discouragement I met with in the management of this Affair, made me desist for that time.

The truth of this statement is not to be doubted, as Waller in putting together Hooke's papers for the *Posthumous Works* discovered a "Draught of an Agreement between the Lord *Brouncker*, Mr. *Boyle*, and Sir *Robert Moray*, with *Robert Hooke* Master of Arts to this purpose." Under the terms of the document, Hooke was to disclose to them his method of measuring "the parts of Time at Sea," and the document also laid out the financial terms of the agreement. But Hooke refused to sign this agreement, which would have been financially beneficial to him, because it contained the following clause: *"That if after I had discover'd my Invention about the finding the Longitude by Watches (tho' in themselves sufficient) they, or any other Person should find a way of improving my Principles, he or they should have the benefit thereof during the term of the Patent, and not I."* One can hardly blame Hooke for not signing a document containing such a clause, it being, as Hooke himself noted in a postscript to his treatise on *Helioscopes*, 1676, *facile Inventis addere* (easy to add to an invention).

In refusing to sign this agreement because of the objectionable clause, Hooke then kept to himself his method for determining longitude. But instead of blaming Hooke for keeping this knowledge secret, as some writers have done, one should wonder why his colleagues did not simply delete the clause to which he objected in order to gain the important technique of accurate determination of longitude. Michael Wright (1989) made a detailed analysis of an unpublished manuscript of Hooke's discovered by A. R. Hall in Trinity College, Cambridge, in which Hooke described his effort in designing a marine timekeeper suitable for determining longitude at sea. Wright went so far as to draw a possible layout of Hooke's timekeeper, cleverly inscribing the words *R. Hooke invenit* and *M.T. Wright fecit*.[18]

Hooke was deeply hurt that the Royal Society failed to support him in this case. He has been accused of being miserly about money because of his meticulous record-keeping of small amounts of money and the discovery of his chest full of money at his death. This incident concerning his watch indicates his willingness to forego a large profit purely on principle. As Robinson and Adams (1935, p. xix) pointed out in their preface to Hooke's diary:

He was full of contradictions. Jealous over trivial money matters, yet almost exceptionally generous, he does not seem to have desired money for his own sake, but to have been greedily strict in claiming what he regarded as his rights. The fuller insight into his character given by the Diary makes it possible to believe that he was capable of abandoning a project if he felt that his personal rights were not fully admitted, even though such a course would be to his disadvantage materially. He can scarcely be blamed for keeping important knowledge secret, since it was a common practice at that time, necessitated by the organization of the scientific world and by the methods of publication.

Hooke was understandably embittered by the dispute over the inventions. As 'Espinasse points out, the Royal Society might have felt some collective guilt in its unfair treatment of their loyal curator, as when Henry Oldenburg, the society's secretary, died in September 1677, Hooke was immediately appointed secretary of the Royal Society along with Nehemiah Grew as co-secretary.[19]

## THE CUTLERIAN LECTURES

Although Hooke had read several papers in 1664 considered by him his first Cutlerian Lectures, only six were published as such, *Lectiones Cutlerianæ*, in one volume in 1679. Even these six were given years before the publication date and based on earlier experiments. Hooke had also given many others as the holder of the Cutler Lectureship. The early (1664) lectures were on the perennial problem of how to find longitude at sea, and records of the Royal Society (Gunther, 1931, 8: 149–150) support Hooke's claims in the matter of the watch, reporting his demonstration of the application of springs to the balance of a watch (controlling its action in a regular manner, thus making it useful for finding longitude.) In the preface to the volume, he wrote in reference to Oldenburg, who transmitted information abroad where it could be used before Hooke could claim credit, that he chose to publish this collection of physical, mechanical, geographical, and astronomical lectures, both to provide variety to the reader and to establish priority for himself—that is, "the securing of Inventions to their first Authors." Hooke explained that "there are a sort of Persons that make it their business to pump and spy out others Inventions, that they may vend them to Traders of that kind, who think they do ingenuously to print them for their own, since they have bought and paid for them."

The first lecture in the volume, originally published in 1674, is entitled "An attempt to prove the Motion of the Earth by Observation." Hooke had made observations and experimentation for the discourse as early as 1665, and later read it at Royal Society meetings in Gresham College in 1670. He had wished to publish this paper then but was dissuaded by others who wished him to do more trials and observations. Somewhat vexed by this delay, he wrote "To the Reader": "I do rather hast[e] it out now, though imperfect, then detain it for a better compleating, hoping it may be at least a Hint to others to prosecute and compleat the Observation, which I much long for."

At the time of his discourse (1670), the question whether the planets moved according to the Copernican system around the stationary Sun or according to the Ptolemaic or Tychonic versions, in which the Earth stands still, had not been completely resolved. The Jesuit astronomer Ricciolus, for example, had 77 arguments against the Copernican view. Hooke was convinced that the failure of discovery of parallax of the moving Earth hitherto was due to the ineffective instruments then used by the astronomers, not that there was no parallax. Most of the attempts to measure parallax were made with huge quadrants but with plain sights requiring naked-eye observations. Hooke knew that the unaided eye is unable to distinguish any angle much smaller than a minute. Earlier, while they were still in Oxford with the Wilkins Group, Hooke and Wren seemed to have experimented with such use of the telescope (Bennett, 1989, p. 31–32). Hooke built the first astronomical instrument using the telescope and eyepiece micrometer.

To avoid the complication of the difference of refraction of the atmosphere between June and December, Hooke made the ingenious decision to observe (Gunther, 1931, 8: 11):

> the passing of some considerable Star near the Zenith of Gresham Colledge, whether it did not at one time of the year pass nearer to it, and at another further from it: for if the Earth did move in an Orb about the Sun, and that this Orb had any sensible Parallax amongst the fixt Stars; this must necessarily happen, especially to those fixt Stars which were nearest the Pole of the Ecliptick.

The discourse describes in detail his equipment, which involved his telescope aimed vertically through two rooms and the roof of his lodging at Gresham College (Fig. 1–25). He chose the bright star in Draco and observed it at various times and found that it was 23 seconds more northerly in July than in October. He wrote:

> 'tis manifest then by the observations of July the Sixth and Ninth: and that of the One and Twentieth of October, that there is a sensible parallax of the Earths Orb to the fixt Star in the head of Draco, and consequently a confirmation of the Copernican System against the Ptolomaick and Tichonick.

Although Hooke thought he detected parallax, what he probably unknowingly discovered was a phenomenon known as stellar aberration. Stellar aberration, which shows the finite velocity of light, also demonstrates the motion of the Earth. The changing direction of the light ray from the star is the result of the combination of the changing motion of the Earth with the steady velocity of the starlight. James Bradley (1693–1726), who is credited with the discovery in 1725 of stellar aberration, explaining it as aberration of light in 1729, constructed an apparatus similar to Hooke's with a telescope made by George Graham. Bradley, with Samuel Molyneux, built the set-up at Molyneux's house and chose the same star in Draco to make their observations. It is entirely possible that Bradley knew of Hooke's experiment.

The importance of this lecture lies in three suppositions mentioned toward the end of the lecture that explain Hooke's "System of the World." This system differed "in many particulars from any yet known, answering in all things to the common Rules of Mechanical Motions":

> 1. "that all Cœlestial Bodies whatsoever, have an attraction or gravitating power towards their own Centers, ... that they do also attract all the other Cœlestial Bodies that are within the sphere of their activity."

> 2. "that all bodies whatsoever that are put into a direct and simple motion, will so continue to move forward in a streight line, till they are by some other effectual powers deflected and bent into a Motion, describing a Circle, Ellipsis, or some other more compounded Curve Line."

3. "that these attractive powers are so much the more powerful in operating, by how much the nearer the body wrought upon is to their own Centers."

It is apparent, therefore, that by this stage (1670) Hooke possessed a sophisticated notion of celestial mechanics.

The second Cutlerian Lecture, also published in 1674, is entitled *Animadversions On the first part of the Machina Cœlestis of the Honourable, Learned, and deservedly Famous Astronomer Johannes Hevelius Consul of Dantzick; Together with an Explication of some Instruments."* Essentially it is Hooke's critique of the instruments Hevelius and others used, and his explication of the superiority of employing telescopic sights and the best way to show divisions of a minute on a scale. He also describes his clock-driven telescope, which is activated by a conical pendulum, an invention generally attributed to Huygens although Hooke had discussed the conical pendulum before the Royal Society in the 1660s. Huygens published his *Horologium Oscillatorium* in 1673.

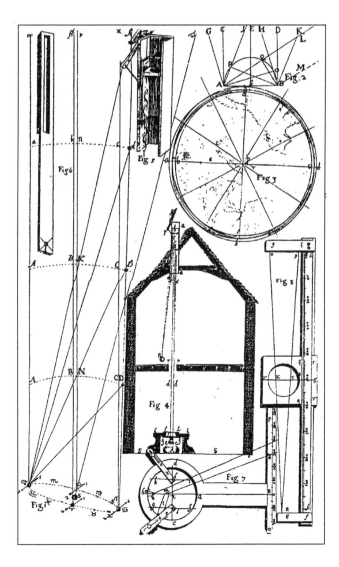

*Figure 1-25. Center of diagram shows Hooke's telescope set up through two floors of his lodgings in Gresham College. Top center shows the trapdoor mechanism on the roof. Other parts of the diagram illustrate apparatus and drawings for astronomical observations. (From Gunther, 1930, vol. 7)*

44  RESTLESS GENIUS

The third lecture, *A Description of Helioscopes, and some other Instruments*, was published in 1676. On the title page is an abbreviated Latin phrase, *Hos ego, &c. Sic vos non vobis ___ ___*. Obviously some words are missing in this quote. ("Those (or them) I, &c. Thus you not to you ___ ___"). The phrase may be loosely translated as "They are mine; thus they are not yours." Or, more specifically: "I invented them; thus you cannot claim them as yours." The reason I believe this last translation is appropriate is because Hooke adds a postscript to this discourse in which he claims priority over Huygens in the invention of the spring-controlled balance wheel of a watch.

The main part of the lecture describes a helioscope for observing the Sun without hurting one's eyes, a way of shortening reflective and refractive telescopes, and other instruments for astronomical or terrestrial measurements. In constructing such an instrument which typically needed to be rotated in various directions without shifting away from its original center, Hooke invented his "Universal Joynt." (Fig. 1–26) He found this universal joint to be most useful "for resolving almost all *Spherical Questions*, makes it of great use in *Navigation*" for taking azimuths and altitudes. Indeed, today we can hardly visualize not having such a useful thing—no car is without universal joints.

*Figure 1-26. Hooke's universal joint. (From Gunther, 1930, vol. 7)*

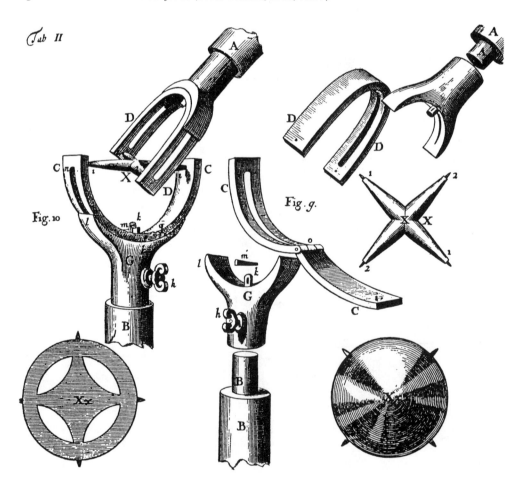

This tract also contains Hooke's observations of an eclipse of the moon that occurred on January 1, 1675, by "making use of a Telescope of eight foot, and my pocket-watch, whose ballance was regulated with springs." He then could not resist to add the following postscript:

> I should have here taken leave of my Reader for this time, but that finding in the *Transactions* a passage inserted out of the *French* Journal de Scavans, about the invention of applying a *Spring to the Ballance of a Watch, for the regulating the motion thereof*, without at all taking notice that this Invention was *first* found out by an *English*-man, and long *since published to the World*: I must beg the Readers patience, whilst I, in vindication of my own right against some unhandsome proceedings, do acquaint him with the state of this matter.

The *English*-man, of course, was Hooke himself; the author of the description of the invention in the French journal was Huygens, and the "unhandsome proceedings" referred to Oldenburg's part in the dispute (see note no. 18). There followed in the *Philosophical Transactions* a vehement but rather lame response from Oldenburg, secretary of the Royal Society and editor of the *Transactions*, claiming (no. 118, October 25, 1675, p. 440–442):

> that the Describer of the *Helioscope* [i.e. Hooke], some years ago, caused to be actually made some Watches of this kind, yet without publishing to the world a Description of it *in print*; but it is as certain, that none of those Watches succeeded, . . . until Monsieur Hugens . . . sent hither a Letter dated January 30, 1675, acquainting us with an Invention of his very exact Pocket-Watches, . . . ."

But Hooke's watch did succeed, as Andrade (1950, p. 443) reasoned: "I do not think that an agreement of this kind, drawn up by Brouncker, Boyle, Moray and Hooke, would have been drawn up unless a going watch had been produced."

The principal interest in Hooke's fourth Cutlerian Lecture (published in 1677), *Lampas: or a Description of some Mechanical Improvements of Lamps,* lies in his discussion of hydrostatics, his illustration of reflection and refraction of light (Fig. 1–27), and his criticism of "the learned Dr. More" [Henry More], and his idea of a "Hylarchick Spirit" in gravity and all other physical motions and effects. Hooke believed, of course, that these effects follow the general rule of mechanics, "that the proportion of the strength or power of moving any Body is always in a duplicate proportion of the Velocity it receives from it; that is, if any Body whatsoever be moved with one degree of Velocity, by a determinate quantity of strength, that body will require four times that strength to be moved twice as fast, and nine times the strength to be moved thrice as fast, "—as is true for the motion of bullets shot from guns or of pendulums moved by gravity and "all other Mechanical and Local motions," allowing only for friction. More, who was a Platonist at Cambridge, rooted in the classics and the teachings of the ancients, exemplifies a segment of the intellectual circles that strongly persisted in the 17th century. It was a period of a strange mixture of medievalism and rationalism, with this dichotomy often existing even within one individual, as with Newton himself. The perception that the 17th century represented an explosion of rationalism among the learned, free from supernaturalism and mysticism, is not accurate. Of all the thinkers of the era, however, Hooke seemed the least hampered by medievalism.

Hooke went on in this discourse to discuss such diverse topics as a waterclock, seeds of moss, and sunspots. He depicted several designs of his "New

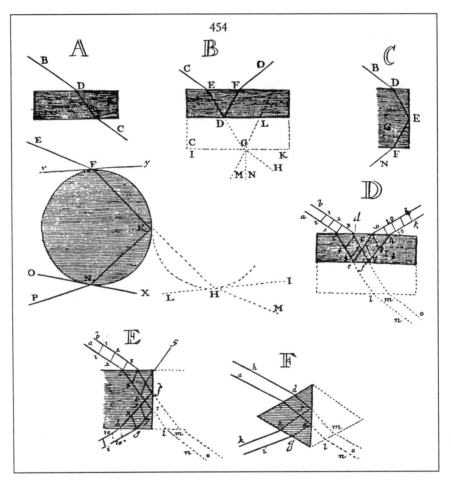

*Figure 1-27. Hooke's diagrams showing the principles of reflection and refraction. (From Cutlerian Lecture* Lampas, *1677; Gunther, 1931, vol. 8.)*

Principle for Watches"—that is, the use of balance wheels which he had shown to the Royal Society more than a decade before (Fig. 1–28). He again ended with a Postscript, here a tirade against Oldenburg's assertion in the *Philosophical Transactions* that Hooke had not published a description of his spring-controlled balance-wheel watch and did not possess a watch that worked. The intensity of Hooke's justifiable anger shows through in this tract:

> The Publisher of Transactions in that of *October 1675.* indeavours to cover former injuries done me by accumulating new ones, and this with so much passion as with integrity to lay by discretion; otherwise he would not have affirmed, that it was as certain that none of my Watches succeeded, as it was that I had made them several years ago: For how could he be sure of a Negative? Whom I have not acquainted with my Inventions, since I looked on him as one that made a trade of Intelligence.

Of interest to the history of geology is a point Hooke made in answer to Oldenburg's claim that "several discoveries of the Accuser [Hooke] had been vindicated from the usurpation of others." With barely suppressed fury Hooke responded, "the clean contrary is upon good grounds suspected from the Publication of a Book about Earthquakes, Petrifactions, &c. Translated and Printed by H. O. the manner of doing which is too long for this place. Such ways this mis-informer hath of vindicating discoveries from the usurpation of others."

It was Oldenburg who translated Steno's writings and published the English version. Hooke strongly suspected Oldenburg's perfidy in transmitting Hooke's ideas on geology to Steno, thus once again undermining Hooke's claims of priority. But as we shall see later, Hooke's ideas were far in advance of what Steno proclaimed, and Hooke's suspicion of Steno's plagiary thus was not well-founded in this case–although as Andrade points out (1950, p. 470) there is clear proof that Oldenburg did his best to discredit Hooke's work. As secretary of the Royal Society, Oldenburg's duty was to record the proceedings, but when it came to Hooke's lectures or publications, Oldenburg often failed to enter them in the records. Notable examples include Hooke's lectures on felt-making (Fig. 1–29) and on the comet of 1665 in which Hooke suggested, for the first time, periodic orbits for comets. Neither of these discourses was recorded but both were referred to by Samuel Pepys in his diary [2/21/1666; 3/1/1665.] Oldenburg also conveniently neglected to write an account of Hooke's book *Lampas* in the *Transactions*.[20]

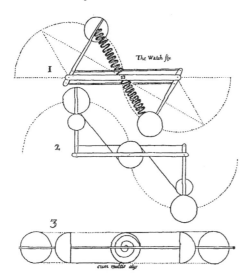

*Figure 1-28. Hooke's drawing of balance wheels for watches. (From* Lampas, *1677; Gunther, 1931, Vol. 8.)*

*Figure 1-29. Hooke's sketch of felt-making, illustrating a lecture he gave to the Royal Society that Oldenburg neglected to record in the* Transactions. *(From Andrade, 1950.)*

**48** RESTLESS GENIUS

The fifth tract, published in 1678, *Lectures and Collections made by Robert Hooke, Secretary of the Royal Society,* is in two parts, *Cometa* and *Microscopium*. *Cometa* contains Hooke's observations on comets of 1664, 1665, and 1677 (Fig. 1-30), and an assortment of discussions on the subject by Boyle, Cassini, and Halley. In *Microscopium* are two letters from Leeuwenhoek, the well-known Dutch microscopist, concerning such small life forms as protozoa and bacteria in pepper-water. Hooke confirmed Leeuwenhoek's observations by making many of his own, concluding that 20,000 of these tiny creatures could reside in the space of the cross-section of a human hair. Fortunate indeed must have been the assistants who learned from such an imaginative mind and at the same time acquired the meticulous skills of experimentation on which Hooke insisted. His correspondence shows that he trained his assistants well, and they went on to succeed in their vocations. In the same lecture he included a "Discourse and Description of Microscopes, improved for discerning the nature and texture of Bodies," in which he also discussed the importance of applying proper lighting to the apparatus.

The sixth and last Cutlerian lecture, published in 1678, contains Hooke's famous and only eponymous law (Fig. 1-31), "Lectures *De Potentia Restitutiva,* or of Spring Explaining the Power of Springing Bodies." Hooke had discovered his law years earlier and announced it in an anagram at the end of his lecture on Helioscopes, published in 1676, where he listed a number of his inventions not yet published (Gunther, 1931, 8: 151). Item no. 3 is "The true Theory of Elasticity or Springiness, and a particular Explication thereof in several Subjects in which it is to be found: And the way of computing the velocity of Bodies moved by them." This statement is followed by the anagram: ceiiinosssttuu, which, he explained, represents the Latin phrase *ut tensio sic vis,* with the letters arranged in alphabetical order, thus, "The Power of any Spring is in the same proportion

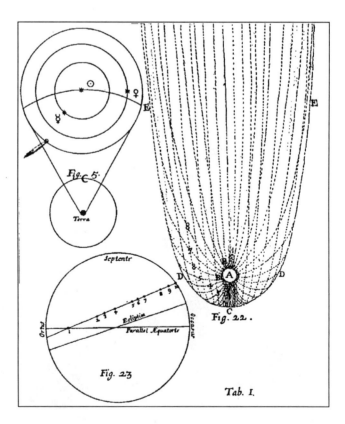

*Figure 1-30. Hooke's diagram of the Comet of 1677. (From Cutlerian Lecture* Cometa, *1678; Gunther, 1931, vol. 8.)*

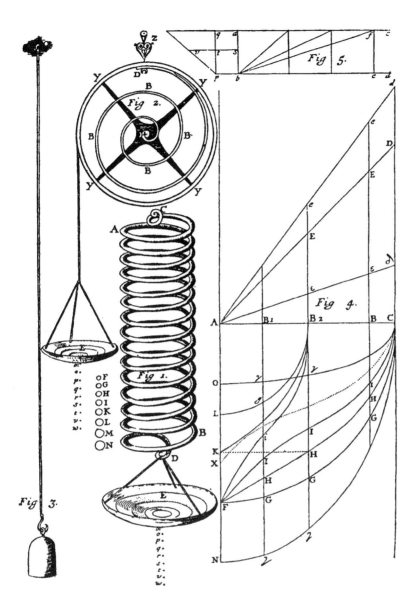

*Figure 1-31. Hooke's Law. The diagram shows three different kinds of springs and the effect of varying weights on them. On the right he shows the consequent motions of the suspended weights, illustrating his idea of harmonic motion. (From lecture* Of Spring, *1678; Gunther, 1931, vol. 8.)*

with the Tension Thereof." As a consequent of this law, Hooke also showed with exceptional perspicacity that in the absence of any energy dissipation, such as friction, any vibration in which the restoring force is proportional to the displacement must have a constant period—that is, simple harmonic motion.

Appended to this lecture is a discussion on his meaning of congruity and incongruity of bodies, an idea first expressed in Hooke's early paper on capillary action published in 1660 and found by him useful in chemistry, as well as in his kinetic theory of matter. To Hooke things that are congruous can more easily be mixed or dissolved in certain fluids. When he referred to bodies and motions it is clear that he was talking about matter at the atomic or molecular level; for example, he spoke of the "particles of matter." To him all things in the sensible world are composed of body and motion, and a particle of matter owes its sensible extension to motion. This hypothesis on congruity and incongruity of bodies has been misinterpreted and erroneously applied to Hooke's ideas on his theory of gravitation by Newton scholar Richard Westfall to discredit Hooke's priority over Newton on some of the basic concepts associated with the Universal Law.[21]

## HOOKE'S LAST TWO DECADES

Throughout the 1670s and 1680s Hooke continued his many activities. He invented a calculator for multiplying and dividing and devised a "weather-wiser" that recorded the time, temperature, air pressure, humidity, amount of rainfall, and strength and direction of wind. He made astronomical observations from the Monument on Fish Street Hill but found that the oscillations of the structure were too great for accuracy The Royal Observatory was founded in 1675, and John Flamsteed was made the first Astronomer Royal in spite of Hooke's own profound interest and achievements in astronomy (Murden, 1985). In fact, Hooke helped to design and construct the Greenwich Observatory, lent his instruments for Flamsteed's use, and because of the many improvements he made for astronomical observations, contributed to the increased accuracy achieved at the Royal Observatory.

In 1677, Oldenburg died, and Hooke was appointed to succeed him as secretary of the Royal Society. He also retained his position as curator of experiments. The death of his nemesis must have been somewhat of a relief to him, as he wrote to Martin Lister on January 5, 1678 (Bodleian Mss. Lister 34, 35, fol. 201, 65):

> I doubt not but you have heard of the changes that have been made in our Society. It hath I'll assure you very much revived us and put a new spirit in all our proceedings which I persuade myself will not only be beneficial and delightful to the members of the Society, but to the whole learned world.

In addition to serving in both positions, Hooke initiated a new publication series, the *Philosophical Collections.* His co-secretary, Nehemiah Grew, was to carry on the *Philosophical Transactions* started by Oldenburg, but they were suspended in January 1679 with no. 142. Hooke's *Collections* began in October 1679 and ended in April 1682, with seven issues having been published.

By 1679, Hooke was so busy that he employed an amanuensis for writing letters, Monsieur Papin, who was paid 18 pence per letter. Papin turned out to be more than a letter writer; he also later demonstrated experiments and invented some "engines" of his own. On September 30, Hooke gained for Papin an annual salary of £20 for writing any letter that needed to be written. It was at the end of 1679 that Newton wrote to Hooke concerning the path of a falling body and Hooke responded by correcting Newton, rekindling the latter's animosity which had been initiated with Hooke's criticism of Newton's optics in 1672. As earlier discussed, it was during the correspondence that followed that Hooke generously apprised Newton of his ideas concerning celestial mechanics.

By now Hooke's life was so jammed with activity that he requested and was granted a paid assistant to prepare experiments. The idea of today's "graduate research assistant" probably started with Hooke, as hitherto under other "natural philosophers," assistants were much more servant than student, unless they had extraordinary talents. Hooke had hired "boy assistants" before this time, as by 1667 he certainly needed help and the society had agreed to pay him £15 toward keeping such an assistant. He, in fact, had the idea of training and teaching his assistants so that they could "graduate" to better employment. In the Bodleian Library is a letter Hooke wrote to his friend John Aubrey, who recommended such an assistant to him. Hooke's reply of August 24, 1675, reveals his philosophy about educating these young men: "I should be very glad to have him live with me if he be a sober virtuous young man and Diligent in following such things as I shall imploy him about which I doubt not will be much for his good hereafter . . . though he doe me service yet I shall assist him much more." He also said that he would provide food, lodging, and instruction to the young man, but that the latter must stay with him for at least seven years, as "it will not be reasonable that after I have been at the pains to fit him for the doing my business. he should presently leave me to take a new one." Indeed, a former assistant trained by him had been hired by another employer at an excellent salary of £150 or £200 per annum, and the employer was so pleased with the assistant's skills that he was sending him to Italy to improve himself further.[22]

In addition to his long-time interest in celestial mechanics and the law of gravitation throughout this period, Hooke experimented with different metals and alloys, continued his astronomical and microscopical observations, recorded differences in terrestrial magnetism, and theorized on the causes and process of digestion—this owing to his own preoccupation with the daily effect of digestion. As secretary he received and wrote letters to many interested people, including Newton, Leewenhoeck, Leibniz, and others.

During the years 1680–1682, Hooke delivered a number of lectures on light that were printed posthumously in Waller's volume. While his contributions on light are important, such as his explanation of interference colors and descriptions of refraction and diffraction, and his anticipation of the wave theory, he could never visualize the tremendous, yet finite, velocity of light that the Danish scientist Olaus Roemer discovered (Wrøblewski, 1985, p. 625). The practical Hooke believed that if the speed of light is "so exceeding swift that 'tis beyond Imagination," then it might as well be instantaneous as far as we on Earth are concerned. The evidence shows, however, that the issue remained open in his mind, as he referred to the velocity of light as "almost instantaneous" [e.g., see *Earthquakes*, Discourse no. 26 in this volume]. Ironically, he was probably the first person to observe the effect of the finite velocity of light, known as stellar aberration, but without recognizing it as such. He had difficulty, therefore, imagining the enormity of the finite velocity of light and, as we shall see, while he knew the age of the Earth had to be much longer than anyone could imagine, he also found it difficult to visualize the billions of years of geological time. In these respects Hooke was indeed a product of his time, although in many other ways he was far in advance of his contemporaries.

Hooke was more interested in practical matters at hand that would facilitate this age of travel and exploration. He undertook cartographical work at this time, devising a projection which was considered superior to Mercator's in that the distances represented were more accurate. Contrary to what many writers say of his personality, however, he did not "take the benefit of a patent, but desired, that the use and benefit thereof might be free" (Gunther, 1930, 7: 574). Even Flamsteed who had earlier impugned Hooke's projection as false, now acknowledged that Hooke's method was "true and real." Hooke continued to improve his telescopes and invented the iris diaphragm that opened and closed "just like the pupil of a man's eye, leaving a round hole in the middle of the glass of any size desired." He showed this invention to the society on July 27, 1681. The iris diaphragm, of course, became indispensable in many instruments, notably the modern camera.

For the explorers, Hooke devised a number of navigational instruments for taking azimuths, altitudes, and other measurements. He made maps, compasses, weather instruments (constantly improving his barometer; it was he who noted the change to lower pressure before a storm), and sampling devices. Decades before Hadley, Hooke had hypothesized on the circulation of air, stating that warm air rising at the Equator moves poleward and cool air at the poles sinks and moves toward the Equator [see *Earthquakes*, Discourse no. 8].

In the 1680s, as a result of his experience in rebuilding London, Hooke lectured on the classification of different types of building stones, noting that clay is best for making bricks, that loam is a mixture of clays and sands, that stones that are flaky "ought to be laid in the building as they lie in their bed," and that some stones that calcine into lime dissolve in the air.

After Hooke passed his secretaryship on to Robert Plot in 1682, the publication of *Philosophical Collections* discontinued and that of the *Philosophical Transactions* was resumed. Hooke continued as curator, but the seemingly ungrateful society changed the condition of his employment from a salary to "Payment by Results." The Royal Society records for June 6, 1683, show (Gunther, 1930, 7: 611):

> It was resolved, that Mr. Hooke shall receive every meeting-day order for the bringing in two experiments at the next meeting-day, together with a declaration by word of mouth of the purpose and design of the experiments, and an account in writing of the history thereof, and the purpose as aforesaid, such as may be fit to be entered in the Register: and that at the end of every quarter there shall be a meeting of the council, where his performances shall be considered, and a gratuity ordered him accordingly; and that from this time he have no other salary.

Hooke was quite wealthy by this time, and for the sake of his first love, experimental philosophy, therefore, he "declared his satisfaction therewith, and his resolution to proceed in his office of curator upon those conditions." He conducted experiments on magnetism, ice and snow crystals, and invented a new kind of balance. He devised a way of communication over long distances of 30 and 40 miles by sight (with a telescope) from high places. He gave a lecture on this subject on May 21, 1684, (Derham, 1726, p. 142). He devised symbols and characters that can be seen distinctly with a telescope (the apparatus on tall stilts is shown in Fig. 1–32) whereby the characters and letters are stored behind the screen, D, and are drawn into view to the right as the message is spelled out. In addition, certain shorthand symbols are used to mean various commonly used phrases such as: ) = I see plainly what you show; O = I am ready to communicate; )( = I am ready to observe; ( = I shall be ready presently, etc.

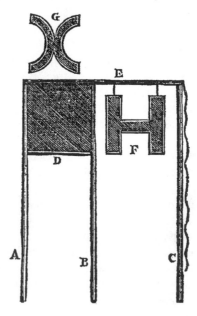

*Figure 1-32. Hooke's optical telegraph, 1684. (From Gunther, 1930, vol. 7.)*

He also continued his long-time interest in modes of transportation such as improving carriages, both for land and sea. In July 1685 he even gave a lecture on Chinese characters and an account of the Chinese use of the abacus. Amazingly, with the use of an ancient book and a Chinese dictionary, Hooke taught himself the basic strokes of the written language and recognized the three styles of writing—regular, cursive, and ceremonial, the last used in epitaphs and other inscriptions. He learned the numbers and was able to reproduce them (Fig. 1-33). The pronunciation phoneticized next to the characters seems to be that of the Cantonese dialect, spoken probably by the only contacts available to him in London then.

He continued to improve on, and made observations with, his barometer (Derham, 1726, p. 169). Apparently, all this time, contrary to the "Payment by Results" plan, Hooke was not paid at all by the society. On December 1, 1684, Hooke himself was elected to the council, and it must have been an embarrassment to them that one of their members had been put in such an untenable position. On June 2, 1686, "it was ordered, that the Council Book be searched as to what had been done about Mr. Hooke's salary." Accordingly, the decision of the council of 1683 was reversed, and on June 16, 1686, "it was ordered, that Mr. Hooke be allowed his arrears for the years 1684 and 1685; and that the Treasurer pay him sixty pounds in full till Lady Day last." So, in spite of all his dedication to the society over a quarter of a century, he never had a raise from his original salary of £30 per annum, which had been decreased to that level on the false promise of Cutler's £50 for the lectureship. Still, he remained a central figure around whom much intellectual activity revolved.

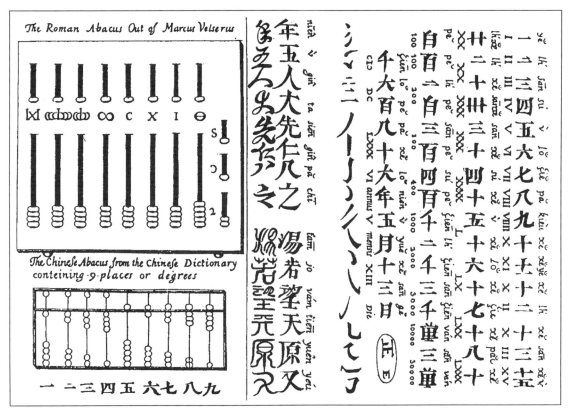

*Figure 1-33. Illustration of Hooke's lecture on the Chinese abacus and the Chinese language, 1686. (From* Philosophical Transactions, *no. 180, 1686.)*

Early in 1687 Hooke's beloved niece, Grace, who had lived with him for many years, died—"the concern for whose death he never wore off, but was observed from that moment to grow less active, more melancholy & cynical than before" (Waller, 1705, p. xxiv). Indeed, Grace's welfare had always been of prime concern to him, as shown by entries in his diary. He stayed up worrying when she was sick; he was angry when she ran off with some beau; he bought silks and pretty trinkets for her; and he acquainted her with certain books he felt young ladies should read. Among these he recorded were *The Accomplisht Woman* and *The Gentlewoman's Companion*. Grace Hooke's death and the publication of Newton's *Principia* with no acknowledgment of his own contributions must have been two heavy indelible hurts from which Hooke never fully recovered. He apparently became visibly quieter, less quarrelsome, with "an extraordinary reservedness of temper." Like the great spirit that he was, however, he was creative, inventive, and productive to the end. He continued to construct new instruments, such as a mechanism for deep-sea sounding and sampling, and to lecture—especially on fossils, on geology, earthquakes, and volcanoes; on the figure of the Earth and on finding the true meridian; on exact measurement of time; on the anatomy of the eye; on interpretation of ancient fables; on the structure of the nautilus shell; on a history of inventions of microscopes and telescopes; on a way to measure weights and distances at sea; on the formation of amber; and on many other topics.

Yet, having been a very sociable man all his life, interested in so many things, he could hardly turn into a recluse. Almost daily entries in both his earlier and later diaries show that he had morning teas or coffeehouse get-togethers with friends and associates, albeit often to the distress of his gastrointestinal tract. Coffeehouses like Jonathan's were frequented by many of his contemporaries as the best means to receive the latest news of domestic and foreign politics. And in Hooke's case, of course, these meetings with friends like Christopher Wren, Richard Waller, Edmond Halley, and others afforded him the opportunity to discuss matters of scientific interest as well.

As Hooke's health, never very robust, deteriorated, he was plagued with more headaches, stomachaches, dizziness, fainting spells, and colds. Yet, he habitually stayed up late reading or working; e.g., the note "t.b.3" in his diary means "to bed at 3." He loved to read and frequented auctions to buy books for himself or to augment the library of the Royal Society. Waller confirms that Hooke never went to sleep till two, three, or four o'clock in the morning, often continuing his studies all night. His habitual ingestion of mercury in self-doctoring necessarily exposed him to the harmful vapors of the element. Mercury fumes cause cumulative damage to one's body, especially to one's nervous system. By 1689, Hooke was almost blind in his right eye. But he continued to lecture, although he no longer felt the need to prepare papers for publication.

On December 7, 1691, at the recommendation (actually on the "warrant") of John Tillotson, the archbishop of Canterbury, Hooke was conferred with the degree of "Dr. of Physick" "as he is a Person of a prodigious inventive Head, so of great Virtue & goodness: and as exceedingly well vers'd in all mathematical & mechanical, so particularly in astronomical knowledge" (Wood, 1721, p. 1041) It would have been interesting to know Hooke's reaction to this singular honor; unfortunately this period of his life is missing from his diaries. The year 1691 was also when Hooke made his last attempt to seek justice in his dispute with Newton over the Law of Gravitation. Could the missing part of this diary have been damaging to the memory of the great Newton? Could this section have been censored by unseen hands whose presence was ever felt in other matters that might have been interpreted to be embarrassing to the memory of Newton? These and other questions come to mind.

Another significant event during this time for which we have no record of Hooke's reaction, was the death of his patron and long-time friend, Robert Boyle (Fig. 1–34). But his sense of loss must have equalled that he felt when Wilkins died, which is recorded. When Wilkins was dying, Hooke had visited him daily. His association with Boyle was possibly at an even deeper level, as, while he was protégé of Wilkins, Hooke's association with Boyle developed from being an employee to a respected friend. He was a frequent guest at Boyle's house for dinner.[23]

On his 61st birthday, July 18, 1696, Hooke's suit against Sir John Cutler's estate was finally settled in his favor. In his inimitable style he poured out his heart in his diary in gratitude to God.[24] Since he was a rich man by then, it could hardly have been the receipt of the money long due him that made him so happy and thankful, especially after 30 years of living without that stipend. Hooke lived by a strict code of honesty and the golden rule; in all his writings and lectures, he never failed to give credit where due; as the text in *Discourse of Earthquakes* shows throughout, he always mentioned the names of people from whom he might have received an idea, some information, or a specimen of rock or fossil. He expected others to do the same and was always hurt and disappointed when they failed to do so. He had a strong sense of what was right, and the right thing had finally happened in the business with the Cutler Lectureship.

*Figure 1-34. Robert Boyle (1627-1691). (From an engraving by R. Woodman in* Gallery of Portraits, *1833.)*

## HOOKE'S LAST DAYS AND AFTERMATH

According to Waller, by about the middle of 1697, Hooke began to complain of swelling and pain in his legs. For one who moved about London at a rapid pace, mostly on foot, this situation must have been intolerable. His eyesight also grew worse, until he was almost totally blind. As late as 1700, however, when he was almost bedridden, he invented his marine barometer which Halley demonstrated to the Royal Society and recorded in the *Philosophical Transactions* in July 1700. At the end of 1702, three months before his death, his mind still active, he wrote a memo to himself about a way to measure the diameter of the Sun to the tenth of a second. But the last year of his life was one of physical suffering. In Waller's words (1705, p. xxvi):

> Thus he liv'd a dying Life for a considerable time, being more than a Year very infirm, and such as might be call'd Bed-rid for the greatest part, tho' indeed he seldom all the time went to Bed but kept in his Cloaths, and when over tir'd, lay down upon his Bed in them, which doubtless brought several Inconveniences upon him, so that at last his Distempers of shortness of Breath, Swelling, partly of his Body, but mostly of his Legs, increasing, and at last Mortifying, as was observ'd after his Death by their looking very black, being emaciated to the utmost, his Strength wholly worn out, he dy'd on the third of March 1702/3. being 67 Years, 7 Months, and 13 Days Old.
>
> His Corps was decently and handsomely interr'd in the Church of St. *Hellen* in *London*, all the Members of the Royal Society then in Town attending his Body to the Grave, paying the Respect due to his extraordinary Merit.

The symptoms as described here point to the condition of congestive heart failure in which the heart is so weak that circulation is impaired to the extent that the body, especially the extremities, swells with retention of fluid, constricting circulation even further—hence, the black color of his legs.

In spite of the decent interment described, but in line with Hooke's usual bad luck, his grave is now lost. His body was originally buried on March 6, 1703, at "about ye Midle of the South Eyle in ye Church." His beloved niece, Grace, was buried in the "South Ile over against the first Piller from the Quier." In 1875, the vicar of St. Helen's, Reverend John Cox, "intimated to the Vestry that the West Window of the Nun's Choir is considerably in want of repair and that in order to effect the necessary repair and to place therein 10 stained glass figures of worthies interred within the precincts of this Church," he needed funds from the parish.[25] Either Hooke's grave lacked an inscription, which is highly unlikely, or the 19th century parish failed to recognize the importance of Hooke's name. When the church floor was dug up to be replaced, 10 individuals or families, with obviously distinguished names, including that of Sir Thomas Gresham, founder of Gresham College, were exempted from removal, but Hooke's name is conspicuously absent from this list.[26] The remains of all except these 10 family names were carted off to the Ilford Cemetery during May to September 1892. Again, the records of those removed do not list either Robert or Grace Hooke, but they do note that there were several coffins of bones that could not be identified. Most likely the zealous 19th century workmen started demolishing the floor from the south side of the church and smashed up inscriptions before anyone told them to match bones up with inscriptions. Since Robert Hooke himself must have seen to the burial of Grace, it is again unlikely that he would have allowed her to be put to rest without an inscription. So Hooke's remains, along with those of his beloved niece Grace, are gone.

Curiously, it was a foreign concern, a German company, Messrs Meyer and Co. of Munich, that donated money to the church to include Hooke as one of the "Worthies" in the west window of St. Helen's. Except for the lost painting/drawing of Hooke referred to earlier, this stained glass window is probably the only representation of Hooke extant. Although the picture does not agree with most descriptions of him, it is appropriately decorated with a shield above his head that includes pecten shells, possibly a symbol to depict his interest in fossils (Fig. 1–35).[27]

Hooke died intestate, although for many years he had intended to leave his money to the Royal Society, in particular to found and endow a "Physico-Mechanick" Lectureship. Perhaps at the end, he became embittered toward the society that had never really treated him very well, but most likely, it simply became too much of a chore to attend to it. A woman named Elizabeth Stephens, whose maiden name might have been Hooke, was Hooke's sole heir. She was his nearest relative, but the exact relationship is not known. Elizabeth, who was apparently illiterate, signed her documents with a mark resembling either a hook or a lower case h. She later married Joseph Dillon, the person who handled the transactions in behalf of her inheritance, which totaled £9,580, 4 shillings and 8 pence, around £8,000 of which were in cash in a trunk – a great fortune for the time.

*Figure 1-35. Representation of Hooke in the "Worthies" window in St. Helen's Church, Bishopsgate, London. (Photograph by June Wilson.)*

Hooke also left a library of about 500 volumes in folio, 1,310 volumes in quarto, 845 volumes in octavo, and 393 volumes in duodecimo, all bound and in quires, along with several bundles of pamphlets (Hunter and Schaffer, 1989, p. 287–294). Among other items found in the cellar was "a picture of Naked woman without a frame," which apparently had been acquired from Ralph Montagu, later duke of Montagu, for whom Hooke had designed and built a great mansion. Because of the discovery of this item, never framed and carelessly relegated to the cellar and probably forgotten by Hooke, Shapin (1989) concludes that Hooke "shared a minor interest in pornography" with Montagu who lent the picture to Hooke, in Shapin's words, "for home consumption." To me, it is revealing indeed that Shapin would consider the discovery of an unframed picture of a nude in a cellar and two terse references to it in Hooke's diary as evidence that Hooke had an interest in pornography. This is another case of unbridled extrapolation from the slimmest of evidence so typical of many skewed accounts of Hooke. Hooke seems to have enjoyed a normal sex life as indicated by entries in his diary and had no need to stimulate his imagination in this respect.[28] Considering his preoccupation with science and his staying up till all hours of the night working or reading because there was not enough time during the day, it is highly improbable that Hooke would have wasted any time gazing upon this picture in order to satisfy any pornographic tendencies. Had pornographic material been found among the over 3,000 items in his library, one might have cause to make such a judgment. As it is, it only adds to the accumulation of slurs and innuendoes that result in the bad press from which Hooke has always undeservedly suffered.

The totality of Hooke's contributions to the world's knowledge and technological improvements is astronomical in terms of one man's life. This restless and indefatigable genius possessed supreme confidence that he and his circle of intellectual friends and trusted technicians could change the world, not just in terms of technological advances in time-keeping, in navigation, in transportation and communication, but also in terms of a rational, global view that is free from superstition, mysticism, and other forms of medievalism. This confidence is exemplified in a small, undated piece of paper I found in the archives of the Royal Society that had been carefully folded, and probably carried about in a coat pocket, either to remind him of the things to do or of the things he had already done (see Fig. 1–36 for the items on the list). Few scientists today, or in any age, would have attempted such an ambitious program.

In comparison to the glorious tributes to Newton at Westminster Abbey, at the Royal Society, and elsewhere, Hooke has been shamefully and negligently treated by history, by most of his countrymen, and especially by the society that could not have survived in its early years without him. Hooke has not been totally forgotten, though. Besides the stained-glass window of "worthies" at St. Helen's Church [the window was damaged, unfortunately, during the Irish Republican Army bombing of the financial district in April 1992], the undergraduate library and an institute for oceanographic studies at the University of Oxford are named in his honor. The parish at Willen proudly acknowledge that Hooke designed their church. A stone block with a plaque on Hook Hill at Freshwater, the Isle of Wight, commemorates his birthplace (Fig. 1–37). These are modest tributes, however, for a man who was among the greatest and most prolific of geniuses who worked in the service of science and humanity.

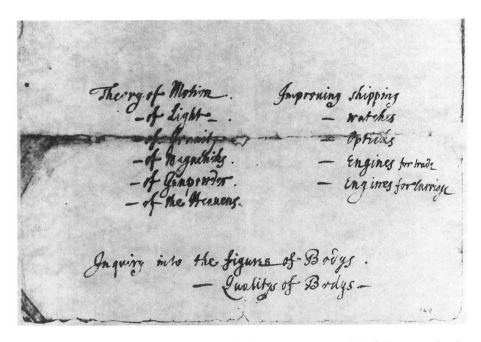

*Figure 1-36. Facsimile reproduction of Hooke's list of things accomplished, from an undated manuscript in the archives of the Royal Society. (Reproduced with the kind permission of the President and Council of the Royal Society, London.)*

*Figure 1-37. Commemorative stone on Hook Hill, Isle of Wight, near All Saints Church and about at the site of his birthplace. (Photograph by C. W. Drake.)*

# 2

# *The Isle of Wight and Its Influence on Hooke's Earthly Thoughts*

*An Observation of my own nearer Home.*

—Robert Hooke, 1668

Most biographical sketches of Robert Hooke mention that he was born on July 18, 1635, in a village called Freshwater on the western end of the Isle of Wight, England (Fig. 2–1). Then, except for noting that his father, John Hooke, had been curate of the Church of All Saints in Freshwater and that his father's house was located on what is now known as Hook Hill, no more information is given of this geologically fascinating island that so strongly influenced Hooke in developing his hypothesis of the terraqueous globe.

Hooke's hypotheses on the origin of fossils and terrestrial features (discussed in detail in Chapters 4, 5, 6 and in Part II, Hooke's *Discourse of Earthquakes*) stemmed from his intimate knowledge of his birthplace. Many of his descriptions of sediments, rocks or fossils are based directly on those features occurring on the Isle of Wight. Also, in *Discourse of Earthquakes,* he tells us he "often" explored the coastal cliffs of this island.[1]

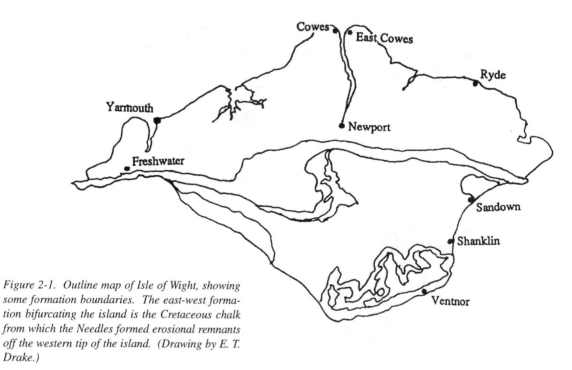

Figure 2-1. Outline map of Isle of Wight, showing some formation boundaries. The east-west formation bifurcating the island is the Cretaceous chalk from which the Needles formed erosional remnants off the western tip of the island. (Drawing by E. T. Drake.)

Robert Hooke roamed his island as a child, and even then made astute observations which he later related. In 1668, when he was 33 and already famous as the author of the book *Micrographia* and as city surveyor for the rebuilding of London after the Great Fire of 1666, he wrote in reference to some of Pliny's writings (*Earthquakes*, p. 297):

> The Second is an Observation of my Own, which I have often taken notice of, and lately examined very diligently, which will much confirm these Histories of *Pliny*, and this my present Hypothesis; and that is a Part of the Observation I have already mentioned, which I made upon the western Shore of the Isle of *Wight*.

This passage shows that he not only often noticed the geology of his birthplace as a child but that he returned to his beloved island as an adult and "lately examined" it diligently. Part of the reason for his return visits to the Isle of Wight must have been because his mother Cicely Giles, who was John Hooke's second wife, was alive until 1665. His father had died in 1648, after several years of illness, when Hooke was 13 years old. Most accounts then relate Hooke's being sent to London to be apprenticed to the portrait painter Peter Lely, whom he had to quit because he couldn't bear the smell of oil paints—and the Isle of Wight is never again mentioned as having any effect on the rest of Hooke's life.

It is clear, however, that as far as the development of geology is concerned, the western and southern shores of the Isle of Wight should perhaps stand in importance as much as Scotland's Glen Tilt of Huttonian fame. An account of the geology of the Isle of Wight, therefore, is not only appropriate but fundamental in understanding how Hooke arrived at his hypotheses.

The Isle of Wight was separated from the Purbeck Peninsula of the Dorset mainland by marine erosion and is part of the geological structure called the Hampshire-Dieppe Basin. The island's river system shows that the land once extended much farther south. The upper Cretaceous to Oligocene rocks on the island have been folded and faulted by tectonic activity in the Cenozoic in several episodes from early Paleocene to the Oligocene or early Miocene. The oldest Cretaceous and oldest rocks on the island are exposed in the cores of two east-west trending, monoclinic, asymmetrical anticlines, the Brightstone or Brixton Anticline in the west and the Sandown Anticline in the east, arranged *en echelon*. The most intense of the Alpine orogeny (mountain-building forces) in Britain, in fact, occurred on the southernmost fringe of England, from the Isle of Wight to the Isle of Purbeck and the Weymouth Peninsula. The violent tectonic movement on the island resulted in some overturned strata and overthrust faulting. Little wonder that Hooke attributed the changes in the "superficies" of the globe to "Earthquakes."

The northern limbs of both folds are almost vertical, as shown by the Chalk pinnacles known as the Needles and often mentioned by Hooke in his *Discourse of Earthquakes* (Fig. 2–2). Vertical beds are also exhibited by the colorful Eocene beds in Alum Bay (Fig. 2–3).

During the earliest Cretaceous times the Isle of Wight was low-lying land across which a large river flowed. Gradually, however, the area was submerged, and by late Cretaceous times almost all of the British Isles were inundated. The teeming microscopic calcareous life-forms in this sea, such as the foraminifera, deposited a great thickness of their calcareous shells (carbonate ooze) which when solidified became the characteristic chalk cliffs of southern Britain. Hooke's theory of petrifaction, described in Observ. 17 of *Micrographia*, will be discussed later. He was the first to observe, describe, and draw a foraminifera shell (Fig. 2–4), marveling at the perfect shellform which he could see only through

*Figure 2-2. Western end of the Isle of Wight showing the Needles, often mentioned by Hooke in his* Discourse of Earthquakes. *(Photograph by C. W. Drake.)*

*Figure 2-3. Vertical Eocene beds in Alum Bay, Isle of Wight, showing darker lignite streaks. (Photograph by C. W. Drake.).*

his microscope. His description of this shell is worth quoting here as it exemplifies the wide-eyed wonder at nature of that age (*Micrographia*, Observ. no. 11, p. 80–81):

> I was trying several small and single Magnifying Glasses, and casually viewing a parcel of white Sand, when I perceiv'd one of the grains exactly shap'd and wreath'd like a Shell, but endeavouring to distinguish it with my naked eye, it was so very small, that I was fain again to make use of the Glass to find it; then, whilest I thus look'd on it, with a Pin I separated all the rest of the granules of Sand, and found it afterwards to appear to the naked eye an exceeding small white spot, no bigger than the point of a Pin. Afterwards I view'd it every way with a better *Microscope* and found it on both sides, and edge-ways, to resemble the Shell of a small Water-Snail with a flat spiral Shell: it had twelve wreathings, *a, b, c, d, e,* &c. all very proportionably growing one less than another toward the middle or center of the Shell, where there was a very small round white spot. I could not certainly discover whether the Shell were hollow or not, but it seem'd fill'd with somewhat, and 'tis probable that it might be *petrify'd* as other larger shels often are, such as are mention'd in the seventeenth *Observation*.

*Figure 2-4. Hooke's drawing of a foraminifer. (From* Micrographia, *1665.)*

One can almost feel the excitement of discovery Hooke felt in seeing this microscopic shell representing such a tiny life-form, and he was the first person to document the existence of microorganisms. From this description and the image it invokes, one sees that Hooke, the 17th-century scientist looking down his microscope, is not much different from a modern-day micropaleontologist bent over his or her microscope. The difference is that the modern scientist does not have to design and build the microscope first!

Other types of fossils abound on the Isle of Wight, such as large ammonites and bones of extinct vertebrates. One can hardly walk along the beaches of any of the western bays, such as Totland, Alum, or Freshwater, without seeing fossil shells in the loosely consolidated strata (Figs. 2–5 and 2–6). Hooke's *Discourse of Earthquakes*, for example, has the following passage (*Earthquakes*, p. 334–335):

> I shall add an Observation of my own nearer Home, which others possibly may have the opportunity of seeing, and that was at the West end of the Isle of *Wight*, in a Cliff lying within the *Needles* almost opposite to *Hurst-Castle*, it is an Earthy sort of Cliff made up of several sorts of Layers, of Clays, Sands, Gravels and Loames one upon the other. Somewhat above the middle of this Cliff, which I judge in some parts may be about two Foot high, I found one of the said Layers to be of a perfect Sea Sand filled with a great variety of Shells, such as Oysters, Limpits, and several sorts of Periwinkles, of which kind I dug out many and brought them with me, and found them to be of the same kind with those which were very plentifully to be found upon the Shore beneath, now cast out of the Sea. This Layer is extended along this Cliff I conceive near half a Mile, and may be about sixty Foot or more above the high Water mark.

Hooke logically and perceptively concluded that these sediments with their teeming marine life had been deposited at the bottom of the sea and were raised to become land more than 60 feet above the high-water mark.

*Figure 2-5. Part of the cliff on the southwest coast of Isle of Wight showing strata of loosely consolidated conglomerates of varying sizes eroded from the chalk and other pre-existing rocks. (Photograph by E. T. Drake.)*

*Figure 2-6. A typical fossiliferous rock along the southwestern coast of Wight, chock full of fossils. (Photograph by E. T. Drake.)*

The Cretaceous transgression of the seas deposited white chalk widely. Some 450 meters of chalk occur south of the Isle of Wight. The Middle Chalk reaches 80 to 90 meters and is largely soft white chalk, with flints appearing near the top (Fig. 2–7). The flints continue throughout the Upper Chalk. Toward the end of the Cretaceous the seas slowly retreated, leaving great areas of chalk that immediately started to erode. The effect of dissolution of the calcareous rock can be seen in Fig. 2–8.

*Figure 2-7. Massive white chalk showing flint nodules or holes from which the flints fell, Isle of Wight. (Photograph by E. T. Drake.)*

*Figure 2-8. The effect of dissolution on the chalk, Isle of Wight. (Photograph by E. T. Drake.)*

The Mesozoic Era ended around 65 million years ago. The principal sections of the Paleogene sediments (which include the Paleocene, Eocene, and Oligocene series) are found on the west and east coasts of the Isle of Wight and in east Dorset. Throughout this time there appeared to be a competitive interplay between marine and non-marine environments, producing sediments showing cycles of transgression and regression of the eastern sea. The latest Eocene and early Oligocene sediments in Britain are preserved only on the Isle of Wight-Hampshire area where, by the end of the Eocene, the sea again retreated and the marine beds were replaced by sediments deposited in freshwater ponds containing the molluscs *Viviparus* and *Unio*. In the Oligocene epoch that followed, however, this area was again inundated by the sea from the east. The sediments exposed on the island indicate that the marine influence generally decreased toward the west in spite of the marine and non-marine cycles of deposition. The Oligocene rocks on the Isle of Wight illustrate well the seesaw effects of these environments. Freshwater snails like *Galba* and *Planorbis* abound in some beds, while other beds contain oysters, barnacles, and marine worms. Still others encase terrestrial faunas, including mammals. Throughout this period, episodes of tectonic activity and uplift occurred that preceded episodes of erosion in the higher grounds in the west, bringing sediments for deposition on the island.

The Isle of Wight was much larger in area in former days, becoming smaller through attrition by erosion. Hooke was very much aware of the foundering of the cliffs into the sea. In *Discourse of Earthquakes*, he wrote (*Earthquakes*, p. 335):

> The place I mentioned before near the *Needles* in the Isle of *Wight* afforded a most evident and convincing one [i.e. proof of petrifaction] as could well be desired, which was from the following Observation. I took notice that the aforesaid Earthy Cliff did founder down and fall upon the Sea-shoar underneath, . . . . I observed several great lumps of the said Founderings lying below.

Fig. 2–9 shows that a big area of land had been eroded and weathered away, as the depression representing a Roman road in the near view can be traced to a corresponding depression across the wide gap of the Freshwater Bay. Fig. 2–10,

*Figure 2-9. Roman road shows extent of erosion in Freshwater Bay, Isle of Wight. The road shown in the foreground was once connected to its extension across the bay (notch in the cliff on left). (Photograph by E. T. Drake.)*

also photographed at Freshwater Bay, shows huge blocks that broke off the main island. The nearest block is reported to have crashed down in one night. What dramatic demonstration of erosion, weathering, breaking down into sand, redeposition, and then re-hardening again into rock!

Another of Hooke's perceptive observations is based on his knowledge of fossil evidence in Britain as a whole, especially that from the southern shore. He speculated that Britain may have been situated in the "Torrid Zone" at one time and had since moved north to its present position. He based this idea particularly on the giant ammonites found in the quarry stone of the Isle of Portland, to the west of and not far from the Isle of Wight. His plate of drawings and accompanying descriptions of ammonites in *Discourse of Earthquakes* are the most exquisite that can be found in any modern treatise on invertebrate paleontology [see Part II, *Earthquakes*, Discourse no. 1]. He reasoned that these *Cornu Ammonis* (horns of Ammon) as he called them, were much greater in size than any shell-fish known today, and he had the testimonies of travelers that certain animals, such as sea turtles, grow much larger in size in the tropics than similar species of colder regions. He asked (*Earthquakes*, p. 343): "whether it may not have been possible that this very land of *England* and *Portland*, did, at a certain time for some Ages past, lie within the *Torrid Zone*; and whilst it there resided, or during its Journying or Passage through it, whether it might not be covered with the Sea."

Hooke's tectonics consists essentially of vertical movements, but his suggestion here of possible *horizontal* displacement of land due to polar wandering was a highly original concept—a radical idea even as late as the first half of the 20th century. He was certainly not advocating continental drift here, only polar wandering, an idea proposed by Hooke that is still useful today and will be discussed in Chapter 5. Since the advent of plate tectonics, however, it has been

*Figure 2-10. Huge blocks of the chalky cliffs that broke off the main island, Isle of Wight; they have been given names by the local inhabitants: (from left) Mermaid, Arch, and Stag. Since the taking of this photograph by the author during the summer of 1990, the second block from the left, known as Arch Rock and known to have existed in the 17th century, toppled into the sea on the night of October 24, 1992. (Photograph by E. T. Drake.)*

shown that Britain had indeed been part of a hot, tropical belt in latitudes around 10° N in the continent of Laurasia during Permo-Triassic times, near the beginning of the Mesozoic Era. During the next period, the Jurassic, Britain had moved possibly to around 30° N and was partly covered by shallow seas—especially along the Dorset coast, where 65 meters of Corallian beds are exposed. To the south, in mid-Jurassic times, the Atlantic had begun to open along its rift. The British Isles continued to move northward during the Cretaceous and, as stated earlier, almost all of the British Isles were inundated by the sea in the late Cretaceous. By the beginning of the Tertiary, Britain had moved to about 40°N.

Several other of Hooke's ideas have also proved to be startlingly on target. These will be discussed in Chapter 4 where Hooke's system of the Earth will be summarized. The essential theme here is that the geology of the Isle of Wight must have served as the source of many of Hooke's ideas announced in his hypothesis. Hooke's observations of his island and his conclusions about its geological history are remarkably accurate. But before presenting his system, we should first examine in Chapter 3 some other prevalent theories of the Earth espoused by highly intelligent and influential men of the 17th century, thereby underscoring the extent that Hooke's system, presented in Chapter 4, was ahead of his time.

# 3
## Other Theories of the Earth

*Ye first 6 revolutions or days might
containe time enough for ye whole Creation.*

—Isaac Newton, 1680

Fossils, or "figured stones," posed a difficult problem to thinkers in the 17th century. To some, nature as created by God is perfect and therefore cannot change; how can perfection be anything different from itself? Yet some fossils, if they represent real animals or plants, are discernibly different from their corresponding living members, while others seem to have no living counterparts at all. Where fossils came from and what they represented were questions that stirred much debate. Many fossils are clearly hardened shells of marine animals, but they can be found on the highest mountains. If these figured stones represented once living animals, could the ocean have covered the Earth to such great heights? Had there been a different distribution of land and sea in the past? Or had these shells formed there through some trick of nature or been left behind by Noah's flood?

The Royal Society of London became a center for much of the discussion and debate regarding these questions. The issue of fossils became an essential element in various theories about the Earth postulated to explain the distribution of continents and oceans and its superficial terrestrial features. Members of the Royal Society were scattered all over Britain, and even when they could not attend the weekly meetings, they often communicated their ideas through correspondence. They eagerly read books speculating about the Earth—in spite of the motto upon which the society was founded, *Nullius in Verba*, (nothing in words) expressing its dedication to the Baconian idea of collection and repeated observation.

Imbued with the Baconian sense of information gathering (which did little to impede them from speculation), therefore, these founders of modern science diligently collected field specimens or wrote letters to correspondents in other countries requesting them to search for certain forms of animal or mineral; they collected other information by means of questionnaires sent to all parts of the world. As the astonishing accounts and samples were brought to the attention of the Royal Society at their weekly meetings, many members began to be concerned with bringing observations of nature into concordance with the biblical accounts of creation, the deluge, and the final conflagration.

Formulation of theories of the Earth became a popular activity among the virtuosi of the 17th century. Not only did some of these theories generate lively debate among the scholars, they also sparked interest among the general public. The motivation underlying these theories was to show that scientific laws based on empiricism are natural laws, and therefore God's laws, and must synchronize with the processes recorded in scripture. In rationalizing the biblical chronology, however, one must relegate the teachings of the ancients to the realm of fantasy even though most of these theorists of the 17th century were educated in the classical mode. As Yushi Ito (1988) points out, many of the ancients, like Aristotle, Plato, Ovid and Seneca, read by the learned in their original Greek or Latin, ascribed to a cyclical view of terrestrial changes which was in diametric opposition to the biblical account, if one interpreted *Genesis* literally. One can understand, then, the dilemma in which many of these scholars found themselves, their determination to rationalize to their own satisfaction, and, if they were good Christians, to proselytize. The formulation of theories, therefore, was not just an amusing pastime; it was a moral necessity.

Some of the intelligentsia, such as Robert Plot (1640–1696), first keeper of the Ashmolean Museum at Oxford, or zoologist Martin Lister (1638?–1712), believed that fossils were *Lapides sui generis*, stones that were just formed that way to look like real animals through a "Plastick Virtue," an inherent characteristic of some soils. But could God have created forms so much like live animals with no function, just as a *lusus naturæ*, a trick of nature? Could the perfection of nature be viewed in this way? On the other hand, if they did represent real animals, then some of these, like the giant ammonites, referred to as "serpentine" or "helical stones" among other names, no longer existed. Could perfection lead to extinction? Others, like Thomas Sherley (1638–1678), physician-in-ordinary to Charles II, were interested in the causes of formation of stones in the kidneys and bladders of humans. Sherley (1672) wrote an essay inquiring into the origin of stones in general, questioning the Aristotelian doctrine of the four elements, and noting that fire and air were not sufficient to produce these bodies. To Sherley, all bodies consist of matter and seed; their universal matter is water, and their invisible seeds contain the "ideas" of the bodies. These ideas depend on their "Exemplars" which are the decrees of God. These seeds act with design. In this latter respect he followed the philosophy of Plato. God and the universal deluge, therefore, are the essential elements. In advocating his principle of water and seeds, however, Sherley used examples from chemistry, especially experiments with precipitation and crystallization.

John Woodward (1665–1728), physician and naturalist, thought that (1695, p. 47):

> although there do indeed happen some Alterations in the Globe, yet they are very slight and almost imperceptible . . . ; but that the bounds of the Sea and Land have been more fix'd and permanent: and in short, that the terraqueous Globe is to this day nearly in the same condition that the Universal Deluge left it: being also like to continue so till the time of its final ruin and dissolution, preserved to the same end for which 'twas first formed, and by the same Power which hath secured it hitherto.

In *An Essay toward a Natural History of the Earth* (1695), Woodward claimed that fossils were the "real spoils" of once-living beings. He even recognized that the rocks containing the fossils are in layers, or strata, but he attributed the demise of the animals they represented to the deluge. The great churnings of the waters of the flood mixed all the animals in a helter-skelter mélange; the creatures then settled out according to their individual specific gravities along with the sediment. If the fossils were found on high places, like the tops of mountains, then of course they were brought there by Noah's flood and settled out in situ. It would have been inconceivable that there had been any exchange of land and sea areas. In the spirit of the age, Woodward amassed a collection of fossils which he bequeathed to his alma mater, the University of Cambridge, and which formed the nucleus of the present Woodwardian Museum.

As we shall see later, Robert Hooke alone held a theory of organic origin of fossils, one not specifically connected with Noah's flood. He argued against

Woodward's scheme, favoring a cyclic theory of terrestrial changes. He also criticized Woodward's idea that sediments settled out by specific gravity after being mixed by the universal deluge. From his own observations Hooke maintained that the order of strata is not correlative with what one would expect if the sediments had settled out from the flood.

To John Ray (1627–1705), highly respected zoologist, botanist and theologian, nature was a continuous demonstration of the wisdom of God. Except for his preoccupation with his religion, Ray's ideas for the formation of the present configuration of the Earth were very similar to Hooke's. The latter had hypothesized that if fossil marine animals are found at high places, then that land must have been at one time under the sea and was later raised up. There is some evidence, in fact, that Ray might have helped himself to parts of Hooke's system of the Earth (see Chapter 7). Ray (1713) supported the organic origin of fossils, stating that the animals either died as a result of the flood or they had lived there in the first Earth before the flood when the bottom of the sea had been elevated, carrying with it the shellfish. He denied that animals could become extinct, however, rationalizing that species unknown in his day may still exist somewhere on this Earth.

Thomas Burnet (1635?–1715), master of the charterhouse and later chaplain-in-ordinary to William III, was convinced that "Truth cannot be an enemy to Truth" and that "God is not divided against Himself." He sought a "natural" explanation for what was observed in concordance with the biblical account. Burnet's *Sacred Theory of the Earth*[1] generated much discourse during that time. Unlike those who claimed that God's creation must be perfect and remain so, Burnet believed that as God could create so He could also destroy. To him, the present uneven surface of the Earth was the result of the universal deluge and God's wrath against the degradation of mankind. In other words, what we have now are the ruins of an originally perfectly smooth and beautiful Earth, the water having been inside the globe underneath a crust emanating forth vapors to provide the moisture needed in paradise (Fig. 3–1).

According to Burnet, not only were man and all living creatures destroyed by the deluge, but the entire "Natural world, and the frame of it . . . and consequently, the present Earth is of another form and frame from what it had before the Deluge" (Burnet, 1691, Book I, p. 50–51). If one accepts this theory, the difficulty posed by fossils disappears: everything was destroyed save what was on Noah's ark. But Noah, after all, could not have gathered a pair of all living beings that dwelled on the first Earth. Burnet wrote (Book I, p. 27) that the deluge "brought upon the Earth a general destruction and devastation . . ., and all things in it, Mankind and other living Creatures; excepting only *Noah* and his Family, who by a special Providence of God were preserv'd in a certain Ark, or Vessel made like a Ship, and such kinds of living Creatures as he took in to him." The implication is that although he took in what he could of the animal and vegetable kingdoms one can certainly conclude that much was left behind and therefore no longer exists today. The idea of extinction, then, also becomes nonexistent. Fossils representing forms of which none exists today simply meant that Noah overlooked them, either accidentally or by design, and they did not get on board.

To explain the distribution of continents and ocean and the uneven surface of the Earth, Burnet provided a hypothesis that was reasonable for his age and highly imaginative, but ultimately and inexorably tied to his religious convictions. He recounted that when Earth began to crack, as the result of the ire of God, and the waters underneath overflowed the surface, great blocks of the crust started to break off and plunge into the water. He observed that "the shores and coasts of the Sea are no way equal or uniform, but go in a line uncertainly crooked

*Figure 3-1. Frontispiece of Thomas Burnet's* The Sacred Theory of the Earth from Its Origin to the Consummation of All Things *(1691), showing the stages of development of Earth from chaos, to paradise, to the Flood (Noah's ark can be seen in the globe to the lower right being pushed by angels), to the present-day Earth, to the final conflagration.*

and broke; indented and jag'd as a thing torn, as you may see in the Maps of the Coasts and the Sea-charts" (Burnet, 1691, p. 103). He illustrated his ideas with figures (see Fig. 3-2). One can imagine that he might well have considered piecing the broken land together to show that the blocks were at one time contiguous, an idea that has occurred to many who ever studied an atlas—including Alfred Wegener (Drake, 1976).

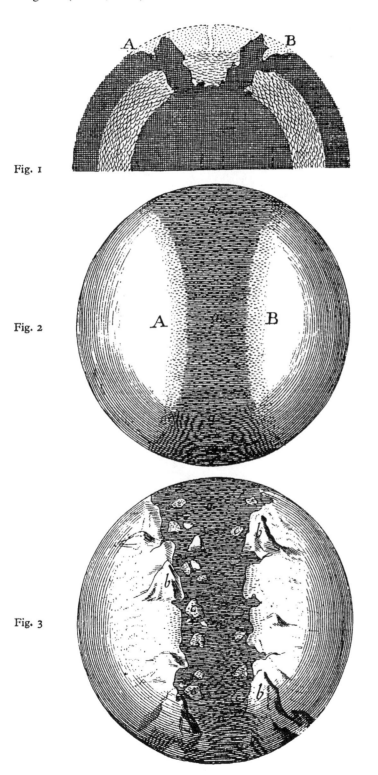

*Figure 3-2. Burnet's figures illustrating break-up and collapse of Earth's crust to create continents. (From* The Sacred Theory of the Earth, *1691.)*

Another creative account was written in 1696 by William Whiston (1667–1752), who by then was under Newtonian influence. The title page of his book summarizes his thesis: *A New Theory of the Earth, From its Original, to the Consummation of all Things. Wherein The Creation of the World in Six Days, The Universal Deluge, and the General Conflagration, As laid down in the Holy Scriptures, Are shewn to be perfectly agreeable to Reason and Philosophy*. True to the Newtonian mechanical age, Whiston attempted to explain the Mosaic account of creation on principles of gravitation and motion of the planets. The deluge, then, was caused by a collision of the Earth with a gigantic comet (Fig. 3–3).[2]

These writers go to great lengths in order to concord with *Genesis*, yet even Isaac Newton approved of the Burnet theory. Newton wrote to Burnet in January 1681 after the latter consulted him regarding his *Sacred Theory*: "Of our

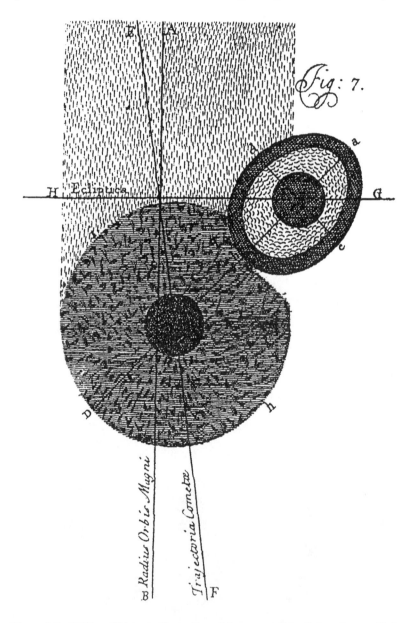

*Figure 3-3. William Whiston's illustration of the cause of the Deluge by a collision of a gigantic water-filled comet with Earth. (From* A New Theory of the Earth, *1696.)*

present sea, rocks, mountains &c I think you have given the most plausible account" (Turnbull, 1960, 2: 329). The great Newton himself had no more sophisticated a view of the creation of the world. In his letter to Burnet of December 24, 1680 (Turnbull, 1960, 2:319), he wrote the following [Note: ye = the; yt = that]:

> Ye heat of ye Sun rarefying yt side of ye Chaos yt ley next it, or by ye pressure of ye vortex or of ye Moon upon ye Waters, some inequalities might bee made in ye Earth, & then ye waters flowing to those lower parts or cavities would make ye seas there, & ye upper parts of ye Earth towards ye poles wch they flowd from, would bee dry land. And all this might ye rather bee, because at first wee may suppose ye diurnal revolutions of ye Earth to have been very slow, soe yt ye first 6 revolutions or days might containe time enough for ye whole Creation, & ye Sun in yt time might contort & shrinke ye parts of ye Earth about ye Æquator more then towards ye poles, & make them hollower.[3]

Newton's use of the word vortex indicates Cartesian influence even at this date, only seven years before the publication of *Principia,* but he was also thinking in terms of gravitation—for example, the gravitational pull of the moon on terrestrial waters, making inequalities on the Earth. Even he, however, believed in the veracity of the biblical account of the six days of Creation and saw no contradiction between scripture and nature, if one allowed that the initial diurnal revolutions of the Earth to have been very slow. The Sun, then, could bring about the shape of the Earth by contorting it and shrinking it. To this letter Burnet replied that Newton was forgetting his Moses, whose Earth formed with land and seas divided before the sun or moon existed; he wrote (Turnbull, 1960, p. 319): "These were made ye fourth day according to Moses, & ye Earth was finisht ye 3d day, as to ye inanimate part of it, sea & land, & even ye plants alsoe; you must then according to Moses bring ye Earth into irregular forme it hath by other causes, & independently upon ye Sun or Moon."

Seeking laws in nature for what was related in the Bible was serious business. Geologists have usually lightly dismissed this early literature; Frank Dawson Adams (1954, p. 210) advised geologists to read these works only if they have time and "a certain sense of humor" and are "in need of mental recreation." To us these theories may seem fanciful, naive, and contorted to accord with the Bible, yet the men who conceived of them were the leading intellects of the day. As Roy Porter (1977, p. 62) pointed out, their philosophies of nature made conflicting demands on them and produced difficulties that were hardly resolvable at that time. Such conflicting demands can occur in any age to impede the development of science; such history, therefore, could perhaps give insights to modern scientists.

Not all scholars of that age, however, were impressed by these hypotheses that were essentially exegeses of the Bible. Edward Lhwyd (1660–1709), Celtic scholar and naturalist and the second keeper of the Ashmolean, was a field man, not a theorist. He saw nothing good in these theories and hypotheses, scoffing at both Burnet's *Sacred Theory* and Woodward's *Essay*. He expressed the hope that reading these narratives would "make man preferre Natural History to these romantic theories" (Gunther, 1945, 14: 269).

Another direction of objection came from the clergy itself. Erasmus Warren, rector of Worlington in Suffolk, strongly objected to Burnet's hypothesis, bluntly calling it false. In 1690 he argued that "the Flood could not be caused by the Dissolution of the Earth, and its falling into the Abyss, . . . because it would be inconsistent with Moses's Description of Paradise" (Warren, 1690, p. 284–285) His objections were not based on physical and mechanical grounds but purely on a literal interpretation of the biblical account. Refuting the idea of a smooth globe, totally without blemishes, described by Burnet as the first pre-Deluge Earth, Warren cited topographic descriptions from the Bible referring to rivers, gardens, trees, and countries. It is ironical that the first appearance of the word "Geologia" or "Geology" is in the title of his book .

Burnet's *Sacred Theory* attracted more attention and debate than other theories of the Earth, not only because of the nature of the subject but also because of Burnet's stature. He was a bishop, and his ideas were supported by Newton. If Newton, the ultimate mechanical philosopher of the age, supported the theory, then, save for objections on Mosaic grounds as expressed by Warren, there should be no scientific objections. While Hooke referred to Burnet as the "ingenious Author" (*Earthquakes*, e.g. p. 371, 395, 402) in his *Discourse of Earthquakes*, his true feelings were quite evident; in a letter to a Mr. Yonge he wrote: "The ingeneous representation of things in Mr. Burnet's Book make many things to be swallowed more glibly than they would in a more common dress, and I generally find that most readers leave off very well pleased with the perusing of it, but I fear there may be many queries thereupon made pretty difficult to be answered" (Unpublished papers of Robert Hooke, Royal Society Library, dated January 9, 1685).

John Beaumont, a supporter of Hooke's views having been persuaded by the latter of the organic theory of fossils, wrote a natural history of some underground caverns in Somersetshire which Hooke published in his *Philosophical Collections* , a series Hooke initiated in 1677 on becoming secretary of the Royal Society. Beaumont's position was in opposition to Burnet's theory, and in fact, he wrote a critique of the Burnet's *Theory* in 1693 entitled *Considerations on a Book, entitled the Theory of Earth, published by Dr. Burnet.* Much to Hooke's pleasure, Beaumont dedicated this book to him.[4] Another supporter was Hooke's friend, John Aubrey, who wrote a natural history of Wiltshire. Aubrey's story in connection with Hooke's system is told in Chapter 7. Hooke's ideas, it would seem, were shared by more people than historians would have us believe; As Ito (1988, p. 308) stated, "the very fact that Hooke repeatedly and tirelessly gave lectures on geological change suggests that his lectures had a great appeal to the audience at the Royal Society, even if some of them were not in agreement with him." A letter written by Tancred Robinson to Edward Lhwyd February 4, 1699 (Bodleian MS Ashmole 1817 a.f. 343), about four years before Hooke died, intimated that while those who were disturbed by Hooke's ideas "do not fail to snarl at him," Hooke's followers constituted a sufficient mass among the audience for these lectures to continue week after week.

In the following chapters, we shall discuss Hooke's own system of the Earth—not only showing in what ways Hooke differed from these theories and was far ahead of his time but also in which aspects he could not totally escape from the tenets of his age.

# 4
# *Hooke's System of the Earth*

*All Cœlestial Bodies whatsoever have an attraction
or gravitating power towards their own Centers.*

—Robert Hooke, 1670

In searching for the causes of the natural processes that produce changes to the terrestrial surface, Hooke hypothesized a number of extraordinary ideas that in hindsight seem almost clairvoyant. Over many years he spent much thought on celestial mechanics and the theory of gravitation, and their effects on Earth. In all his thinking about the Earth, Hooke always kept in mind that it is a planet in the solar system. Some of his ideas were freely borrowed by Newton, either without acknowledgment of debt or grudgingly mentioned with other names only after much urging by contemporaries.[1]

On May 23, 1666, Hooke read a paper to the Royal Society concerning "the Inflexion of a direct Motion into a Curve by a Supervening Attractive Principle," in which he says (Royal Society ms. no. RBO.RBC.2.242):

> I have often wondered, why the Planets should move about the Sun according to Copernicus his Supposition, being not included in any solid Orbs (which the Antients possibly for this reason might embrace) nor tyed to it, as their Center, by any visible strings; and neither depart from it above such a degree, nor yet move in a streight line, as all bodies, that have but one single impulse ought to doe:... But all the Celestiall bodies, being regular solid bodies, and moved in a fluid, and yet moved in Circular or Elliptical Lines, and not streight, must have some other cause, besides the first imprest Impulse, that must bend their motion into that Curve.

Hooke recognized the usefulness of the principle of multiple hypotheses and always tried to exhaust all ways of looking at an issue. For the effect of a straight motion bent into a curve, he could not "imagine any other likely cause besides these two" possible explanations: The first has roots in the Cartesian universe in that the medium through which the bodies move are of unequal density, the outer space being more dense, the motion of the bodies would be deflected inward. His second possible cause of deflecting a body into a curve is from "an attractive property of the body placed in the centre; whereby it continually endeavours to attract or draw it to itself." Hooke obviously favored the second explanation and deduced that if his idea was correct—that there was an attractive force at the center of the planetary system—then "all the phenomena of the planets seem possible to be explained by the common principle of mechanic motions, and . . . we may be able to calculate them to the greatest exactness and certainty, that can be desired." He concluded that not only can the motions of the planets and satellites be solved but also those of comets.

This day, May 23, 1666, when Hooke was only 31 years old, in my opinion, was the birth of the mechanical philosophy that dominated the world not long afterward. In a lecture Hooke gave in 1670 and published in 1674 as one of the Cutlerian Lectures, he expressed the first of his three suppositions regarding his principle of universal gravitation (Gunther, 8:27–28):

> That all Cœlestial Bodies whatsoever, have an attraction or gravitating power towards their own Centers, whereby they attract not only their own parts, and keep them from flying from them, as we may observe the Earth to do, but that they do also attract all the other Cœlestial Bodies that are within the sphere of their activity; and consequently that not only the Sun and Moon have an influence upon the body and motion of the Earth, and the Earth upon them, but that [here Hooke uses the symbols for Mercury, Venus, Mars, Saturn and Jupiter—i.e. all other planets known at that time] by their attractive powers, have a considerable influence upon its motion as in the same manner the corresponding attractive power of the Earth hath a considerable influence upon every one of their motions also.

As physicist Robert Weinstock (1992a) intimates, if the reader had not been informed that Hooke made this statement in 1670, seventeen years before the publication of Newton's *Principia*, one could easily attribute this "far-reaching all-embracing gravitational principle" to the great Sir Isaac Newton, because most of us have been successfully indoctrinated in our schools by the Newton legend, apple and all.

Hooke, like some of his contemporaries, notably his friend Christopher Wren, with whom he had many fruitful discussions (Bennett, 1975), tended to strive for a cosmic view within which such theories as that of gravitation constitute only one element. He perceived the universe as evolving, so that on Earth both the organic and the inorganic components are forever changing. His cosmic view had at its base the fundamental law of the conservation of energy and matter, that gravitation, though applied universally, was not exclusive; other forces were to be reckoned with, such as the magnetic force discussed by Gilbert in 1600 and the "force," that the Sun exerts in repelling, not attracting, the comet's tail—an observation he himself made. The latter "force," it is now known, is an effect of radiation pressure and solar wind.

Naively, Hooke shared many of his original and fertile ideas with others. In a letter to Newton dated December 9, 1679, Hooke corrected Newton in the latter's notion that a body falling toward the center of the Earth would go in a spiral to the east. Hooke asserted correctly that the line of fall would not be a spiral line but an "excentrical elliptoïd" and that the fall of the body would not be directly to the east as Newton claimed, but to the southeast, and more to the south than east. In making this correction, Hooke seemed to have taken into consideration not only the centrifugal effect but also the Coriolis effect (as observed relative to the Earth's surface at the latitude of London). During this same period, Hooke also informed Newton of his concept of centripetal force, being a force "compounding the celestial motions of the planets of a direct motion by the tangent and an attractive motion towards the central body." Before receiving this essential idea from Hooke in 1679, Newton was unproductively focused on the opposite effect, that of the *vis centrifuga*, (centrifugal force), which is not a true force. Without the concept of centripetal force, the theory of universal gravitation was, to use Richard Westfall's word (1967, p. 260), "inconceivable."

## THE EARTH IN SPACE

The law of gravitation was part of Hooke's great scheme, and placing Earth in the scheme was one of Hooke's greatest achievements in geology. He summarized his ideas concerning his system of the Earth in its orbit in a lecture he delivered to the Royal Society on January 26, 1687. To eliminate some repetition, his 15 propositions in his lecture are here summarized into the following nine points (*Earthquakes*, p. 346–348):

1. The axis of rotation in its journey on the ecliptic around the sun keeps very nearly parallel to itself, but over time the direction in which the Earth's axis points in the heavens would vary with respect to the fixed stars. This motion is the precession of the Equinoxes.
2. The axis of the diurnal rotation of the Earth has a progressive motion and in time it changes in position within the body of the Earth; consequently, "the Polar points upon the Surface of the Earth have alter'd their Situation; so that the present Polar Points have formerly been distant from those Poles that were then; and consequently that those former Polar Points are now remov'd to a certain distance from the present, and move in Circles about the present." This is Hooke's concept of polar wandering.
3. Because of the Earth's rotation, the shape of the Earth is an oval whose "longest Diameters lye in the Plain of the Equinoctial, and whose shortest is the Axis." In other words, it is an oblate spheroid.
4. In time there is a change in the "gravitating Power and Tendency" of the different solid and fluid parts of the Earth.
5. Depending on the positions of those parts with respect to the polar points, the changed center of gravity would cause "sliding, subsiding, sinking and changing of the Internal Parts of the Earth, as well as External, tho' the latter will be more powerful, as being more affected by the Rotation."
6. Such change would also cause "an alteration in the Magnetical Power and Vertue of the Body of the Earth, especially of such Parts as are more loose and of a more fluid Nature."
7. The same principle would cause swelling of the sea near the Equator and sinking near the Poles so that "many submarine Regions must become dry Land, and many other Lands will be overflown by the Sea."
8. The "many places which by degrees are made Submarine, will be cover'd with Various Coats or Layers of Earth" because the land is continually washed down and "by Rivers carried into the Sea, and there deposited in the Submarine Regions" into "Layers or Stratifications of divers kinds of Substances according to the nature of those which are this or that way brought thither, and there deposited."
9. Such major changes in the environment would cause changes in the flora and fauna but "preserving in the mean time the Characterisks and Marks of the former Qualifications, when in another Condition."

The concept of polar wandering implicit in the second point in this list is truly original. He proposed that as a result of this relative displacement of the poles with respect to the surface of the Earth, England must have been in the Torrid Zone at some time in the past. Polar wandering is not continental drift, and it is also different from the motion of the spin axis relative to the orbital plane, such as the precession of the Equinoxes. It is offered by him as an additional possible explanation for the redistribution of land and water. If the equatorial bulge occurs at a different circumference, he surmised, land there would be inundated by water, and land would appear near the new positions of the poles. Hooke's concept of polar wandering and his hypothesis concerning the shape of the Earth are discussed in greater detail in the next chapter.

As for the distribution of matter within the Earth, Hooke evoked gravity as the essential element in the makeup of the terraqueous globe, which he likened to an onion (*Earthquakes*, p. 326). The heaviest and most dense materials are concentrated nearest the center or

> at least nearest to that part which is attractive and the cause of gravitation, and the next lighter in the second place, and so on to the third, fourth, fifth and according to their several degrees of Gravity and Density...so Water would always have covered the Face of the Earth, and the lightest Liquor would always have been at the top, and the Air above that, and Æther above that...till disturbed by Earthquakes.

Although the term isostatic compensation was unknown to Hooke, he seemed to have had some idea of its action. He wrote (*Earthquakes*, p. 320):

> tis very probable, that whensoever an Earthquake raises up a great part of the Earth in one place it suffers another to sink in another place; for Gravity is a Principle that will not long suffer a space to remain unfill'd under so vast a pile of Earth as a Mountain, unless the Substances, so thrown up, be of very hard, close and vast Stones that may, as it were, vault it.

Hooke's use of the term "vault" indicates that he might have had an idea of the mass deficiency beneath high mountains as was later observed.

Having placed the Earth in the context of the solar system, Hooke proceeded to discuss the Earth itself and the dynamic processes that effect changes on its surface. He arrived at the crux of his hypotheses concerning the Earth long before he found himself amidst the all-pervasive biblical atmosphere of the time described in the last chapter. Despite his early religious training, he was always more influenced by what he observed and experienced than by what was handed down to him. As for his ideas about the Earth, the shores of his birthplace, the Isle of Wight, had a greater effect on him than any received opinion (see Chapter 2). Throughout his life, Hooke was against being "tied up to the Opinions we have received from others" if evidence from observation and experimentation opposed them. *Jurare in verba Magistri* (to judge in the words of the masters) to Hooke would be contrary to the Baconian motto of the Royal Society, *Nullius in Verba*. He urged (*Earthquakes*, p. 450) that "sensible Evidence and Reason may at length prevail against Prejudice, and that Libertas Philosophandi may at last produce a true and real Philosophy."

## HOOKE'S METHODOLOGY

Unlike Lhwyd, who adhered to Baconianism to the letter, Hooke's approach to epistemology was much more sophisticated. In D. R. Oldroyd's opinion (1972, p. 110), Hooke had a much clearer idea of the importance of "hypothetico-deductive" methods in science than Bacon himself, and therefore Hooke's methodology represents an advance over that of Bacon.

In the *Discourse of Earthquakes* (pre–1668), Hooke opens by making an "ecological" statement—that the Earth we inhabit and everything about it, its air, its water, its land, and all species of life, should concern us as humans. It behooves us therefore to know as much about the nature of things on Earth as possible. To do so, we must collect all kinds of data through observation, testing, and experimentation. Furthermore, all the data collected must be marshaled into some kind of order to be useful. He asks (*Earthquakes*, p. 280): "When this mighty Collection is made, what will be the use of so great a Pile? Where will be found the Architect that shall contrive and raise the Superstructure that is to be made of them, that shall fit every one for its proper use?"

Until such an architect could be found, this collection would indeed be "a heap of confusion." He therefore advocated that there should be predesigned models or theories to provide purpose to experiment and aim to observation. In this respect he goes beyond Bacon and the motto of the Royal Society, as he believes that a collection of facts should not be an end in itself but that it should form the beginning of a "solid, firm and lasting Structure of Philosophy." This aimed-at structure, then, becomes a "true and certain knowledge of the Works of Nature." In a later lecture (1686), he expanded on his philosophy (*Earthquakes*, p. 330):

> The methods of attaining this end may be two, either the Analytick or the Synthetick. The first is the proceeding from the Causes to the Effects. The second from the Effects to the Causes: The former is the more difficult, and supposes the thing to be already done and known, which is the thing sought and to be found out; this begins from the highest, most general and universal Principles or Causes of Things, and branches itself out into the more particular and subordinate.
>
> The second is the more proper for experimental Inquiry, which from a true information of the Effect by a due process, finds not the immediate Cause thereof, and so proceeds gradually to higher and more remote Causes and Powers effective, founding its Steps upon the lowest and more immediate Conclusions.

Hooke's writing in *Discourse of Earthquakes* and elsewhere is a literary as well as scientific accomplishment. His imageries aided by a free use of metaphors are wonderfully strong. He likens the *Analytick* method to that used by an Architect "who hath a full comprehension of what he designs to do and acts accordingly." But the *Synthetick* method is "more properly resembled to that of a Husbandman or Gardener, who prepares his Ground and sows his Seed, and diligently cherishes the growing Vegetable, supplying it continually with fitting Moisture, Food, Shelter, &c. observing and cherishing its continual Progression, till it comes to its perfect Ripeness and Maturity, and yields him the Fruit of his Labour" (*Earthquakes*, p. 330). This explanation is as clear a definition as can be found of the methods of deduction and induction respectively, the meanings of which have been confounding generations of students.

### THE FORMATION OF ROCKS AND MINERALS

As early as 1663, when Hooke was 27 years old, he had already built and exhibited his new microscope to the Royal Society. On March 25th of that year, he was "solicited by the Royal Society to prosecute his microscopical observations in order to publish them" (Gunther, 1938, Preface to *Micrographia*, p. vii). Thereafter he was charged to demonstrate at least one microscopical demonstration at every meeting of the society. On May 27th, he exhibited the pores of petrified wood, the subject of Observ. no. 17 in *Micrographia*. True to his methodology, he subjected his specimen of petrified substance, which "resembled" wood, to a battery of tests, determining its weight, specific gravity, hardness, solubility, combustibility, friability, and so on, concluding that it was indeed a piece of wood that had become petrified. Although he had to use the inexact terminology of a science not yet born, his description of the process of petrifaction is reminiscent of the lit-par-lit molecular replacement process in which the host material is replaced by a different substance (e.g., $SiO_2$) molecule by molecule, so that even the surface texture (like that of wood in this instance) is retained in the petrified specimen. He wrote (*Micrographia*, Observ. no. 17, p. 109) that the wood must have lain in some place where it was "well soak'd with *petrifying* water," and

> did by degrees separate, either by straining and *filtration*, or perhaps, by *precipitation, cohesion* or *coagulation*, abundance of stony particles from the permeating water, which stony particles, being by means of the fluid *vehicle* convey'd, not onely into the *microscopical* pores, and so perfectly stoping them up, but also into the pores or *interstitia*, which may, perhaps, be even in the texture or *Schematisme* of that part of the Wood, which, through the *Microscope*, appears most solid, do thereby so augment the weight of the Wood, as to make it above three times heavier then Water, and perhaps, six times as heavie as it was when Wood.

While most British writers were influenced by continental European views of ore genesis, considered the mainstream by Rachel Laudan (1987), Hooke's ideas involving these petrifying waters, were wholly original. They were neither like those of his contemporaries (Boyle, for instance) nor were they constrained by the all-pervasive biblical account of the Creation. Boyle (1680) in his *The Sceptical Chemist* (first published in England in 1661) argued against the Aristotelian view that natural substances are composed of air, fire, earth and water, but he presented evidence that minerals grow like vegetables *in situ*. Boyle's ideas influenced many writers, such as Sherley (1672), Short (1734) and Whitehurst (1778). As Firman (1986) points out, Hooke possessed "very sophisticated opinions on the origin of ores and gangue minerals." Noah's flood figured prominently in most of the other principal ideas of ore genesis, in which the basic process, with variations upon the theme, involved water—flood waters that fractured, destroyed, and dissolved rocks and reprecipitated in the fractures.

Hooke believed that petrifying waters were an important agent in the formation of stalactites and stalagmites in caves and in other crystallization and precipitation processes that take place in rock fissures. But other types of ores, he holds in his *Discourse of Earthquakes*, formed deep in the Earth and were later exposed by uplift and erosion; the natural place of minerals is "very deep under the Surface of the Earth" and the reason we find it "sometimes near the Surface of the Earth, as in Mountains, is not because it was there generated, but because it has been by some former Subterraneous Eruption (by which those Hills and Mountains have been made) thrown up towards the Surface of the Earth" (*Earthquakes*, p. 317). These highly perceptive ideas had profound influence over some later writers, notably Hutton, as we shall see later.

## THE ORIGIN OF FOSSILS

Hooke conceived of the process of petrifying waters in order to explain how organic matter could be changed into stone—and to explain the occurrence of fossils he had observed on the Isle of Wight and elsewhere. The prevailing explanation for the occurrence of fossils was that somehow the Earth possessed latent "virtues" that could generate such plant- or animal-like forms, and that these forms resembled real organic beings only by coincidence. These virtues could be "seminal" in nature—that is, left there in the Earth by God during creation, as Sir John Pettus, deputy governor of British mines, believed, or they could be "particular" and "plastick," in that they mimicked real forms through latent terrestrial forces. The term "figured stones" implied both fossils and minerals in crystalline form. Hooke, however, always distinguished between the two. Minerals, he claimed, possessed their crystalline structure as a result of the internal arrangement of particles; he makes this point very clear in his lectures, as we shall see later. As early as 1665 he had concluded, in *Micrographia*, that the figured stones we recognize as fossils "owe their formation and figuration, not to any kind of *Plastick virtue* inherent in the earth, but to the Shells of certain Shel-fishes, which, either by some Deluge, Inundation, Earthquake, or some such other means, came to be thrown to that place, and there to be fill'd with some kind of Mudd or Clay, or *petrifying* Water, or some other substance, which in tract of time has been settled together and hardned in those shelly moulds into those shaped substances we now find them" (*Micrographia*, p. 111). The German Jesuit Athanasius Kircher who wrote *De Mundo Subterreano*, which was reviewed in the *Philosophical Transactions* in 1665 and cited by Hooke in *Discourse of Earthquakes*, first invoked terrestrial magnetic forces to explain the idea of plastic virtue for the formation of fossils in situ, and the plastic-virtue idea was then adhered to by many of Hooke's colleagues in the Royal Society (Schneer, 1954). Hooke was convinced that "*Nature does nothing in vain*" and that it would be contrary to the wisdom of Nature to generate these figured stones "for no higher end then onely to exhibit such a form" as a *lusus naturæ* (trick of nature). He urged that as many varieties of fossils be collected and studied as possible (*Micrographia*, p. 112).

After that, the origin of fossils became a subject of debate among the members of the Royal Society, many of whom espoused the plastick-virtue idea. A few supported the organic origin of fossils but attributed their occurrence to Noah's flood—fossil marine shells found on high places, such as the tops of mountains, were brought and left there by the deluge. The Danish cleric-scholar Nicholas Steno in Italy also was of this opinion. Hooke alone was steadfast in insisting that fossils were petrified organic remains and, if they represented marine species, the rocks in which they were imbedded once must have been at the bottom of the ocean. He wrote (*Earthquakes*, p. 314, 320), "'tis a very cogent Argument that the superficial Parts of the Earth have been very much chang'd since the beginning, that the tops of mountains have been under the Water, and consequently also, that divers parts of the bottom of the Sea have been heretofore Mountains." While he accepted the possibility of Noah's flood, "several other Floods we find recorded in Heathen Writers" could also make great alterations on the Earth's surface and, most important, that it was "not very probable" that the shells of fishes found at the tops of mountains like the Alps or Andes were carried there by the Flood which lasted "but a little while."

## TECTONICS AND CYCLICITY

Hooke was not an atheist, however; he did not deny the Creation of the world and "the eternal Command of the Almighty, that the Waters under the Heaven should go to their Place," so that the oceans and dry land were created, but he refers to this act of God as "that extraordinary Earthquake" (*Earthquakes*, p. 314). Earthquakes, to

Hooke, are the cause of much of the uneven surface of the Earth. By earthquakes, he meant every manner of movement of the terrestrial crust, whether violently by slipping, sliding, or subsiding (faulting) or by volcanic eruptions; or slowly and imperceptibly by degrees, such as those changes effected by wind, water, waves or ice. The terrestrial crust could be raised into dry land and mountains, either suddenly or by degrees, or it could subside to form valleys or "vast Vorages and Abysses" of the deep sea. Hooke's tectonics, therefore, are represented essentially by vertical movements, although horizontal displacements are also intimated in the faulting mechanisms and his discussion about England having at one time resided within the Torrid Zone. This latter thought, however, is more related to his idea of polar wandering, discussed in the next chapter, than large-scale horizontal and relative movements of land as in continental drift.

While Hooke believed earthquakes to have been more violent in the past as, for example, in the case of the disappearance of Atlantis, he also regarded them as part of the general cyclic dynamic phenomena of uplift, erosion, and sedimentation. He clearly recognized that changes could occur "by degrees," and in this respect, he cannot be categorized solely as a catastrophist.[2] He was keenly aware that land is constantly subject to such erosional agents as wind, rain, snow, or tides. Just as floods from rivers carry away all things that stand in their way and cover the land with mud, "levelling Ridges and filling Ditches," so tides and currents in the sea wash away cliffs and waste the shores. As the Nile delta is enlarged by sediments, the Gulf of Venice "choak'd with the Sand of the Po," and the mouth of the Thames grows shallow from sand brought down with the stream, so "Cliffs that Wall this Island do Yearly founder and tumble into the Sea." As noted earlier, Hooke was keenly aware of the dramatic demonstration of the foundering of such cliffs near his boyhood home on the Isle of Wight (see Fig. 2–10).

To Hooke, all things in nature "almost circulate and have their Vicissitudes." He expressed this conviction in his poetic prose as follows (*Earthquakes*, p. 313):

> Generation creates and Death destroys;
> Winter reduces what Summer produces:
> The Night refreshes what the Day has scorcht,
> and the Day cherishes what the Night benumb'd.

The words "almost circulate" signify that he did not believe these "vicissitudes" to be completely closed cycles which could never progress toward an end. These "almost" circulating changes have been going on "as long standing as the World." Neither did he, therefore, adhere to uniformitarianism in the strict Lyellian sense of the word—that is, a steady-state that allows no progression. In fact, towards the end of his life, in speaking of the fuel that caused "the Subterraneous Flame or Fire, or Expansion," and without knowledge of radioactivity (more than two centuries in the future), he wrote that this fuel becomes spent and is "converted to another Substance, not fit to produce any more the same Effect" (*Earthquakes*, p. 425). The almost prophetic Hooke sensed in the whole universe a perpetual change that is progressive and teleological, even though the cyclic phenomena were superimposed on a general progression.

Hooke's grasp of the cyclic process of sedimentation, consolidation, uplift, erosion, and denudation was much in advance of his time. But he often found corroboration in the writings of such ancients as Seneca and Fabianus whose descriptions of natural processes fit his own hypothesis. In *Earthquakes*, (p. 314), he wrote:

> This Part being thus covered with other Earth, perhaps in the bottom of the Sea, may by some subsequent Earthquakes, have since been thrown

up to the top of a Hill, where those parts with which it was by the former means covered, may in tract of time by the fall and washing of Waters, be again uncovered and laid open to the Air, and all those Substances which had been buried for so many Ages before, and which the devouring Teeth of Time had not consumed, may be then exposed to the Light of the Day.

Hooke's natural cycles include the hydrologic cycle in which the sea water is cycled through the atmosphere as vapor and rain, and back to the sea through rivers. By the same token, if an earthquake causes one part of the Earth's surface to be raised, another will sink in compensation. Thus, the Earth renews itself.

## *BIOLOGICAL EVOLUTION*

Hooke then tied into his system of the Earth, with its causal principle of "earthquakes," his notion of biological evolution and the usefulness of fossils. He recognized the extinction of species and the generation of new ones. For a discussion of Hooke's ideas on evolution and his attitude toward religion, God, and time, see Chapter 6.

## *FASCINATION WITH THE CLASSICS*

Naturally enough, Hooke was bothered by criticism aimed at his hypotheses. If his critics were fools he could have ignored them, but most were the philosophers of his age. In Oldroyd's opinion (1989), because of the objections of Wallis and his cohorts in the Oxford Philosophical Society against Hooke's axial displacement (polar wandering) idea (see Chapter 5), Hooke started to examine classical writings and myths to try to determine the verisimilitude of such ancient narrations to see if they might provide historical support for his own hypotheses as to changes in latitudes and longitudes. Yushi Ito (1989) proposed a different and more plausible interpretation of Hooke's preoccupation with the classics. In Ito's opinion, because Hooke was critical of Burnet's non-cyclic theory of the Earth, he undertook a search through the ancient writings to find evidence for his own cyclical theory. Thus began his fascination with Hanno's *Periplus*, Plato's *Timæus,* Ovid's *Metamorphoses* and Aristotle's *Meteorologica.*

Being a sponge for knowledge, however, he became so fascinated by the ancient writers that he seemed to become mired in his classical studies. While he never wavered in the fundamental aspects of his earthly thoughts, he became bogged down by exactly those elements, the absence of which had freed his imaginative mind and allowed the formulation of brilliant ideas which from today's vantage point seem alarmingly on target. While he had earlier espoused distrust of received opinion, he now was diligently looking among the ancient writings for corroboration.

Hooke's lectures cited the legend of Atlantis in Plato's *Timæus* as supporting his hypothesis of exchange of land and sea areas—in this case, a catastrophic one. Repeatedly quoted, Ovid's *Metamorphoses* is presented as corroboration of his ideas on the formation of the Earth and its features; these poems also seemed to him to correlate with the account of Creation in *Genesis.* Faced with the work of distinguished clerics like Thomas Burnet (*Sacred Theory of the Earth*) and John Ray (*Three Physico-Theological Discourses* ) which sought to bring natural phenomena into concordance with the Bible, Hooke was obviously aware of the negative effect he could produce by advocating a theory of the Earth without God and the biblical account as its central theme. John Ray, in fact, as we shall see later, was quite taken by Hooke's hypothesis of the terraqueous globe, so much so that evidence suggests he might have "borrowed" Hooke's ideas and adapted them to his physico-theological treatise.

Some of Hooke's closest and most respected friends were distinguished churchmen. John Wilkins, Warden of Wadham College and later Bishop of Chester, for example, had worked with him in constructing flying machines and devising a universal language. It was Wilkins who first encouraged Hooke to produce his book *Micrographia*. During the plague in 1665 Hooke stayed at Durdans, Lord Berkeley's seat near Epsom, in constant association with Wilkins and Sir William Petty. Later, in 1672, when Wilkins was dying, Hooke visited him daily. It did not hurt his image with these close religious friends, therefore, that he also saw in the classics corroboration of the biblical story as well.

Nevertheless, Hooke never wavered from his essentially earthly thoughts involving the organic origin of fossils; polar wandering; subterraneous eruptions and earthquakes leading to the exchange of land and sea areas; the cyclical nature of sedimentation, uplift, and erosion, and his many other ancillary ideas such as petrifaction. He felt that none of his ideas was in real conflict with anything religious, or mythological for that matter, if one allowed free interpretation. Where he was deeply disturbed was regarding the concept of time. Rationally he knew that the Earth could not have been as young as the scriptural chronology would have it, and he was in fact looking for corroboration from some of the ancient writings, including those of the Chinese, for a much older Earth. As noted earlier, he accepted Noah's flood as possible but it lasted for only 200 days and therefore could not have emplaced the fossils on high mountains where they are found or piled up such thick layers of sands and consolidated sediments like those on the shores of his native Isle of Wight. See Chapter 6 for a discussion of Hooke's concept of time.

In many important ways, therefore, Hooke's system of the Earth was well in advance of his time and very close to the modern view. He should be honored as a true founder of the science of geology. His ideas gave not only impetus but also substance to later earth scientists, notably to the "Father of modern geology," James Hutton, discussed in Chapter 8.

# 5

# Hooke's Concept of Polar Wandering on an Oblate Spheroid Earth

*The Poler points upon the Surface of the Earth have alter'd their Situation.*

—Robert Hooke, 1687

In speaking of England being in the Torrid Zone at some time in the past, based on his notion that the huge fossil ammonites found on the southern coast were species that must have produced in hotter climates, Hooke was not describing continental drift. Rather, he was referring to the relative displacement of the poles with respect to the entire surface of the Earth which could have caused England to be successively covered by the sea or to emerge as land. Because this process, known as polar wandering, is referred to as axial displacement, it has been confused with the motion of the spin axis relative to the orbital plane, such as the precession of the Equinoxes. Hooke (*Earthquakes*, p. 347) offered polar wandering as an additional possible explanation for the redistribution of land and water, because there is

> a more than ordinary swelling or rising of the Sea in those Parts which are near the Æquinoctial, and a sinking and receeding of the Sea from those which are near the Poles; so that as any Parts do increase in their Latitudes, so will the Sea grow shallower, and as their Latitudes decrease, so must the sea swell and grow high; by which means many submarine Regions must become dry Land, and many other Lands will be overflown by the Sea, and these variations being slow, and by degrees will leave very lasting Remarks of such States and Positions, in the superficial Substances of the Earth.

The figure of the Earth, therefore, must be that of an oblate spheroid. Thus if the equatorial bulge occurs at a different circumference, land there would be inundated while land would appear near the new positions of the poles—as a result of the "*Oval Figure of the Sea and Body of the Earth in some measure*" (*Earthquakes*, p. 349). He is saying here, therefore, that not only does the water bulge out as a result of the centrifugal effect but the "Body of the Earth" as well, realizing that over geologic time, the solid Earth under a constant force also behaves like a fluid. But apparently he also believed that the "Body of the Earth" would bulge only "in some measure" indicating that he might have been considering the difference in viscosity between water and rock. Although we know now that this difference is insignificant over the geologic timescale and therefore would not create the flooding and receding of the water as described in the previous quote, Hooke's description of polar wandering is correct and its effect on climatic zones is accurate.

In his 1687 lecture discussed in chapter 4, Hooke's summary concerning the Earth included the following five points: First, that the Earth revolves around the Sun in the plane of the Ecliptic, making a revolution once in 12 months. Second, the Earth rotates on its own axis $365\frac{1}{4}$ times during the year, and the axis is an imaginary line passing through the the Earth center and that is inclined to the plane of the Ecliptic at $23\frac{1}{2}°$. Third, this axis is parallel to itself, or nearly so, in its trip around the Sun and, at present, points to a spot in the heavens not far from "the last Star of the Tail of the little Bear call'd the *Pole-star*, but heretofore 'twas at a greater distance from it." Fourth, the axis varies with respect to the pole star, "and by degrees proceed nearer towards it, not directly, but in a Circle parallel to the Ecliptick, or whose Center is the Pole of the Ecliptick." This last item refers, of course, to the precession of the Equinoxes, a complete revolution of which is 26,000 years. Hooke added at this point (*Earthquakes*, p. 346):

> Thus far I take the same with the Hypothesis of *Copernicus* and his Followers. But Fifthly, I suppose yet further, that the *Axis of Diurnal Rotation* of the Earth had also had a progressive motion, and hath, in process of time, been chang'd in position within the Body of the Earth, and consequently that the Poler points upon the Surface of the Earth, have alter'd their Situation; so that the present Polar Points have formerly been distant from those Poles that were then; and consequently that those former Polar Points are now remov'd to a certain distance from the present, and move in Circles about the present.

This is Hooke's statement of the polar wandering concept. Notice he said the *Polar points upon the Surface of the Earth*—that is, the geographic poles. But how can the pole positions change with respect to the Earth's surface without changing the orientation of the Earth's axis with respect to the plane of the Ecliptic? He had expressly iterated that the axis is always nearly parallel to itself in its journey around the Sun and is "kept in an Inclination to the said Plain of $23\frac{1}{2}°$." The answer must be that the Earth's surface has shifted with respect to the pole positions.

Hooke coupled this concept with the hypothesis of the oblate spheroid shape of the Earth to explain his observations of the occurrence of tropical species in England and also the redistribution of land and sea areas. Newton has long been credited with describing the true figure of the Earth. But in a series of lectures Hooke delivered in 1686 and 1687 Hooke delved into the problem of the figure of the Earth, pointing out that this subject had already been discussed by him 10 or 12 years earlier, or around 1675, which date is years before Newton published on the Earth's shape in *Principia* (1687). Hooke asked (*Earthquakes*, p. 343):

> whether the Superficies of the Ocean be equally distant from a Central Point in the Bowels of the Earth, and whether any other perpendiculars to the Surface thereof, besides those of every single Parallel, and its Poles, do tend to any other Point of its Axis; and if there should be found more than one Point, then what are the limiting or terminating Points of a Line of such Points; that is, at what distance they must be from one another, or from a Central Point? This I mention'd in two of my preceding Lectures, the one read about ten or twelve Years since, and in the other about two Years since; in both which I indeavour'd to shew that the form of the Earth was probably somewhat flatter towards the Poles than towards the Equinoctial, since which I have met with some Observations that do seem to make a probability in my Conjecture and Hypothesis.

He asked whether lines normal to the surface of the Earth intersect the axis at more than one point and if so, what are the *limiting* points of a line of such points, "that is, at what distance they must be from one another, or from a Central Point?" The description unquestionably depicts the figure of the Earth as an oblate spheroid, and there can be no doubt as to what Hooke intended. Unfortunately, he sometimes referred to the figure as a prolated spheroid when he meant an oblated one, as, according to the Oxford English Dictionary, the term "oblate" did not come into use to mean "flattened at the poles" until 1705. This confusion of terminology might have contributed partially to the confusion of later writers as to the author of the true shape of the terrestrial globe. The content of Hooke's lectures concerning the issue, however, clearly demonstrates his meaning.

The two experiments that he performed in 1687 to prove his hypothesis also could only give one conclusion: that the Earth bulged at the Equator and was depressed at the Poles. In the first one, while he blew a bubble of glass in a pipe at its molten temperature he twirled it in a circular motion around the pipe, and the shape of the bubble quickly assumed that of an "Oval." In the second experiment, he showed that water naturally recedes from the Poles toward the Equinoctial. He set a round dish of water on a stand that could be moved in a circular motion around an axis passing through the center of the dish. He noted that as the spinning increased, the water began to sink in the middle and to rise toward the edge of the dish. He wrote (*Earthquakes*, p. 351):

> This last Experiment doth hint, that the Convexity of the Sea near the Poles of the Earth must necessarily be much flatter than elsewhere, and not only less Spherical than the rest of the Sea, but possibly plain, nay, beyond a plain, possibly Concave, for that the Water cannot but have or receive from the vertiginous Motion, an endeavour to recede from the Center of that Motion, and the Gravity of the Earth working there more powerfully and freely . . . . But which seems more material, I conceive that a Degree of Latitude, if there measured would be very much longer than a Degree of Latitude under the Æquinoctial.

Hooke, therefore, was quite aware that the way to prove the shape of the Earth was to measure the distance along the Earth's surface near the Equator that corresponds to a degree of latitude and then compare that with the distance of a degree near the poles. On the surface of an oblate spheroid the distance subtended by one degree between two normals would be greater near the poles than near the Equator, as is shown in Fig. 5-1.

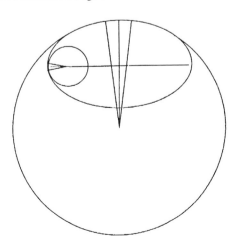

*Figure 5-1. Diagram showing that on the surface of an oblate spheroid (the oval shape), the ground distance subtended by one degree between two radii of curvature is greater near the poles than near the Equator. (Drawing by E. T. Drake.)*

On the other hand, Newton, to whom faithful followers have attributed, and continue to attribute, all sorts of wonderful ideas whether they were truly his or not, definitely did *not* believe that the Earth was an oblate spheroid until possibly after he learned of Hooke's experiments and shortly before publication of his *Principia*. As late as 1681, Newton wrote that he believed the Earth to be spherical in shape in response to a letter of Thomas Burnet asking Newton what was his opinion of the figure of the Earth. Newton gave his reasoning as follows (Turnbull, 1960, 2:329):

> My chief reason for that opinion is ye analogy of ye Planets. They all appear round so far as we can discern by Telescopes, & I take ye earth to be like ye rest. If it's diurnal motion would make it oval that of Jupiter would much more make Jupiter oval; the *vis centrifuga* at his equator caused by his diurnal motion being 20 or 30 times greater then the *vis centrifuga* at our equator caused by the diurnal motion of our earth, as may be collected from the largeness of his body & swiftness of his revolutions. The sun also has a motion about his axis & yet is round.

In the late 17th and early 18th centuries, France and England were competing with each other in trying to determine the shape of the Earth, and at one juncture, the French insisted that they proved that the Earth was a prolate spheroid while the English adhered to the oblate spheroidal figure. Edmund Halley of England and Jean Richer of France were among the first to discover that a pendulum clock set to an exact time in the northern latitudes would lose time near the Equator unless the pendulum were shortened at the Equator to adjust for the loss in time. This phenomenon remained a mystery until Hooke explained the shape of the Earth and therefore the decrease in gravity at the Equator. Hooke wrote (*Earthquakes*, p. 349), "this Theory [which he had discussed publicly as early as 1675], has been verify'd, first by Mr. *Hally* at St. *Helena*, and since by the *French* in *Cayen*, and now lately in *Siam*, in all which places it is affirmed, that 'twas necessary to shorten the Pendulum to make it keep its due Time." This statement perhaps reflects Hooke's nationalistic attitude, as Halley's expedition to St. Helena to chart the southern skies took place from 1676–1678 while that of Jean Richer to Cayenne in French Guiana at 5°N was in 1672, although it is true that Richer's observations were not published until 1679 in the *Mémoires de l'Académie des Sciences*.

Newton, who seemed often to plant evidence in his writings to establish priority especially for the sake of posterity, after he was convinced of the true shape of the Earth, tried to give the impression in the *Principia* that he had come to his conclusion shortly after Richer's trip to Cayenne in 1672. He wrote (Book III, Proposition 20, of the third edition of *Principia* ):

> And, first of all, in the year 1672, M. *Richer* took notice of it in the island of *Cayenne*; for when, in the month of *August*, he was observing the transits of the fixed stars over the meridian, he found his clock to go slower than it ought in respect of the mean motion of the sun at the rate of $2m.28s.$ a day.

If one did not look into his private correspondence, one certainly would get the impression that as soon as he was apprised of Richer's discovery, he was convinced of the oblate spheroid shape of the Earth. This false impression has been carried forth in the literature to this day.[1] His 1681 letter to Burnet, quoted earlier, however, proves that even after the publication of Richer's observations in 1679 (much less in 1672), Newton still believed in the roundness of the Earth and gave what he felt to be cogent argument for his belief. Here is another example of history allowing Newton to usurp the credit that rightfully belongs to Hooke.

As for the concept of polar wandering, it is apparent that many of Hooke's contemporaries failed to understand it fully. Hooke stated clearly that "there may be in the Rotation of the Body of the Earth, a change of the Axis of that Rotation, by a certain slow Progressive Motion thereof, whereby the Poles of the said Motion appear to be in superficial parts of the Earth, which heretofore were at some distance from the then polar Points or Parts" (*Earthquakes*, p. 357). He was referring to the *relative* positions of the poles with respect to the *surface*, "superficial parts," of the Earth, not advocating a wholesale flip-flop of the rotational axis with respect to the orbital plane. Polar wandering, or the displacement of the poles *relative* to the continents and oceans of the surface of the Earth, is a totally distinct idea from axial displacement with respect to the heavens. But Hooke's contemporaries as well as later and recent writers and editors have not understood this distinction. Hooke's theories of a nonspherical Earth and axial displacement with respect to the Earth's surface (polar wandering) were postulated by him to explain interchanges of land and sea areas, as evidenced in the previous chapter in the summary list of his lecture of January 26, 1687.

Edmund Halley communicated Hooke's ideas to John Wallis in Oxford in a letter dated February 15, 1687. Wallis, a quarrelsome mathematician of note and notoriety, had been embroiled in several disputes ranging from solving mathematical puzzles to the cure of a deaf-mute child (Chapter 7); he read the letter to the Oxford Philosophical Society, then replied to Halley on March 4 and April 26, 1687. He admitted that "some little alteration may have been in the Earth's axis from that different obliquity of it to the Plain of the Ecliptick," but that "so vast a change as is now suggested, could not possibly have been" without some historical record of it. In connecting these two statements in proximity to each other suggests that Wallis and his colleagues confused an *astronomical* axial displacement with respect to the heavens with Hooke's polar wandering concept. In the first instance, in referring to the obliquity of the Earth's axis to the plane of the ecliptic, Wallis obviously meant the orientation of the Earth's axis in the solar system resulting in a precession. At the same time, Wallis and his Oxford group concluded, "sure we are, that there is no evidence in history that ye top of ye Alps was ever sea; Except in Noah's Floud," and the notion of the Earth changing its axis "to turn ye world upside down ... seems too extravagant for us to admit." For supporting evidence Wallis, with words dripping with sarcasm and ridicule, related that a "Dr. of Physick of good credit" showed the group what seemed the shell of a fish taken by the good doctor from the kidney of a woman. It was more likely, the society concluded, that the fishshell was formed in the woman's kidney "than that this kidney had once been sea."

Hooke read his reply to Wallis at a meeting of the Royal Society on April 27, 1687. It was apparent to Hooke that not only had he been misrepresented possibly out of some malice, but that also he had been misunderstood; his letter reads: "I find it [Wallis' letter] to be made up partly of misrepresentation, partly of designed Satyr. Arising as I conceive partly from a misunderstanding of what I have here deliverd but cheifly from some prejudice conceived against me and my performances."[2] This confusion, I believe, has been carried down through the centuries and still plagues the minds of many today.

That the Wallis group also opposed the oblate spheroid form of the Earth emerges as strong indicaton that its members were yet unaware of Newton's eventual conversion to that view. Newton was very well known and respected by the date of Wallis' letter to Halley of March 4, 1687, attacking Hooke's hypotheses. As mentioned, citing the results of Richer's expedition to Cayenne at

5°N in 1672, Newton in his *Principia* tried to give the impression that he was the first to discover the true form of the Earth. His belief in the oblate spheroidal form for the Earth must have come so late in the course of events that the Oxford group at this time (1687) was still not privy to it, not to mention Halley himself who had been urging Newton to publish. Had they been aware of Newton's conversion, given Newton's reputation by this time, the group would also have believed in the true shape. Yet, in his 1687 letter Wallis wrote (Oldroyd, 1989, p. 210–211):

> The Earth's figure (whether a Sphere, or Spheroid,) was here [meaning at Oxford] discoursed about a year ago. (& at other times, before & since.) When it was doubted by some, whether we have convincing evidence, from Observation, that it is a perfect sphere (& not somewhat flat, or oblong;) Which seemed not certainly to be adjusted without an accurate measure of a Degree in a great circle, at many & far distant places.
>
> And, if a sphereoid; it was thought more likely to be an oblong (on it[s] longest Axe from Pole to Pole) That being better fitt for motion, & more likely to preserve ye Axe in its proper place . . . .
>
> And it was judged to be, if not a true sphere, at least very neer it; Both from ye Figure of ye Earths shadow in ye Moons Eclipse; (which appears circular, & is so supposed to have ever been:) And from ye falling of heavy bodies, at all places, in a perpendicular to ye Earths surface, in (what we call) planes: which (supposing all tending to the same quarter) would not be, in all places, if the Earth be not (as to sense) spherical.

If the shape of the Earth had to be a spheroid, this group thought, then they would favor a prolate spheroid with its "longest Axe from Pole to Pole."

It is rather strange that Edmund Halley, who must have thought Hooke's hypotheses of sufficient interest and merit to transmit them to Oxford for comments, seemed to have been intimidated by such rejection. Halley had been interested in the secular variations in the geomagnetic declination since 1683. Perhaps because he was 20 years younger than Hooke and seeing the disastrous airing of Hooke's hypotheses, especially with respect to polar wandering, Halley felt it politic to drop the subject of his own research. In spite of the fact that polar wandering has apparently nothing to do with the observed, relatively rapid wandering of *magnetic* poles, Halley waited five years before coming forth, in 1692, with his own, rather complicated, scheme to explain westward secular variation of the geomagnetic field. He proposed that the Earth has a solid shell separated from a solid inner core by a liquid layer and each of the solid parts has a pair of poles, with the European North Pole and the American South Pole attached to the inner core, while the American North Pole and Asian South Pole are attached to the shell. In this way, a westward drift occurs when the inner part rotates more slowly than the outer. The lag would require a period of about 700 years for the core to fall one complete revolution behind the shell (Evans, 1988). Surely polar wandering is not more complicated or less palatable than this idea.

Stranger still that Halley apparently did not react to the Wallis group's assertion concerning the shape of the Earth. It was Halley (and Hooke names him in his lecture), among others, who verified the fact that gravity is stronger at the Poles than at the Equator and that, in Hooke's words, it "was necessary to shorten the Pendulum to make it keep its due Time" near the Equator. Halley's close association with Newton and the latter's taking credit for being the first to arrive at the true shape of the Earth, something that both Newton and Halley must have known was Hooke's idea, make one wonder what really took place behind the scenes.

Strangest of all is the fact that today it is still not generally acknowledged that the deification of Newton over the centuries has precluded a true account of this history.

As for Hooke's experiment showing the recession of water from the poles toward the Equator, it is interesting to note that Alfred Wegener in the 20th century describes an almost identical experiment (attributed to U. P. Lely, 1927) in *The Origin of Continents and Oceans*. Wegener and his associate, J. Letzmann, repeated the demonstration. Figure 5–2 illustrates the equipment they used.

Wegener observed that when the nail is pointing upwards so that the float's center of gravity is displaced upward, the centrifugal effect decreases and the float drifts to the center. Conversely, when the nail points downward, the float drifts to the edge because its center of gravity is farther from the axis than that of the water it displaced and the centrifugal effect dominates the pressure gradient. Wegener termed this phenomenon the "pole-flight force," invoking it as the means of displacing continental blocks through the sima. The reason the motion seems opposite to what he wished to demonstrate for the Earth, of course, is because of the reversed curvature of the fluid in the experiment to that of the planet (Wegener, 1929, p. 173).

Because the concept of polar wandering is a highly imaginative one, requiring the abstract thinking with which few are so abundantly endowed as Hooke, its significance is still elusive to many and its mechanism is not clear. Its effect and manifestation, however, are evident. Hooke cited as evidence the occurrence of tropical and subtropical flora and fauna in England which showed that England must have been in a hot climate zone some time in the past. And since climatic belts have a strong relationship to the positions of the poles, then the North Pole must have been at a different place relative to the surface of the Earth in the past.

Polar wandering was very important to Wegener in explaining his hypothesis of continental drift. Wegener reasoned that polar wandering must be defined as a surface phenomenon because only the top crust of the Earth is accessible to geologists and the location of former poles can be deduced only from fossil and climatic evidence—both essentially surface features. He defined polar wandering, then, as "a rotation of the system of parallels of latitude relative to the whole surface of the globe, or otherwise as a rotation of the whole surface relative to the system of parallels (which amounts to the same thing because all movements are relative)" (Wegener, 1929, p. 148). Continental drift and polar

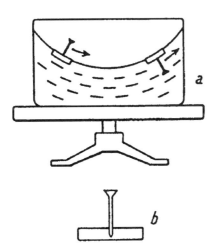

*Figure 5-2. Wegener's equipment used in demonstrating his "pole-flight" force; the center of gravity of the float is above or below the displaced water depending on the direction the nail is oriented as the table is spinned. Hooke had performed an identical experiment to show this effect; i.e., when the nail points downward, the float drifts to the edge because its center of gravity is farther from the axis than that of the displaced water, and the centrifugal effect outweighs the pressure gradient, and the opposite is true when the nail points upward. It must be remembered, of course, that the curvature of the Earth is convex while the curvature of the fluid is in the opposite sense. (Reproduced with the kind permission of Friedr. Vieweg und Sohn, from the 4th revised edition of* Die Entstehung der Kontinente und Ozeane, *1929, Braunschweig.)*

wandering are two distinct and separate ideas, but this distinction has not been understood by modern historians of Earth science. Large-scale horizontal movements of the continents relative to each other do not constitute a form of polar wandering as some have claimed (see Oldroyd, 1989, p. 232), although it is true that a large shift in mass within the mantle due to convection could cause continental drift and a shift of the spin axis with respect to the surface of the Earth because of a change in its moment of inertia (Goldreich and Toomre, 1969).

Another point of clarification is that although Hooke did not rule out any possible movements of the terrestrial crust, such as his mention of "slipping, sliding and subsiding," he did not advocate largescale horizontal movements of the continents with respect to each other. Because his tectonics was essentially vertical, he can not be touted as a proponent of drift theory. But he was the originator of the idea of polar wandering which is a necessary component in continental drift theory and especially in biogeography.

Probably because of the ridicule that Hooke suffered at the hands of Wallis and his Oxford group, the idea of polar wandering, or axial displacement, as it was known, did not prevail in the ensuing centuries. This concept was unfathomably unacceptable to most of Hooke's colleagues. Hooke himself complained to Wallis in a letter read to the Royal Society dated April 27, 1687, why should the idea of a slow variation of the axis of rotation frighten people when "whoever admitts the Copernican Hypothesis allowes 10000 more and yet we doe not find them frighted with the Bugbear of turning the world upside down"—that is, if we believe that the planets revolve around the Sun, it doesn't seem to bother us when we find ourselves turned 180° every 12 hours, to say nothing of our racing around the Sun once a year. John Ray, a religious man and noted naturalist, citing Henry More, the neoplatonist, wrote in 1691: "Speaking of the *Parallellism* of the *Axis* of the *Earth* he saith, I demand whether it be better to have the *Axis* of the *Earth* steady and perpetually parallel to itself, or to have it carelessly tumble this way and that way as it happens, or at least very variously and intricately: And you cannot but answer me, it is better to have it steady and parallel" (Ray, 1691, p. 142) Besides showing Ray's familiarity with the Hooke hypothesis (discussed in Chapter 7 on plagiarism), this quote is evidence again that somehow Hooke's idea of axial displacement, or polar wandering, has been misconstrued to mean that the axis of rotation itself actually tumbles around with respect to the orbital plane. As we shall see later, others in history, including James Hutton, made a point of attacking the axial displacement idea (Drake, 1981). Even in the early decades of the 20th century, Wegener (1929, p. 129) bemoaned, "some opponents still rejected this view . . . with a severity which is hard to understand." A partial explanation must be that this concept is too difficult for many to comprehend.

Polar wandering is evidenced by paleoclimatic arguments regarding the distribution of flora, fauna, and sediment types. Continental drift, on the other hand, was found to be a complication in deciphering these distribution patterns at various geologic periods, during which the climatic belts changed with respect to the varying polar positions. As Wegener pointed out (1929, p. 130), "if one starts out from the standpoint of drift theory, and if one maps the fossil evidence for climates on a chart developed for the relevant period by means of the theory, these contradictions completely vanish, and all the climatic evidence arranges itself to form the pattern of climatic zones which is familiar to us today." While the two concepts, continental drift and polar wandering, are distinct from each other, the distribution patterns could not be interpreted coherently unless the continents were assembled more or less the way Wegener had proposed as Pangæa in his drift theory. The connection between these ideas is diagrammatically demonstrated in figure 5–3.

*Figure 5-3. Schematic diagram showing the relationships among continental drift, polar wandering, and biogeography. Diagram 1 shows the position of the original continent with its assemblage of flora and fauna (x's and y's). Diagram 2 shows the beginning of continental drift. Diagram 3 shows continued drifting and, simultaneously, polar wandering (direction shown by arrow). Diagram 4 shows the way the continent and the biogeography (geographical distribution of the biota) looks to us today. In order to reconstruct the original scene, therefore, one must take into consideration both drift and polar wandering. (Drawing by E. T. Drake.)*

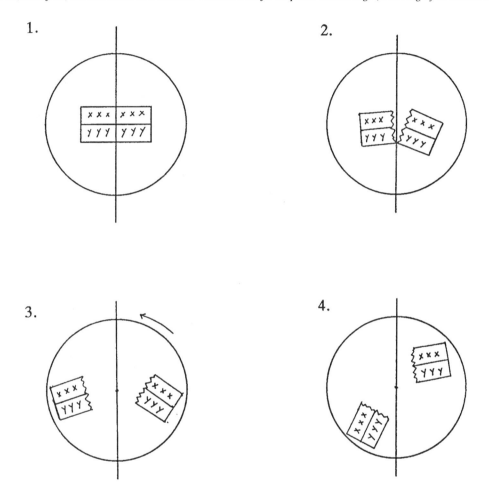

Today, the vindication of Hooke's polar wandering concept can be found in its usefulness in the plate tectonics paradigm. Paleomagnetism, like the paleoclimatic zones, can only be explained by displaced poles. Goldreich and Toomre in 1969 lent mathematical support to the concept of large angular displacements of the earth's rotation axis relative to the entire mantle having occurred on a geological time scale. Courtillot and Besse (1987) deduced what they consider the "true polar wander" path by comparing the apparent polar wander path as described by plate motions over fixed hotspots (i.e., hotspots are coupled to the mantle) with that indicated by paleomagnetic data. The rate of such a whole "mantle roll" averages about 5 cm/yr, which is of the same order of magnitude as seafloor spreading. Hooke's idea of three centuries ago, rejected until only recent times, is now seriously discussed. Not only that, polar wandering is now actually measured. Robert Duncan (pers. comm.) and his coworkers (Duncan and Richards, 1991) on a voyage of the drillship JOIDES *Resolution*, using their data on the Réunion hotspot in the Indian Ocean, measured the amount of true polar wander, or a whole Earth shift of 8° to 12° within the last 66 million years around a pole on the Equator at 60°E.

# 6

# *Hooke's Theory of Evolution and Attitude toward God and Time*

*There have been many other Species of Creatures in former Ages of which we can find none at present, and . . . there may be divers new kinds now, which have not been from the beginning.*

—Robert Hooke, 1667

Over the thirty years during which Hooke discoursed on "Earthquakes," he never wavered from his fundamental concepts on (1) the origin of fossils, (2) petrifaction, (3) evolution, (4) the shape of the Earth, (5) polar wandering, (6) universal gravitation, (7) cyclicity of many terrestrial processes and (8) that subterraneous eruptions and earthquakes cause changes to the terrestrial surface and the exchange of land and sea areas. What crept into his later discourses was a need to bring God into his picture of the system of the Earth—not to show the presence of God's hand in all phenomena but to show that natural causes, and natural explanations, do not gainsay God. From the start he questioned the validity of the narration in *Genesis* and asserted that Noah's flood was nothing special, not because he was a disbeliever but because he distrusted the truthfulness and accuracy of the biblical account, which, after all, he reasoned, was written by men about things that happened before the invention of writing. Still, God and the Bible entered into his lectures as the general political and religious climate of England became more steeped in ecclesiastical doctrines around the time of the ascension to the throne by William III of Orange and his wife Mary II in 1689.

## NOAH'S FLOOD

Even so, Hooke's practical attitude toward the story of the Flood was brought out in many passages of his discourses. The duration of the Flood, he noted, was about 200 "Natural Days." Half a year, therefore, could not have been long enough to produce and perfect "so many and so great and full grown Shells, as these which are so found do testify." His writing also indicates that he had some sense of the length of time it takes for great thicknesses of sediment to accumulate. He had witnessed weathering, erosion, and deposition along the coastal cliffs of his birthplace, the Isle of Wight. He wrote, "the quantity and thickness of the Beds of Sand with which they [the fossils] are many times found mixed, do argue that there must needs be a much longer time of the Seas Residence above the same, than so short a space can afford" (*Earthquakes*, p. 341).

This statement was made during the first group of lectures on earthquakes delivered before September 1668, when he was totally unhampered by the need to consider religion or to reconcile nature with the Bible. And even later, around 1686–1687, he still saw the Bible as being, after all, written by men, and he distrusted any record of things that happened before the invention of writing; he wrote (*Earthquakes*, p. 334):

> The great transactions of the Alterations, Formations, or Dispositions of the Superficial Parts of the Earth into that Constitution and Shape which we now find them to have, preceded the Invention of Writing, and what was preserved till the times of that Invention were more dark and confused, that they seem to be altogether Romantick, Fabulous, and Fictious, and cannot be much relied on or heeded, and at best will only afford us Occasions of Conjecture.[1]

Hooke appears to be questioning the traditional scriptural chronology—that is, how can we trust the veracity of something before recorded history? No mention is made, in fact, of the oft-quoted phrase in relation to the Bible, that these were words inspired by God Himself! One passage in particular reveals his sense of humor about the legend of Noah even though it was from a lecture dated February 22, 1699, when he was already trying hard to consider the biblical account seriously. In it he related an account of a ship found on the bottom of a lake in Italy that was determined to have been that of the Roman Emperor Tiberius. Other accounts told of ships found buried deep in sediments in mines—vessels as technologically advanced as those of Hooke's own time "whose Anchors were of Iron and the Sails of Linnen." He wrote, "'tis reasonable to think that the *Antediluvians* were as ingenious, if not much more, than the *Postdiluvians*, for the inventing of Ships, and for the use of them, for the transplanting of Colonies, for Trading and for War" (*Earthquakes*, p. 443). Hooke's admiration of the Antediluvians (who were supposed to be so bad in the eyes of God that they had to be destroyed), stating they were much more ingenious than the Postdiluvians, indicates somewhat his attitude toward religion: the bunch saved by Noah must not have been very bright. Hooke continued, if "Ships of that Bigness and perfection of Anchors, Sail and Rigging" were being built at the time of Noah, the latter should have known about it. Furthermore, others besides those in the Ark could have survived the Flood. He wrote (*Earthquakes*, p. 440):

> For the Hypotheses of *Noah*'s Flood, 'tis not said in any History, that Navigation, especially on the Ocean, was grown to such a perfection in *Noah*'s time as to make Ships of that Bigness and perfection of Anchors, Sail and Rigging, as this by this short Description seems to have been; and 'tis very likely if any such Navigation had been, it would have been taken notice of in the History of the Bible; for it cannot be suppos'd that *Noah* should not be inform'd of it, if any such Art had been then practiced in any part of the World how remote soever from the place of his Abode. Next, if such should have been, it might have happen'd that some other Men or Creatures might have escap'd with Life besides those in the Ark.

## BIOLOGICAL EVOLUTION

More important for the history of science than his attitude toward Noah was Hooke's notion of biological evolution and his conviction of the usefulness of fossils. He recognized the extinction of species and the generation of new ones: "there have been many other Species of Creatures in former Ages, of which we can find none at the present; and that 'tis not unlikely also but that there may be divers new kinds now, which have not been from the beginning" (*Earthquakes*, p. 291). As Cecil Schneer (1954, p. 267) so aptly put it, "This 'crooked man of a melancholy and jealous disposition,' . . . could, in a single subordinate clause, revise creation and, at least by implication, read God out of nature." Hooke believed environmental change to be "the reason why we now find the shells of divers Fishes Petrify'd in Stone, of which we have now none of the same kind." And, "Since we find that there are some kinds of Animals and Vegetables peculiar to certain places, and not to be found elsewhere; if such a place have been swallowed up, 'tis not improbable but that those animal Beings may have been destroyed with them." By the same token, "there may have been divers new varieties generated of the same Species, and that by the change of the Soil on which it was produced; for since we find that the alteration of the Climate, Soil and Nourishment doth often produce a very great alteration in those Bodies that suffer it"; and "'tis not to be doubted but that alterations also of this nature may cause a very great change in the shape, and other accidents of an animated Body. And this I imagine to be the reason of that great variety of Creatures that do properly belong to one Species" (*Earthquakes*, p. 327–328). Rossiter (1935) thought it "strange that these striking anticipations of later biological theory have passed unnoticed." Hooke's writings, however, might not have been as unnoticed as has been thought. It is difficult to prove positively the influence of any writings on any later thinkers, but in Hooke's case it is clear that his writings were widely known by the latter half of the 18th century and during the 19th century (see Chapter 8). While no specific mention of Hooke was made by Darwin in his writings, we do know that Darwin owned a copy of Hooke's *Micrographia* even before his voyage on the *Beagle* (Digregorio, 1990, p. 388).

In a fairly well-known passage, Hooke stated clearly that he believed fossils could be used as records of the past (*Earthquakes*, p. 321):

> There is no Coin can so well inform any Antiquary that there has been such or such a place subject to such a Prince, as these will certify a Natural Antiquary, that such and such places have been under the Water, that there have been such kind of Animals, that there have been such and such preceding Alterations and Changes of the Superficial Parts of the Earth.

More startling still, and more than 100 years before William "Strata" Smith, Hooke wrote (*Earthquakes*, p. 411):

> And tho' it must be granted, that it is very difficult to read them, and to raise a *Chronology* [Hooke's own italics] out of them, and to state the intervals of the Times wherein such, or such Catastrophies and Mutations have happened; yet 'tis not impossible, but that, by the help of those joined to other means and assistances of Information, much may be done even in that part of Information also.

These quotes undoubtedly illustrate Hooke's perceptive mind; unfortunately, he was far ahead of his time.

Several major objections to Hooke's "Earthquake" theories including his ideas on evolution and extinction emerged to complicate Hooke's ever busy life. Extinction was an especially difficult concept for his audience to swallow. But Hooke asked them what else could be concluded when (*Earthquakes*, p. 433):

> There have been, former times, certain Species of Animals in Nature, which in succeeding and in the present Age have been and are wholly lost; for neither have we in Authors any mention made of such Creatures, nor are there any such found at present, either near the places of their position (as on the Shores or Sea about this Island) nor in any other part of the World for ought we yet know.

Hooke's rhetoric in these discourses is especially effective and persuasive because he was able to verbalize his critics' concerns. He recognized that they looked upon the inference of extinction as "absurd and extravagant" because "it would argue an imperfection of the first Creation, which should produce any one Species more than what was absolutely necessary to its present and future State, and so would be a great derogation from the Wisdom and Power of the Omnipotent Creator" (*Earthquakes,* p. 433).

Another of his critics' arguments was that if some animals were not found alive, that did not necessarily mean that such animals did not exist somewhere, yet to be discovered. This was John Ray's argument against Hooke's idea of extinction, and it was here that the churchman Ray departed from the rest of Hooke's theories (see Chapter 7 on "Plagiarism or Paranoia"). Furthermore, his detractors, probably in desperation, countered his persuasive logic with the argument that no written history existed to document the changes he claimed. Hooke's response was: granted, we are not familiar with all the numerous varieties of animal and plant life in the world, so for now we won't insist that they do not exist somewhere. He made the analogy that the absence of written history detailing the building of the pyramids in Egypt should not prompt us to conclude that humans did not build them. It would be just as ridiculous, he continued, for us to consider the extinct ammonites as a trick of nature, not representing living animals "because he could not tell what Creature they were." He reemphasized that species change to adapt to the environment, that "We see what variety of *Species,* variety of Soils and Climates, and other Circumstantial Accidents do produce; and a *Species* transplanted habituated to a new Soil, doth seem to be of another kind, tho' possibly it might return again to its first Constitution, if restored to its first former Soil" (*Earthquakes*, p. 450). The odds against Hooke's ideas being accepted were great indeed, because even as late as 1694, when he had been lecturing on fossils for 30 years, some people were still skeptical that fossils represented once living forms. It is hardly strange, therefore, that to many his ideas on evolution spoke against the perfection of God.

In 1689, Hooke asked his listeners to simply assume that the animals represented by their fossils were extinct, regardless of their personal convictions. He wrote (*Earthquakes,* p. 435):

> We will, for the present, take this Supposition to be real and true, that there have been in former times of the World, divers *Species* of Creatures, that are now quite lost, and no more of them surviving upon any part of the Earth. Again, That there are now divers *Species* of Creatures which never exceed at present a certain Magnitude, which yet, in former Ages of the World, were usually of a much greater and Gygantick Standard; suppose ten times as big as at present; we will grant also a supposition that several *Species* may really not have been created of the very Shapes they now are of, but that they have changed in great part their Shape, as well as dwindled and degenerated into a dwarfish Progeny; that this may have been so considerable, as that if we could have seen

both together, we should not have judged them of the same Species. We will further grant there may have been, by mixture of Creatures, produced a sort differing in Shape, both from the Created Forms of the one and other Compounders, and from the true Created Shapes of both of them.

The principle that certain species are of a certain size now but were much bigger in a former age has certainly been expressed by modern paleontologists. Furthermore, some forms have changed so much that they cannot now be identified as the same species. Mixing, or interbreeding, of creatures can also produce forms different from their ancestors. He did not see how this should offend God in anyway. He wrote (*Earthquakes*, p. 435):

And yet I do not see how this doth in the least derogate from the Power, Wisdom and Providence of God, as is alledged, or that it doth any ways contradict any part of the Scripture, or any Conclusion of the most eminent Philosophers, or any rational Argument that may be drawn from the Phænomena of Nature; nay, I think the quite contrary Inferences may, nay, must, and ought to be made.

He is obviously speaking with much passion here, indicating that he is not an irreligious man—rather, that the mysteries and grandeur of Nature affirm his religious beliefs. To Hooke, God designed the model and the conditions allowing the world with all its creatures to evolve. While God might have been the ultimate cause—"That extraordinary Earthquake"— the immediate causes are always natural ones. I think that God might also have been a very personal entity to Hooke, one to whom he gave extravagant praise and thanks in the privacy of his *Diary* when something happened in his favor, such as when he won his case against Cutler's estate (see Chapter 1, note 24).

## HOOKE'S CONCEPT OF TIME

Thus Hooke's attitude toward God and the Bible were both rational and practical, without denying the existence of one or denigrating the importance of the other. To him, the concept of evolution was inescapable, given the fossil evidence, and despite objections, he was not going to retract any of his conclusions. Besides, it in no way derogated from God's perfection. He would defend his right to articulate hypotheses based on facts and evidence as the way humans can approach the truth. The one element that on the surface appears to have escaped his full comprehension was the duration of time since the Creation. Here was a date for the beginning of time that was so firmly established by Archbishop Ussher (1581–1656) from the Bible that it was hardly subject to interpretation: at 9:00 a.m. on October 26, 4004 B.C. Yet, evidence throughout Hooke's discourses shows him to be much disturbed by the insufficiency of the scriptural chronology; he rationally felt that the Earth had to have been older than 6,000 years.

The most convincing evidence for him was the geology of the Isle of Wight, along whose shores Hooke, both as a child and as an adult, spent many hours studying the strata and processes of erosion and sedimentation. It seems inconceivable that this mathematically minded man would not have made mental calculations on the rate of erosion and sedimentation. He advocated a process of cyclicity for the surface features of the Earth, including exchanges of land and sea areas by denudation, uplift, erosion, and sedimentation. He also believed these cyclic changes to have undergone "several vicissitudes," and knew that these processes took time—a much longer time than is allowed by the Bible.

He not only sought out passages in the classics to corroborate his cyclic theory, but he also searched for any mention of real time. Occasionally, when periods of time mentioned by the ancient writers exceeded the biblical allowance, he tried to mollify his audience by explaining, for example, that the Earth might have rotated and revolved faster in its earlier days, hence a day at the beginning of the world would have been shorter than now (*Earthquakes*, p. 322). The number itself, therefore, was not to be taken too seriously. Because of these comforting reassurances he supplied to his audience, critics claim that Hooke really adhered to the scriptural chronology. Yet, one can easily read between the lines. He repeatedly referred to Egyptian and Chinese histories, which told of events of "many thousand Years more than ever we in Europe heard of by our Writings." But to placate his audience, he quickly added (*Earthquakes*, p. 327), "if their Chronology may be granted, which indeed there is great reason to question." Often, however, especially during his earlier discourses, he failed to supply this balm to his audience and, in fact, revealed his contempt for those who pretended to be authorities on all matters. For example, in reference to his idea of polar wandering, that England might have been in the torrid zone at some time in the past, he wrote (*Earthquakes*, p. 319):

> And those Persons that will needs be so over confident of their Omniscience of all that has been done in the World, or that could be, may, if they will vouchsafe, suffer themselves to be asked a Question, who inform'd them? Who told them where *England* was before the Flood; nay, even where it was before the *Roman* Conquest, for about four or five thousand Years, and perhaps much longer.

Although we see here that Hooke himself did not visualize the unbelievably long geological timescale as we know it, yet in relation to the chronology of the Bible, the Roman conquest around the beginning of the first millenium A.D., plus 5,000 years and "perhaps much longer" before that, would certainly outstrip Bishop Ussher's date of 4004 B.C. for the beginning of the world. If he were skeptical of the duration of time as enunciated by "Heathen Historians" as some claim because he adhered to the scriptural chronology, why did he repeatedly bring forth the "space of Time" allowed by these historians like the Chaldeans, the Egyptians and the Chinese, who "do make the World 88,640,000 years old"? (*Earthquakes*, p. 395). It is too facile, thus, to say that Hooke was bound by the biblical chronology.

The evidence shows Hooke searching desperately through the literature, and *wanting* the Earth to be much older than was allowed by the Church in order to have the "space of Time" to accomplish all the terrestrial changes he saw continually taking place, including the violent earthquakes and volcanic eruptions, as well as those processes that advance slowly "by degrees." Add to these cycles his concepts of polar wandering, biological change, extinction of species and generation of new ones (all of which require a very long time) and his skepticism of biblical accounts shown by his repeated denials that Noah's flood could accomplish so much, it is reasonable to assume that he believed in a much older Earth than 6,000 years.

Yet his writings show his ambivalence on this subject. The question is whether Hooke was deliberately equivocal, assuring his audience when necessary but giving free rein to his imagination at other times. I believe he was deeply disturbed by the question of time. Critics have used certain passages in Hooke's *Discourse of Earthquakes*, such as that regarding the disappearance of

Atlantis, as proof that he was a true believer and accepted the scriptural 6,000 years as the age of the Earth. One such passage is from a lecture, dated February 15, 1688, more than 20 years after he started his discourses. By then Hooke had launched into his study of the classics for confirmation of some of his ideas as cyclic theory, raising of land from the sea or subsidence, and so on. People were still not convinced of the organic origin of fossils. For a full understanding of this passage, one should read the entire lecture, which is No. 12 in the present transcribed version and starts on p. 403 in the original *Posthumous Works*. His style is argumentative; he is trying to convince his listeners of the verisimilitude of some ancient legends, that islands sometimes are born by rising out of the sea and they disappear by subsidence, and that many things may seem unlikely to them and yet are true. The stumbling block in the Atlantis story, however, is that the collapse of Atlantis happened more than 9,000 years ago, which is a direct contradiction of the biblical time constraint. But he forged ahead with an analogy, saying if he saw oysters and other fish in the rocks of the Alps and "I confess it seemed to me a little hard, because I could not give the Pedigree of the Fish [that is, I didn't recognize it because it was an extinct species], therefore I should not be allowed to believe it a Fish, when I saw all the sensible marks of a Fish"? Note the phrase, "I confess it seemed to me a little hard," meaning certain preconceived notions about where bodies of marine animals should be found might prevent one from seeing the truth.

In the same vein, he argued that the story of Atlantis could very well be true, saying "I confess the account of nine thousand Years is Argument enough to make the whole History to be suspected as a Fiction"—that is, from the point of view of the preconceived ideas of time of the audience. Hooke went on to say that he did not know what that 9,000 years signified in terms of time and, yes, he agreed, it is hard to accept the truth of Atlantis given what we've all been led to believe about the age of the Earth. His purpose, however, was to convince the audience of the truth of the legend of Atlantis—he did not want them to disregard it out of hand because of the biblical time constraint; does that prove, as critics claim, that he himself was so constrained?

My interpretation is that while he considered the time frame as described in the Bible to be insufficient to produce all that he saw on the Earth, his own conception of time had not crystallized to the extent that he could express a definite stand about its duration. To claim that Hooke was constrained by the scriptural chronology because of this passage, however, is to fail to understand his thinking throughout the 30 years of his lectures on Earth science. Being a product of his time but endowed with a vision far beyond that of most of his contemporaries, Hooke obviously found it necessary to profess adherence to the contemporary dicta. Readers should look behind these words and study everything else he said to arrive at his true feelings on the issue of time. In his search through the classics he found corroboration for many of his ideas; Ovid's *Metamorphoses*, for example, talks of land rising up from the sea and other great changes to the surface of the Earth as a result of upheavals caused by earthquakes or volcanic activity. The problem for Hooke was how to convince those listeners who were apt to reject his theses if the time component opposed their basic religious beliefs. In making an assessment of Hooke's concept of time, one should always remember that his earlier series of lectures on "earthquakes," before September 15, 1668, was notable for its lack of reference to any religious consideration. Only later, when religion became more of a political issue, that time also became an issue and Hooke found it necessary to contort to accommodate the general politico-religious attitude.

Hooke did not live to witness the terrible devastation and tremendous energy released by such events as the birth of the new volcanic island near Santorini in the Ægean Sea in 1707 and the Lisbon earthquake in 1755. But he believed that such catastrophic occurrences had taken place in the history of the Earth and went to great lengths in his later lectures to convince his listeners that Plato's legend of Atlantis, for example, was really a historical account of an actual event that took place as a result of an explosion of hot, subterranean heat, or "fires." By the time of these later lectures, he was older and more versed in the ways of the world and especially in the prejudices of human nature. He found the mention of time in these classics a distraction in his effort to convince his listeners. And for that reason whenever a period of time much longer than the biblical limit is mentioned, he had to tell his audience to disregard it in order to persuade them to accept the rest of the narrative. Plato, in *Timæus*, for example, wrote about great alterations in the world taking place for 9,000 years before Solon, who lived more than 2,000 years before Hooke's time (*Earthquakes*, p. 372–373). The story Hooke was telling his audience, then, happened more than 11,000 years ago, and he felt compelled to add "bating only his number of Years," but the rest of the story, he urged, should be accepted as true, because it had "so much of Reason and Agreement with the State of things." The evidence suggests, therefore, it would be erroneous to interpret Hooke's personality to be so steeped in biblical edict that he could not imagine the duration of Earth's time to be longer than the scriptural account.

But the irony is that while he easily accepted time to have been thousands of years longer than the biblical limit, there is no explicit proof that he conceived of the billions of years of the geological timescale. Even though all of the evidence and argument he had amassed over 30 years were pointing the way to a world of almost unfathomable longevity, he realized that he could not resolve this question without committing heresy. If he thought the Earth was at least millions of years old, which seems likely because he admired the wisdom of the Chinese, he was prudent enough to keep quiet about it. Some of the men for whom he had the greatest respect were churchmen like his old sponsor John Wilkins, the bishop of Chester. The conferment of his own doctoral degree at Oxford in 1691 was recommended and "warranted" by Archbishop of Canterbury John Tillotson. Being a loyal friend and protégé, Hooke could not have bitten the proverbial hand by flaunting an idea that would have offended just about everyone in the 17th century.

The explicit breaking away from the 6,000-year age of the Earth, however, was essential for humans to understand the Earth in all its complexity. Even if we believe that Hooke must have had a much more profound sense of the true age of the Earth, he could not have served the future in this respect. To say his thinking was limited by the scriptural chronology, however, is not to understand Hooke at all. His insistent but discrete questioning of the prescribed duration of time and his insightful accounts of the time-consuming "vicissitudes" of the cyclic terrestrial processes turned out to be very important in the development of Earth science, and, as we shall see in a later chapter, they led to the formulation of the Huttonian world-without-end and the beginning of modern geology.

# 7

# *Plagiarism or Paranoia?*

*'Tis a right Presbyterian trick.*

—John Aubrey, 1692

Hooke has been singled out among all the great scientists of the 17th century and characterized by some writers as "cantankerous" in his claims of priority and disputes over ideas. Yet if one studies the intellectual milieu of the time, the controversies and rivalries of the type in which he was involved seem almost to be the rule rather than the exception. And Hooke's reaction to such controversy involving his own discoveries and inventions seems mild in comparison to the behavior of some of his contemporaries. Because Hooke was involved in more experiments and observations and articulated more hypotheses than many, it follows that perhaps more of these controversies surrounded his person than they did others. But I have shown several notable incidents involving friends and colleagues such as Boyle and Mayow in which Hooke clearly had priority, but he never disputed their claim. His attempts to defend himself against his detractors or those who pirated his ideas, however, have earned him the labels "paranoid," "suspicious," "jealous," and "unusually vitriolic" by modern writers. Furthermore, his behavior is sometimes considered "notorious." As unfair as these labels seem to me, it is the argument that his behavior associated with these disputes is "antithetic to genteel values" of the time and that such quarrels are more associated with tradesmen than intellectuals (Shapin, 1989, p. 276) that is a matter of debate.

As British author Anthony Powell has pointed out, it is not easy to define social position in the latter half of the 17th century, and this dilemma is illustrated by the ramifications of the Pepys family. The famous diarist and man-about-town, Samuel Pepys, was the impecunious son of a London tailor; his father was the third son of a third son and married a washmaid, Margaret Kite, and Pepys was the fifth of eleven children. But Pepys's grandfather's sister happened to marry Sir Sidney Montagu and became the mother of Edward Montagu, later First Earl of Sandwich. When Pepys was 23 years old, he entered into service in the family of Sir Edward Montagu as a secretary, or domestic steward. And because Sir Edward took command of the fleet that brought Charles II back to England, Pepys's place under the sun was secured. His fast rise to fame as secretary of the admiralty to Charles II and James II—and to fortune—was achieved both by virtue of his connections and by taking bribes: the February 2nd, 1664, entry in his diary, for example, relates his taking 40 pieces of "good gold" hidden in a package of gloves for his wife from Sir W. Warren, timber merchant. As secretary of the admiralty, of course, Pepys had great influence in deciding which merchants were chosen to supply materials for shipbuilding. According to Powell (1988, p. 116), "The prejudice against 'trade,' as such, was a later development, and at this period younger sons from many great and ancient families became not only merchants but apprentices to skinners, malsters, and such like." The argument that Hooke's problems sprang from the accident of his birth and that he was discriminated against because he was not a "gentleman" simply has no substance.

Hooke's friends, colleagues, and associates apparently did not find his conduct in the defense of his reputation as anything remarkable, indicating that they did not think it offensive; in fact, many found it justifiable. In the controversy between John Wallis and Hooke, discussed earlier concerning the apparent shift in the Earth's rotational axis, there is no evidence that anyone *at that time* thought Hooke was over-reacting. Some present-day writers, however, have judged Hooke's behavior unseemly for a "gentleman." Hooke suffers from a bad press today, in spite of genuine attempts to restore his reputation. The tendency to carry on an erroneous and misguided tradition is often as strong among social scientists as among some subjects of their studies—for example, scientists with a pet idea.

Controversies were commonplace among the intelligentsia of the 17th century and probably much enjoyed by bystanders including the Royal Society audience. The scene might perhaps be analogous to the shoutings and rantings of members of the British Parliament during session, in contrast to the more boring goings-on in the U.S. Congress. Just because the MPs yell at each other does not necessarily mean that they are not gentlemen and ladies or that they are mortal enemies outside of Parliament. It is more likely that modern scholars have over-reacted to the records of these disputes than that Hooke and his contemporaries were behaving uncharacteristically of their age. Nevertheless, it is still strange that Hooke would be singled out for censure by modern writers when he was only reacting (when he did react) in a way that was normal and characteristic of the time, especially because there was just cause for his reaction.

Hooke's adversary in the rotational-axis dispute was John Wallis, who was a passionate opponent of the well-known philosopher, Thomas Hobbes. In 1655, Hobbes had produced solutions of some ancient puzzles, and in an unsparing manner Wallis, the mathematician, showed the absurdities of these solutions. Thereafter, in the words of the 1908 edition of the *Dictionary of National Biography*, the two disputants "rivalled each other in abuse and verbal quibbling"—a feud that lasted 20 years. Hobbes resented his exclusion from the Royal Society and attributed it to the malice of Wallis. Wallis was further involved in another seemingly acrimonious dispute with William Holder, whose name occurs often in Hooke's *Diary*. As rector of Bletchington, Oxfordshire, Holder had taught a deaf-mute boy to speak. This was the son of Colonel Edward Popham, admiral for the Parliament. Later, young Popham had a relapse, and Wallis was called in to restore his speech. Thereafter, Holder and Wallis each claimed credit for having cured the young man.

John Aubrey, a notable defender of Hooke and a "gentleman," considered Wallis "a capable but envious and contentious man" (Powell, 1988, p. 267). He was apparently equally contentious in the axial-shift controversy with Hooke. Yet, some modern writers would attach these adjectives to Hooke rather than to his opponent. It has become positively fashionable among some authors to picture Hooke in a negative light, regardless of the circumstances involved. Hooke's honest and straight-forward statements made publicly or in letters directed to his critic are taken by later writers as "bellicose." As for occasional complaints one finds in a diary, it is patently unfair to judge that person's character by querulous remarks made in the privacy of his own diary. But this tradition of characterizing Hooke as notoriously cantankerous, a chronic claimer of credit and paranoid complainer, passes on from generation to generation of writers. In addition, some of this genre are curiously reluctant to give Hooke credit for much of anything, glossing over his great scientific achievements as not much out of the ordinary of the time. This phenomenon is perhaps an attempt to rationalize the careless treatment of his memory by the Royal Society and others who have built their reputations on the winning (i.e., Newtonian) side.

Compared to some others involved in controversy, Hooke's behavior was most "gentlemanly." A famous example is the fracas that resulted from John Woodward's attack on the work of Dr. John Freind in Woodward's publication entitled the *State of Physic* (1718). The colleagues of these two took sides in the issue to the extent that one of Freind's champions, Dr. Richard Mead, assaulted Woodward as the latter was entering Gresham College, and the two fought a duel with their swords right there and then at the gates of Gresham. Woodward fell to the ground and had to be helped by onlookers.

Numerous accounts of such quarrels over disagreement or piracy can be related of the period, and cries of plagiarism were frequent. Aubrey, for example, also seemed to have suffered a case of shady mistreatment at the hands of someone he respected. He had written a biography of Thomas Hobbes, and in 1680 he sent his manuscript to Dr. Richard Blackbourne of Trinity College, Cambridge, for advice and editorial comments. Aubrey specifically wished to know whether he should publish it in Latin or English. Blackbourne did not reply and apparently claimed later that he had lost Aubrey's paper, but a year later, in 1681, Blackbourne published *Vitæ Hobbianæ Auctarium*—in Latin. Aubrey complained to no avail that Blackbourne used him poorly; he wrote, "I suffer the grasse to cutt under my feet; for Dr. Blackbourne will have all the Glory" (Powell, 1988, p. 178–180). I suppose there could have been some misunderstanding that Blackburne thought he could use the Aubrey manuscript with the latter's consent, but Aubrey's reaction certainly speaks differently of the affair. Curiously, even Aubrey's experience with such carelessness, if not lack of ethics, on the part of a colleague, his mistrust of others thereafter, and his fear of losing his manuscripts, all have been blamed on Hooke's influence (see Hunter, 1975, p. 83).

In view of such widespread violation of trust among colleagues, Aubrey's respect and implicit trust of Hooke should speak in favor of Hooke's integrity. Instead, Aubrey is pictured as somehow tainted by Hooke. In 1686, Aubrey wrote a draft will and bequeathed his manuscript *Memoires of Naturall Remarques in the County of Wilts* to "my worthy friend Mr. Robert Hooke of Gresham College" to publish the manuscript in case he should die before he had a chance to do it himself. He worried, however, that Hooke "hath so much to doe of his owne that he will not be able to finish," a rather typical circumstance with Hooke. Eventually this manuscript was published as *Natural History of Wiltshire*, but not until 1847, more than a century and a half later, by the Wiltshire Topographical Society.

What concerns us here in regard to this manuscript is that it represents the starting point of a potentially serious case of plagiarism of Hooke's hypothesis of the terraqueous globe. And the probable reason the matter did not become explosive is because Hooke, contrary to what some would expect of him, chose not to make a case against the offender, someone the world respected. The alleged perpetrator seems to have been the distinguished and highly regarded botanist-zoologist and theologian John Ray (Fig. 7–1). Powell (1988, p. 217) considers him a "brilliant savant" who was perhaps "the greatest of the early naturalists." Ray, the son of a blacksmith, was pre-eminent in the fields of plants, birds and insects. In 1691, Aubrey had sent Ray a copy of his Wiltshire manuscript for his comments. One chapter entitled "An Hypothesis of the Terraqueous Globe. A Digression," was clearly identified by Aubrey as based on one of Hooke's discourses on his system of the Earth. Aubrey wrote:

I have heard *S<sup>r</sup> Will Davenant* say, that Witt did seem to be the easiest thing in the world for when it is delivered, it appeares so naturall, that everyone thinkes he could have sayd the same: this of his may also be applied to Inventions and Discoveries; so after the Discovery of America by Columbus, then every Navigator could have donne as much. So now (me thinkes) I could be angry with my selfe for my stupidity, that have so often trod upon and rid over these Remaines of the old world mentioned in the former Chapter [Aubrey is referring to the chapter on "Formed Stones" or fossils], without making a due Reflexion, which deserves so much a greater admiration and research than the Ruines of the August Buildings of the Greekes & Romans; as this Globe excells in magnitude those ancient Monuments: But this Hypothesis was never heard of till *Mr. Rob. Hooke* in a Discourse of his before the Royall Societie at Gresham Colledge An°. Dom. 1663 or 1664, did *first discover* it: and now by many perhaps forgotten, and which I doe here handdowne to Posterity (if this Essay of mine live) with a due acknowledgment of his great Discovery and Witt. But I hope this excellent Person will bee pleased to write and publish this *Hypothesis himselfe*.

This new Notion of M<sup>r</sup> Hookes was no sooner, or not long after delivered; but that intelligence of it was sent by the Post to Nich. Steno, who printed much of this Hypothesis in Latin in 8° translated into English by *M<sup>r</sup> H. Oldenburgh*; whereas it appears by the number of Transaction, that M<sup>r</sup> Hooke was the Author of it before.[1]

This passage not only refers to the dates of the first public expression of Hooke's hypothesis, 1663 or 1664, but also refers to the role that Henry Oldenburg played in transmitting many of Hooke's ideas abroad to be used by others. Steno is widely recognized as a founder of geology based on his one publication on the

*Figure 7-1. John Ray (1627-1705). (From an engraving by H. Meyer in* Gallery of Portraits, *1833.)*

subject, his *Prodromus* (1669). There is some suspicion, at least on the part of Hooke and Aubrey, that Oldenburg might have deliberately sent Steno Hooke's ideas and championed Steno's book, to the detriment of Hooke (see Oldroyd, 1989, p. 217, n. 46 and Gunther, 1931, 8:208). In the next chapter we will compare the contributions of Hooke and Steno. Now we must examine the case involving John Ray.

After carefully perusing Aubrey's manuscript, Ray wrote to Aubrey on September 22, 1691, from his home, Black Notley. He expressed his "great pleasure & satisfaction" in reading Aubrey's manuscript, but added: "I find but one thing that may give any just offence & that is ye Hypothesis of ye Terraqueous Globe, wherewith I must confesse my self not to be satisfied. But that is but a Digression, & aliene from your subject, & so may very well be left out" (Gunther, 1928, p. 169) Aubrey annotated this letter adding, "This Hypothesis is Mr. Hooks. I say so, and it is the best thing in the Book; it (indeed) does interfere w[it]h ye 1 chap. of Genesis" (Gunther, 1928, p. 171). Aubrey's annotation suggests that perhaps it was Hooke's departure from the Mosaic account that might have offended Ray. Indeed, while Hooke toward the end of his life sought corroboration of his ideas both from the classics and from the Bible, his early expression of his hypothesis was devoid of religious considerations. Ray's account, on the other hand, is replete with demonstrations of the power and wisdom of God, but the similarity to Hooke's hypothesis on the basic ideas about the causes of terrestrial changes and the origin of land and sea is quite evident.

As sketchy as Aubrey's chapter on Hooke's hypothesis is, clearly Ray did not want it published. It must also be remembered that at this time Hooke had been discoursing on the same subject for over 20 years, so that his views should have been familiar to Ray anyway. Ray's expressed reason to Aubrey for deleting the chapter is that it was, after all, only a "Digression." Not a year later, however, Ray published *Miscellaneous Discourses Concerning the Dissolution and Changes of the World*, dated January 6, 1692, with two "Digressions" of his own. In a letter Ray wrote on November 25, 1691, to Edward Lhwyd, Ray announced that the manuscript for this book was now finished and in press. In addition to the body of the book, which consists of a sermon he had preached some 30 years earlier at St. Mary's Church, Cambridge, he wrote (see Keynes, 1976, p. 107–108):

> there are two large Digressions, one concerning the general Deluge in the days of Noah: another concerning the Primitive chaos & creation of the World. In the former of those at the instance & importunity of some friends I have inserted something concerning formed stones [i.e. fossils] as an effect of the Deluge, I mean their Dispersion all over the Earth. Therefore you will find all I have to say in opposition to their opinion, who hold them to be primitive productions of Nature in imitation of shels.

The circumstances under which this volume was so hastily sent to the publishers seem at least suspicious. Only a short while earlier, in 1691, Ray had published *The Wisdom of God Manifested in the Works of the Creation,* a book the sole purpose of which was to show that the entire universe "the Sun and Moon, and all the Heavenly Host" and "the vast multitude of Creatures" are proof of the power and wisdom of God; the section on the Earth gave no indication that at this date he had a hypothesis on the formation of its mountains, valleys, and seas. Yet the main body of the new book of 1692, *Miscellaneous Discourses*, contained a 30-year-old sermon, which apparently he had not been in a hurry to publish. This book, with its two "Digressions," was dedicated to John

Tillotson, archbishop of Canterbury. In the preface Ray apologized "for being too hasty in huddling up and tumbling out Books." He continued,

> The longer a Book lies by me, the perfecter it becomes. Something occurs every day in reading or thinking, either to add, or to correct and alter for the better; but should I defer the Edition till the Work were absolutely perfect, I might wait all my life-time, and leave it to be published by my Executors. But I see that Posthumous Pieces generally prove inferiour to those put out by the Authors in their lives.

Ray "doth protest too much, methinks." Surely it could not have been the 30-year-old sermon that suddenly spurred him on to publication. His tagging on to this sermon the hypothesis on the formation of terrestrial features is rather analogous to the U. S. Congress pushing through an unpopular piece of legislation on the tailcoat of a popular one—with the President not having the line-item veto.

Geoffrey Keynes, the bibliographer, makes the following excuse for Ray: that the reason for his haste was "due in fact to his having been at death's door with pneumonia in March 1691, and his consequent fear that he might not live to see his book through the press if he delayed" (Keynes, 1976, p. 108) Ray, in fact, was fully recovered by the time he was writing to Aubrey about the latter's manuscript. He lived to be 78 years old; Hooke had died at age 68, two years before Ray. It is more likely that Ray's rush to print was prompted by reading chapter 8 of Aubrey's Wiltshire manuscript which contained Hooke's hypothesis. Expressing his dissatisfaction with Aubrey's chapter, then publishing something quite similar, should certainly be cause for suspicion.

Indeed, Ray's account, while replete with demonstrations of the power and wisdom of God, is similar to Hooke's hypothesis in the basic ideas about the causes of terrestrial changes and the origin of land and sea. In describing Hooke's ideas on land rising from the sea and the creation of mountains, Ray cites some of the same lines from Ovid's *Metamorphoses* that Hooke does. He also cites similar passages from Pliny, Strabo, and Kircher the Jesuit, the same authors cited by Hooke. Ray was apparently careful about giving credit to his sources; yet we know that he was familiar with Hooke's hypothesis and he did not mention Hooke. Aside from having read Hooke's hypothesis in Aubrey's manuscript, Ray had been a Fellow of the Royal Society since 1667 and therefore should have been thoroughly imbued with Hooke's earthly ideas since he was so interested in the subject matter. The incensed Aubrey wrote to Anthony Wood on February 13, 1692, shortly after the appearance of Ray's book (Powell, 1988, p. 219):

> Your advice to me was prophetique, *viz.* not to lend my Mss. You remember Mr. J. Ray sent me a very kind letter concerning my *Naturall History of Wilts*: only he misliked my Digression, which is Mr. Hooke's Hypothesis of the terraquious Globe whom I name with respect. Mr. Ray would have me (in the letter) leave it out. And now lately is come forth a booke of his in 8° which all Mr. Hooke's hypothesis in my letter is published and without any mention of Mr. Hooke or my booke. Mr. Hooke is much troubled about it. 'Tis a right Presbyterian trick.

Another indication that Ray was quite familiar with Hooke's ideas is evidenced in a small treatise he published in 1691 entitled *The Wisdom of God Manifested in the Works of the Creation,* dedicated to the Lady Letice Wendy. In discoursing about the Creation he borrowed the words of the neoplatonist Henry More who had also objected to Hooke's axial shift idea. Some of More's words quoted by Ray were "I demand whether it be better to have the *Axis* of the *Earth* steady and perpetually parallel to it self, or to have it carelessly tumble this way and that way as it happens, or at least very variously and intricately: And you cannot but answer me, it is better to have it steady and parallel." Without mentioning Hooke, therefore, Ray attacked Hooke's axial shift idea through someone else's words, while also demonstrating the confusion of Hooke's contemporaries between axial shift with respect to the heavens and Hooke's idea of polar wandering. It is not known, of course, whether Ray's motive was simply to beat Hooke to print via Aubrey or whether he actually stole the material. Ray must have known that Hooke had been discoursing on his earthly ideas for 20 years and had not got around to publishing much about them; Hooke was widely known to be overly busy, having more to do than could be done. Ray was therefore not particularly worried until he saw the Aubrey manuscript. The unpleasant fact remains that the circumstances seem suspicious and unbecoming to such a religiously fervent savant—that is, urging Aubrey to delete his "Digression" while adding "Digressions" on the same subject in his own book.

Ray clearly felt that his own digressions were very important, because in the second edition (1693), he had changed the title to *Three Physico-Theological Discourses* and his "Digressions" became major parts of the book. The title page shows the extent of Hooke's hypothesis covered by Ray. After the main title and three subtitles are these words: "Wherein Are largely Discussed the Production and Use of Mountains; the Original of Fountains, of Formed Stones, and Sea Fishes Bones and Shells found in the Earth; the Effects of particular Floods and Inundations of the Sea; the Eruptions of Vulcano's; the Nature and Causes of Earthquakes: With an Historical Account of those Two Late Remarkable Ones in Jamaica and England." He also added four engraved plates of ancient coins and fossils. It was Hooke, of course, who had made the analogy between coins and fossils as both indicators of past history. In the preface of the second edition, Ray wrote, "But now this Treatise coming to a second Impression, I thought it more convenient to make these several Discourses upon these Particulars, substantial Parts of my Work, and to dispose them according to the Priority and Posteriority of their Subjects, in Order of Time, beginning with the Primitive *Chaos*."

Ray's glorious reputation, however, makes him rather invincible. Much like Newton's situation, there are plenty of writers, even today, who are willing to pick up his shield. Gunther, the editor of Ray's correspondence, writes, "Ray's work on *Chaos and Creation* published in 1692, a year after the date of the letter to Aubrey, was a far more valuable contribution to the science of geology"—a judgment not considered so by geologists since this group generally do not connect Ray's name with any contribution in geology. Michael Hunter (1975, p. 58) writes that Ray's theory differs from Hooke's and presumably, even if it did not differ, "undoubtedly" Ray's work was independent. Anthony Powell points out in a footnote to this whole affair that C. E. Raven, Ray's biographer, considered "the charge of plagiarism (which Hooke was always inclined to scent) most improbable" (Powell, 1988, p. 219, footnote). Note the positive tones regarding Ray—i.e., almost infallible—in contrast to the derogatory tone of this statement against Hooke when the circumstantial evidence is so weighted on the side of unethical conduct on the part of Ray. A curious phenomenon!

The unlucky Hooke never had the kind of press that Ray, like Newton, has always enjoyed. Ironically, when Aubrey's *Natural History of Wiltshire* was finally published in 1847 by the Wiltshire Topographical Society, the editor, John Britton, in an age that still worshipped Newton like a god, saw fit to delete chapter 8 on Hooke's hypothesis altogether. Britton's rationale was that the substantial and brilliant discoveries of Newton induced many of his less gifted contemporaries to pursue inquiries into the arcana and profound mysteries of science; but where rational inferences and deductions failed, they too frequently had recourse to mere unsupported theory and conjectural speculation. . . . The chapter of Aubrey's work which bears the above title [i.e. An Hypothesis of the Terraqueous Globe, A Digression] is, to some extent, of this nature.

As troubled as Hooke was in Ray's apparent perfidy, as reported by Aubrey, he did nothing in retaliation, either publicly or privately that we know of; he made no claims of priority against Ray. Hooke's doctorate from the University of Oxford was granted just a short while before by the authority of John Tillotson, archbishop of Canterbury. Ray's book was dedicated to none other than the archibishop. Hooke, then, was not insensitive to prudent silence. Was he wrong to "scent" plagiarism? Was he "notoriously cantankerous" and a paranoid so that he would be "always inclined" to "cry" plagiary? Or do the facts and circumstantial evidence speak for themselves in this age of free-for-all piracy of ideas, seemingly perpetrated by the most distinguished members of the intellectual community?

# 8
# Final Assessment

*Facile Inventis addere.*

—Robert Hooke, 1676

As sociologist Robert Merton (1970) noted, it was not so much a sudden biologic emergence of gifted people in the 17th century that caused the flowering of science but rather the interest of extraordinary people in the same things. Hooke was not alone in his scientific interests and pursuits. The spectrum of his interests was so broad, in fact, that his activities converged with those of many others. His colleagues in the Royal Society devoted much thought and debate to geology, especially to the question of the origin of fossils. Scholars on the continent were also interested in geology. Historian Rachel Laudan (1987) argues eloquently that geology as a science actually developed from the European (continental) tradition of the Wernerian School at Freiberg, which emphasized a chemical-mineralogical approach to the study of the Earth, while the British tradition, centering on the general principle of plutonism, was insular, isolated, and of lesser importance. It is perhaps true that English-speaking geologists have long ignored the European, continental development, but what they consider the mainstream, the British tradition, was neither minor nor insular. The spread of knowledge was quite effective even in the 17th century, as evidenced in the correspondence of the Royal Society of London. By the 18th century such communication between Britain and the continent had become routine among the ever-increasing numbers of learned societies all over Europe. British ideas, therefore, were central to the intellectual debate of the time.

In this chapter I will show that Hooke's ideas surpass Steno's in depth by comparing their geological contributions and I will provide the textual proof that Hooke had a profound influence on the thinking of Hutton. While Steno is widely acclaimed as "founder" of the the science of geology and Hutton is considered "father" of modern geology, Hooke's geological contributions, like so many of his accomplishments, have suffered undeserved obscurity. I am hopeful that the evidence supplied may allow historians of science to reassess their received perception with regard to the foundation of geology and to include Hooke in that roster of luminaries in the history of geology. After being apprised of the depth of Hooke's thinking and the extent of its subsequent transmission, one can hardly avoid the conclusion that Hooke had laid the foundation of the pre-continental-drift paradigm in geology.

The ancient Greeks and Romans believed in an organic and marine origin for fossil shells, but among the philosophers of the Middle Ages and the Renaissance, only a few held such enlightened beliefs; one was Leonardo da Vinci (1452–1519). Another, the German Konrad Gesner (1516–1565), suggested that tonguestones, or *glossopetræ* resembled fossil sharks' teeth. The French potter Bernard Palissy (1510–1581) claimed that fossil shells represented once-living forms that had been petrified. But most people in the Middle Ages thought of fossils as self-generated in the soil by a trick of nature. By the 17th century, some philosophers were convinced that fossils had an organic origin, but even most of these people believed them to be only relics of Noah's flood. As for minerals and crystalline material, they were long believed to hold supernatural, mystical, or magical properties. Nevertheless, because of strong mining interests in the 16th century, many regions, especially in Saxony, were explored and some valuable information was collected. The German physician Georg Bauer (1494–1555), known by his Latin name, Agricola wrote treatises on geology and attempted a classification of minerals. He wrote *De re Metallica,* a comprehensive survey of the mining industry. This book is famous not only for its facts about mining but also for its highly informative woodcuts by Hans Deutsch. But geology as a subject of study had not emerged in a coherent sense although there was much interest in the objects found in rocks.

## NICOLAUS STENONIS (1638-1686)

Roughly a century after Agricola, around 1666, Niels Stensen (1638–1686), a Danish physician, better known as Nicolaus Stenonis or Steno, was engaged in anatomical studies in Italy. By then, Hooke's *Micrographia* with his explanations for petrifaction and ideas on the formation of terrestrial features, had been published and he was actively defending his hypothesis that fossils were the "exuviæ" of former living things. He had also been lecturing on his system of "this terraqueous globe," the Earth. Steno was attached to the Court of the Medicis in Florence where, in the autumn of 1666, the head of a large great white shark, caught in the Ligurian Sea, was brought to him to be dissected by order of Ferdinand II grand duke of Tuscany. Ferdinand was extraordinary in that he was intensely interested in scientific investigation and sponsored the activities of the Accademia del Cimento, the learned society founded by his brother Leopold in 1657 (Albritton, 1980, p. 22). Steno's report on his dissection of the shark's head was published as a supplement to a larger treatise on muscles. The supplement, *Canis carchariæ dissectum caput* (The head of a shark dissected) is considered the first geologic treatise because it rekindled interest in the origin of tonguestones (*glossopetræ*), demonstrating unequivocally that tonguestones were actually petrified sharks' teeth (Fig. 8-1). Tonguestones, which are found embedded in rock in many parts of the world, are especially abundant on

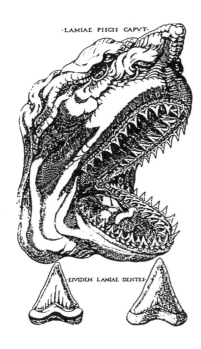

*Figure 8-1. Illustration of a shark's head and teeth used by Steno in his report on dissecting a shark. (From* Canis Carcharia Dissectum Caput, *1667.)*

the island of Malta. Explanations for their origin ranged from objects fallen from the heavens to things grown in the ground *sui generis* or as a *lusus naturæ* (self-generated; as a joke of nature). Legends were also associated with them. One involved the wrath of the Apostle Paul who, bitten by a snake on Malta, put a curse on all snakes on that island and turned their teeth into tonguestones (Adams, 1938, p. 115; Albritton, 1980, p. 25). Magical powers were attributed to the stones if they were worn around one's neck or taken internally in powder form.

The main interest for Steno however, was how these objects had become encased in rock. In 1669 he published his only substantial contribution to geology, *De Solido intra Solidum Naturaliter Contento Dissertationis Prodromus* (Predecessor of a dissertation of a solid naturally contained within a solid). He had intended to write a more extensive treatise on this subject, but after converting to Catholicism in 1667, he vowed to devote his life in the service of God. He had become more and more disaffected by science, which he relegated to only an intermediate stage in the hierarchy of things. He expressed this conviction in the following words: "Beautiful is that which we see, more beautiful that which we know, but by far the most beautiful that which we do not comprehend" (Scherz, 1960, p. 17).

Steno was ordained a priest in 1675 and two years later was consecrated in Rome as bishop. He led the remainder of his life in a constant state of self-denial, giving all his personal wealth to the poor. In 1686, at age 48, suffering from fasting and poverty, he died. One wonders why Steno dropped his scientific activities so precipitously, in the middle of things and at the height of his career. Not only that, he transformed himself from a darling of royal courts, living lavishly, to a religious ascetic whose own parishioners abhorred his extreme self-deprivation so much that some even threatened to assault him (Albritton, 1980, p. 39). One explanation for his abandonment of science is the speculation that he might have been frustrated trying to conform nature to scripture, and in a Kuhnian manner rejected science because he could not tolerate the crisis of overturning his conceptual paradigm—that of his faith (Albritton, 1980, p. 40–41). This explanation seems thin since practically all the philosophers of the age were steeped in the same religious traditions, the same paradigm, and, in spite of contradictions, they were able to rationalize that nature was a manifestation of God.

Cosimo III ordered the interment of Steno's body in the famous Chapel of the Medicis of San Lorenzo. Today, on the cloister wall of San Lorenzo is a medallion portrait of Steno, surrounded by a marble wreath, with a Latin inscription under it. This carving was provided by scientists who convened in Bologna at the International Geological Congress in 1881. More than a thousand scientists, mostly geologists and anatomists, from many parts of the world contributed to it and then made a pilgrimage to his tomb from Bologna to Florence to pay homage to this man "of surpassing distinction." He had written 24 papers in anatomy, 14 on theological subjects, and one in geology, his *Prodromus*. In spite of Steno's relatively short scientific career and the paucity of his geological writings, scientists and historians generally agree that his geological contributions have been recognized as fundamental.

One cannot help but note the respect and admiration history has accorded Steno in contrast to that received by his contemporary, Hooke. The difference is all the more striking and incomprehensible when the contents of the *Prodromus* are compared to ideas expressed earlier by Hooke. As Gordon Davies (1964) correctly pointed out, "the neglected Hooke had an understanding of Earth-history which is in some respects superior to that of the much lauded Steno."

Steno is credited with expressing three basic principles in geology (*Prodromus*, p. 30):

1. The superposition of strata: "at the time when any given stratum was being formed all the matter resting upon it was fluid, and therefore at the time when the lowest stratum was being formed, none of the upper strata existed."
2. Original horizontality of strata. That a stratum was initially horizontal and continuous at the time of its formation. In fact, Steno conceived that the stratum would cover the whole Earth unless existing solid bodies prevented its deposition.
3. The constancy of interfacial angles. This principle, called Steno's Law, concerns crystallography. In examining the six-sided columnar crystal of quartz, he noted that regardless of the size of the crystal the angles of corresponding faces were always the same.

The first two principles are illustrated in the *Prodromus* in "The history of Tuscany," six cross sections (Fig. 8-2) representing a schematic geologic history

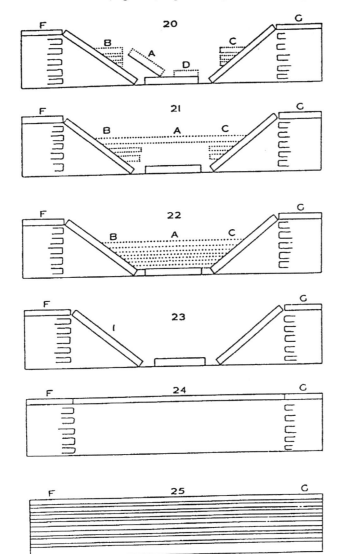

*Figure 8-2. Steno's diagrams showing the six aspects of Tuscany. (From Prodromus, 1669.)*

of the province. The sections depict a sequence of events involving an initial sea and deposition; formation of an underground cavern; collapse of the overlying strata, causing tilting; a great flood and more deposition; formation of more cavern, and a second collapse of strata. Interestingly, as a testimony to Steno's influence on modern geology, his cross sections of Tuscany are read from the bottom up, the oldest stage being at the bottom. It is widely accepted that the expression of such seemingly simple geological truths established the science. A comparison of the ideas of Steno and Hooke, however, reveal that Hooke's contributions were more extensive and more profound (Drake and Komar, 1981).

**On Fossils**

Both Steno and Hooke stood apart from their contemporaries in expressing an organic origin for fossils. But Hooke, who had expressed many of his innovative ideas on fossils and geology by the early 1660s, went much further. In lectures given prior to 1668, he discussed the extinction and transmutation of species and the idea that fossils might provide a key to the chronology of natural events in the past—just as ancient coins and monuments might provide information on the history of a region. Fossils could (*Earthquakes*, p. 321)

> certify a Natural Antiquary, that such and such places have been under the Water, that there have been such kind of Animals, that there have been such and such preceding Alterations and Changes of the superficial Parts of the Earth; And methinks Providence does seem to have design'd these permanent shapes, as Monuments and Records to instruct succeeding Ages of what past in preceding. And these written in a more legible Character than the Hieroglyphics of the ancient Egyptians, and on more lasting Monuments than those of their vast Pyramids and Obelisks.

In the famous passage already quoted (see Chapter 6) in which he predicts that fossils could "raise a *Chronology* [Hooke's italics]," it is uncertain whether he understood the fundamental principle of relating different fossils to corresponding levels of strata that enclose them, thus establishing an orderly succession and a relative timescale. Nevertheless, his sense that fossils were useful in determining a sequence of events was quite evident and outstrips Steno's ideas on fossils. Hooke has also been considered an early evolutionist in that he recognized the evolution of species and generation of varieties and the concept of extinction (Edwards, 1963). For example, the giant ammonites found in Portland quarry stone represented to him extinct species, and that there existed "shells of divers Fishes Petrify'd in Stone, of which we have now none of the same kind," because of the destruction of their environment. By the same token, diverse new varieties of the same species could be generated by a change in the environment, because "there have been many other species of Creature in former Ages, of which we can find none at present; and that 'tis not unlikely also but there may be divers new kinds now, which have not been from the beginning" (*Earthquakes*, p. 291; see Chapter 6).

Dom Remacle Rome (1956), writing in French about the correspondence between Steno and the Royal Society of London, states that Hooke held fast to his ideas concerning the organic origin of fossils despite the limited success in convincing his colleagues and therefore "il ne pouvait manquer de voir en Sténo un puissant allié" (he could not help but see in Steno a powerful ally). There is no evidence, however, that Hooke felt particularly supported by Steno's assertions. In fact, Hooke had accused Henry Oldenburg, secretary of the Royal

Society, of sending his ideas to Steno, thereby allowing the latter to plagiarize him, writing:

> I must now add in my own vindication that I did long since prove Steno had much of his treatise from my Lectures which some time before that I had read in G[resham] C[ollege] which Lecture[s] Mr Old[enburg] Borrowed and transcribed and by Divers circumstances I found he had transmitted the substance if not the very Lectures themselves. And he did as good as own it, and upon my challenging him with it he did in two of his transactions publish that I had Read A great part of that Doctrine & hypothesis in my Lectures in Gresham Colledge Some time before Mr Steno had published his Booke.[1]

There is no reason to dispute Hooke's charges against Oldenburg in this case, as the latter acknowledged his action only after Hooke challenged the secretary on his role in transmitting Hooke's ideas abroad. In making this acknowledgment, Oldenburg established Hooke's priority. Hooke had referred to the matter enigmatically earlier, in a postscript to his Cutlerian treatise *Lampas* (1677, p. 208):

> Whereas he [Oldenburg] asserts that several discoveries of the Accuser [Hooke] had been vindicated from the usurpation of others. It is answered, the clean contrary is upon good grounds suspected from the Publication of a Book about Earthquakes, Petrifactions, &c. Translated and Printed by H. O. [Henry Oldenburg] the manner of doing which is too long for this place. Such ways this mis-informer hath of vindicating discoveries from the usurpation of others.

These passages highlight Hooke's frustration in attempting to gain credit for his freely expressed ideas and which led to his later penchant for secrecy and anagrams. Hooke never challenged Steno directly on the matter; the latter had already entered his state of self-denial and any further accusation would have been superfluous. Hooke's anger, therefore, was directed only toward Oldenburg. Had it not been for Oldenburg's support of Steno's *Prodromus* by not only bringing it to the attention of the Royal Society but also translating it into English, geologists, at least, might have recognized their debt to Hooke.

## On Strata

It is not so much the substance of the *Prodromus* as the way Steno expressed it that appeals to modern geologists. His ideas on sedimentary strata are clearly, concisely, and pedagogically stated. He defined his terms and stated the simple general truths in a distinctly modern fashion. Following his expression of superposition, for example, he states that all strata except the lowest are bounded by two planes parallel to the horizon and that "strata either perpendicular to the horizon or inclined toward it, were at one time parallel to the horizon." These statements seem self-evident to us because we are so familiar with their meaning; geology students learn them as freshmen. But for geology to become a science, such principles needed to be expressed. And as the distinguished geologist/historian, the late Claude Albritton (1980), noted, while Steno's ideas seem simple, to infer anything about time, even relative time, from a solid mass of strata is not obvious. I am constantly surprised, for example, at students' difficulty in unraveling a sequence of events from a simply drawn cross section containing such features as tilted beds, an unconformity, some horizontal beds, and maybe an intrusion and a couple of dikes. Steno's contributions, therefore, should not be underestimated.

As modern as Steno's approach was, however, Hooke actually approached more closely our thinking today. He was one step ahead of such fundamental truths, assuming them as self-evident. His ideas on strata were implicit rather than explicitly expressed as in Steno's case. Hooke spoke of sediments as layers: the "Layers of several Earths, Sands, Clays, Stones, Minerals, etc. that are met with in digging Mines and wells" (*Earthquakes*, p. 325). He assumed superposition and original horizontality of sediments: "many parts are cover'd and rais'd by Mud and Sand that lye almost level with the Water" (*Earthquakes*, p. 313). Hooke also realized that water was not the only agent of deposition; he recognized burial by volcanic ash and that the motion of air could transport dust from place to place. Instead of setting some of his assumptions down systematically as Steno had done, his mind raced ahead. He discussed at length not only depositional features, but also erosional forces. Thus, the concept of a geological unconformity is implicit in his cyclic theory—that is, deposition following denudation and uplift of the land. It was typical of Hooke's mental *modus operandi* to penetrate quickly beyond what was apparent to him to a more profound understanding. So he proceeded from the formation of layers of loose sediments to the processes of consolidation of the sediments, to erosional forces and all agents that cause alterations to the Earth's surface, resulting in the denudation that leads to another cycle of exchange of land and sea areas. This cyclic theory of Hooke is very important in geology and is discussed in Chapter 4. Steno mostly ignored or did not recognize these processes; on how the loose sediments were consolidated into rock, for example, he was silent.

### On Formation of Mountains

Hooke and Steno did not differ on the basic ideas of mountain building, but again Steno stopped short of Hooke whose mind reached original and productive concepts. Steno clearly understood the notion of faulting, as his diagrams show that strata could change their position by slipping or collapsing as a result of "the withdrawal of the underlying substance, or foundation," thereby forming underground caverns. He recognized that mountains could form by "eruption of fires which belch forth ashes and stones together with sulphur and bitumen." He also thought that mountains could be formed by "the violence of rains and torrents," meaning presumably the piling up of debris washed down by water from higher places. Yet Hooke recognized not only these processes, but uplift and subsidence as essential characteristics of tectonics. His all-inclusive term "earthquakes" refers not only to strictly seismic disturbances but also to diastrophic movements. Not stopping with sedimentation and tilting of the strata through faulting, as Steno did, Hooke placed the dynamic processes of the Earth in a universal context. He took into account the motion of the Earth around the Sun in the plane of the ecliptic, the shape of the Earth, and the tilt of the axis of diurnal rotation. He invoked the highly original idea of polar wandering and possible shifts in the Earth's center of gravity (see Chapter 5). Such shifts could cause "sliding, subsiding, sinking and changing of the internal parts of the earth as well as external," thereby causing earthquakes and changes in sea level, altering the positions of sea and dry land. Such shifts could also cause changes in the "Magnetical Power and Verture of the Body of the Earth." On such cosmic ideas Steno was silent.

### On Crystallography

Steno's Law on the constancy of interfacial angles is illustrated by diagrams in his *Prodromus*. He believed that because these angles never vary, in spite of changes in length and number of sides of crystals, a crystal must grow by accretion—that is, "new crystalline matter is being added to the external planes of the crystal already formed." Hooke, on the other hand, delved deeper. His diagrams published in his book *Micrographia* (1665), four years earlier than the *Prodromus*, clearly show his conviction that the external form of crystals are an expression of the internal arrangement of particles, which he referred to as "bullets" or "globules" (Figs. 8-3 and 8-4). In contrast, Steno had rejected the idea that the shape of the crystal should resemble the shape of the arrangements of particles. And, as mineralogist/historian Cecil Schneer (1960) points out, Hooke's diagrams express not only Steno's Law but Hauy's law of rational axial intercepts as well. Schneer suggests that Steno may well have contemplated Hooke's diagrams and redrawn them omitting the corpuscles.

**On Scriptural Chronology and the Deluge**

The most striking difference between Steno and Hooke, however, was in their attitude toward scriptural chronology and the deluge. Even before he published the *Prodromus* Steno had converted to Catholicism, and as a deeply religious man, he could not have denied the scriptures. The Earth's history then must be confined to the biblical time limit. And although he recognized the organic origin of fossils, he considered them relicts of the deluge. He divided geological events into six "aspects" corresponding to the six days of creation. And he presented the geological history of Tuscany in terms of the same six aspects; hence, there are six cross sections. He equated the universal deluge with the fourth

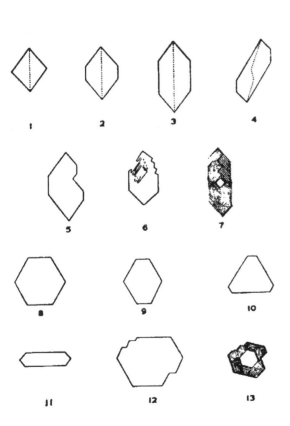

*Figure 8-3. Steno's diagrams showing that crystal forms grow by surface accretion. (From* Prodromus, *1669.)*

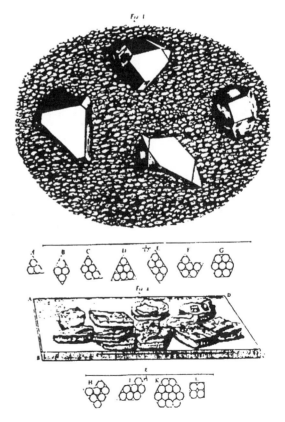

*Figure 8-4. Hooke's diagrams of crystals showing that the external form is an expression of the internal arrangement of particles. (From* Micrographia, *1665.)*

aspect, shown in his sections as a period of flooding and deposition. He considered the remains of extinct Tertiary animals, including elephants found in the Arno Valley, pack animals of Hannibal's army that died after being mired in mud (*Prodromus*, p. 64–65). He freely alluded to *Genesis* for confirmation of his arguments. For example, he wrote (*Prodromus*, p. 69), "In regard to the first aspect of the earth, Scripture and Nature agree in this, that all things were covered with water." The second and third aspects of his history were a dry plane and a surface of great relief, respectively. With regard to the time of the flood, the fourth aspect, Steno wrote (*Prodromus*, p. 72):

> Secular history is not at variance with sacred history, which relates all things in detail. The ancient cities of Tuscany, of which some were built on hills formed by the sea, put back their birthdays beyond three thousand years; in Lydia, moreover, we come nearer to four thousand years so that it is possible thence to infer that the time at which the earth was left by the sea agrees with the time which Scripture speaks.

Huge plains again appeared after the flood for the fifth aspect, and "Nature proves that those plains existed, and Scripture does not gainsay it" (*Prodromus*, p. 74). The sixth aspect of the Earth, corresponding to the topmost cross section of Tuscany, is "evident to the senses," as it represents the present configuration of the surface of the Earth.

Hooke's attitude toward scriptural chronology and the deluge was, by contrast, very practical. As the son of a curate, he was exposed to religious education as intense as anyone's in the 17th century, but his curious mind was constantly questioning. He made innumerable statements questioning a literal interpretation of the Bible. The story related in *Genesis* could only have been written down after the invention of writing, and he mistrusted the account of anything before that invention. Until then, it was passed down by word of mouth and therefore was "dark and confused . . . and cannot be much relied on or heeded." He accepted Noah's flood as possible but knew that "several other Floods we find recorded," albeit by "Heathen writers," also could have altered terrestrial surface features. Certainly, he insisted, the fossils in rocks could not have been placed there either by "Men's Hands" or "the General Deluge which lasted but a little while." His own knowledge of the hundreds of feet of sediments exposed on the Isle of Wight and the occurrence of giant fossils of extinct creatures was more convincing than the words from scripture.

The consensus among writers who refer to Hooke's geology is that his geological writings were forgotten or neglected by later scientists (e.g., Lyell, 1830; Carozzi, 1970; Turner, 1974). True, his writings on geology seem to have had little influence for 50 years or so after the publication of his *Posthumous Works* (1705). Newton's stature during the period after Hooke's death must have had something to do with this situation.

### JAMES HUTTON (1726-1797)

By the middle of the 18th century, however, it is clear that Hooke's *Discourse of Earthquakes* was a well-known piece of literature. It was cited, quoted, and even adopted by geologists. In particular, James Hutton (1726–1797, Fig. 8-5), considered the "founder" of modern geology, was keenly aware of Hooke's legacy, although he never credited Hooke or mentioned his name in any context. Hutton's theory of the Earth, as announced to the Royal Society of Edinburgh in 1785, was almost identical to Hooke's system of the Earth. This similarity, recognized

*Figure 8-5. An eighteenth-century caricature of James Hutton. (From Wright, 1865.)*

by Gordon Davies (1964), in my opinion was not a coincidence. Davies stated that Hooke deserves to be remembered as a precursor of Hutton, and David Oldroyd (1972) interprets Davies as asserting that "in a number of instances Hooke should receive the credit for ideas which are usually believed to have originated in the work of James Hutton." The suggestion that Hutton was familiar with Hooke's hypotheses is compelling.

**Hutton's Reading Material**

The spread of geological knowledge was extensive during the 18th century (Eyles, 1969; Porter, 1977). Contrary to E. B. Bailey's (1967) claim that Hutton was probably ignorant of what others had written up to that date and therefore did not mention the names of many other writers in *Theory of the Earth* (1788), Playfair (1803) tells us that Hutton had "carefully perused almost every book of travels from which anything was to be learned concerning the natural history of the earth." A number of well-known naturalists wrote about Hooke's ideas in the 18th century, some in such a matter-of-fact manner that the reader has the impression Hooke was well known and therefore needed no further explication. John Whitehurst (1786), for example, famous for his geologic work in Derbyshire, in his treatise of 1786, *An Inquiry into the Original State and Formation of the Earth*, quotes freely from "Dr. Hooke's Post." throughout the first part of his book before he finally gave the citation on p. 132 as "the posthumous work of the learned Doctor Robert Hooke." Whitehurst must have had reason to assume that his readers would be familiar with Dr. Hooke and the publication referred to as "Post."

One person in the 18th century both championed Hooke's hypotheses and adopted them as his own. This colorful character, Rudolf Erich Raspe, famous as the author of the *Travels of Baron Münchausen*, was so convinced of the validity of Hooke's ideas that he published in 1763 *Specimen Historiæ Naturalis Globi Terraquei, ...*, displaying the words *Hookiana Telluris Hypothesi* prominently in the long title: (*A Model of the Natural History of the Terraqueous globe, especially about new Islands born out of the sea, and from their exact descriptions and observations, further corroborating, the Hookian hypothesis of the Earth, on the orgin of mountains and petrified bodies*). [My own translation of the word *Specimen* is "model" while others translate the word as "introduction."] On the strength of the publication of this volume, Raspe was elected a Fellow of the Royal Society of London in 1769.[2] Clearly, the society to which Hooke had devoted most of his life accepted Raspe's *Specimen* as a significant publication. Furthermore, it is solid evidence that Hooke's *Discourse of Earthquakes* was alive and well.

Raspe was not unimportant in the history of geology. Although Rachel Laudan (1987) considers him a Wernerian, Raspe was part of the rebellion that helped finally to break away from such Wernerian teachings as the chemical precipitation of basalt from an aqueous solution. He was among the first to recognize the significance of the Frenchman Nicholas Desmarest's (1725–1815) discovery of the volcanic origin of basalt in the Auvergne published in 1774 in the memoirs of the Académie des Sciences in Paris, but which had been communicated to the Académie in 1765. The speed with which Raspe took up this new line of research testifies to the effectiveness of the transmission of ideas in the 18th century. He quickly realized that the basalts in the vicinity of Kassel in Germany must also be of volcanic origin and communicated his idea to the Royal Society of London in a letter which was read on February 8, 1770. In 1774, he published his findings and interpreted the Habichtswald as the remains of an ancient volcano. This work, which introduced the idea of the volcanic origin of basalt in Germany, was hailed by no less a personage than Goethe as epochal (Raspe, 1771, 1774; Iversen and Carozzi, 1970).

If we can believe Playfair that Hutton indeed "carefully perused almost every book of travels from which anything was to be learned concerning the natural history of the earth," then he could hardly have missed seeing Raspe's *Specimen*. Furthermore, Raspe who was fluent in Latin, Italian, German, French, and English, had translated the travels and observations of some important continental naturalists, which were published under the aegis of the Royal Society of London. One of the translations was of the travels through Italy by J. J. Ferber published in 1776, and another was of the travels of I. von Born through the Bannat of Temeswar, Transylvania, and Hungary published in 1777. In both books Raspe took the opportunity to review Hooke's ideas on earthquakes and volcanoes, albeit sometimes zealously expressing them as his own, nevertheless referring to Hooke by name in several instances. The publication dates of these books were in advance of the 1785 publication date of Hutton's *Abstract* and his communication to the Royal Society of Edinburgh.

At least one other book of which Hutton should have been aware was *The History and Philosophy of Earthquakes, from the Remotest to the Present Times: Collected from the best Writers on the Subject* (1757). The author was given as "a member of the Royal Academy of Berlin," but was probably the astronomer John Bevis. After the great Lisbon earthquake of 1755, much attention was centered on the problem of earthquakes, and the publication of this volume resulted from the widespread interest generated by the event. Of 334 pages of text, this book devotes 106 to Hooke's *Discourse of Earthquakes*. An indication of the high regard in which Hooke was held in the 18th century is that on the title page were two quotations on the scientific method, one in Latin by Verulam (Francis Bacon) from the *Novum Organum* and the other from "Dr. Hooke's *Method of improving Natural Philosophy*" (see Fig. 8-6). Hooke, therefore, was placed on the same level as the greatly respected Lord Verulam.

Another publication of note following the Lisbon earthquake was a long article by John Michell, Fellow of Queen's College, Cambridge, published in the *Philosophical Transactions* in 1761 and read to the Royal Society in February

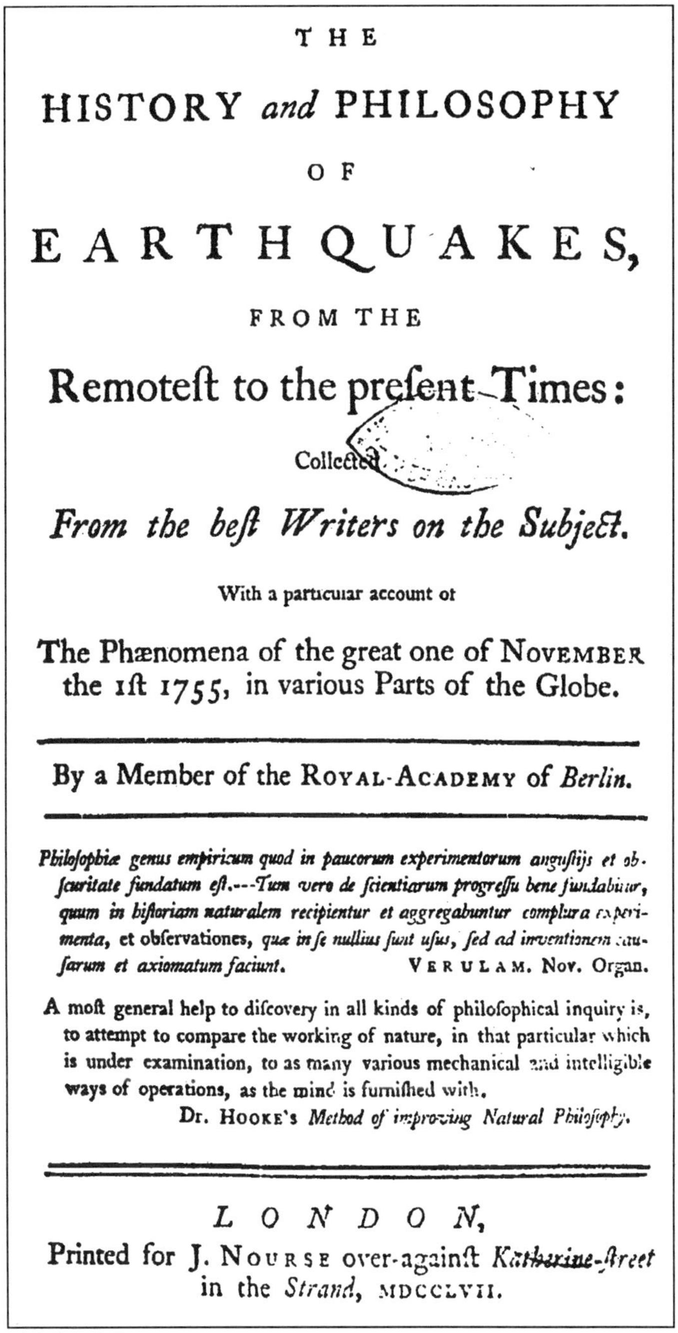

THE

HISTORY *and* PHILOSOPHY

OF

EARTHQUAKES,

FROM THE

Remoteſt to the preſent Times:

Collected

*From the beſt Writers on the Subject.*

With a particular account of

The Phænomena of the great one of NOVEMBER the 1ſt 1755, in various Parts of the Globe.

By a Member of the ROYAL-ACADEMY of *Berlin.*

*Philoſophiæ genus empiricum quod in paucorum experimentorum anguſtijs et obſcuritate fundatum eſt.---Tum vero de ſcientiarum progreſſu bene ſundabitur, quum in hiſtoriam naturalem recipientur et aggregabuntur complura experimenta,* et obſervationes, *quæ in ſe nullius ſunt uſus, ſed ad inventionem cauſarum et axiomatum faciunt.* VERULAM. Nov. Organ.

A moſt general help to diſcovery in all kinds of philoſophical inquiry is, to attempt to compare the working of nature, in that particular which is under examination, to as many various mechanical and intelligible ways of operations, as the mind is furniſhed with.

Dr. HOOKE's *Method of improving Natural Philoſophy.*

*LONDON,*

Printed for J. NOURSE over-againſt *Katharine-ſtreet* in the *Strand*, MDCCLVII.

*Figure 8-6. Title page of* The History and Philosophy of Earthquakes *(1757).*

and March of the previous year, 1760. Michell states from the start that he relied mostly on the book mentioned, *The History and Philosophy of Earthquakes* recommending it as "a work well worth the perusal of those who are desirous of being acquainted with this subject, . . . a very judicious abridgment of ten of the most considerable writers." Since Hooke's ideas occupied a third of the volume, it should be concluded that Hooke was a highly respected authority. Michell himself was quite taken by many of Hooke's ideas and adopts the latter's theory that earthquakes were caused by "vapours raised from waters suddenly let out upon subterraneous fires" (Michell, 1761, p. 594).

Hutton was not a reclusive scholar shut away at his farm in Scotland miles away from other scientists. He was a close friend of the chemist Joseph Black, and both men had met Raspe in Edinburgh in 1787, although Raspe's reputation probably had reached Scotland many years before. Hutton was also in constant communcation with members of the active Lunar Society of Birmingham, which promoted the importance of knowledge about the Earth in industrial, technological, agricultural, and transport developments. As a visiting geologist, Hutton was entertained by the Lunar Society members, among whom were John Whitehurst and Rudolf Raspe. Historian of British geology, Roy Porter (1977), reports that the membership carried on "extensive correspondence with other devotees at home and abroad." Among other visiting geologists was J. J. Ferber, whose book on travels Raspe translated, adding his Hookiana. As stated before, Hutton made scant mention of others in his work in 1785, but in later volumes of *The Theory* he did cite several writers, including Raspe.

**Textual Evidence**

But the most convincing evidence that Hutton was aware of Hooke's ideas and might have freely borrowed them is in Hutton's own writings. The following comparison of Hooke's and Hutton's texts shows their theories being so similar that it is difficult to find where they differ, but where they do differ is precisely where Hutton's style becomes polemical. As one reads Hutton, one might wonder with whom is Hutton arguing; the evidence presented here shows him to be arguing with Hooke. Fundamental to both theses is that marine fossils found in high places indicate that the rocks in which they are imbedded were once loose sediment at the bottom of the sea.

Hooke (*Earthquakes*, p. 298):

> Many Parts which have been Sea are now Land, and others that have been Land are now Sea; many of the Mountains have been Vales, and the Vales Mountains, etc.

Hutton (1785, *Abstract*, p. 5):

> The solid parts of the present land appear, in general, to have been composed of the productions of the sea ... that, while the present land was forming at the bottom of the ocean, the former land maintained plants and animals.

Hooke assigns four "species" to the category of raising of parts of the land from their former level, and conversely he cites four species of lowering or sinking of the level of the land (see Introduction to Part II and *Earthquakes*, p. 298–299).

Hutton agrees with Hooke's four "species" of raising of the land but has difficulty accepting the action of sinking. He thus argues (1788, *Theory*, p. 265):

the sinking the body of the former land into the solid globe, so as to swallow up the greater part of the ocean after it, if not a natural impossibility, would be at least a superfluous exertion of the power of nature. Such an operation as this would discover as little wisdom in the end elected, as in the means appropriated to that end; for, if the land be not wasted and worn away in the natural operations of the globe, why make such a convulsion in the world in order to renew the land? If, again, the land naturally decays, why employ so extraordinary a power, in order to hide a former continent of land, and puzzle man?

In light of plate tectonics and the "sinking" of land as a result of subduction, the lowering of the topography as a result of thermal contraction, and other sinking processes, Hooke seems indeed more clairvoyant than Hutton. The latter favors denudation and lowering of the land by erosion, something that Hooke adheres to most strongly in his cyclic theory. If physical sinking is superfluous to Hutton's theory, why did he mention it and further argue against it? He must have been aware, therefore, that some authors accepted the sinking idea originated by Hooke. Why did Hutton not simply say, "in this respect I disagree with Hooke"?

Another interesting Huttonian argument against Hooke concerns petrifaction and consolidation and introduces the issue of neptunism versus plutonism long before the geological community divided themselves into Huttonians or Wernerians. To explain how loose sediment turns into rock Hooke proposes four causes: (1) "fiery Exhalations arising from subterraneous Eruptions or Earthquakes"—that is, heat and fusion; (2) a "Saline Substance, whither working by Dissolution and Congelation, or Crystallization, or else by Precipitation and Coagulation"—that is, aqueous solution; (3) a glutinous or bituminous matter, "which upon growing dry or settling grows hard, and unites sandy bodies together into hard stone"; and (4) "a very long continuation of these Bodies under a great degree of Cold and Compression" (Earthquakes, p. 290).

Hutton (1788, p. 223) generally agrees that,

> Besides an operation, by which the earth at the bottom of the sea should be converted into an elevated land, or placed high above the level of the ocean, there is required, in the operations of the globe, a consolidating power, by which the loose materials that had subsided from water should be formed into masses of the most perfect solidity.

Instead of the four species of causes of consolidation that Hooke enumerated, however, Hutton allows only two possibilities—heat and fusion and aqueous solution. But within the category of aqueous solution, Hutton includes "congelation from a fluid state by means of cold," which is reminiscent of Hooke's fourth species.

In showing that water could indeed be "petrifying," Hooke uses the example of the formation of stalactites and stalagmites in subterraneous caverns of England (*Earthquakes*, p. 293):

> The water itself does, by degrees, produce several conical pendulous Bodies of Stone, shap'd and hanging like Icicles from the Roof of the Vault; and dropping on the bottom, it raises up also conical Spires, which, by degrees, endeavour to meet the former pendulous Stiriæ.

But Hutton, after admitting to aqueous solution as a possible process, argues against it, stating that water could only consolidate substances that are soluble in water, and, "having found strata consolidated with every species of substance, . . . concluded, that strata in general have not been consolidated by means of aqueous solution" (1785, p. 11). Then in his *Theory* (p. 229) he states,

> We have strata consolidated by calcareous spar, a thing perfectly distinguishable from the stalactical concretion of calcareous earth, in consequence of aqueous solution.

Here Hutton, therefore, is referring specifically to Hooke's example of stalactite formation which, in Hutton's thinking, has nothing to do with strata consolidated by "spar" or calcite formed by crystallization from a melt.

If Hutton were simply presenting a new and original theory, why should he be compelled to argue against straw men of his own making? He could just state, "the loose sediment is consolidated by heat and fusion." If he was aware that others might have been advocates of aqueous solution, then he should have been also aware that it was Hooke who presented the idea. If so, then why didn't he refer to Hooke by name?

It has been generally supposed that Hutton's dismissing aqueous solution as a process of consolidation is an attack against neptunism. But at the time of the publication of his *Abstract* in 1785, neptunism versus plutonism as we know it had hardly become an issue. Werner's paper on the aqueous origin of basalt was not written until 1788, the same year that Hutton's *Theory* was published. Geologist R. H. Dott, Jr. (1969) rightly reminds us that the issue of neptunism versus plutonism predated both Hutton and Werner and therefore "the neptunism attacked by Hutton himself was largely pre-Wernerian." My interpretation is that these Huttonian arguments against aqueous solution were simply directed against Hooke. Hutton compacted Hooke's theory into a Huttonian theory and had to rationalize throwing out those parts of Hooke's hypothesis that were unacceptable to him. It is a bit of historical irony that Hutton's "debate" with Hooke regarding aqueous solution as one of Hooke's four processes of consolidation should have been inadvertently and fortuitously assumed by all who followed as an attack on neptunism. Hutton might have been more surprised than anyone that his argument with a man who lived a century earlier should have become the point around which the plutonists rallied their forces.

**Axial Shift**

But the point of Hutton's contention that most strongly supports my interpretation is Hooke's idea of shifts in the Earth's axis of rotation, his polar wandering concept discussed in Chapter 5. Similar to his unnecessary objection to the sinking of the terrestrial surface, which did not need to be mentioned at all in the Huttonian theory, much less argued against, Hutton superfluously brings out the point of axial displacement. The question asked by Hutton is "How such continents, as we actually have upon the globe, could be erected above the level of the sea?" To Hooke, of course, earthquakes, with his broad definition of the word, including any movements of the terrestrial crust whether violently by faulting and volcanic eruptions or slowly and imperceptibly "by degrees," is his general answer. It is also not impossible, Hooke added, that there have been shifts in the center of gravity of the Earth so that the position of the axis of rotation on *the surface of the Earth* had been different in the past. If there is a change in the rotational axis, there would be a change in the distribution of land and sea, because the Earth's shape is an oblate spheroid as a result of the centrifugal effect created by the spinning axis. If the equatorial bulge occurs in some other belt around the Earth, Hooke postulated, water there would rise and cover the land, while land would appear near the new poles (see Chapter 5 and Hooke's words in *Earthquakes*, p. 346–347).

Hutton's answer to Hooke across a century is that (1788, p. 222),

> no motion of the sea, caused by this earth revolving in the solar system, could bring about that end; for let us suppose the axis of the earth to be changed from the present poles, and placed in the equinoctial line, the consequence of this might, indeed, be the formation of a continent of land about each new pole, from whence the sea would run towards the new equator; but all the rest of the globe would remain an ocean. Some new points might be discovered, and others, which before appeared above the surface of the sea, would be sunk by the rising of the water; but on the whole, land could only be gained substantially at the poles. Such a supposition as this, if applied to the present state of things, would be destitute of every support, as being incapable of explaining what appears.

Hutton's argument here seems totally confused. On the one hand, he seems to agree with Hooke that land could "be gained substantially at the poles," if the rotational axis were to be placed in the equatorial plane," and on the other hand he seems adament that Hooke's supposition would be "destitute of every support." Hutton's contradictory statement here clearly shows the confused state of mind he was in. He seems to want this issue just to disappear by not accepting it. But having demolished this thought with such finality, Hutton seems at a loss as to what to do with it since he must have realized that no one except Hooke himself would debate this issue. Certainly Hooke's contemporaries, led by John Wallis, felt that they had already put the issue to rest (see Chapter 5). Even if Hutton did not read Hooke directly but only read Raspe's summaries of Hooke's hypothesis, he still could have omitted the whole issue of axial displacement, because even Raspe had dismissed this Hookian idea in his Sp*ecimen* as too philosophical and therefore to be ignored and, indeed, he never mentioned it again after this dismissal. Hutton, therefore, weakly continues his argument as follows (1788, p. 223):

> But even allowing that, by the changed axis of the earth, or any other operation of the globe, as a planetary body revolving in the solar system, great continents of land could have been erected from the place of their formation, the bottom of the sea, and placed in a higher elevation, compared with the surface of that water, yet such a continent as this could not have continued stationary for many thousand years—

without, he continues, also the process of consolidation of the loose sediment—a point that is not only not denied by Hooke but originated with him. The idea of axial shift is so superfluous to Hutton's general thesis that, having brought it up, he must retreat from it by side-stepping the issue. Hooke had no support from his contemporaries on this concept and the idea was not prevalent in the 18th century.[3] Regardless of the physical validity of this idea in raising or lowering landmass, that Hutton should refer to it at all, much less argue so vehemently against it, indicates his thorough familiarity with Hooke's theory. His desire to rationalize a theory that contained disturbing elements is apparent.

### Hooke's Importance

At the risk of being labeled an iconoclast, I maintain that the evidence presented strongly suggests that Hooke's hypotheses were so ingrained in Hutton's mind, because of his careful perusal of them, that he found it necessary to bring up the Hookian points with which he disagreed—not because these were necessary to the Huttonian Theory, but because they were, after all, products of a brilliant mind and had to be rationalized away. The parts of Hooke's system with which he agreed, which include practically everything else in his theory, especially the cyclic nature of sedimentation and denudation and including such important concepts as the unconformity, Hutton left intact or expanded with great skill, illustrating them with astute field observations in his later volumes. To Davies (1969) Hooke's theory is actually superior to Hutton's, as his understanding of the processes of consolidation of loose sediments is much more sophisticated. And his cyclic theory was "too advanced for his age."

Hooke was important in the history of Earth science and it is time this fact should be recognized by geologists and historians alike. Not only do his writings reveal the extent of pre-Hutton and pre-Werner geological knowledge, they contain the fundamental concepts of pre-continental-drift geological paradigm. Further, in spite of a seeming hiatus of about fifty years during which the figure of Newton dominated the intellectual scene, Hooke's ideas *were* transmitted by later writers, demonstrating the continuity of the development of geological thought. Hutton's greatest contribution to geology, however, is not his adoption and expansion of Hooke's system, important as it is to Hooke's standing in the history of geology. In the concept of time Hutton, almost a century later, was able to make the leap of imagination that Hooke was never willing to make. Brilliant, original and imaginative as he was, Hooke could not have conceived of a world without either a beginning or an end. Just as he felt that no experiments of his day could demonstrate to him a finite velocity for light, so his working hypotheses could not include an infinite time. But for someone in the 17th century to consider that light traveled so fast that for all practical purposes it was infinite or that the biblical timescale was not long enough for all that happened on Earth is quite extraordinary. The most remarkable quality about this remarkable man, however, is that the totality of his contributions in so many areas of learning is almost beyond the scale of human endeavor and a gift to all generations to come.

# NOTES

Complete bibliographic information on all references can be found in the Bibliography. The author, year, and page numbers, of sources of specific ideas or direct quotes are given in parentheses near the referenced text. Entries of Pepys's diary are identified in the text by their entry dates. A superscript number denotes an extended note with comments and/or additional information. The italicized word *Earthquakes* refers to Hooke's *Lectures and Discourses of Earthquakes*, and the page references refer to the original page numbers of the *Discourses* as contained in the 1705 edition of Hooke's *Posthumous Works* edited by Richard Waller and not to the page numbers in the transcribed version in the present volume which follows the chronology established by Rappaport (1986). The original 1705 page numbers, however, are given in brackets throughout the transcribed version.

*INTRODUCTION*

1. John Aubrey's *Brief Lives* compiled by the late Oliver Lawson Dick first published by Martin Secker & Warburg, Ltd., in 1949, and reprinted in Penguin Classics in 1987, p. 242-245. Mr. Dick researched the original Aubrey manuscripts from some sixty-six volumes of manuscripts reposited at the Bodleian Library, Oxford, and the libraries of Wiltshire Archæological and Natural History Society, The Royal Society, the Corporation of London, and the British Museum.
2. Rappaport, 1986. Rhoda Rappaport has done a great service to the Earth science history community by working out a chronology for Hooke's *Discourses of Earthquakes*. While there may still be some discrepancies in dates, I have followed Rappaport's chronology in transcribing Hooke's lectures.

*PART I. ROBERT HOOKE'S LIFE AND WORK*

### Chapter 1. The Life of Robert Hooke

1. Previous studies include 'Espinasse, 1956, Andrade, E. N. da C., 1950, Gunther, R. T., 1930. All of these accounts naturally draw from the works of Richard Waller, 1705, John Ward, 1740, and from Aubrey, 1949.
2. Some residents of the Isle of Wight, led by Mr. Trevor Clarke, retired engineer, are attempting to resurrect the reputation of the island's distinguished native son, Robert Hooke, by establishing a museum in Hooke's honor on the property of the All Saints' Church.
3. Note, for example, entries for August 1, 2, and 3, 1672, all begin with "Drank [Fe] and [Hg]." (Hooke used the astronomical symbols for Mars and Mercury—i.e., iron and mercury.)
4. Hooke's £100 was not insignificant; it is probably equivalent today to two years' salary for a young Oxbridge graduate, or about $50,000. It is generally

assumed that the entire £100 were left him by his father. Mr. Trevor Clarke of the Isle of Wight has kindly sent me a copy of John Hooke's will in which it is clearly stated that only £40 were left him by his father, along with all of John Hooke's books. The source of the remaining £60 is unknown, although Hooke's maternal grandmother, Ann Giles, also left him some money. Even £40, however, was not considered too poor. The income from Newton's father's estate was £150 per annum and was considered a very respectable amount; that of Newton's stepfather, Barnabas Smith, £500 per annum and was termed a "plentiful estate" by Conduitt, the husband of Newton's niece. Westfall considered Conduitt's assessment an understatement. See Westfall, 1980, p. 47ff.

5. Hooke's *Diary* entries for both April 20, 1674, and June 8, 1675, report going to Mrs. Mary Beale (1632-1697), portrait painter. She is thought to have been a pupil of Sir Peter Lely and therefore could have known Hooke at Sir Peter's studio during his apprenticeship at painting. The April 20th entry in his diary indicates that while he succeeded in talking his friend Boyle into sitting for Mrs. Beale some time in the future, Hooke himself went that very day, as he had also shaved and cut his hair that day, presumably to look his best for the sitting. The second date, June 8, 1675, was apparently a sitting for Boyle, and Hooke accompanied his friend to Mrs. Beale's. See also 'Espinasse, 1956, p. 13.

6. For example, to this day, the Magdalen College School at Oxford, founded in the 15th century, gives scholarships to students who can sing in the Magdalen College choir. Our son attended Magdalen College School in 1972-1973. The school stood ready to grant him a scholarship had he been able to sing. It is not considered demeaning today to be a chorister, nor was it in Hooke's day.

7. For accusation of Parliamentary "Visitors" that Oxford was hostile to science, see Shapiro, 1969. p. 109.

8. There is a replica of this pump in the Ashmolean Museum, Oxford.

9. Furthermore, in the 1662 edition of Boyle's book, *New Experiments*, pp. 63-64 of *A Defence*, 2nd. edition, Oxford, where the Law is stated, Boyle added a "Defence of the Authors Explication of the Experiments, against the Objections of Franciscus Linus and Thomas Hobbes," revealing that when Hooke first heard about the hypothesis, he told Boyle that he had, a year before, "made observations to the same purpose, which he acknowledged to agree well with Mr. Townley's Theory." Also, Boyle had such poor eyesight at the time that he could not have executed the experiment without help and often used the pronoun *we* in his description; for example, he noted that some discrepancies, "we guessed," were probably caused by some tiny air bubbles trapped in the mercury. Gunther notes that, "with his poor sight he would hardly have noticed the air-bubbles himself, and no ordinary assistant would have perceived their significance." Hooke was clearly no ordinary assistant. Additionally, as L. T. More (1944) stated in his biography of Boyle, p.95, the latter never did any more quantitative experimental work after Hooke ceased to work for him.

10. Butler became famous for his quotable sayings, such as "spare the rod and spoil the child," in his book *Hudibras*, which pleased Charles II so much that the king sent Butler £300.

11. On June 2, 1676, Hooke records in his *Diary* that he saw the play with Godfrey and Tompion: "met Oliver there. Damnd Doggs. Vindica me Deus. people almost pointed." See also 'Espinasse, 1956, p. 25.

12. John Robison, 1803, pp. 535-537. Similar statements can be made and repeated and evidence can be produced again and again, but somehow they do not make an impact on the general attitudes passed on to historians from

generation to generation. One explanation is that Hooke was "ahead of his time," as the phlogiston theory that claimed the loss of something to the air by the burning substance, dominated the scene in the 18th century and illustrates the power of fashion and the bandwagon syndrome in science as well as in the humanities. Hooke's undeserved obscurity *today*, however, must be attributed to the unfortunate situation of having been made an enemy of the great Newton by Newton himself. Few writers in the ensuing centuries have dared to champion Hooke's cause, while at the same time a whole phalanx of writers over the years have allied themselves with the glorious "winning side" which dictates the opinion to be received.

13. Gunther, 1930, vol. VI, pp. 221, 237, 256. The description of Hooke's instrument in Royal Society records of February 1965 (Gunther, vol. VI, p. 237) is as follows: "Mr. Hooke produced a new small quadrant contrived by himself, to make, by the means thereof, both celestial and terrestrial observations with more exactness than by the largest instruments, that had been hiterto publicly known. This quadrant was only of 17 inches radius, being by the contrivance of a small roller, that moved upon the limb of it, made so accurate, that each degree was actually distinguished into 60 minutes, each of which minutes being about one-third of an inch long, was actually divided into six parts, denoting every 10 seconds in a minute. The sights were likewise so contrived, though but short, as to be no less curious in distinguishing the parts of a minute in the visible object. The perpendicular also of the quadrant was so contrived, that, though it exceeded not much three feet in length, yet it could be adjusted, by the means of an index, so exactly, as if it were 60 feet long."

14. In Olmstead's article (1949) the author berates scientist-historians in general and astronomer Delambre in particular for making a "categorical assertion" without systematic search and "rigorous historical criticism." The astronomer Delambre's assessment that the application was in use not later than 2 October 1667 and *probably even earlier*, however, turned out to be slightly more accurate than Olmstead's own painstaking conclusions considering that the latter leaves out Hooke's earlier application altogether. This story again illustrates the unnecessary tension that exists between historians of science and scientists who write history of their own science.

15. For an account of the controversy over the origin of Meteor Crater, see Ellen T. Drake, 1985, "The Coon Butte Crater Controversy," in: Ellen T. Drake and William M. Jordan, eds., *Geologists and Ideas: A History of North American Geology*, Geological Society of America Centennial Special Volume No. 1, Boulder, Colorado: Geological Society of America, 1985.

16. The Latin sentence represented by this anagram is *Ut pendet continuum flexile, sic stabit contiguum rigidum inversum,* which according to Waller (1705) is the catenary line. Aside from the many references to the rebuilding of St. Paul's in Hooke's diary, there are also letters extant in Hooke's hand referring to his work at the cathedral. For example, in a letter to Lord Conway for whom Hooke had designed Ragley Hall, Hooke warned that the mortar and brick work should be completed well before winter sets in so that the mortar could be thoroughly dry before the winter, for otherwise a great part of the walls and stone work would have to be taken down again, and he adds, "as I have found twice in the building of St Paules and in a staircase at Mountacue house and severall other places." (Batten, 1936-37, p. 102).

17. A letter by Wren dated July 28, 1675, shows that he decided to reject the phoenix idea on the basis of cost and that the phoenix symbol not only would not be understood at that height but that the spread wings might be dangerous as they catch the wind.

18. Wright, 1989, p. 101. The unpublished Trinity College manuscript is numbered Ms. 0.11a.1. 'Espinasse, 1956 p. 63-65. In the matter of the dispute with Huygens who also claimed the invention of the pocket-watch, 'Espinasse strongly suspects Henry Oldenburg's perfidy. As secretary of the Royal Society Oldenburg was certainly privy to whatever was going on and it was his duty to record the proceedings. Hooke had demonstrated his invention to the Royal Society, but Oldenburg had omitted all reference to it in the records. Furthermore, Oldenburg is suspected of having communicated the ideas for such a watch to Huygens. The fact that Huygens assigned his English patent rights to Oldenburg is circumstantial evidence that Oldenburg's interest was not altogether objective. King Charles II was well pleased with Hooke's version of the watch and probably would have granted him a patent in 1675 if it were not for the dispute and the unpleasant fact that Lord Brouncker, president of the Royal Society at that time, though present in 1660 when Hooke showed his watch and in fact was one of the "Persons of Honour" in the discussion regarding the patent document, did not back up Hooke and he lost his claim. Brouncker, unfortunately, was not the person of honor that Hooke thought he was. In the opinion of Samuel Pepys, and apparently others as well, Brouncker was "a rotten-hearted false man." It was small satisfaction to Hooke that the King saw fit also not to grant a patent to Oldenburg for the Huygens watch. In the privacy of his diary Hooke referred to Oldenburg as the "Lying Dog."
19. Huygen's *Œuvres*, vol. 4, p. 427; in: Hesse, 1966b, pp. 439-440. In fairness to Oldenburg, however, it must be said that any number of people could have communicated Hooke's inventions abroad. One such person was Robert Moray, one of the "Persons of Honour" who actually drew up the patent agreement Hooke refused to sign. In a letter dated August 1st, 1665, Moray wrote to Huygens "Until now I have never told you another thing he [Hooke] has put forward in his lectures on Mechanics...." This very sentence is evidence that Moray habitually wrote to Huygens to inform him of one thing or "another" Hooke lectured about. Moray goes on to say in his letter: "It is a quite new invention or rather a score for measuring time more exactly than with your pendulum clocks, either on sea or land, not being at all disturbed as he says by changes of position or even of wind. It is, in a word, to apply to the balance, in place of the pendulum, a spring which can be done in a hundred different ways, and he has even given us a discourse in which he has undertaken to prove that there are means of adjusting the excursions so that the small and great are isochronous."
20. In the opinion of the Halls (A. R. and M. B., 1962), Oldenburg was not to be blamed, that he only carried out his duty as Secretary of the Royal Society in a responsible manner, that Hooke should not have feuded with him. But Hooke was not the only one who distrusted Oldenburg. Christopher Wren, apparently, also disliked and distrusted Oldenburg, as revealed in *Parentalia* ['Espinasse, 1956, p. 127]. Even Newton did not like him much in spite of the latter's obsequiousness toward the great man [Andrade, 1950, pp. 470-471]. The Halls' article criticizes Andrade for asserting in the Introduction of Turnbull's *Correspondence of Isaac Newton* that Oldenburg stirred up trouble "as always" on the grounds that there is no evidence for such an assertion. They then produce their evidence to the contrary consisting of dutiful and well-meaning letters written by Oldenburg, but at a very early date, before Hooke and Oldenburg became avowed enemies. The Halls then add, "We leave others more skilled in esoteric arts than we are to determine the poisonous contents of other letters which Oldenburg wrote at this time, which have failed to survive." This sarcasm directed toward scholars like Andrade

who hold a different opinion respecting the accomplishments of Hooke and his relationship with some of his contemporaries is an example of the extent of partisanship of some Newton scholars. The revealed attitude seems to be that any study giving credence to Newton's adversaries, especially Hooke, must be discredited. In this same article the authors cite the age difference between the older more famous Hooke (in 1672) and the younger Newton, thus implying what a cad Hooke was to criticize the newcomer Newton's first paper (on optics). It is possible, of course, that the Halls made a simple arithmetic error, but they claim a difference in age of 15 years between the two men when in actuality it was only seven.

21. Westfall (1969) seizes upon these rather archaic-sounding words, uttered in Hooke's first paper when he was only 25 years old, to convince the reader not to take Hooke's startling pronouncements on gravitational law too seriously, on the grounds that these ideas really show him to mean "particular" gravities rather than universal gravitation when he discusses celestial mechanics. To Westfall, Hooke was thus worlds behind Newton on the universal law. It is my opinion that Westfall misinterpreted Hooke's concept of congruity and incongruity. Hooke is referring to bodies and motions at the atomic or molecular level rather than at the planetary level. When Hooke speaks of vibrations he is referring to the oscillations of the "particles of matter," in his development of a kinetic theory of matter.

22. Hooke letter to John Aubrey August 24, 1675, Bodleian Mss. Aubrey 13. fol. 185. Although the letter seems to read that a Mr. "Mountaine" hired Harry and sent him to Italy, Hooke's diary entries indicate that "Mountaine" must be misspelled, that the employer was probably Montague for whom Hooke had designed a mansion.

23. It has been remarked that the fact it was always Hooke who went to Boyle's house instead of vice versa underscores the great social gap between the two men. But it should be remembered that Hooke was a bachelor who lived in Gresham College rooms and except for his niece Grace and a maid Nell (who also served as a sexual partner) and other maids who took their places after Grace died and Nell left his employ, Hooke was not equipped well for entertainment. Boyle, on the other hand, lived in a great mansion run by his sister Lady Ranelagh, who, with the help of many servants, footmen and such, was in the habit of entertaining and having guests for dinner. It makes sense, therefore, if the two men wanted to see each other to discuss subjects of common interest, they could more easily do it at Boyle's house, and especially if they are good friends, it is more pleasant to do so over dinner. Regardless of the age, an intelligent and interested person like Boyle, will always enjoy and appreciate association with a brilliant and creative mind irrespective of social differences. After all, how many Hookes can one know in one's lifetime?

24. Waller, 1705, p. xxv. Waller writes, "On the 18th of July 1696. being his Birth Day, his Chancery Suit for Sir John Cutler's Salary, was determin'd for him, to his great satisfaction, which had made him very uneasy for several Years. In his *Diary* he shews his sense of it in these Terms DOMSHLGISS:A. which I read thus *Deo Opt. Max. summus Honor, Gloria in secula secularum, Amen. I was Born on this Day of July 1635. and God has given me a new Birth, may I never forget his Mercies to me; whilst he gives me Breath may I praise him.*"

25. Publications of the Harleian Society, 1864, and the Registers of St. Helen's Church, Bishopsgate, London, vol. 31, 1904, edited by W. Bruce Bannerman, p. 347.

26. The list of distinguished names buried at St. Helen's and exempted from removal is as follows: Sir John & Lady Crosby, merchant (grocer); Sir John

Spencer, clothworker; Sir Wm. Pickering (father & son), ambassador to Spain under Elizabeth I; Sir Thomas Gresham, mercer; Sir John & Dame Abigail Lawrence; Frances Bancroft, Founder of Bancroft's School & charity administration by the Drapers' Co.; Alderman Thom. Robinson; Sir Julius Cæsar, Privy Counsellor to James I; Alexander Macdougall; and the Buddington family. Only six of these names, however, are featured as "Worthies" in the west window along with Robert Hooke.

27. Trevor Clarke, native of the Isle of Wight and who is attempting to commemorate Hooke in various ways, interprets the pecten symbolism along with a staff Hooke is shown carrying, as representing a pilgrim. But given Hooke's generally non-churchgoing nature, I prefer to associate the pecten shells with his interest in fossils. As for the staff, I interpret that as a walking stick, because Hooke apparently did carry one, especially when he was investigating the cliffs near his home of Freshwater on the Isle of Wight. See his own discription of poking at the loosely consolidated sediments with his stick, *Earthquakes,* p. 297.

28. For example, entries of "Nell ♓" for April 7, 8, and 9 of 1673 represent "intercourse with Nell." Sometimes he records the double symbol ♓ ♓ which apparently means he performed twice that night, as with the entries of June 19th or July 2nd, 1673. Mostly he did not bother to record this activity, being more concerned with other matters, including his successful elimination.

## Chapter 2. The Isle of Wight and Its Influence on Hooke's Earthly Thoughts

1. Information on the geology of the Isle of Wight was mostly gleaned from my own field work on the island during the summer of 1990; I was assisted by my geologist daughter, Linda Drake Neshyba. Photographs in this chapter were taken by myself or by my husband, Charles Drake. Additional information on the geology of the island and on British geology as a whole is obtained from the following sources: Bennison and Wright, 1969; Raynor, 1981; and Wells & Kirkaldy (1967).

## Chapter 3. Other Theories of the Earth

1. The Latin edition of Burnet's book, *Telluris Theoria Sacra,* was published in 1681. Burnet consulted Newton regarding his theory before printing this edition. Subsequently it was translated into English, the first part of which was published in 1684 and the second in 1689. A later edition appeared in 1691.
2. Whiston, 1696. It is interesting to note that Whiston succeeded to Newton's Lucasian Professorship at Cambridge in 1703, the year that Hooke died.
3. Turnbull, 1960, vol. II, p. 319. The original of this letter seems to be missing, but Burnet quoted this passage from Newton's letter in his reply to Newton of January 13, 1681.
4. Hooke records in his diary on Feb. 24, 1693, "Mr Jo. Beaumont presented me his Book (Considerations on Dr. Burnet's Theory etc.): When he was gone I found he had Dedicated it to me." Gunther, 1935, vol. X, p. 217.

## Chapter 4. Hooke's System of the Earth

1. See Patterson 1949 and 1950. The suspicion of plagiarism on the part of Newton was extant even in Newton's time; see, for example, the diary of Thomas Hearne, *Reliquiæ Hearnianæ* published in 1857 in Oxford, p. 585 and p. 659. The April 7th, 1726, entry reads: "I was told last night by Mr.

Whiteside, and I suppose 'tis what others think and say also, that Sir Isaac Newton took his famous book called *Principia Mathematica*, another edition whereof is just come out, from hints given him by the late Dr. Hook (many of whose papers cannot now be found) as well as from others that he received from Sir Christopher Wren, both of which were equally as great men as Sir Isaac ... ." And again on the 27th March, 1727: "It is remarkable, that Sir Isaac owed much to some papers he had got of Dr. Hooke's. " The tone of these entries also reveals the degree of respect that people seemed to regard Hooke during that era.
2. Some writers have misinterpreted his meaning and taken the word "earthquake" at face value to mean exclusively occurrences of catastrophic dimensions; see Westfall, 1972, p. 486, who calls Hooke "the first catastrophist." It is clear that Hooke's writings show him to employ both philosophical approaches to the study of the Earth, uniformitarianism and catastrophism.

## Chapter 5. Hooke's Concept of Polar Wandering on An Oblate Spheroid Earth

1. See Fernie, 1991; except for giving Newton credit for the true shape of the Earth, Fernie's article is an excellent account of the history of the determination of the shape of the Earth. The original latin version of the Newton quote regarding Richer's trip is as follows: "Et primo quidem D. *Richer* hoc observavit anno 1672 in insula Cayennæ. Nam dum observavet transitum fixarum per meridianum mense *Augusto*, reperit horologium suum tardius moveri quam pro medio motu solis, existente differentia 2'.28" singulis diebus."
2. Quotes are from letters reposited at the Royal Society no. ClP.XX.75. Transcriptions in Oldroyd, 1989; and Turner, 1974.

## Chapter 6. Hooke's Theory of Evolution and Attitude toward God and Time

1. According to the Rappaport chronology of Hooke lectures, this one is probably one of six lectures delivered between 8 December 1686 and 19 January 1687.

## Chapter 7. Plagiarism or Paranoia?

1. Two copies of this Aubrey manuscript exist, one is reposited at the Bodleian Library and the other is at the Royal Society. The text differs somewhat from each other.

## Chapter 8. Final Assessment

1. Answer of Hooke to Wallis, read to the Royal Society April 27, 1687; Wallis referred to Steno in his letter to Halley of March 4, 1687, and Hooke suspected that Wallis had mentioned the name with design, intimating that Hooke's ideas had originated with Steno. Royal Society manuscript ClP.XX.75. Also see Oldroyd, 1989, p. 217.
2. Raspé was a fascinating and colorful character whose financial state of indebtedness led him into embezzlement and disgrace, from which he was able not only to extricate himself but to restore his standing in polite society, albeit never again in his native country of Germany. At age 30 he had gained a reputation in science as well as in poetry. Further, the Landgrave of Hesse-Cassel had, in 1767, appointed him to the Curatorship of the Landgrave's

Collections, Chair of Antiquity at Cassel's University, and a seat on the Hessian Privy Council. But being constantly in debt, he pawned some medals and coins in the Collections for which he was caring and for which he had meticulously compiled a catalogue. The missing pieces were discovered when the Landgrave appointed him to a diplomatic post as Hessian Resident at Venice and, before Raspé's departure, ordered an inventory with the aid of Raspé's own catalogue. He was ignominiously arrested. But the clever fellow was able to escape from jail, leave Germany and travel to England. The unfortunate author of *Baron Münchausen*, then, was expelled by the Royal Society of London in December 1775 and divorced by his wife in 1778. It is testimony to his ingenuity, however, that in spite of his expulsion from the Royal Society, the latter contrived to publish his volumes including his translations of J. J. Ferber and Baron I. Born. Later in 1787, Raspé moved to Scotland where he managed to be introduced to the intellectual circles of Edinburgh Society which included such men as James Hutton and Joseph Black. See Iversen & Carozzi (1970) and Carswell (1950).

3. Thomas Burnet, Hooke's contemporary, famous for his book *Sacred Theory of the Earth* (1691) also wrote of the displacement of the Earth's axis but in an entirely different context. Burnet's axis-tilt happened as a result of God's wrath; that is, in its paradisaical days the Earth had no tilt in its rotational axis, and therefore there were no seasons to complicate the lives of Man. This version of axial displacement, with respect to the ecliptic, of course, is different from Hooke's idea of polar wander.

# BIBLIOGRAPHY

Adams, Frank D. (1938). *The Birth and Development of the Geological Sciences.* New York: Dover, 1954 edition, 506 pp.

Adamson, Ian (1978). "The Royal Society and Gresham College, 1660-1711." *Notes and Records of the Royal Society* 33, 1-21.

Agassi, Joseph (1977). "Who Discovered Boyle's Law?" *Studies in History and Philosophy of Science* 8, 189-250.

Albritton, Claude C., Jr. (1980). *The Abyss of Time: Changing Conceptions of the Earth's Antiquity after the Seventeenth Century.* San Francisco: Freeman, Cooper, 251 pp.

Anderson, D. L. (1982). "Hotspots, Polar Wander, Mesozoic Convection and the Geoid." *Nature* 297, 391-393.

Andrade, E. N. da C. (1935). "Robert Hooke and His Contemporaries." *Nature* 136, 358-361.

——— (1950). "Robert Hooke." Wilkins Lecture delivered December 15, 1949, *Proceedings of the Royal Society*, London, ser. A., 201, 439-473; also printed in "Robert Hooke," *Proceedings of the Royal Society*, ser. B, 137, 153-187.

——— (1951). "Robert Hooke, Inventive Genius." *The Listener*, 8 February 1951.

——— (1953). "Robert Hooke 1635-1703." *Nature* 171, 365-367.

——— (1954). "Robert Hooke." *Scientific American* 191, 94-98.

——— (1960). "Robert Hooke, F. R. S." *Notes and Records of the Royal Society* 15, 137-145 (reprinted in Hartley 1960).

Ariotti, P. E. (1971-1972). "Aspects of the Conception and Development of the Pendulum in the Seventeenth Century." *Archive for History of Exact Sciences* 8, 329-410.

Armitage, Angus (1947). "The Deviation of Falling Bodies." *Annals of Science* 5, 342-351.

——— (1950). "Borell's Hypothesis and the Rise of Celestial Mechanics." *Annals of Science* 6, 268-282.

Aubrey, John (1949). *Brief Lives and Other Selected Writings.* New York: Scribner's, 409 pp.

Badcock, A. W. (1962). "Physical Optics at the Royal Society, 1660-1800." *British Journal for the History of Science* 1, 99-116.

Bailey, E. B. (1967). *James Hutton - The Founder of Modern Geology.* Amsterdam: Elsevier, 161 pp.

Baillie, G. H. (1951). *Clocks and Watches: An Historical Bibliography.* London: N.A.G. Press, 414 pp.

Ball, W. W. Rouse (1893). *An Essay in Newton's Principia.* New York: Johnson Reprint, 1972, 175 pp.

Bardell, David (1988). "The Discovery of Microorganisms by Robert Hooke." *American Society for Microbiology News* 54, 182-185.

Batten, M. I. (1936-1937). "The Architecture of Dr. Robert Hooke, F.R.S." *Walpole Society* 25, 83-113.

Baumann, R. (1917). "Wissenschaft, Geschäftgeist und Hookesches Gesetz." *Zeitschrift der Zereines Deutscher Ingenieure* 61, 117-124.

Beaumont, John (1693). *Considerations on a Book, Entituled The Theory of the Earth*, publist by Dr. Burnet. London. Printed for the author, to be sold by Randal Taylor.

Bechler, Zev (1974). "Newton's 1672 Optical Controversies: A Study in the Grammar of Scientific Dissent." In: *The Interaction between Science and Philosophy*, Y. Elkana (ed.). Atlantic Highlands, NJ, pp. 115-142.

Bekhterev, P. (1925). *Analytische Untersuchung des verallgemeinerten Hookeschen Gesetzes*. Leningrad.

Bell, Walter G. (1923). *The Great Fire of London in 1666*, 3rd ed. London: John Lane The Bodley Head, 387 pp.

Bennett, J. A. (1973). "A Study of *Parentalia*, with Two Unpublished Letters of Sir Christopher Wren." *Annals of Science* 30, 129-147.

——— (1975). "Hooke and Wren and the System of the World: Some Points Towards an Historical Account." *British Journal for the History of Science* 8, 32-61.

——— (1980-1981). "Robert Hooke as Mechanic and Natural Philosopher." *Notes and Records of the Royal Society* 35, 33-48.

——— (1981). Cosmology and the Magnetic Philosophy, 1640-1680. *Journal for the History of Astronomy* 12, 165-177.

——— (1989). "Hooke's Instruments for Astronomy and Navigation." In: *Robert Hooke: New Studies*, M. Hunter and S. Schaffer (eds.). Woodbridge, England: Boydell, pp. 21-32.

——— (1986). "The Mechanics' Philosophy and the Mechanical Philosophy." *History of Science* 24, 1-28.

Bennison, George M., and Wright, Alan E. (1969). *The Geological History of the British Isles*. New York: St. Martin's, 406 pp.

Birch, Thomas (1756-1757). *The History of the Royal Society of London for Improving of Natural Knowlege from its First Rise*. London: A. Millar. Reprint edition 1968 in 4 vols. and index: Johnson Reprint, *The Sources of Science no. 44*.

Birrell, T. A. (1963). *The Cultural Background of Two Scientific Revolutions: Robert Hooke's London and James Logan's Philadelphia*. Utrecht: Dekker and Van de Vegt.

Blay, Michel (1981). "Un exemple d'explication mécaniste au 17e siècle: l'Unité des théories Hookiennes de la couleur." *Revue de l'Histoire des Sciences* 34, 97-121.

von Born, I. (1777). *Travels through the Bannat of Temeswar, Transylvania and Hungary in the Year 1770; described in a Series of Letters to Professor Ferber, on the Mines and Mountains of these Different Countries by Baron Inigo Born, Counsellor of the Royal Mines, in Bohemia; to which is added John James Ferber's Mineralogical History of Bohemia, translated from the German with some Explanatory Notes, and a Preface on the Mechanical Arts, the Art of Mining, and its Present State and Future Improvement by R. E. Raspe*. London: G. Kearsley.

Boston Medical Library (1935). "Robert Hooke, 1635-1703." *New England Journal of Medicine* 213, 1040-1042.

Bradbury, Savile (1967). *The Evolution of the Microscope*. Oxford: Pergamon, 357 pp.

——— (1968). *The Microscope: Past and Present*. Oxford: Pergamon.

Brewster, David (1831). *The Life of Sir Isaac Newton*. London: Wm. Tegg,, 372 pp.

——— (1855). *The Memoirs of Sir Isaac Newton*, 2 vols. Edinburgh: Constable.

British Museum (Natural History) (1931). *Guide to an Exhibition Illustrating the Early History of Palaeontology*. Special Guide, No. 8. London.

Britton, John, ed. (1969). "Aubrey's Natural History of Wiltshire," a reprint of "The Natural History of Wiltshire" by John Aubrey, originally published by the Wiltshire Topographical Society in 1847. London: J. B. Nichols. Newton Abbot, Devon: David & Charles Reprints, 1969, 132 pp. Introduced by K. G. Ponting.

Brown, T. M. (1971). "Introduction to Second Edition." In: *The Posthumous Works of Robert Hooke*, Richard Waller (ed.). London.

Buchdahl, Gerd (1957). "Robert Hooke." *Scripta Mathematica* 23, 77 (review of 'Espinasse).

Burnet, Thomas (1691). *The Sacred Theory of the Earth.* The Theory of the Earth: Containing an Account of the Original of the Earth, and of all the General Changes which it hath already undergone, or is to undergo till the Consummation of All Things. The Two Fisrt (sic) Books Concerning the Deluge, and Concerning Paradise. The Second Edition. Dedicated to the King. London: Printed by R. Norton, for Walter Kettilby, at the Bishops-Head in S. Paul's Church-Yard. Reprinted as a Centaur classic, 1965, by Southern Illinois University Press, Carbondale, IL, 412 pp.

Carozzi, A. V. (1970). "Robert Hooke, Rudolf Erich Raspe and the Concept of Earthquakes." *Isis* 61, 85-91.

Carswell, J. (1950). *The Prospector, Being the Life and Times of Rudolf Erich Raspe (1737-1794).* London: Cresset Press, 277 pp. (New York edition has title: *The Romatic Rogue; Being the Singular Life and Adventures of Rudolf Erich Raspe, Creator of Baron Munchausen.* New York: Dutton, 277 pp.)

Centore, F. F. (1968). "Copernicus, Hooke and Simplicity." *Philosophical Studies* 17, 185-196.

——— (1970). *Robert Hooke's Contributions to Mechanics.* The Hague: Martinus Nijhoff, 135 pp.

Chapin, Seymour (1953). "A Survey of the Efforts to Determine Longitude at Sea, 1660-1760." *Navigation* 3, 247.

Clark, G. N. (1949). *Science and Social Welfare in the Age of Newton,* 2nd ed. Oxford: Clarendon, 159 pp.

Clay, Reginald S. and Court, Thomas H. (1932). *The History of the Microscope.* London: Charles Griffin, 266 pp.

Cohen, I. Bernard (1964). "Newton, Hooke, and 'Boyle's Law.'" *Nature* 204, 618-621.

Colvin, Howard (1978). *A Biographical Dictionary of British Architects 1600-1840.* London: John Murray, pp. 428-431.

Courtillot, Vincent, and Besse, Jean (1987). "Magnetic Field Reversals, Polar Wander, and Core-Mantle Coupling." *Science* 237, 1140-1147.

Crombie, A. C. (1967). "The Mechanistic Hypothesis and the Scientific Study of Vision." In: *Historical Aspects of Microscopy*, G. L'E. Turner and S. Bradbury (eds.). Cambridge: Heffer, pp. 3-12.

Davies, Gordon L. (1964). "Robert Hooke and His Conception of Earth-History." *Proceedings of the Geologists' Associations* 75, 493-498.

——— (1969). *The Earth in Decay, A History of British Geomorphology 1578-1878.* London: Macdonald, 390 pp.

De Milt, Clara (1939). "Robert Hooke, Chemist." *Journal of Chemical Education* 16, 503-510.

Deacon, Margaret (1965). "Founders of Marine Science in Britain: The Works of the Early Fellows of the Royal Society." *Notes and Records of the Royal Society* 20, 28-50.

——— (1971). *Scientists and the Sea, 1650-1990: A Study of Marine Science.* London and New York: Academic Press, 445 pp.

Dear, Peter (1985). "Totius in Verba: Rhetoric and Authority in the Early Royal Society." *Isis* 76, 145-161.

Derham, W. (1726). *Philosophical Experiments and Observation of the Late Eminent Dr. Robert Hooke.* London: Cass Library of Science Classic no. 8 (1967), 398 pp.

Dick, Oliver Lawson, ed. (1949). *Aubrey's Brief Lives.* London: Martin Secker and Warburg. Reprinted in Penguin Classics (1987), 388 pp.

DiGregorio, Mario A. (1990). *Charles Darwin's Marginalia,* Vol. I. New York: Garland Publishing Co., Inc.

Dilworth, C. (1986). "Boyle, Hooke and Newton: Some Aspects of Scientific Collaboration." *Memorie di Scienze Fisiche e Naturali* 103, 329-331.

———, and Sciuto, M. (1983). "Hooke, Grimaldi and the Diffraction of Light." In: *Atti dell 4 Congresso Nazionale di Storia della Fisica*, Pasquale Tucci (ed.). Como, pp. 31-37.

Dostrovsky, S. (1974-1975). "Early Vibration Theory: Physics and Music in the Seventeenth Century." *Archive for History of Exact Sciences* 14, 169-218.

Dott, R. H., Jr. (1969). "James Hutton and the Concept of a Dynamic Earth." In: *Toward a History of Geology*, C. J. Schneer (ed.). Cambridge: Massachusetts Inst. Technology Press, pp. 122-141.

Drake, Ellen T. (1976). "Alfred Wegener's Reconstruction of Pangea." *Geology* 4, 41-44.

——— (1981). "The Hooke Imprint on the Huttonian Theory." *American Journal of Science* 281, 963-973.

——— (1983). "Robert Hooke and the Huttonian Theory: A Discussion." *Journal of Geology* 91, 231-232.

——— (1985). "The Coon Butte Crater Controversy." In: *Geologists and Ideas: A History of North American Geology*, GSA Centennial Special Volume 1, Ellen T. Drake (ed.). Boulder, CO: Geological Society of America, pp. 65-77.

——— (1988). "Review of *From Mineralogy to Geology: The Foundation of a Science, 1650-1830* by Rachel Landau." *Geology* 16, 766.

———, and Komar, Paul D. (1981). "A Comparison of the Geological Contributions of Nicolaus Steno and Robert Hooke." *Journal of Geological Education* 29, 127-134.

——— (1983). "Speculations about the Earth: The Role of Robert Hooke and Others in the 17th Century." *Earth Science History* 2, 11-16.

——— (1984). "Origin of Impact Craters: Ideas and Experiments of Hooke, Gilbert, and Wegener." *Geology* 12, 408-411.

Duncan, Robert A. (1991). "Ocean Drilling and the Volcanic Record of Hotspots." *GSA Today* 1, 213-219.

———, and Richards, M. A. (1991). "Hotspots, Mantle Plumes, Flood Basalts, and True Polar Wander." *Review of Geophysics* 29, 31-50.

——— et al. (1990). "1. The Volcanic Record of the Réunion Hotspot." *Proceedings of the Ocean Drilling Program, Scientific Results* 115, 3-10.

Edwards, W. N. (1963). "Robert Hooke as Geologist and Evolutionist." *Nature* 137, 96-97.

'Espinasse, Margaret (1937). "Robert Hooke on His Literary Contemporaries." *Review of English Studies* 13, 212-216 (under name of M. Wattie).

——— (1956). *Robert Hooke.* London: William Heinemann, 192 pp.

Evans, Michael E. (1988). "Edmond Halley, Geophysicist." *Physics Today* 41, 41-45.

Eyles, V. A. (1969). "The Extent of Geological Knowledge in the Eighteenth Century, and the Methods by Which it was Diffused." In: *Toward a History of Geology*, C. J. Schneer (ed.). Cambridge, MA: Massachusetts Inst. Technology Press, pp. 159-183.

Ezell, M. J. E. (1984). "Richard Waller, S. R. S.: 'In the Pursuit of Nature.'" *Notes and Records of the Royal Society* 38, 215-233.

Faul, Henry, and Faul, Carol (1983). *It Began with a Stone.* New York: Wiley, 270 pp.

Feisenberger, H. A. (1966). "The Libraries of Newton, Hooke and Boyle." *Notes and Records of the Royal Society* 21, 42-55.

——— (1975). *Sale Catalogues of Libraries of Eminent Persons:* Vol. 11, *Scientists.* London: Mansell and Sotheby Parke-Bernet.

Fernie, J. Donald (1991). "The Shape of the Earth." *American Scientist* 79 (part 1), 108-110.

Firman, R. J. (1986). "Some British Ideas about Ore Genesis from Hooke and Whitehurst 1668-1786." *Bulletin Peak District Mines Historical Society* 9, 404-412.

Frank, R. G. (1975). "Institutional Structure and Scientific Activity in the Early Royal Society." *Proceedings of the 14th International Congress of the History of Science* No. 4, 82-101.

―――― (1980). *Harvey and the Oxford Physiologists: A Study of Scientific Ideas and Social Interaction.* Berkeley: University of California Press, 368 pp.

Gaposchkin, Sergei (1963). "Newton and Hooke." *Sky and Telescope* 25, 14-15.

Goidanich, Gabriele (1941). "I primi documenti dell'esistenza dei funghi Microscopici." *Bolletino Sta. Patol. Vegetale*, new ser. 21, 318-331.

Goldreich, Peter, and Toomre, Alar (1969). "Some Remarks on Polar Wandering." *Journal of Geophysical Research* 74, 2555–2567.

Gough, H. J. (1935). "Robert Hooke (1635-1703); A Short Appreciation." *Iron and Steel Industry* 8, 373-375.

Gouk, Penelope (1980). "The Role of Acoustics and Music Theory in the Scientific Work of Robert Hooke." *Annals of Science* 37, 573-605.

―――― (1982). "Acoustics in the Early Royal Society 1660-1680." *Notes and Records of the Royal Society* 36, 155-175.

Gould, Rupert Thomas (1923). *The Marine Chronometer, Its History and Development.* London: Holland Press, 287 pp.

Greene, J. C. (1959). *The Death of Adam.* Ames: The Iowa State University Press, 388 pp.

Gunther, R. T. *Early Science in Oxford*, vols. 6 and 7 (*Life and Work of Robert Hooke*, 1 and 2) (Oxford, 1930); vol. 8 (*The Cutler Lectures of Robert Hooke*) (Oxford, 1931); vol. 10 (*Life and Work of Robert Hooke*, 4) (Oxford, 1935); vol. 13 (*Micrographia: Life and Work of Robert Hooke*, 5) (Oxford, 1938).

Gunther, Robert W. T. (1928). *Further Correspondence of John Ray.* London: Printed for the Ray Society, 332 pp.

―――― (1945). *Early Science in Oxford:* Vol. 14, *Life and Letters of Edward Lhwyd.* Oxford: Clarendon, 576 pp.

Hall, A. Rupert (1951). "Robert Hooke and Horology." *Notes and Records of the Royal Society* 8, 167-177

―――― (1951). "Two Unpublished Lectures of Robert Hooke." *Isis* 42, 219-230.

―――― (1966). *Hooke's Micrographia, 1665-1965.* London.

―――― (1966). "Mechanics and the Royal Society 1668-1670." *British Journal for the History of Science* 3, 24-38.

―――― (1978). "Horology and Criticism: Robert Hooke." *Studia Copernicana* 16, 261-281.

――――, and Hall, Marie B. (1962). "Why Blame Oldenburg?" *Isis* 53, 482-491.

――――, and Westfall, R. S. (1967). "Did Hooke Concede to Newton?" *Isis* 58, 403-405.

Hall, D. H. (1976). *History of the Earth Sciences during the Scientific and Industrial Revolutions.* Amsterdam: Elsevier Scientific Pub, 297 pp.

Hall, Marie B. (1952). "The Establishment of the Mechanical Philosophy." *Osiris* 10, 412-541.

―――― (1976). "Science in the Early Royal Society." In: *The Emergence of Science in Western Europe*, M. P. Crosland (ed.). New York: Science History Publications, pp. 57-77.

―――― (1981). "Solomon's House Emergent: The Early Royal Society and Cooperative Research." In: *The Analytic Spirit*, H. Woolf (ed.). Ithaca: Cornell University Press, pp. 177-194.

——— (1985). "Oldenburg, the *Philosophical Transactions* and Technology." In: *The Uses of Science in the Age of Newton*, J. G. Burke (ed.). Berkeley: University of California Press, pp. 21-47.

Hallo, R. (1934). *Rudolf Erich Raspe, ein Wegbereiter von deutscher Art und Kunst*. Stuttgart: W. Kohlhammer.

Hambly, Edmund C. (1987). "Robert Hooke, the City's Leonardo." *City University* 2, 5-10.

Hannay, J. B. (1892). "Formation of Lunar Volcanoes." *Nature* 47, 7-8.

Hartley, H., ed. (1960). *The Royal Society: Its Origins and Founders*. London: Royal Society, 275 pp.

Haswell, J. Eric (1951). *Horology: The Science of Time Measurement and Construction of Clocks, Watches and Chronometers*. London: Chapman and Hall, 288 pp.

Hearne, Thomas (1857). *Reliquiae Hearnianae*. Oxford: The Editor.

Henry, John (1983). "Matter in Motion: The Problem of Activity in 17th Century English Matter Theory." Ph.D. thesis, Open University, London.

——— (1986). "Occult Qualities and the Experimental Philosophy: Active Principles in Pre-Newtonian Matter Theory." *History of Science* 24, 335-381.

Hesse, M. B. (1962). "Hooke's Development of Bacon's Method." In: *Proceedings of the Tenth International Congress of the History of Science*. Ithaca, 2 vols.; vol. 1 pp. 265-268, Paris, 1964.

——— (1966a). "Hooke's Philosophical Algebra." *Isis* 57, 67-83.

——— (1966b). "Hooke's Vibration Theory and the Isochrony of Springs." *Isis* 57, 433-441.

*History and Philosophy of Earthquakes, From the Remotest to the Present Times: Collected from the Best Writers on the Subject, 1757*. London: J. Nourse, 351 pp.

Hooke, Robert (1665). *Micrographia: Or Some Physiological Descriptions of Minute Bodies Made by Magnifying Glasses, with Observations and Inquiries Thereupon*. London: Jo. Martyn and Ja. Allestry, printers to the Royal Society. Reprinted by Bruxelles, Culture et Civilisation, 1966, 246 pp.

——— (1677). *Lampas, or Descriptions of Some Mechanical Improvements of Lamps and Waterpoises, Together with Some Other Physical and Mechanical Discoveries*. London: John Martyn, printer to the Royal Society, 54 pp. In: *Early Science in Oxford:* Vol. 8, *The Cutler Lectures of Robert Hooke*, R. T. Gunther (ed.) (1931) Oxford: Oxford University Press, pp. 154-208, postscript.

——— (1678). *Lectures. De Potentia Restitutiva or of Spring, Explaining the Power of Springing Bodies*. London: John Martyn, printer to the Royal Society, 25 pp. In: *Early Science in Oxford:* Vol. 8, *The Cutler Lectures of Robert Hooke*, R. T. Gunther (ed.) (1931). Oxford: Oxford University Press, pp. 331-356.

——— (1705) (posthumously), 1668-ca. 1699. "Lectures and Discourses of Earthquakes, and Subterraneous Eruptions, Explicating the Causes of the Rugged and Uneven Face of the Earth; and What Reasons May be Given for the Frequent Finding of Shells and Other Sea and Land Petrified Substances, Scattered Over the Whole Terrestrial Superficies." In: *The Posthumous Works of Robert Hooke, M.D. S.R.S. Geom. Prof. Gresh. &c. Containing his Cutlerian Lectures, and Other Discourses, Read at the Meetings of the Illustrious Royal Society*, Richard Waller (ed.), Royal Society Secretary. London: Smith & Walford, printers to the Royal Soc., pp. 279-450. Reprinted in *The Sources of Science*, no. 73, New York and London: Johnson Reprint, 572 pp.

Hoskin, M. A. (1982). "Hooke, Bradley and the Aberration of Light." In: *Stellar Astronomy: Historical Studies*. Chalfont St. Giles, Bucks: Science History Publications, pp. 29-36.

Howse, Derek (1980). *Greenwich Time and the Discovery of the Longitude.* Oxford: Oxford University Press, 254 pp.

Hult, Jan (1977). "Robert Hooke Och Hans Fjäderlag: Ett 300-Arsminne." *Daedalus,* 39-57.

Hunter, Michael (1975). *John Aubrey and the Realm of Learning.* New York: Science History Publications, 256 pp.

——— (1981). *Science and Society in Restoration England.* Cambridge: Cambridge University Press, 233 pp.

——— (1982). "Reconstructing Restoration Science: Problems and Pitfalls in Institutional History." *Social Studies of Science* 12, 451-466.

——— (1982). *The Royal Society and its Fellows, 1660-1700: The Morphology of an Early Scientific Institution.* Chalfont St. Giles, Bucks: Published by the British Society for the History of Science, 270 pp.

——— (1984). "A 'College' for the Royal Society. The Abortive Plan of 1667-1668." *Notes and Records of the Royal Society* 38, 159-186.

——— (1989). *Establishing the New Science: The Experience of the Early Royal Society.* Woodbridge, England: Boydell, 382 pp.

——— and Schaffer, Simon, eds. (1989). Robert Hoo*ke: New Studies.* Woodbridge, England: The Boydell Press, 310 pp.

———and Wood, P. B. (1986). "Towards Solomon's House: Rival Strategies for Reforming the Early Royal Society." *History of Science* 24, 49-108.

Hutton, J. (1785). Abstract of a Dissertation Read in the Royal Society of Edinburgh, Upon the Seventh of March, and Fourth of April, 1785, Concerning the System of the Earth, its Duration, and Stability. Reproduced in Facsimile. In: *Contributions to the History of Geology.* G. W. White (ed.), vol. 5. Darien, Conn., Hafner, 1970, 30 pp.

——— (1788). "Theory of the Earth, or an Investigation of the Laws Observable in the Composition, Dissolution, and Restoration of Land Upon the Globe." Transactions Royal Society of Edinburgh I, pt. II, pp. 209-304.

Huygens, Christian (1629-1645). *Oeuvres complètes,* Vol. 4. La Hague: Nijhoff, 1888.

Ito, Yushi (1988). "Hooke's Cyclic Theory of the Earth in the Context of Seventeenth Century England." British *Journal for the History of Science* 21, 295-314.

Iversen, A. N., and Carozzi, A. V. (1970). Editors' introduction to Raspe, R. E., *An Introduction to the Natural History of the Terrestrial Sphere, Principally Concerning New Islands Born from the Sea and Hooke's Hypothesis of the Earth on the Origin of Mountains and Petrified Bodies to be Further Established from Accurate Descriptions and Observations.* New York: Hafner, cx pp.

Jeneman, H. R. (1985). "Robert Hooke und die frühe Geschichte der Federwäge." *Berichte zur Wissenschaftsgeschichte* 8, 121-130.

Jones, E. L. (1951). "Robert Hooke and *The Virtuoso.*" *Modern Language Notes* 66, 180-181.

Jones, Richard F. (1961). *Ancients and Moderns: A Study of the Rise of the Scientific Movement in Seventeenth Century England,* 2nd ed. St. Louis: Washington University, 354 pp.

Jourdain, Philip E. B. (1913). "Robert Hooke as a Precursor of Newton." *Monist* 23, 353-384.

——— (1914). "The Principles of Mechanics with Newton from 1666-1679 and 1679-1687." *Monist* 24, 188-224 and 515-564.

——— (1915). "Newton's Hypothesis of Ether and Gravitation from 1672 to 1679." *Monist* 25, 76-106.

Kargon, R. H. (1971). "The Testomony of Nature: Boyle, Hooke and the Experimental Philosophy." *Albion* 3, 72-81.

Kassler, J. C., and Oldroyd, D. R. (1983). "Robert Hooke's Trinity College 'Musick Scripts': His Music Theory and the Role of Music in His Cosmology." *Annals of Science* 40, 559-595.

Keynes, Geoffrey (1966). *A Bibliography of Robert Hooke.* Oxford: Clarendon, 115 pp.

——— (1976). John R*ay. A Bibliography 1660-1970.* Amsterdam: Gérard Th. van Heusden, 184 pp.

King, Henry C. (1955). *The History of the Telescope.* New York: Dover, 456 pp.

Knowlson, James (1975). *Universal Language Schemes in England and France 1600-1800.* Toronto: University of Toronto Press, 301 pp.

Koyré, Alexandre (1950). "A Note on Robert Hooke." *Isis* 41, 195-196.

——— (1952). "An Unpublished Letter of Robert Hooke to Isaac Newton." *Isis* 43, 312-337.

——— (1955). "A Documentray History of the Problem of Fall from Kepler to Newton." *Transactions of the American Philosophical Society* 45, 329-395.

——— (1965). *Newtonian Studies.* London: Chapman and Hall, 288 pp.

Kubrin, D. C. (1968). "Providence and the Mechanical Philosophy: The Creation and Dissolution of the World in Newtonian Thought." Ph.D. Thesis, Cornell University, chapter 6.

Laudan, Rachel (1987). *From Mineralogy to Geology: The Foundations of a Science, 1650-1830.* Chicago and London: University of Chicago Press, 278 pp.

Lehmann-Haupt, H. (1973). "The Microscope and the Book." In: Fest*schrift für Claus Nissen.* Wiesbaden: Pressler, pp. 471-502.

Lely, U. P. (1927). Een Proef die de Krachten demonstreert, welke de Continentdrift kan veroorzaken. "Physica," *Nederlandsch Tijdschrift voor Natuurkunde* 7, 278-281.

Leopold, J. H. (1976). "Hooke's and Huygen's Balance-Spring Inventions." *Antiquarian Horology*, 672-675.

Lilley, Steven (1948). "Robert Hooke and the Modern Music Hall." *Discovery* 9, 189.

Lister, M. (1671). A Letter of Mr. Martin Lister. *Philosophical Transactions* 6, 2281-2285.

Lloyd, Claude (1929). "Shadwell and the Virtuosi." *Publications of the Modern Language Association of America* 44, 472-494.

Lohne, J. A. (1960). "Hooke versus Newton: An Analysis of the Documents in the Case on Free Fall and Planetary Motion." *Centaurus* 7, 6-52.

Lyell, C. (1830). *Principles of Geology, Being an Attempt to Explain the Former Changes of the Earth's Surface, by Reference to Causes Now in Operation.* London: John Murray, pp. 31-35.

Lyons, Sir Henry G. (1944). *The Royal Society 1660-1940: A History of Its Administration under Its Charters.* New York: Greenwood, 1968, 354 pp.

Lysaght, D. J. (1937). "Hooke's Theory of Combustion." *Ambix* 1, 93-108.

Macintosh, J. J. (1983). "Perception and Imagination in Descartes, Boyle and Hooke." *Canadian Journal of Philosophy* 13, 327-352.

MacPike, Eugene Fairfield, ed. (1932). *Correspondence and Papers of Edmond Halley.* Oxford: Clarendon, 300 pp.

Maddison, R. E. W. (1951). "Studies in the Life of Robert Boyle, 1." Notes *and Records of the Royal Society* 9, 1-35.

Manuel, Frank E. (1968). *A Portrait of Isaac Newton.* Cambridge: Harvard University Press, 478 pp.

Matzke, E. B. (1943). "The Concept of Cells Held by Hooke and Grew." *Science* 98, 13-14.

McKie, Douglas (1953). "Fire and the Flamma Vitalis: Boyle, Hooke and Mayow." In: *Science, Medicine and History*, 2 vols. E. A. Underwood (ed.). London: Oxford University Press,. 1: 469-488.

Merton, R. K. (1938). *Science, Technology and Society in Seventeenth-Century England.* London: First published as vol. IV, part 2 of *Osiris*. New edition, New York: H. Fertig, 1970.

Michell, John (1761). "Conjectures Concerning the Cause, and Observations Upon the Phænomena of Earthquakes; Particularly of that Great Earthquake of the First of November, 1755, Which Proved so Fatal to the City of Lisbon, and Whose Effects were Felt as Far as Africa, and More or Less Throughout Almost all Europe. *Philosophical Transactions of the Royal Society of London*, Vol. 51, Pt. II, pp. 566-634. (Read February 28; March 6, 13, 20, 27, 1760).

Middleton, W. E. Knowles (1964). *The History of the Barometer.* Baltimore: Johns Hopkins University Press, 489 pp.

——— (1965). "A Footnote to the History of the Barometer: An Unpublished Note by Robert Hooke." *Notes and Records of the Royal Society* 20, 145-151.

——— (1966). *A History of the Thermometer and Its Use in Meteorology.* Baltimore: Johns Hopkins University Press, 249 pp.

——— (1969). *Invention of the Meteorological Instruments.* Baltimore: Johns Hopkins Universiity Press, 362 pp.

Middleton, W. S. (1927). "The Medical Aspect of Robert Hooke." *Annals of Medical History* 9, 227-243.

Millington, E. C. (1945). "Theories of Cohesion in the Seventeenth Century." *Annals of Science* 5, 253-269.

Mills, A. A. (1969). "Fluidization Phenomena and Possible Implications for the Origin of Lunar Craters." *Nature* 224, 863-866.

Montagu, M. F. Ashley (1941). "A Spurious Portrait of Robert Hooke." *Isis* 33, 15-17.

More, Louis T. (1934). *Isaac Newton, a Biography.* New York: Scribner's, 675 pp.

——— (1944). The Life and *Works of the Honourable Robert Boyle (1627-1691).* New York: Oxford University Press, 313 pp.

Morgan, J. R. (1930-1931). "Robert Hooke (1635-1703) Part I." *Science Progress* 25, 282-284.

Moyer, A. E. (1977). "Robert Hooke's Ambiguous Presentation of 'Hooke's Law.'" *Isis* 68, 266-275.

Mulligan, Lotte, and Mulligan, G. (1981). "Reconstructing Restoration Science: Styles of Leadership and Social Composition of the Early Royal Society." *Social Studies of Science* 11, 327-364.

Murden, Lesley (1985). *Under Newton's Shadow: Astronomical Practices in the Seventeenth Century.* Bristol and Boston: Adam Hilger, 152 pp.

Nakajima, H. (1984). "Two Kinds of Modification Theory of Light: Some New Observations on the Newton-Hooke Controversy of 1670 Concerning the Nature of Light." *Annals of Science* 41, 261-278.

Neufeld, M. W. (1927). "Ein Vorschlag zur Herstellung Künstlicher Seider a.d.j. 1666 von Robert Hooke." *Kunstseide* 9, 268.

Neville, R. G. (1962). "The Discovery of Boyle's Law, 1661-1662." *Journal of Chemical Education* 39, 356-359.

Newton, Sir Isaac (1726). *Mathematical Principles of Natural Philosophy and System of the World*, 3rd ed. Translated by Andrew Motte in 1729. Translation revised and supplied with an historical explanatory appendix by Florian Cajori. Cambridge: Cambridge University Press, and Berkeley: University of California Press, 1934.

Nicolson, Marjorie H. (1935). "The Microscope and English Imagination." *Smith College Studies in Modern Languages,* Vol. 16, no. 4.

——— (1956). *Science and Imagination.* Ithaca: Great Seal Books, 238 pp.

——— (1965). *Pepys's Diary and the New Science.* Charlottesville: University Press of Virginia, 198 pp.

———, and Mohler, N. M. (1937). "The Scientific Background of Swift's Voyage to Laputa." *Annals of Science* 2, 299-334.

———, and Rodes, David S., eds. (1966). *The Virtuoso: A* Comedy, by Thomas Shadwell. Lincoln: University of Nebraska Press, 153 pp.

Oldroyd, David R. (1972). "Robert Hooke's Methodology of Science as Exemplified in His 'Discourse of Earthquakes.'" *British Journal for the History of Science* 6, 109-130.

——— (1980). "Some 'Philosophical Scribbles' Attributed to Robert Hooke." *Notes and Records of the Royal Society* 35, 17-32.

——— (1987). "Some Writings of Robert Hooke on Procedures for the Prosecution of Scientific Inquiry, Including His 'Lectures of Things Requisite to a Natural History.'" *Notes and Records of the Royal Society* 41, 145-167.

——— (1989). "Geological Controversy in the Seventeenth Century: 'Hooke vs Wallis' and Its Aftermath." In *Robert Hooke, New Studies*, Michael Hunter and Simon Schaffer (eds.), Woodbridge, England: Boydell, pp. 207-233.

Olmstead, J. W. (1949). "The Application of Telescopes to Astronomical Instruments, 1667-1669: A Study in Historical Method." *Isis* 40, 213-225.

Palter, Robert (1972). "Early Measurements of Magnetic Force." *Isis* 63, 544-558.

*Parentalia* (1750). Or Memoirs of the Family of the Wrens; viz., of Mathew, Bishop of Ely, Christopher, Dean of Windsor, etc. but chiefly of Sir Christopher Wren Late Surveyor General of the Royal Buildings, President of the Roylal Society etc., etc. in which is contained besides his works, a great number of original papers and records on Religion, Politicks, Anatomy, Mathematicks, Architecture, Antiquities; and most Branches of Polite Literature. Cmpiled by his son Christopher; Now published by his grandson, Stephen Wren, Esq. in the care of Joseph Ames, London. Printed for T. Osborn in Gray's Inn and R. Dodsley in Pall Mall, 1750. (Farnborough, Hants, Eng., Gregg Press, Ltd., 1965), xii, 368 pp.

Patterson, Louise Diehl (1948). "Robert Hooke and the Conservation of Energy." *Isis* 38, 151-156.

——— (1949). "Hooke's Gravitation Theory and Its Influence on Newton, I: Hooke's Gravitation Theory." *Isis* 40, 327-341.

——— (1950). "Hooke's Gravitation Theory and Its Influence on Newton, II: The Insufficiency of the Traditional Estimate." *Isis* 41, 32-45.

——— (1950). "A Reply to Professor Koyré's Note on Robert Hooke." *Isis* 41, 304-305.

——— (1951). "Thermometers of the Royal Society." *American Journal of Physics* 19, 523-555.

——— (1952). "Pendulums of Wren and Hooke." *Osiris* 10, 277-321.

——— (1953). "The Royal Society's Standard Thermometer, 1663-1709." *Isis* 44, 51-64.

Patterson, T. S. (1931). "John Mayow in Contemporary Setting." *Isis* 15, 47-96.

Pavlov, A. P. (1928). "Robert Hooke, un évolutionniste Oublié due 17e siècle." *Palaeobiologica* 1, 203-210.

Pelseneer, Jean (1929). "Une lettre inédite de Newton." *Isis* 12, 237-254.

——— (1931). "Un journal inédit de Hooke." *Isis* 15, 97-103.

Pepys, Samuel (1633-1703). *Diary.* Deciphered and with a life and notes by Baron Richard G. Braybrooke and by Reverend Mynors Bright. 10 volumes. New York: Dodd, Mead, 1889.

Playfair, J. (1803). "Biographical Account of the Late James Hutton, F.R.S." *Transactions of the Royal Society of Edinburgh* V, 39-99.

Porter, R. S. (1977). *The Making of Geology: Earth Science in Britain 1660-1815.* Cambridge: Cambridge University Press, 288 pp.

Pourciau, Bruce H. (1991). "On Newton's Proof that Inverse-Square Orbits Must Be Conics." *Annals of Science* 48, 159-172.

Powell, Anthony (1988). *John Aubrey and His Friends.* London: Hogarth, 342 pp.

Power, M. E. (1945). "Sir Christopher Wren and the Micrographia." *Transactions of the Connecticut Academy of Arts and Sciences* 36, 37-44.

Price, D. J. de S. (1957). "The Manufacture of Scientific Instruments." In *History of Technology*, C. Singer (ed.), 6 vols. Oxford: Clarendon, 3: 620-647.

Pugliese, Patri J. (1982). "The Scientific Achievement of Robert Hooke: Method and Mechanics." Ph.D. thesis, Harvard University, 729 pp.

——— (1989). "Robert Hooke and the Dynamics of Motion in a Curved Path." In *Robert Hooke: New Studies*, Michael Hunter and Simon Schaffer (eds.). Woodbridge, England: Boydell, pp. 181-205.

Purver, Margery (1967). The Royal *Society: Concept and Creation.* London: Routledge and Kegan Paul, 246 pp.

Rahman, A. (1957). "Biography and Science (review of 'Espinasse)." *Science Culture* 22, 551-553.

Raman, V. V. (1972). "Background to Newtonian Gravitation." *Physics Teacher* 10, 439-442.

Ranalli, G. (1982). "Robert Hooke and the Huttonian Theory." *Journal of Geology* 90, 319-325.

Rappaport, Rhoda (1986). "Hooke on Earthquakes: Lectures, Strategy and Audience." *British Journal for the History of Science* 19, 129-146.

Raspe, Rudolf Erich (1763). Specimen Historiae Naturalis Globi Terraquei, Praecipue de Novis e Maris Natis Insulis, et ex his Exactius Descriptis et Observatis, Ulterius Confirmanda, Hookiana Telluris Hypothesi, de Origine Montium et Corporum Petrefactorum. Amsterdam and Leipzig: J. Schreuder and P. Mortier, 191 pp. Translated and edited by A. N. Iversen and A. V. Carozzi (1970), New York: Hafner.

——— (1763). An Introduction to the Natural History of the Terrestrial Sphere. Translated and edited by Audrey Notvik Iversen and Albert V. Carozzi, Editor's Introduction xvii-cx. New York: Hafner Publishing, 1970.

——— (1771). "A Letter Containing a Short Account of Some Basalt Hills in Hassia." *Royal Society of London Philosophical Tranactions.* 61, 580-583.

——— (1774). Beytrag zur alterältesten und natürlichen Historie von Hessen oder Beschreibung des Habichtswaldes und verschiedner andern Niederhessischen alten Vulcane in der Nachbarschaft von Cassel. Cassel: Johann Jacob Cramer.

Ray, John (1691). *The Wisdom of God Manifested in the Works of the Creation.* London: Printed for Samuel Smith, at the Princes Arms in St. Paul's Church Yard. Facsimile reproduced from a copy in the possession of the Calvörsche Bibliothek, Clausthal-Zellerfeld. Hildesheim: Georg Olms Verlag, 1974 in Anglistica and Americana, a series of reprints selected by Bernhard Fabian, Edgar Mertner, Karl Schneider, and Marvin Spevack.

——— (1692). *Miscellaneous Discourses Concerning the Dissolution and Changes in the World.* Wherein the Primitive Chaos and Creation, the General Deluge, Fountains, Formed Stones, Sea-Shells, Vulcanoes, the Universal Conflagration and Future State, are Largely Discussed and Examined. London: Printed for Samuel Smith, at the Prince's Arms in St. Paul's Church Yard, 144 pp.

——— (1713). *Three Physico-Theological Discourses, Concerning: I. The Primitive CHAOS, and Creation of the World; II. The General DELUGE, its Causes and Effects; III. The Dissolution of the WORLD, and Future Conflagration.*

Wherein are largely discussed, The Production and Use of Mountains; the Original of Fountains, of Formed Stones, and Sea-Fishes Bones and Shells Found in the Earth; the Effects of Particular Floods, and Inundations of the Sea; the Eruptions of Vulcanos; the Nature and Causes of Earthquakes. Also an Historical Account of those Two Late Remarkable Ones in Jamaica and England. With Practical Inferences, 3rd ed. London: Printed for William Innys, at the Prince's Arms in St. Paul's Church Yard.

Rayner, Dorothy H. (1981). *The Stratigraphy of the British Isles*, 2nd Edition. Cambridge: Cambridge University Press, 460 pp.

Reddaway, T. F. (1940). *The Rebuilding of London after the Great Fire.* London: Jonathan Cape, 353 pp.

Richards, P. W. (1981). "Robert Hooke on Mosses." *Occasional Papers of the Farlow Herbarium* 16, 137-146.

Robinson, H. W. (1938). "Gleanings from a Library. 1. A Note on the Early Minutes." *Notes and Records of the Royal Society* 1, 92-94. Reprinted as "Hooke's Pocket Watch" (1939). *Annals of Science* 4, 322-323.

——— (1949). "Robert Hooke as a Surveyor and Architect." *Notes and Records of the Royal Society* 6, 48-55.

Robinson, Henry W. and Adams, Walter, eds. (1935). *The Diary of Robert Hooke M.A., M.D., F.R.S., 1672-1680.* Transcribed from the Original in the Possession of the Corporation of the City of London (Guildhall Library). London: Taylor and Francis, 527 pp.

Robison, John (1803). Lectures *on the Elements of Chemistry.* Delivered in the University of Edinburgh by the Late Joseph Black, M.D., vol. I. Notes and Observations, pp. 535-537. Edinburgh: Printed by Mundell for Longman and Rees, London, and William Creech, Edinburgh.

Rome, D. R. (1956). "Nicolas Sténo et la 'Royal Society of London.'" *Osiris* 12, 244-268.

Rosen, Edward (1975). "Richer, Jean." *Dictionary of Scientific Biography.*, New York: Scribner's, 11: 423-424.

Rosenfeld, L. (1927). "La théorie des couleurs de Newton et ses adversaires." Isis 9, 44-65.

Rossi, Paolo (1984). *The Dark Abyss of Time: A History of the Earth and the History of Nations from Hooke to Vico.* Translated by Lydia G. Cochrane. Chicago: University of Chicago Press, 338 pp.

Rossiter, A. P. (1935). "The First English Geologist: Robert Hooke (1635-1703)." *Durham University Journal* 27, 172-181.

——— (1936). "Comment on W. N. Edwards' Article 'Hooke as Geologist.'" *Nature* 137, 455.

Rostenberg, Leona (1989). *The Library of Robert Hooke: The Scientific Book Trade of Restoration England.* Santa Monica, CA: Madoc Press, 257 pp.

Rozet (M. le capitaine) (1846) "Mémoire sur la selénologie." *Comptes Rendus* 22, 470-473.

Rudwick, M. J. S. (1972). *The Meaning of Fossils: Episodes in the History of Palaeontology*, chap. 2. London: Macdonald and New York: American Elsevier, 287 pp.

Ruffner, J. A. (1966). "The Background and Early Development of Newton's Theory of Comets." Ph.D. thesis, Indiana University, chapter 6.

Sabra, A. I. (1981). *Theories of Light from Descartes to Newton.* Cambridge: Cambridge University Press, 365 pp.

Salmon, Vivian (1974). "John Wilkins's Essay (1668): Critics and Continuators." *Historiographic Linguistica* 1, 147-163. Reprinted in *The Study of Language in Seventeenth Century England* (1979). Amsterdam: Benjamins, pp. 191-206.

Scherz, Gustav (1960). "Niels Stensen in Copenhagen." *Geotimes* 5, 10-17, 56-57.

Schneer, Cecil J. (1954). "The Rise of Historical Geology in the Seventeenth Century." *Isis* 45, 256-268.

——— (1960). "Kepler's New Year's Gift of a Snowflace." *Isis* 51, 531-545.

——— (1971). "Steno: On Crystals and the Corpuscular Hypothesis." In: *Dissertations on Steno as a Geologist*, G. Scherz (ed.). Odense University Press, pp. 293-307.

Schwartzenburg, Dewey (1980). "The Great Red Spot." *Astronomy* 8, 6-14.

Shapin, Steven (1989). "Who Was Robert Hooke?" In: *Robert Hooke: New Studies*, Michael Hunter and Simon Schaffer (eds.). Woodbridge, England: Boydell, pp. 253-285.

———, and Schaffer, Simon (1985). *Leviathan and the Air Pump: Hobbes, Boyle and the Experimental Life*. Princeton: Princeton University Press, 440 pp.

Shapiro, A. E. (1973). "Kinematic Optics. A Study of the Wave Theory of Light in the Seventeenth Century." *Archive for History of Exact Sciences* 11, 134-266.

Shapiro, Barbara J. (1969). *John Wilkins, 1614-1672: An Intellectual Biography*. Berkeley: University of California Press, 333 pp.

Sherley, Thomas (1672). *A Philosophical Essay: Declaring the Probable Causes Whence Stones are Produced in the Greater World*. London. Reprint edition (1978) by Arno Press.

Sherwood, Taylor F. (1947). "An Early Satirical Poem on the Royal Society." *Notes and Records of the Royal Society*, October, pp. 37-46.

Short, T. (1734). *A History of the Mineral Waters of Derbyshire, Lincolnshire and Yorkshire*. London.

Simpson, A. D. C. (1984). "Newton's Telescope and the Cataloguing of the Royal Society's Repository." *Notes and Records of the Royal Society* 38, 187-214.

Singer, B. R. (1976). "Robert Hooke on Memory, Association and Perception." *Notes and Records of the Royal Society* 31, 115-131.

Singer, Charles (1955). "The First English Microscopist, Robert Hooke." *Endeavour* 14, 12-18.

Slaughter, M. M. (1982). *Universal Languages and Scientific Taxonomy in the Seventeenth Century*. Cambridge: Cambridge University Press.

Solis, Carlos, ed. (1989). *Robert Hooke, Micrografía*. Spanish edition with notes and apparatus. Madrid.

St. Clair Humphreys, A. (1891). "An Hypothesis of Lunar Formations." *British Astronomical Association Journal* 3, 132-136.

Steno, Nicolaus (1667). *Canis carcharia dissectum caput*. Translated by A. Garboe (1958) as *The Earliest Geological Treatise*. London: Macmillan, 51 pp.

——— (1669). *De solido intra solidum naturaliter contento dissertationis prodromus*. Florentiae, 76 pp. English version with notes by J. G. Winter, 1916. New York: Macmillan, pp. 169-283. Reproduced in facsimile in G. W. White, 1968.

Stimson, Dorothy (1948). *Scientists and Amateurs: A History of the Royal Society*. New York: H. Schuman, 270 pp.

Suess, E. (1904). *Das Antlitz der Erde* [The Face of the Earth]: Vol. I. Translated by H. B. C. Sollas. Oxford: Clarendon, 604 pp.

Symonds, R. W. (1952). "Robert Hooke and Thomas Tompion." In: *Illustrated Catalogue to British Clockmaker's Heritage Exhibition*. London: Science Museum.

Tait, P. G. (1885). "Hooke's Anticipation of the Kinetic Theory." *Proceeding of the Royal Society of Edinburgh*. [See also: Tait, 1898-1900, *Scientific Papers*, vol. 2, Cambridge University Press, p. 122.]

Taylor, E. G. R. (1937a). "The Geographical Ideas of Robert Hooke." *Geographical Journal* 89, 525-538.

―――― (1937b). "Robert Hooke and the Cartographical Projects of the Late Seventeenth Century." *Geographical Journal* 90, 529-540.

―――― (1940). "The English Atlas of Moses Pitt, 1680-1683." *Geographical Journal* 95, 292-299.

―――― (1950). "The Origin of Continents and Oceans: A Seventeenth Century Controversy." *Geographical Journal* 116, 193-198.

―――― (1954). *The Mathematical Practitioners of Tudor and Stuart England.* Cambridge: Cambridge University Press for the Institute of Navigation, 442 pp.

Turnbull, H. W., ed. (1960). The *Correspondence of Isaac Newton:* Vol. 2, *1676-1687.* Published for the Royal Society. Cambridge: Cambridge University Press, 552 pp.

Turner, A. J. (1974). "Hooke's Theory of the Earth's Axial Displacement: Some Contemporary Opinions." *British Journal for the History of Science* 7, 166-170.

Turner, Gerard L'Estrange (1980). *Essays on the History of the Microscope.* Oxford: Senecia, 245 pp.

―――― (1981). *Collecting Microscopes.* New York: Mayflower, 120 pp.

Turner, H. D. (1956). "Hooke and Theories of Combustion." *Centaurus* 4, 297-310.

―――― (1959). "Robert Hooke and Boyle's Air Pump." *Nature* 184, 395-397.

Waller, Richard (1705). "The Life of Robert Hooke." In: *The Posthumous Works of Robert Hooke, M.D., S.R.S., Geom. Prof. Gresh. and Containing His Cutlerian Lectures, and Other Discourses, Read at the Meetings of the Illustrious Royal Society.* London: Sam. Smith and Benj. Walford, p. xxvi-xxvii.

Warren, Erasmus (1690). *Geologia: Or, A Discourse Concerning the Earth Before the Deluge.* Wherein the Form and Properties ascribed to it, in a Book entitled The Theory of the Earth, Are Excepted Against: And it is made appear, That the Dissolution of that Earth was not the Cause of the Universal Flood. Also A New Explication of that Flood is attempted. London: Printed for R. Chiswell, at the Rose and Crown in St. Paul's Church Yard, 359 pp.

Webster, Charles (1965). "The Discovery of Boyle's Law and the Concept of Elasticity of Air in the Seventeenth Century." *Archive for History of Exact Sciences* 2, 441-502.

――――, ed. (1974). *The Intellectual Revolution of the Seventeenth Century.* London" Routledge and Kegan Paul, 445 pp.

Wegener, A. (1929). *Die Entstehung der Kontinente und Ozeane* [The origin of continents and oceans]. 4th revised German ed. Translated by J. Biram. New York: Dover, 240 pp.

Weil, Ernest (1946). "Robert Hooke's Letter of December 9, 1679, to Isaac Newton." Nature 158, 135.

Weinstock, Robert (1982). "Dismantling a Centuries-Old Myth: Newton's *Principia* and Inverse-Square Orbits." *American Journal of Physics* 50, 610-617.

―――― (1989). "Long-Buried Dismantling of a Centuries-Old Myth: Newton's *Principia* and Inverse-Square Orbits." *American Journal of Physics* 57, 846-849.

―――― (1992a). "Problem in Two Unknowns: Robert Hooke and a Worm in Newton's Apple." *The Physics Teacher* 30, 282-288.

―――― (1992b). "Newton's 'Principia' and Inverse-Square Orbits: The Flaw Reexamined." *Historia Mathematica* 19, 60-70.

Wells, A. K., and Kirkaldy, J. F. (1967). *Outline of Historical Geology.* London: George Allen and Unwin, 503 pp.

Westfall, Richard S. (1962). "Newton and His Critics on the Nature of Colours." *Archives Internationales d'Histoire des Sciences* 15, 47-58.

―――― (1963). "Newton's Reply to Hooke and the Theory of Colours." *Isis* 54, 82-96.

——— (1966). "Newton Defends His First Publication." *Isis* 57, 299-314.

——— (1967). "Hooke and the Law of Universal Gravitation: A Reappraisal of the Reappraisal." *The British Journal for the History of Science* 3, 245-261.

——— (1969). "Introduction." In: *The Posthumous Works of Robert Hooke*, Richard Waller (ed.). New York and London: Johnson Reprint Corp., pp. ix-xxvii.

——— (1971). *Force in Newton's Physics: The Science of Dynamics in the Seventeenth Century.* London: Macdonald, and New York: American Elsevier, 579 pp.

——— (1972). "Robert Hooke." In: *Dictionary of Scientific Biography*, Vol. 6. New York: Scribner's, pp. 481-488.

——— (1973). "Newton and the Fudge Factor." Science 179, 751-758.

——— (1980). *Never at Rest: A Biography of Isaac Newton.* Cambridge, Cambridge University Press, 908 pp.

——— (1983). "Robert Hooke, Mechanical Technology and Scientific Investigation." In: *The Use of Science in the Age of Newton*, J. G. Burke (ed.). Berkeley: University of California Press, pp. 85-110.

Wheatley, Henry B., F.S.A. (1891). *London Past and Present, Its History, Associations and Traditions.* London: John Murray, Abemarle Street, pp. 171-177.

Whiston, William (1696). *A New Theory of the Earth From Its Origin to the Consumation of All Things.* London. Reprint edition, 1978, New York: Arno Press, 388 pp.

White, G. W., ed. (1968). *Contributions to the History of Geology, Vol. 4, Nicolaus Steno: Prodromus of a Dissertation Concerning a Solid Body.* New York: Hafner, 169-283.

Whitehurst, J. (1786). *An Inquiry into the Original State and Formation of the Earth, Deduced From Facts and the Laws of Nature*, 2nd ed. London: W. Bent, 283 pp. Reprinted 1978, New York: Arno Press.

Whiteside, D. T. (1964). "Newton's Early Thoughts on Planetary Motion: A Fresh Look." *British Journal for the History of Science* 2, 117-137.

Whittaker, Edmund (1951). *A History of the Theories of Aether and Electricity. Vol. I: The Classical Theories.* New York: Harper and Brothers, 434 pp.

Whitrow, G. J. (1938). " Robert Hooke." *Philosophy of Science* 5, 493-502.

Wilchinsky, Z. (1939). "The Theoretical Treatment of Hooke's Law." *American Physics Teacher* 7, 134.

Williams, E. (1956). "Hooke's Law and the Concept of the Elastic Limit." *Annals of Science* 12, 74-83.

Willmoth, Frances (1987). "John Flamsteed's Letter Concerning the Natural Causes of Earthquakes." *Annals of Science* 44, 23-70.

Wilson, C. A. (1970). "From Kepler's Laws, So-Called, to Universal Gravitation: Empirical Factors." *Archive for History of Exact Sciences* 6, 89-170.

Wilson, L. G. (1960). "The Transformation of the Ancient Concept of Respiration in the Seventeenth Century." *Isis* 51, 161-172.

Wood, Anthony (1721). *Athenae Oxonienses, an exact history of all the writers and bishops who have had their education in the University of Oxford.* Oxford, 4 vols.; New York: Johnson Reprint Corp., 1967.

Wood, P. B. (1978). "Francis Bacon and the 'Experimentall Philosophy': A Study in Seventeenth Century Methodology." M. Phil. Thesis, London University, chap.5.

Woodruff, L. C. (1919). "Hooke's *Micrographia*." *American Naturalist* 53, 247-264.

Woodward, John (1695). *An Essay Towards A Natural History of the Earth: and Terrestrial Bodies, especially Minerals: As also of the Sea, Rivers, and Springs. With an Account of the Universal Deluge: And of the Effects that it had Upon the Earth.* London: Printed for Ric. Wilkin at the Kings-Head in St. Paul's Church Yard, 277 pp. Reprint Edition 1978 by Arno Press, Inc.

Wright, Michael (1989). "Robert Hooke's Longitude Timekeeper." In: *Robert Hooke: New Studies,* Michael Hunter and Simon Schaffer (eds.). Woodbridge, England: Boydell, pp. 63-118.

Wright, T. (1865). *A History of Caricature and Grotesque in Literature and Art.* London: Virtue Brothers, 494 pp.

Wrøblewski, Andrzéj (1985). "De Mora Luminis: A Spectacle in Two Acts with a Prologue and an Epilogue." Ameri*can Journal of Physics* 53, 620-630.

# II

**Robert Hooke's**

Lectures and Discourses

OF

# EARTHQUAKES,

AND

**Subterraneous Eruptions.**

EXPLICATING

The Cause of the Rugged and Uneven Face
of the EARTH;

AND

What Reasons may be given for the frequent
findings of Shells and other Sea and Land
Petrified Substances, scattered over the whole
Terrestrial Superficies.

**Transcribed, Annotated, and with an Introduction
by Ellen Tan Drake**

# *Introduction*

These lectures on "earthquakes" are here arranged according to the chronology worked out by Rhoda Rappaport (1986) through her careful research into the records of the Royal Society. They are transcribed into modern type to facilitate reading for those who may have eyesight difficulties or young scholars who might be put off at first sight. Purists might look upon this practice with disdain who may feel a certain masochistic pride in being able to read 17th century texts in their original type and with all their creative font styles, spelling and punctuation. I have retained the spelling, punctuation, capitalization, font styles (i.e., whether italics or roman) of the 1705 Waller edition of the *Posthumous Works of Robert Hooke*.

The original folios are indicated in brackets throughout the text so that the reader can find the corresponding untranscribed passages with relative ease. I have eliminated most of Waller's brief marginal comments and notes because I find them not very helpful, substituting, instead, my own, more extensive, annotations in the margins wherever I feel qualified to make comments in clarifying the text. Hooke expresses his most perceptive ideas about geology in the lectures that ended on September 15, 1668, and in the series of lectures between May 28th, 1684, and toward the end of 1687. After that, he launches into his study of the classics to find corroborative statements about some of his ideas such as cyclicity and islands grown from volcanic eruptions. They are nevertheless important in revealing a progressive change in the tone of his lectures toward a greater need to include the Mosaic account while he struggled to find the verisimilitude of legends. During the last few years of his life, Hooke waxed more philosophical; he seemed then to be more inclined to bring the biblical account into concordance with his scientific ideas. But a healthy skepticism seems always to be present underlying his words. As for Greek and Latin passages, Hooke himself usually makes their meanings rather clear, but a few notes are supplied through the kindness of friends with classics degrees from the University of Oxford. Because only the Greek Symbol font was available to me for transcription, Greek scholars are urged to consult the original Greek passages in Waller's *Posthumous Works of Robert Hooke*.

As Waller noted, these discourses were read on several different occasions to the Royal Society, and therefore were not "so methodically digested" as they would have been had Hooke himself compiled them for publication. In fact, they span some thirty years. As a result, Hooke's organization is difficult to follow. Also, there is much repetition, in addition to changes in tone and attitude between his earlier and later lectures. In general, he did not waver from his basic ideas. To aid the reader in understanding some of his ideas, in addition to my marginal notes, I have summarized here several major groups of ideas in the first series of lectures.

After giving examples of fossils and their illustrations and descriptions, Hooke makes seven generalizations (Propositions) concerning fossils:

1. Stony objects shaped as shells are found in most countries, even far from the sea.
2. These objects, resembling parts of animals and plants, are usually found buried below the surface.

3. Whole bodies of fishes, etc., have been found as stone, with no inner parts.
4. Many sites where these objects were found are far from any sea.
5. Many of these objects are found at great heights above sea level.
6. Many are found below the Earth's surface and below sea level.
7. Some of these bodies are found in very dense rock.

Eleven propositions explain the substance of fossils and why they occur where they do:

1. Fossils represent either organic matter itself turned into stone, or are impressions of the living thing.
2. Some extraordinary event caused their petrifaction.
3. The three causes of petrifaction are:
   a) petrifying waters
   b) saline or sulfureous mixture plus subterranean heat or earthquake
   c) congelation by cold and compression
4. Various liquids can become solid; crystals are formed by crystallizing from a fluid.
5. Various other fluid substances can also harden.
6. Land and sea have exchanged places.
7. Much of Great Britain could have been under water in the past.
8. Sea water receded by changes in the Earth's center of gravity or by earthquake.
9. The tops of some of the highest mountains were under water in the past.
10. The present uneven surface and irregularities of the Earth were caused by earthquakes in the past.
11. Species have become extinct and new ones have been generated which did not exist from the beginning.

The organization for his four effects of earthquakes can perhaps be best illustrated in a chart (Fig. 9-1).

Hooke's Royal Society audience must have enjoyed listening to these lectures on "Earthquakes," as they were allowed to be given over a period of so many years. His contemporaries mostly were not ungenerous in their praise of him and the evidence shows that he was widely respected. The very fact that he was so often plagiarized is additional proof that people liked and appreciated the originality of what he presented. There were apparently sufficient numbers of reasonable men in the 17th century who, by honoring Hooke, went out of their way to counteract the wrongs done to him.

Hooke's obscurity today, however, is undeserved. In the Hooke-Newton controversy, Newton, of course, emerged the winner. The latter was keenly aware of his own historical image and went about during his lifetime to shape this image for posterity—in planting letters to friends, for example, years after the supposed event, thus establishing matters of priority (see Chapter 1 under A Most Productive Idea). But some of this campaign was perpetrated anonymously by Newton's followers and admirers who went as far as cutting words out of his letters in an effort to censor whatever might have been damaging to his reputation. These obsequious throngs have always surrounded persons of power, and they are a formidable force because they are always on the winning, powerful side. Later, unsuspecting researchers, then, partly because of their own selectivity and subjectivity, accept the planted or carefully censored letters as historical "truths." Scholars are usually discerning of the authenticity of evidence, but in the case of Newton, they are often blinded by his god-like stature. Many then who have built their reputations writing about this man and his achievements naturally are resistant to any attempt to reappraise the received opinion. Reassigning to others the priority claimed for Newton for three hundred years, or even reapportioning to others a share of the credit, tends to reduce him to his correct size as but one of a line of extraordinary men in the 17th century.

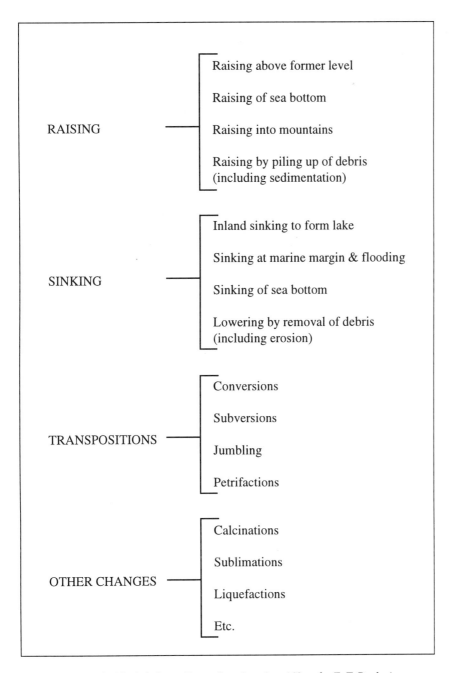

*Figure 9-1. Hooke's four effects of earthquakes. (Chart by E. T. Drake.)*

Decidedly, Robert Hooke should be included in this roster. Enjoy reading Hooke. Discover his remarkable intellect, originality, and sense of humor. Relish his diction and turn of phrasing. Savor his colorful metaphors. Marvel at his keen mind and the extent of his knowledge. His tremendous imagination sometimes carried him off on tangents, but even these are flights of intellectual exercise often to be admired. Hooke sometimes was wrong, of course, but always for the right reasons. A modern geologist, biologist, chemist, or physicist, if he or she takes the time to read Hooke, can usually appreciate his approach, because he seemed to think like one of us.

I am hopeful that the present volume will help to place Hooke in his rightful place of honor in the annals of science.

# A Discourse of Earthquakes

**No. 1.** *A series of lectures that ended September 15, 1668.*

[p.279] I Have formerly endeavour'd to explain several Observations I had made concerning the Figure, Form, Position, Distance, Order, Motions and Operations of the Celestial Bodies, both as to themselves, and one with another, and likewise with respect to the Body of the *Earth* on which we inhabit. But conceiving it may more nearly concern us to know more paricularly the Constitution, Figure, Magnitude and Properties of the Body of the *Earth* itself, and of its several constituent Parts, I have endeavour'd to collect such Observations and Natural Histories of others, as may serve to give some Light toward the making a compleat Discovery of them, so far as the Power, Faculties, Organs, and other helps that Nature has furnish'd Man with, may assist us in performing and perfecting thereof.

*Although Waller says he does not know to what former discourse Hooke is referring here, it is clear that Hooke had indeed lectured on these subjects on "Celestial Bodies." One such lecture, dated May 23, 1666, for example, announces his concept of the centripetal force so important to Newton's gravitational law.*

The Subject is large, as extending as far as the whole Bulk included within the utmost limits of the Atmosphere: And 'tis not less copious and repleat with variety, as containing all the several Parts and Substances included within those Limits, namely, The aerial, watery and earthy Parts thereof, whether Superficial or Subterraneous, whether Exposed or Absconded, whether Supraterraneal, Superterraneal, or Subterraneal, whether Elemental or Organical, Animate or Inanimate, and all the Species and kinds of them, and all the constituent Parts of them, and the Composits constituted of them; of which also there will fall under Consideration, the Artificial as well as the Natural Causes and Powers effective of things; then their Generation, Production, Augmentation, Perfection, Vertue, Power, Activity, Operation, Effect, Conservation, Duration, Declination, Destruction, Corruption, Transformation, and in one word, the motion or progression of Nature sensibly exprest, or any other ways discernable in each of those Species. Which Subject, if we consider as it is thus represented, doth look very like an Impossibility to be undertaken even by the whole World, to be gone through within an Age, much less to be undertaken by any particular Society, or a small number of Men. The number of Natural Histories, Observations, Experiments, Calculations, Comparisons, Deductions and Demonstrations necessary thereunto, seeming to be incomprehensive and numberless: And therefore a vain Attempt, and not to be thought of till after some Ages past in making Collections of Materials for so great a Building, and the employing a vast number of Hands in making this preparation; and those of several sorts, such as Readers of History, Criticks, Rangers and Namesetters of Things, Observers and Watchers of several Appearances, and Progressions of Natural Operations and Perfections, Collectors of curious Productions, Experimenters and Examiners of Things by several Means and several Methods and Instruments, as by Fire, by Frost, by Menstruums, by Mixtures, by Digestions, Putrefactions, Fermentations and Petrifactions, by Grindings, Brusings, Weighings and Measuring, Pressing and Condensing, Dilating and Expanding, Dissecting, Separating and Dividing, Sifting and Streining; by viewing with Glasses and Microscopes, Smelling, Tasting, Feeling, and various other ways of Torturing and Wracking of Natural Bodies, to find out the Truth or the real Effect as it is in its Constitution or State of Being.

*Hooke's environmental attitude: The Earth which we inhabit and everything about it, including all aerial, watery, and earthy parts of it and all species of life on it, concern us as* Homo sapiens.

*Investigations into the nature of things on Earth require the collection of all sorts of data through observation, testing, and experimentation.*

To these may be added Registers or Compilers, such as shall Record and Express in proper Terms these Collections; add to these Examiners and

## DISCOURSE OF EARTHQUAKES

Rangers of Things, such as shall distinguish and marshal them into proper Classes, and denote their Excellencies or Gradations of differing Kinds, their Perfections or Defects, what are Compleat, and what Defective, and to be repeated, and the like.

[p.280] So that we see the Subject of this Enquiry is very copious and large, and will afford Work enough for every Well-willer to employ his Head and Hands, to contribute towards the providing Materials for so large a Fabrick and Structure, as the great quantity of Materials to be collected do seem to denote. However, 'tis possible that a much less number may serve the turn, if fitly qualified and done with Method and Design, and it may be much better and easier.

*Further, all the data collected must be marshaled into some kind of order to be useful. The subject therefore is large and the quantities of materials great; however, if the work is done with a method and design, it may be easier.*

When this mighty Collection is made, what will be the use of so great a Pile? Where will be found the Architect that shall contrive and raise the Superstructure that is to be made of them, that shall fit every one for its proper use? Till which be found, they will indeed be but a heap of confusion. Who shall find out the Experiments, the Observations, and other Remarks, fit for this or that Theory? One Stone is too thick, or too thin, too broad, or too narrow, not of a due colour, or hardness, or grain, to suit with the Design, or with some other that are duly scapled for the purpose: This Piece of Timber is not of a right Kind, not of a sufficient Driness and Seasoning, not of a due length and bigness, but wants its Scantlings, or is of an ill Shape for such a purpose, or was not fell'd in a due time: 'Tis Sap-rotten, or Wind-shaken, or rotten at Heart, or too frow, and the like, for the purpose for which 'tis wanted.

*A huge collection of facts is but a pile of confusion unless they can be fit into this or that theory.*

I mention this, to hint only by the by, that there may be use of Method in the collecting of Materials, as well as in the use of them, and to shew that there may be made a Provision too great, as well as too little, that there ought to be some End and Aim, some pre-design'd Module and Theory, some Purpose in our Experiments, and more particular observing of such Circumstances as are proper for that Design. And though this Honourable Society have hitherto seem'd to avoid and prohibit pre-conceived Theories and Deductions from particular, and seemingly accidental Experiments; yet I humbly conceive, that such, if knowingly and judiciously made, are Matters of the greatest Importance, as giving a Characteristick of the Aim, Use, and Significancy thereof; and without which, many, and possibly the most considerable Particulars, are passed over without Regard and Observation. The most part of Mankind are taken with the Prettiness or the Strangeness of the Phænomena, and generally neglect the common and the most obvious; whereas in truth, for the most part, they are the most considerable. And the greatest part of the Productions of Nature are to be seen every where, and by every one, though, for the most part, not heeded or regarded, because they are so common. I could wish therefore that the Information of Experiments might be more respected, than either the Novelty, the Surprizingness, the Pomp, and the Appearances of them.

*There should be predesigned models or theories to provide purpose to experiment. Although the policy of the Royal Society has been to avoid and prohibit preconceived theories, Hooke feels that if these are made judiciously they can have great relevance and provide aim to observation. For a discussion of Hooke's scientific methodology, see Chapter 4.*

The obviousness and easiness of knowing many Things in Nature, has been the Cause of their being neglected, even by the more diligent and curious; which nevertheless, if well examined, do very often contain Informations of the greatest value. It has been generally noted by common, as well as inquisitive, Persons, that divers Stones have been found, formed into the Shapes of Fishes, Shells, Fruits, Leaves, Wood, Barks, and other Vegetable and Animal Substances: We commonly know some of them exactly resembling the Shape of Things we commonly find (as the

Chymists speak) in the Vegetable or Animal Kingdom; others of them indeed bearing some kind of Similitude, and agreeing in many Circumstances, but yet not exactly figured like any other thing in Nature; and yet of so curious a Shape, that they easily raise both the Attention and Wonder, even of those that are less inquisitive. Of these beautifully shaped Bodies I have observed two sorts: First, some more properly natural, such as have their Figures peculiar to their Substances: Others more improperly so, that is, such as seem to receive their Shape from an external and accidental Mould.

Of the first sort, are all those curiously figured Bodies of Salts, Talks, Spars, Crystals, Diamonds, Rubies, Amethysts, Ores, and divers other Mineral Substances, wherewith the World is adorned and enriched; which I at present omit to describe, as reserving them for a Second Part, they seeming to be, as it were, the Elemental Figures, or the *ABC* of Nature's [p.281] working, the Reason of whose curious Geometrical Forms (as I may so call them) is very easily explicable Mechanically: And shall proceed to the second sort of Bodies.

Of these are two kinds; either first the very Substances themselves converted into Stone, such are Bones, Teeth, Shells, Fruit, Wood, Moss, Mushrooms, and divers Vegetable and Animal Substances: Or secondly, such other Mineral or Earthy Substances, as Clays, Sands, Earths, Flinty Juices, etc. which have filled up, and been moulded in divers other Bodies, as Shells, Bones, Fruits, &c but especially Shells. These, according to the Representations they bear of other Bodies, have received divers Names; of which *Aldrovandus, Bauhinus, Imperatus, Wormius,* and others, reckon a great number: Such are, *Cornu ammonis sine armatura, Helicoides, Hoplites, multiplex obscure lucens, muricatum, cristatum, cristatum pertusum, striatum Campoides, Campoides Echinatum, Caprinum cornu, Cornu Arietinum, Sceleton Serpentis. Conchites bivalvis striatus, Mytulus biforis, cinerius rugosus, Coclites, Chama lapidea, lævis, rugata, Ostracites, Pectenites, Bucardia, Strombites, Belemnitæ Cornu fossile, Glossopetrae, Astroites, Entrochos, Colonetta, Lapis judaicus, Fungites, fungus saxeus, Lapis Indicus, Brontias, Brontias favogineus, Ombria, Ovum anguinum, Lignum petrifactum.* Of these I shall describe some few, because every one has not the Opportunity of seeing and examining them.

I have designed 15 several sorts of Snail rather than Snake-stones, call'd by some Authors *Cornua Ammonis*, or *Sceleta Serpentum*, all of them, both of different Substances and various Shapes; but yet all of them agreeing in these Proprieties, that they were made of a Tapering or Pyramidal Body, coil'd up together, so as that the Tip or Point of it was in the Center, and the Base outmost; next that, in the coiling up, the Axis of this Pyramidal Body kept exactly in the same Plane. 3. That all of them were ridged or furrow'd with Rings, Furrows, or Protuberances and Depressions, which respected the Center of the Spiral, for the most part, but were moulded and rang'd each of them different ways, all of them very regular, and exceedingly ornamental. 4. That in the coiling the lesser and inner Parts sunk, as it were, always into the inside of the greater encompassing Part. 5. That all of them had Diaphragms, or separating Valves, whereby the Parts might oft-times be easily separated. 6. That the *Fimbriae*, or Edges of these Diaphragms, were in most of these Stones very visible; in others of them, where they were somewhat more obscure, they might be made apparent, by scraping or rubbing away the outsides of

---

*Figured Stones, or fossils, resemble the shapes of plant or animal or their parts. Or they are the cast of a mold made by such a shape.*

*The curious geometrical shapes of crystals can be explained easily. Hooke was among the first to assert that the outward expression of a mineral is the result of the inner arrangement of particles, the "Elemental Figures, or the ABC of Nature's Working"—i.e. atoms.*

*Organic matter itself, such as bones, teeth, etc. can be converted into stone, or the shapes of them, especially shells, can be molded into other substances such as clay.*

*Waller's notes inserted at this point explain that the first five tables were drawn by Hooke himself with full description for only the first three tables.*

*Descriptions of ammonites called snail-stones or snake-stones. The Figures in Table I "Diaphragms" refer to the septa of ammonoids.*

them, where they were somewhat more obscure, they might be made apparent, by scraping or rubbing away the outsides of them. 7. That these *Fimbriæ* or Edges appear'd on the Surface, like the Out-lines of some curious Foliage, a Specimen of some of which I have given in the 3d and 10th Figures. This, [p.282] upon Examination of them, I found to proceed from the Fulness of the Edges of the Diaphragms whereby the Edges were waved or plaited somewhat in the manner of a Ruff. 8. That most of them were covered with a very curiously polish'd as well as curiously carv'd Surface, some of them shining like burnish'd Brass, as those of the 1st and 2d Figures; others like Brass, tarnish'd black, but rubb'd smooth; others of them like transparent Horn, as the 12th Figure; others like Coperas-stones; others like a coarser sort of white Marble; others like black Marble. 9. That from these polisht Surfaces one might oftimes easily pick off a Substance exactly resembling the plaited shining Substances of a Shell; and this did very visibly in many of them cover the internal stony Body, with a Coat two or three times as thick as a Snail's Shell. 10. That the biggest end of these Spiral Bodies was always imperfect without any determinate Figure. 11. That many of these Spiral Bodies seem'd, as if they had been broken and shatter'd, and had grown together again in an irregular Posture. 12. That many of them were compounded of several Substances, the Spaces between the Diaphragms being sometimes fill'd with one kind of Substance, sometimes with another, and sometimes they were found empty, only all the sides of the Diaphragm were cover'd with a kind of Tooth-Sparr. 13. There were many of these which were at first included in Stones, out of which they might easily be separated, so as to leave a perfect Impression like themselves; but in most of those incompassing Bodies, the Impression was bigger than the impressing Body, by the thickness of a thin Shell, which seemed to have been heretofore the Cause of both Impressions, but was worn away and decay'd by the Injury of time; they differ'd one from another chiefly in these Particulars: First, That the Bases or Planes supposed to cut these spirall'd Bodies at right Angles, with the Circumference and Plane thro' the Axis, were of different Figures; as that of the first Figure was much like that of a common Nautilus, being somewhat like the Figure of a Turkish Crescent; but the Diaphragms were not smooth and plain like those of a Nautilus, but full and ruffled like the Leaves of Sea-wrack, and several other luxuriant Vegetables, and that (which appear'd by the Foliage visible on the Surface) the Diaphragms were much thicker and closer together: This on the outside was like polish'd Silver, but the inside of a Substance not much unlike blue Slate, but closer, harder, and heavier. That of the 3d Figure was of a Figure, as if the former had been press'd quite flat; so that instead of the round Back in the 2d Figure, this has a Back terminated with a sharp Edge, as in the 4th Figure, 'tis all over almost cover'd with a shining Substance not unlike a Substance we call Alchimy, or whited Brass; on this the Foliage of the Edges of the Diaphragms is very visible, one of which I have described in its posture, as at *a*, and 3 others of them, that the Curiosity of them might be the more visible, I have described by the help of a Lens in the 5th Figure. Somewhat like to this is that of the 6th Figure; but that instead of an edged Back, this is hollowed or furrowed not unlike the Wheel of a Pully, with two protuberant Ridges on either side, as is visible in the 7th Figure; the Surface is undulated like the former two, but somewhat more manifestly, the Fashion of which Waves are not unlike the Ribs of wicker Screens, 2, 3, 4, and sometimes more of them uniting at last together into one more conspicuous Rib which crosses the Center; the other side of this was broken so that the Diaphragms and several

*By "Fimbriæ" Hooke is referring to the suture lines made by the crenulated septa which he likens to a "Ruff," or ruffle. He also describes them as similar to foliage. Further on, he does use the term "suture" to describe the trace of the intersection of septa edges with the surface of the shell, a term still used today.*

hollow Cavities between them were to be seen, 'twas of a Substance somewhat like the Rust of Iron. That of Figure the 8th was of a Substance somewhat like *Portland*-stone, but closer and harder. The transverse Section of it was much like the former, as may be seen in the 9th Figure: The Back of it was gutter'd and knobbed very like a *Japan* Nautilus, one of which I have in Mr. *Colwall's* Gift: The Ribs also, or Furrows of the Side were not much unlike, only they were somewhat finer wrought with Knobs or Buttons, as may be perceived by the Figure. The Knobs and Surface of the 10th was somewhat like this, but that they were a little more gross, as is visible by the Figure; this was of the Colour of a Bone that has been long buried in the Ground, and of a stony Substance almost as hard as a Flint; the outward Shell that seems to have cover'd it, was quite worn away, and all the partitions [p.283] or Edges of the Diaphragms were most conspicuous, the transverse Section you have in the 11th Figure. The Snail-stone describ'd by the 12th Figure was of pellucid Pebble, and look'd almost like Horn; that this had Diaphragms also, is evident by the end of it, which is bounded by one, as may be seen by the 13th Figure, which shews also the transverse Section of it. It had a Spine or Quill *a*, coiled about the back of it, the biggest part of which was broken off, and only the hollow part of the Quill of the Shell was left, as at *b* in the 12th: But at *c* the Quill was intire, and the Substance that fill'd the hollow of it, was transparent Pebble like the rest of the Stone; the incompassing Shell was much worn away, only in some Parts of it 'twas visible enough.

*Portland-stone is an Upper Jurassic limestone used as a building stone. But the substance Hooke is describing here is even "closer and harder," therefore more dense.*

The small one in the 14th Figure was much of the same Make, the Shell that covered it was black like the other, its End was also terminated by a Diaphragm, as is visible in the 15th Figure, where the transverse Section is also describ'd; but the Substance that fill'd it was quite differing, being a kind of Pyrites or Coperas-stone. The small one of the 16th Figure was of a Shape participating of the 1st and 3d, the transverse Section of it shewing it to be thinner than the 1st, but thicker than the 3d, was terminated by a Diaphragm very finely leav'd, and with viewing it very intensely, I could perceive the Sutures very finely wrought, much like those of the 5th Figure: it was fill'd with a black, stony Substance, and on the other Side of it (being a little broken) several of the Cavities between the Diaphragms appear'd empty. The little Part of one describ'd in the 18th Figure, was compounded as it were of the 8th and 14th, as may be seen by the transverse Section of it. Fig. 19. This, as the rest, had visible Diaphragms also, 'twas of a Substance like rusty Iron: These preceding were all of them of a Figure that taper'd very much, so as the Spiral from a very large Circumference, was presently contracted into a little one; but the following were of a Figure more protracted, and made many more Revolutions before they ended. Of this kind the 20th was the plainest, resembling that of the first of the other kind; 'twas ribb'd not much unlike it, and the transverse Section was much the same with that of the 2d Figure, as is visible in the 21st Figure.

*Here Hooke actually uses the term "suture," which is possibly the first time this term was used in describing ammonites.*

The 22d was somewhat like the 12th, but of a smaller Spiral; it was terminated with a Diaphragm, and had a small Spine or Quill which was laid round the Back of it, as may be seen by the 23d Figure; the Shell of it was yet sticking on it, and it look'd very like burnisht Brass; it seem'd to be a very hard Stone. The 24th somewhat resembled the 14th, the Surface of it may be perceiv'd by the 24th, and the transverse Section by the 25th Figure; 'twas a Pyrite, and one part of it was dissolv'd into Salt by the Air:

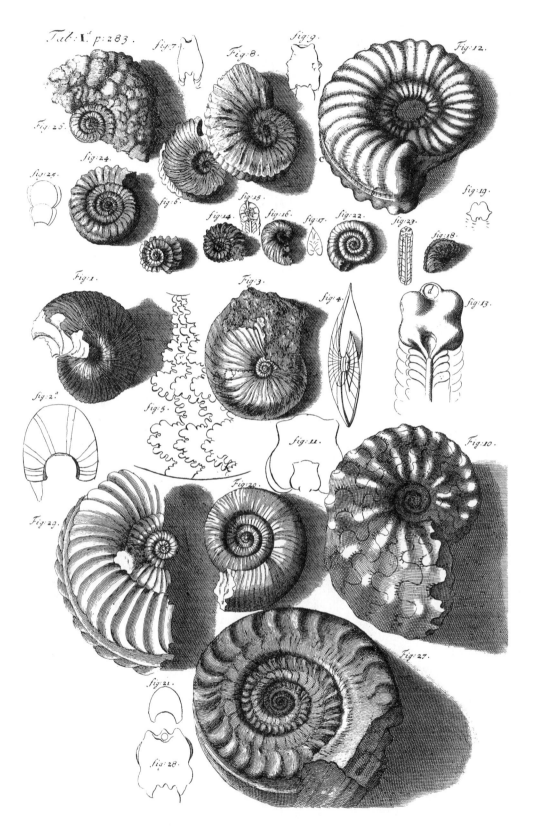

*Table I. Hooke's drawings of ammonites referred to as snail-stones or snake-stones, described on p. 161–163.*

I had two or three others of the same shape, one of a kind of grey Marble, another of a Flint, a third of an iron Stone, which I have describ'd in the 26th Figure, the Stone incompassing it, and the filling of the Shell it self being both of the same Substance: One of these was bruised in the hardning, so that the Cracks of the Shell were very visible; the like Accident was common to many of the rest, and the Shell encompassing was of a distinct Substance from the rest, and was easie to be pick'd off. But the prettiest of all the rest was the 27th, where the several Coats of the Shell and Diaphragms were very distinct, tho' they seem'd to be all petrify'd and turn'd to another Substance. It had been very much broken before the hardning, and all the Cracks of the Shell were very distinct; and which seem'd a little strange, some Parts of it were thrust outward, as if it had been fill'd with Water; and by Congelation, the Ice had swell'd and broken out the sides of it, which probably might be the Cause; for all between the Diaphragms it was fill'd with a transparent Spar or Caulk, such as is usually dug out of Lead Mines, a Substance between Crystal and Talk; the Original of which Substance seems to me not unlikely to be congeal'd petrifying Water; the Voluta of it was curiously ribb'd or moulded, and the Foliage or Edges of the Diaphragms were very Ornamental. It had a Spine or Quill went round the back of it, as several of the former; the transverse Section you have in the 28th Figure. [p.284] The Surface of the 29th was of a peculiar kind of carving, as is visible by the Figure; it was of a reddish Flint, the outside very smooth and polish'd, the side not visible was not so perfect, being broken in several places, and discovering the Diaphragms, and that some of the interjacent Spaces were empty. The greater End of this looking very irregular, I broke with a smart stroke of a Hammer a little piece off from it, and discover'd two small Snakes-stones within it, which probably had been tumbled into the Mouth of it before it was concreted; for they were of the very same Substance, but of a differing Figure from any I have yet describ'd, being ribb'd like the 24th, but only they were bigger and farther distant, and they went quite round the back. This last, and another like the 24th were taken up near *Keinsham*, about 4 or 5 Miles from *Bristol*, and sent by Dr. *Beal*, whence also I suppose several of the other may have come.

Had they not been much too large, I would have describ'd also one or two of those *Cornua Ammonis*, presented to the Repository by the Right Honourable *Henry Howard* of *Norfolk*, which are in Diameter about 2-1/2 Foot, and the concave Impression of one of a greater Magnitude, which I found in a Piece of *Portland*-stone. These large Stones are between 300 and 400 Weight, and of a Stone in all Particulars much like *Portland*-stone, whence I suppose they were at first fetch'd: They are all shap'd much like that of the 20th Figure; but the coiled Cone is not altogher so round, nor so slender, or of so acute an Angle, but the Undulations of the Surfaces are alike, and so are also the Diaphragms. I have been also told by Persons of very good Credit, that they have seen in *Darbyshire* and *Yorkshire* Snail-stones of a much more prodigious bigness, 3 or 4 times as big as these; which I have not had an Opportunity to send to enquire more curiously into, though I have much desired; and so much the rather, because it seems much more to excel in bigness all other Shell-fishes we know, than the Giants (Stories tells us of) did exceed the ordinary Size of Men now living.

I have, to parallel these Snake stones added in Table II. a Description of three several sorts of Nautil-shells, because I had no greater variety by me, though I have seen many other kinds. The 1st Figure represents a large

*Figure 27, "prettiest" of all, shows several coats of the shell and septa all mineralized. Cracks in the shell show that possibly it had been filled with water, which upon freezing expanded and cracked the shell.*

*What geologist has not felt that "smart stroke of a hammer" and found a fossil or two in the broken stone? In this description we see that Hooke was indeed a pioneer geologist.*

*Hooke marvels at the gigantic snail-stones, as they are much bigger in relation to all known living shellfishes than storied giants are to Man.*

*Table II described.*

166  DISCOURSE OF EARTHQUAKES

Nautilus-shell cut *per axin*, and manifests the manner how the Diaphragms are placed in that kind of Shell in the concave Part thereof; and the 2d Figure shews how they are placed up the convex side; this being a small Piece of the middle of a Nautilus-shell, and the wreathed Lines shew where the Diaphragms join'd upon the back thereof. The 3d Figure represents a *Japan* Nautilus-shell, crenated on the sides, and knobbed on the back, much in the manner as several of the Snakes-stones are. The 4th Figure represents a small Piece of a peculiar kind of Nautilus, whose conical Body is divided by small Diaphragms under every of the black circuling Lines, and is coil'd so as its roundnes is kept, and the Parts do not touch one another. The Name of it I know now, being no where describ'd by any Author.

In Table III. I have described some sorts of Helmet-stones, of which I have a very great Variety, and it would have been tedious to have added them all in this Place. I have likewise described three sorts of Button-stones, all, and every of which seem to have been nothing else but the filling up of several sorts of Echini-shells, of which the European Coasts afford a great Variety. The 1st, 2d, and 3d Figures, represent three several sorts of Button-stones, all of them of very hard Flints, two of them, namely the two less, join'd to, or shap'd as 'twere out of irregular Pieces of Flint of the same Substance. They have all of them this in common with all the other sorts of Helmet-stones, that they have two Parts, which seem to have been the fillings up of two Holes or Vents in the Shell, and they are divided into five Parts, though every of them of distinct Shapes, as may be seen by the Figures. That of the 1st Figure also hath this Property, in common with the finer sort of Helmet-stones, that it exhibits the Sutures or Junctures of the Shell, as are more plainly to be seen in the 4th, 5th, and 6th Figures. The 4th Figure represents a Helmet-stone, look'd down upon almost di[p.285]rectly, exhibiting the Impressions of the several Holes, Sutures, and Cracks that appear upon the top of one of these Stones. The 5th and 6th Figures represent another Helmet-stone of the same kind with the former, but less look'd upon against the bottom and side, exhibiting the Impressions of the several Holes, Sutures, and Cracks, that were in the imprinting Shell from which these Stones receiv'd their Shape: These were both of a kind of grey Flint. The 7th Figure represents the bottom of another sort of Helmet-stone, where the Vents *a* and *b* are placed in another manner, than they were in the 1st, 2d, 3d, or 5th Figures. The 8th and 9th Figures represent the bottom and top of another sort of Helmet-stone, which seems to be the filling up of a kind of *Echini*-shell, very like to those found in *Devonshire* and *Cornwal*, one of which I have delineated in the 10th Figure: This last kind was of Chalk. I have several other sorts, which I have not now time to delineate, some of transparent Pebbles, some of Marble, some of a Stone as hard as *Portland*, some of black, red, grey, and other Flints, some of Coperas-stone, some of other kinds of Stone, none of Spar. I would to this have added the Description of a great Variety of *Echini*-shells, divers of which I have by me in the Repository of the Royal Society, and others that I have met with elsewhere, but that I shall do it elsewhere: They are indeed almost infinite, but all concur in these Properties which all Helmet-stones likewise have. First, that they are distinguish'd into five Parts, by Sutures, Ribs, and Furrows. Secondly, that they have two Vent-holes: They have divers of them also little Edges, being the Impressions of the Sutures, and divers little rows of Pins, being the Impressions of the small Holes; and any one that will diligently and impartially examine both the Stones and the Shells,

*Table III.*

*Descriptions of echinoderms. "Helmet-stones" was the general name for echinoderms. "Button-stones" were the smaller varieties.*

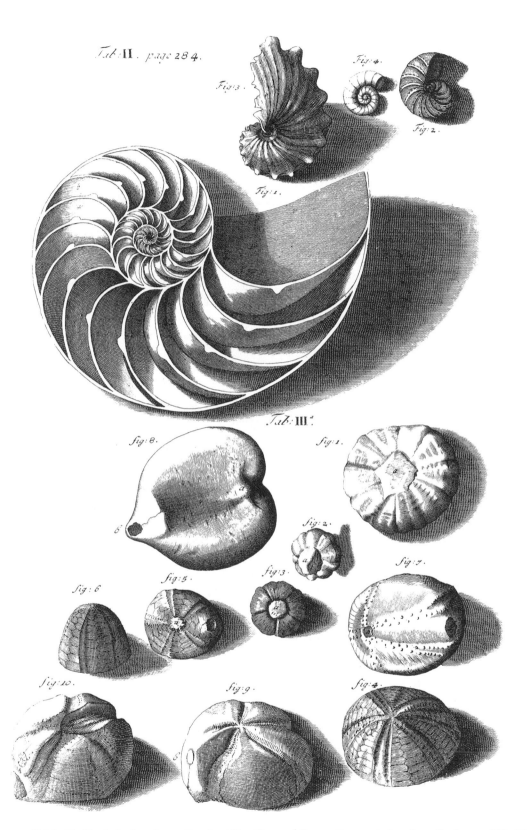

*Table II. Hooke's drawings of a giant Nautilus and three other varieties, described on p. 165–166.*

*Table III. Hooke's drawings of echinoderms known as Helmet-stones or Button-stones for the smaller varieties, described on p. 166.*

and compare the one with the other, will, I can assure him, find greater reason to perswade him of the Truth of my Position, than any I have yet urged, or can well produce in Words; no Perswasions being more prevalent than those which these dumb Witnesses do insinuate.

Table IV.

*The Figures of this and the next Table were left undescribed by Dr. Hooke, which Defect I have endeavoured to suply in some measure. Fig. 1st and 2d represent a sort of Shell, of which I think we have no Species now described; they are very thick and heavy: Of this sort I found one upon the Sand, on the side of the Severn about 8 or 10 Miles from Gloucester; it was a perfect Stone. Fig. the 3d, another unusual shaped Stone exceeding thick and heavy. I know not what the 4th is, it shews something like the Spine of some Fish. Of the 5th Figure I have seen several, and is well represented; they call them in that country Screw-stones. ('tis figured by Dr. Plott, in his Natural History of Oxfordshire; Tab. 4, Fig. I.) The rest of the Figures in this Table shew several sorts of Sharks-teeth, except the 18th and 19th, which seem to be the Shells of some Fish. The 10th Figure shews the Make of the inside of one of those long Teeth, if they are so, of which I think not many now doubt; in this manner of the Fibres radiating from the Center, is very conspicuous. ('Tis true these have by former Writers been thought* Lapides sui generis, *and call'd Belemnites. I shall wave the Dispute.) At the Basis of the 8th Figure, is observable the very great Cavity, as likewise the largeness of it. The 14th Figure represents one of the largest* Glossopetræ *that has been seen. Upon the 16th and 17th Figure Dr. Hooke makes this Remark, that there are 220 of them in the Fishes Mouth. What the 15th is I know not, except it be a petrify'd Grinder bedded in Stone.*

Table V.

*Figure the 1st, the petrified Grinder of some large Animal, possibly of a Whale or Elephant. Dr. Grew, in his Museum, says of a Sea-Animal. Fig. 2. A petrify'd Crab, very much resembling the Fish it self.*

*Figure the 3d, 4th, 5th, 6th, 7th, 8th, and 9th, I take to be Pieces of petrify'd Wood, tho' I know they have been otherwise esteem'd by some Writers. Whether the 14th may not be the same, I will not determine. The 10th, 11th, 12th, and 13th, I take to be some sorts of petrify'd Fruits, or possibly some of them Seed-vessels. The 15th and 16th Figures are the Astroites or Star-stones, of which one is given [p.286] separate in the 16th. These in our Author's Opinion were Pieces broken off from the numerous Legs of that sort of Star-fish, of which one was many Years since sent from New England, if I mistake not, and is now in the Society's Repository: It is described in the Philosophical Transactions (no. 50, p. 1153), publish'd by Mr. Oldenburg by the Name of* Piscis Echino-stellaris Visci formis. *The 17th Figure I take for a sort of Fungus petrify'd.*

*It is a great Misfortune that the Descriptions of these two Tables are wanting amongst the Papers, if they were ever drawn up, which I somewhat question; for the Figures were not number'd: For had he done it himself, he would have made several very considerable Remarks; which if I could, is not so proper for me to attempt; nor do I know my thing of the History concerning the Places where they were found, or the like: However I judged it not convenient they should be lost; and therefore ordered them to be graved, and have ventured this imperfect Description. What follows next, is the Abstract of a Letter I sent him from Bristol, 1687. R. W.*

According to Waller, Tables IV and V were left undescribed by Hooke. Waller, therefore, supplied the brief descriptions that are here printed in italics.

The "Screw-stone" appears to be the petrified filling of a gastropod shell.

Lapides sui generis *literally means self-generated stones.* Waller, having known the controversy over the origin of fossils in which Hooke was involved, prefers now to stay out of the dispute.

These seem to be parts of crinoid stems.

Waller claims he himself drew the two last tables, VI and VII, from fossils he found not far from Bristol and had given the drawings to Hooke, together with an explanation of the figures. The explanation, Waller asserts, is among the lost Hooke papers, but as proof of his part, he supplied excerpts of his letter to Hooke written in 1687, containing descriptions of the figures in Tables VI and VII. Since these are not Hooke material per se, I have placed the letter with the descriptions of Tab. VI and VII in Appendix A.

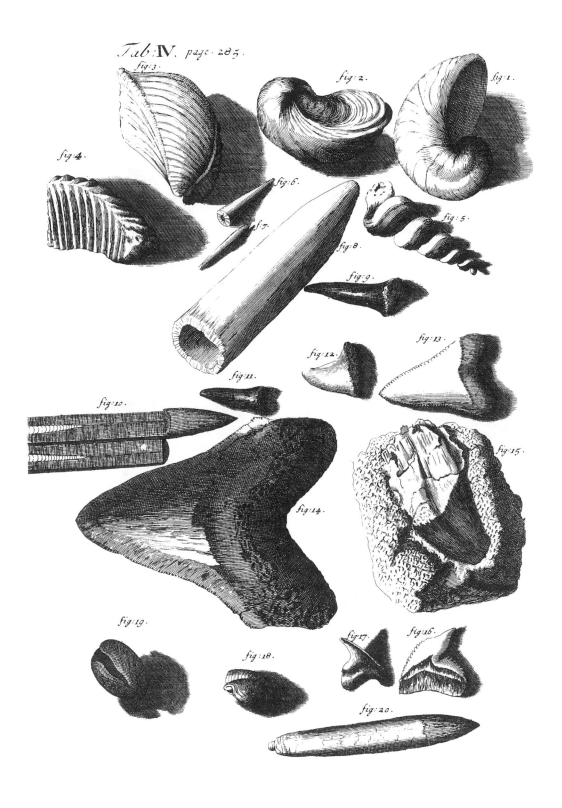

Table IV. Hooke's drawings of a gastropod, bivalves, belemnites and sharks' teeth, described by Waller on p. 168.

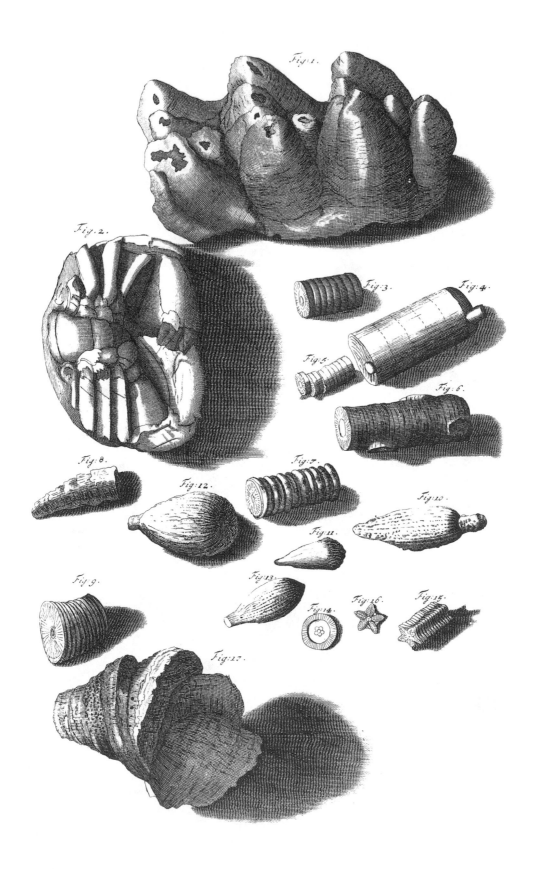

*Table V. Hooke's drawings of various fossils: mammal teeth, a crab, parts of crinoid stems and plant material, described by Waller on p. 168.*

[p.287] These, and the like Shapes, because many of them are curious, have so far wrought on some Men, that they have endeavoured to give us an Explication of the manner of their Formation; in doing of which they have so far rambled from the true and genuine Cause of them, that they have left the Matter much more difficult than they found it. Amongst the rest, *Gaffarel, a* [p.288] *French* Writer, seems not the least mistaken, who has transferr'd them over to the Confirmation, as he thinks, of his Astrological and Magical Fancy; and thinks that as they were produced from some extraordinary Celestial Influence, and that the Aspects and Positions of the fix'd Stars and Planets conduc'd to their Genration, so that they also have in them a secret Vertue whereby they do at a distance work Miracles on things of the like Shape. But these, as fantastical and groundless, I shall not spend time on at present to refute, nor on the Conjectures and Hypotheses of divers others; which though perhaps somewhat more tolerable than that I last recited, yet most of them have recourse to some vegetative or plastick Vertue inherent in the Parts of the Earth where they were made, or in the very parcels of which they consist, which, to me, seems not at all consonant to the other workings of Nature; for those more curiously carved and beautiful Forms are usually bestow'd on some vegetable or animal Body. But my Business at present shall not be so much to confute others Conjectures, as to make probable some of my own; which tho' at the first hearing they may seem somewhat paradoxical, yet if the Reasons that have induced me thereunto be well consider'd and weigh'd, I hope at least they may seem possible, if not more than a little probable.

The particular Productions of this kind that I have taken notice of my self in my own Enquiries, and which I find dispersed up and down in the Writing of others, may be reduced under some one or other of these General Heads or Propositions.

First, That there are found in most Countries of the Earth, and even in such where it is somewhat difficult to imagine (by reason of their vast distance from the Sea or Waters how they should come there) great quantities of Bodies resembling both in Substance and Shape the Shells of divers sorts of Shell-fishes; and many of them so exactly, that any one that knew not whence they came, would without the least scruple firmly believed them to be the Shells of such Fishes: But being found in Places so unlikely to have produced them, and not conceiving how else they should come there, they are generally believed to be real Stones form'd into these Shapes, either by some plastick Vertue inherent in those Part of the Earth, which is extravagant enough, or else by some Celestial Influence or Aspect of the Planets operating at a distance upon the yielding Matter of the Parts of the Earth, which is much more extravagant. Of this kind are all those several sorts of Oyster-shells, Cockle-shells, Muscle-shells, Periwinkle-shells, and the like, which are found in *England, France, Spain, Italy, Germany, Norway, Russia, Asia* and *Africa*, and divers other Places; of which I have very good Testimonies from Authors of good Credit.

Secondly, That there often have been, and are still daily found in other Parts of the Earth buried below the present Surface thereof divers sorts of Bodies, besides such as I newly mention'd, resembling both in Shape, Substance, and other Proprieties, the Parts of Vegetables, having the perfect Rind or Bark, Pith, Pores, Roots, Branches, Gums, and other constituent Parts of Wood, though in another posture, lying for the most

---

*One of the prevailing ideas concerning the origin of fossils is that they were formed by some magic or astrological influence or that the earth where fossils are found had some vegetative or plastic qualities that permitted the purely coincidental formation of animal and plant shapes. Hooke, in contrast, rejects all explanations that invoke "Astrological and Magical Fancy."*

*Hooke's propositions concerning fossils: 1stly, stony objects shaped as shells are found in most countries, even in places far from the sea.*

*Hooke was extremely well read and kept up correspondence abroad. The notion that England at this time was out of touch with the rest of the European intellectual mainstream is therefore untrue.*

*2ndly, these objects resembling parts of both animal and plant have been found in many places buried below the present surface.*

part Horizontal, and sometimes inverted, and much differing from that of the like Vegetables when growing, and wanting also, for the most part, the Leaves, smaller Roots and Branches, the Flower and Fruit, and the like smaller Parts, which are common to Trees of that kind; of which sort is the *Lignum Fossile*, which is found in divers Parts of *England, Scotland, Ireland,* and divers Parts of *Italy, Germany,* the *Low Countries,* and indeed almost in every Country of the World.

Thirdly, That there are often found in divers other Parts of the Earth, Bodies resembling the whole Bodies of Fishes, and other Animals and Vegetables, or the Parts of them, which are of a much less permanent Nature than the Shells abovemention'd, such as Fruits, Leaves, Barks, Woods, Roots, Mushrooms, Bones, Hoofs, Claws, Horns, Teeth, &c. but in all other Pro-[p.289]prieties of their Substance, save their Shape, are perfect Stones, Clays, or Earths, and seem to have nothing at all of Figure in the inward Parts of them. Of this kind are those, commonly call'd Thunder-bolts, Helmet-stones, Serpentine-stones, or Snake-stones, Rams-horns, Brain-stones, Star-stones, Screw-stones, Wheel-stones, and the like.

*3rdly, whole bodies of fishes, other animals, and their parts, as well as parts of plants, are found as stone, with no inner parts.*

*These are the various names that were given to belemnites and ammonites. Helmet-stones refer to echinoids.*

Fourthly, That the Parts of the Earth in which these kinds have been found, are some of them some hundred of Miles distant from any Sea, as in several of the Hills of *Hungary,* the Mountain *Taurus,* the *Alpes, &c.*

*4thly, the locations where these objects have been found are far from any sea.*

Fifthly, That divers of those Parts are many Scores, nay, some many Hundreds of Fathoms above the Level of the Surface of the next adjoining Sea, there having been found of them on some of the most Inland, and on some of the highest Mountains in the World.

*5thly, many of these objects have been found at great heights above the level of the nearest sea.*

Sixthly, That divers other Parts where these Substances have been found, are many Fathoms below the Level both of the Surface of the next adjoining Sea, and of the Surface of the Earth itself, they having been found buried in the bottoms of some of the deepest Mines and Wells, and inclosed in some of the hardest Rocks and toughest Metals. Of this we have continual Instances in the deapest Lead and Tin-mines, and a particular Instance in the Well dug in *Amsterdam,* where at the Depth of 99 Foot was found a Layer of Sea-shells mixed with Sand of 4 Foot thickness, after the Diggers had past through 7 Foot of Garden-mould, 9 Foot more of black Peat, 9 Foot more of soft Clay, 8 of Sand, 4 of Earth, 10 of Potters-clay, 4 more of Earth, 10 Foot more of Sand, upon which the Stakes or Piles of the *Amsterdam* Houses rest; then 2 Foot more of Potters-clay, and 4 of white Gravel, 5 of dry Earth, 1 of mix'd, 14 of Sand, 3 of a Sandy Clay, and 5 more of Potters-clay mix'd with Sand. Now below this Layer of Shells immediately joining to it, was a Bed of Potters-clay of no less than 102 Foot thick; but of this more hereafter.

*6thly, various such objects have also been found below the surface of the earth and of the nearest sea, such as those found in mines and wells. A layer of shells has been found at a depth of 99 feet.*

Seventhly, That there are often found in the midst of the Bodies of very hard and close Stone, such as Marbles, Flints, *Portland,* and Purbeck-stone, &c. which lye upon, or very near to the Surface of the Earth, great quantities of these kind of figured Bodies or Shells, and that there are many of such Stones which seem to be made of nothing else.

*7thly, some of these bodies are found in very dense rock.*

These Phænomena, as they have hitherto much puzled all Natural Historians and Philosophers to give an Account of them, so in truth are they in themselves so really wonderful, that 'tis not easie without making multitudes of Observations, and comparing them very diligently with the

Histories and Experiments that have been already made, to fix upon a plausible Solution of them. For as on the one side, it seems very difficult to imagine that Nature formed all these curious Bodies for no other End, than only to play the Mimick in the Mineral Kingdom, and only to imitate what she had done for some more noble End, and in a greater Perfection in the Vegetable and Animal Kingdoms; and the strictest Survey that I have made both of the Bodies themselves, and of the Circumstances obvious enough about them, do not in the least hint any thing else; they being promiscuously found of any kind of Substance, and having not the least appearance of any internal or substantial Form, but only of an external or figured Superficies. As, I say, 'tis something harsh, to imagine that these thus qualified Bodies should, by an immediate plastick Vertue, be thus shaped by Nature contrary to her general Method of acting in all other Bodies; so on the other side, it may seem at first hearing somewhat difficult to conceive how all those Bodies, if they either be the real Shells or Bodies of Fish, or other Animals or Vegetables, which they represent, or an impression left on those Substances from such Bodies, should be, in such great quantities, transported into Places so unlikely to have received them from any help of Man, or from any other obvious Means.

*So how did these objects get there? Certainly not by a "plastick Vertue." Therefore, if they represent the hardened animals or plants themselves or impressions of them, how did they get transported to such unlikely places?*

[p.290] The former of these ways of solving these Phaenomena, I confess I cannot for the Reasons I now mention'd, by any means assent unto; but the latter, tho' it has some Difficulties also, seems to me not only possible, but probable.

The greatest Objections that can be made against it, are, First, by what means those Shells, Woods, and other such like Substances (if they really are the Bodies they represent) should be transported to, and be buried in the Places where they are found? And,

*Why are these bodies buried in the places where they are found?*

Secondly, Why many of them should be of Substances wholly differing from those of the Bodies they represent; there being some of them which represent Shells of almost all kinds of Substances, Clay, Chalk, Marble, soft Stone, harder Stone, Marble, Flint, Marchasite, Ore, and the like.

*And why are many of them made of substances so different from the bodies they represent?*

In answer to both which, and some other of less Importance, which I shall afterwards mention, give me leave to propound these following Propositions, which I shall endeavour to make probable. Of these in their Order.

*To answer these questions Hooke has 11 propositions:*

My first Proposition then, is, That all, or the greatest part of these curiously figured Bodies found up and down in divers Parts of the World, are either those Animals or Vegetable Substances they represent converted into Stone, by having their Pores fill'd up with some petrifying liquid Substance, whereby their Parts are, as it were, lock'd up and cemented together in their Natural Position and contexture; or else they are the lasting Impressions made on them at first, whilst a yielding Substance by the immediate Application of such Animal or Vegetable Body as was so shaped, and that there was nothing else concurring to their Production, save only the yielding of the Matter to receive the Impression, such as heated Wax affords to the Seal; or else a subsiding or hardning of the Matter, after by some kind of Fluidity it had perfectly fill'd or inclosed the figuring Vegetable or Animal Substance, after the manner as a Statue is made of Plaister of *Paris*, or Alabaster-dust beaten, and boil'd, mixed with Water and poured into a Mould.

*Firstly, that these objects either represent the organic matter itself turned into stone or they are the impressions of the living matter.*

Secondly, Next that there seems to have been some extraordinary Cause, which did concur to the promoting of this Coagulation or Petrification; and

*2ndly, that some extraordinary process occurred that caused the petrification.*

that every kind of Matter is not of it self apt to coagulate into a strong Substance, so hard as we find most of those Bodies to consist of.

Thirdly, That the concurrent Causes assisting towards the turning of these Substances into Stone, seem to have been one of these, either some kind of fiery Exhalation arising from subterraneous Eruptions or Earthquakes; or secondly, a Saline Substance, whither working by Dissolution and Congelation, or Crystallization, or else by Precipitation and Coagulation; or thirdly, some glutinous or bituminous Matter, which upon growing dry or setling grows hard, and unites sandy Bodies together into a pretty hard Stone; or fourthly, a very long continuation of these Bodies under a great degree of Cold and Compression.

*3rdly, four different processes that could accomplish the petrifaction are subterranean heat, chemical processes, hardening from a glutinous substance, or cold and compression.*

Fourthly, That Waters themselves may in tract of time be perfectly transmuted into Stone, and remain a Body of that Constitution without being reducible by any Art yet commony known.

*4thly, various waters can become solid. By the word waters, Hooke obviously is not limiting himself to plain $H_2O$. His arguments further on make this point clear.*

Fifthly, That divers other fluid Substances have after a long continuance at rest, have settled and congealed into much more hard and permanent Substances.

*5thly, various other fluid substances harden also.*

Sixthly, That a great part of the Surface of the Earth hath been since the Creation transformed and made of another Nature; namely, many Parts which have been Sea are now Land, and divers other Parts are now Sea which were once a firm Land; Mountains have been turned into Plains, and Plains into Mountains, and the like.

*6thly, land and sea have exchanged places.*

[p.291] Seventhly, That divers of these kind of Transformations have been effected in these Islands of *Great Britain*; and that 'tis not improbable, but that many very Inland Parts of this Island, if not all, may have been heretofore all cover'd with the Sea, and have had Fishes swimming over it.

*7thly, much of Great Britain could have been under the sea at one time.*

Eighthly, That most of those Inland Places, where these kinds of Stones are, or have been found, have been heretofore under the Water; and that either by the departing of the Waters to another part or side of the Earth, by the alteration of the Center of Gravity of the whole Bulk, which is not impossible; or rather by the eruption of some kind of subterraneous Fires, or Earthquakes, whereby great quantities of Earth have then been rais'd above the former Level of those Parts, the Waters have been forc'd away from the Parts they formerly cover'd, and many of those Surfaces are now raised above the Level of the Water's Surface many scores of Fathoms.

*8thly, the sea waters could have receded by various means, by a change in the Earth's center of gravity, or by earthquakes whereby what had been covered by the oceans is now dry land raised above sealevel.*

Ninthly, it seems not improbable, that the tops of the highest and most considerable Mountains in the World have been under Water, and that they themselves most probably seem to have been the Effects of some very great Earthquake, such as the *Alpes* and Appennine Mountains, *Caucasus*, the Pike of *Tenariff*, the Pike in the *Terceras*, and the like.

*9thly, the tops of some of the highest mountains of the world could have been under water at one time.*

Tenthly, That it seems not improbable, but that the greatest part of the Inequality of the Earth's Surface may have proceeded from the subversion and tumbling thereof of some preceding Earthquakes.

*10thly, the uneven surface of the earth was caused by earthquakes in the past.*

Eleventhly, That there have been many other Species of Creatures in former Ages, of which we can find none at present; and that 'tis not unlikely also but that there may be divers new kinds now, which have not been from the beginning.

*11thly, Hooke's concept of extinction of species and generation of new ones. Hooke's ideas on evolution. are discussed in Chapter 6.*

There are some other Conjectures of mine yet unmention'd, which are more strange than these; which I shall defer the mentioning of till some other time; because tho' I have divers Observations concurring, yet having not been able to meet with such as may answer some considerable Objections that they are liable to, I will rather at present endeavour to make probable those already mentioned, by setting down some of those Observations (for it would be tedious to insert all) I have collected, both out of Authors, and from my own Experience.

*Hooke next attempts to prove his propositions with arguments based on observations.*

The First was, That these figured Bodies dispersed over the World, are either the Beings themselves petrify'd, or the Impressions made by those Beings. To confirm which, I have diligently examin'd many hundreds of these figured Bodies, and have not found the least probability of a plastick Faculty. For first, I have found the same kind of Impression upon Substances of an exceeding differing Nature, whereas Nature in other of her Works does adapt the same kind of Substance to the same Shape; the Flesh of a Horse is differing from that of a Hog, or Sheep, or from the Wood of a Tree, or the like; so the Wood of Box, for Instance, is differing from the Wood of all other Vegetables; and if the outward Figure of the Plant or Animal differ, to be sure their Flesh also differs: And under the same Shape you always meet with Substances of the same kind; whereas here I have observed Stones bearing the same Figure, or rather Impression, to be of hugely differing Natures, some of Clay, some of Chalk, some of Spar, some of Marble, some of a kind of Free-stone, some like Crystals or Diamonds, some like Flints, others a kind of Marchasite, others a kind of Ore. Nay, in the same figur'd Substance I have found divers sorts of very differing Bodies or kinds of Stone, so that one has been made up partly of Stone, partly of Clay, and partly of Marchasite, and partly of Spar, according as the Matter chanced to be jumbled together, and to fill up the Mould of the Shell.

*Arguments for the first proposition: the same shapes can be made of various kinds of materials.*

Another Circumstance, which makes this Conjecture the more probable is, that the outward Surface only of the Body is form'd, and that the inward [p.292] Part has nothing of Shape that can reasonably be referr'd to it; whereas we see that in all other Bodies that Nature gives a Shape to, the figures also the internal Parts or the very Substance of it, with an appropriate Shape. Thus in all kinds of Minerals, as Spars, Crystals, and divers of the precious Stones, Ores, and the like, the inward Parts of them are always correspondent to the outward Shape; as in Spar, if the outward Part be shap'd into a Rhomboidical parallepiped, the inward Part of it is shap'd in the same manner, and may be cleft out into a multitude of Bodies of the like Form and Substance.

*Whereas the outward shape of other bodies, like minerals in crystalline form, correspond to the inner structure (Hooke uses the example of the cleavages of calcite), the insides of fossils do not correspond to the outward shape.*

Another Circumstance is, that I have in many found the perfect Shell inclosed making a concave Impression on the Body that inclosed it, and a convex on the Body that it did inclose; which I have sometimes been able to take out intire, and found it to be both by its Substance and Shape, and reflective shining, and the like Circumstances, a real Shell of a Cockle, Periwinkle, Muscle, or the like.

*Shells, in fact, are often filled with other substances.*

And farther, I have found in the same place divers of the same kinds of Shells, not fill'd with a Matter that was capable of taking the Impression, but with a kind of sandy Substance; which lying loose within it could be easily shook out, leaving the inclosing Shell perfectly intire and empty; others I have seen which have been of black Flint, wherein the Impression has been made only of a broken Shell, which stuck also into it; the other Part of the Surface of that Stone, which are not within the Shell, remaining only form'd like a common Flint.

And which seems to confirm this Conjecture much more than any of the former Arguments, I had this last Summer an Opportunity to observe upon the South-part of *England*, in a Clift whose Bottom the Sea wash'd, that at a good heighth in the Clift above the Surface of the Water, there was a Layer, as I may call it, or Vein of Shells, which was extended in length for some Miles: Out of which Layer I digg'd out, and examin'd many hundreds, and found them to be perfect Shells of Cockles, Periwinkles, Muscles, and divers other sorts of small Shell-Fishes; some of which were fill'd with the Sand with which they were mix'd; others remain'd empty, and perfectly intire: From the Sea-waters washing the under part of this Clift, great quantities of it do every Year tumble or founder down, and fall into the Salt-water, which are wash'd also by several Mineral-waters issuing out at the bottom of those Clifts. Of these founder'd Parts I examined very many Parcels, and found some of them made into a kind of harden'd Mortar, or very soft Stone, which I could easily with my Foot, and even almost with my Finger, crush in Pieces; others that had lain a longer time exposed to the Vicissitudes of the rising and falling Tides, I found grown into pretty hard Stones; others that had been yet longer, I found converted into very hard Stone, not much yielding to the hardness of Flints. Out of divers of these, I was able to break and beat out divers intire and perfect Shells, fill'd with a Substance which was converted into a very hard Stone, retaining exactly the Shape of the inclosing Shell. And in the part of the Stone which had encompass'd the Shell, there was left remaining the perfect Impression and Form of the Shell; the Shell it self remaining as yet of its natural white Substance, though much decay'd or rotted by time: But the Body inclosing and included by the Shell, I found exactly stamp'd like those Bodies, whose Figures Authors generally affirm to be the Product of a Plastick or Vegetative Faculty working in Stones.

*Note that Hooke uses the term "layer" to describe a bed of fossil shells that extends for miles. He initiated the use of a number of geological terms. His observations of seawater eroding the cliffs of the Isle of Wight, breaking them down to sediments which in turn hardened again into rock, led him to formulate his cyclic hypothesis involving exchange of land and sea areas. See Chapter 2*

Another Argument, that these petrify'd Substances are nothing but the Effects of those Shells being fill'd with some petrifying Substance, is this, That among those which are call'd *Cornu-Ammonis*, or Serpentine-stones, (found about *Keinsham*, and in several other Parts of *England*, and in other Countries, as at the *Balnea Bollensia*) which are indeed nothing else but the moulding off from a kind of Shell which is much shap'd like a *Nautilus*-shell, the whole Cavity being separated with divers small Valves or Partitions, much after the same manner as those Shells of the *Nautilus* are commonly observed to be. Among these Stones, I say, I have, upon breaking, found some of the Cavities between those Partitions remain almost quite empty; [p.293] others I have found lined only with a kind of Tartareous, or rather Crystalline Substance, which has stuck to the sides, and been figured like Tartar, but of a clear and transparent Substance like Crystal; whereas others of the Cavities of the same Stone I have found filled with divers kinds of Substances very differing: Whence I imagine those Tartareous Substances to be nought else but the hardning of some saline fluid Body, which might soak in through the Substance of the Shell. Others of these I have, which are quite of a transparent Substance, and seem to be produced from the Petrifaction of the Water that had fill'd them; others I have found fill'd with a perfect Flint, both which I suppose to be the productions of Water petrify'd: And I may perhaps hereafter make it probable, that all kinds of Flints and Pebbles have no other Original.

*Some of these fossil shells are filled with crystals "stuck to the sides" deposited there from solution.*

I could urge many other Arguments to make my first Proposition probable, that all those curiously shaped Stones, which the most curious Naturalists most admire, are nothing but the Impressions made by some real Shell in a Matter that at first was yielding enough, but which is grown harder with time. To this very Head also may be referr'd all those other kinds of petrify'd Substances, as Bones, Teeth, Crabbs, Fishes, Wood, Moss, Fruit, and the like; some of all which Kinds I have examin'd, and by very many Circumstances, too long to be here inserted, judge them to be nothing else but a real petrifaction of those Substances they resemble.

*Fossils therefore are either impressions made by animals or plants or they are the animals and plants or their parts themselves petrified.*

My Second Proposition will not be difficult to prove, That if these be the Effects of Petrifaction or Coagulation, it must be from some extraordinary Cause. And this because we find not many Experiments of producing of them when and where we will; besides we find that most things, especially Animal and Vegetable Substances, after they have left off to vegetate, do soon decay, and by divers ways of Putrefaction and Rotting, loose their Forms and return into Dust; as we find Wood, whether exposed to the Air or Water, in a little Time to waste and decay; especially such as is exposed to the alteration of both, and even in those Places where these petrify'd Substances are to be met with. The like we find of Animal Substances; and we have but some few Experiments of preserving those Bodies, to make them as permanent as Stone, and fewer of making them into a Substance of the like Nature.

*Arguments for the 2nd Proposition: Some extraordinary cause petrified the objects.*

The Third thing therefore, which I shall endeavour to shew, is, That the concurring Causes to these Petrifactions seem to be either some kind of petrifying Water, or else some saline or sulphureous Mixture, with the concurrence of Heat, from some subterraneous Fire or Earthquake; or else a very long Continuation of those Bodies under a great degree of Cold and Compression, and Rest. That petrifying Waters may be able to convert both Animal and Vegetable Substances into Stone, I could, besides several Trials of my own, bring multitudes of Relations out of Natural Historians: But these are so common almost in all Countries, and so commonly taken notice of by the Curious, that I need not instance. *Cambden* and *Speed* will tell you of abundance here in *England*, as the *Peak* in *Derbyshire*, and in several other subterraneous Caverns in *England*. The Water it self does, by degrees, produce several conical pendulous Bodies of Stone, shap'd and hanging like Icicles from the Roof of the Vault; and dropping on the bottom, it raises up also conical Spires, which, by degrees, endeavour to meet the former pendulous *Stiriae*. And indeed I have generally observ'd it, that wherever there is a Vault made with Lime under Ground, into which the Rain-Water soaking through, a pretty thickness of Ground, does at last penetrate through the Arch; I have in several places, I say, observ'd that that Water does incrustate the Roof with Stone, and in many places of it generate small pendulous Icicles. This Water I have found in a little time to incrustate Sticks, or the like Vegetable Substances with Stone, and in some places to penetrate into the Pores of the Wood, filling them up with small Cylinders of Stone. This I have observ'd also in divers of the Arches of St. *Paul's* Church, which have been uncover'd and have lain open to the Rain, though there be [p.294] no Earth for it to soak through. And tho' I have never yet been able to petrify a Stick throughout, yet I have now by me several pieces that retain so perfectly all the Figure of Wood, and are yet so perfectly in all other properties Stone, that I find not the least Reason of doubt to believe that those pieces have been actual Wood, having still the Bark, the Clefts, the Knots, the Grain, the Pores, and even

*Arguments for the 3rd Proposition: Causes of petrifaction: deposition from solution, "petrifying waters," some "saline or sulphureous mixture" together with heat from subterraneous sources or earthquake; or by congelation.*

*Water containing lime in solution deposits stalactites and stalagmites in caverns.*

those too which, for their smalness, I have elsewhere call'd Microscopical; tho' I confess some of these more perfect pieces seem to have been petrify'd from some more subtile and insinuating petrifying Water, than those I newly mention'd; and 'tis not improbable but that some Subterraneous Steams and Heat may have contributed somewhat towards this Effect. But first I shall endeavour to make it probable, that these petrify'd Bodies may have been placed in those Parts where they are found, by some kind of Transformation wrought on the Surface of the Earth, by some Earthquake: And to this end, I shall by and by mention some strange alterations that have been made by Earthquakes, after I have first made probable my fourth Conjecture.

The Fourth Proposition therefore to be explain'd and made probable is, that Waters themselves of divers Kinds, are, and may have been transmuted perfectly into a stony Substance, of a very permanent Constitution, being scarcely reducible again into Water by any Art yet commonly known. And that divers other Liquid or Fluid Substances have in tract of time settled and congealed into much more hard, fixt, solid and permanent Forms than they were of at first.

*Arguments for the 4th Proposition: Waters, liquids, or other fluid substances have been known to congeal into permanent solid forms.*

The probability of which Proposition may appear from these Particulars.

1. That almost in all Streams and running Waters there is to be found great quantity of Sand at the bottom, many of which Sands both by their Figure in the Microscope, and transparently, seem to have been generated out of the Water.

First, I say, That their transparency which they discover in the Microscope is an Argument, because I believe there is no transparent Body in the World that has not been reduc'd to that Constitution by being some ways or other made fluid, nor can I indeed imagine how there should be any. All Bodies, made transparent by Art, must be reduc'd into that Form first; and therefore 'tis not unlikely but that Nature may take the same Course; but this as only probable I shall not insist on. Next, I say, that the Figures of diverse of them in the Microscope discover the same things; for I have seen multitudes of them curiously wrought and figured like Crystal or Diamonds, and I cannot imagine by what other Instrument Nature should thus cut them, save by Crystalizing them out of a Liquid or Fluid Body, and that way we find her to work in the formation of all those curious regular Figures of Salts, and the Vitriols (as I may call them) of Metals and divers other Bodies, of which Chymistry affords many Instances. Sea-Salt and Salgem chrystylizeth into Cubes or four-sided Parrallelipipeds; Niter into triangular and hexangular Prisms. Alume into Octoedrons, Vitriols into various kinds of Figures, according to the various kinds of Metals dissolved, and the various *Menstrua* dissolving them; Tartars also, and Candyings of Vegetables are figured into their various regular Shapes from the same Method and Principle. And in truth, in the formation of any Body out of this mineral Kingdom, whose Origine we are able to examine, we may find that Nature first reduces the Bodies to be wrought on into a liquid or soft Substance, and afterwards forms and shapes it into this or that Figure. But this Argument drawn from the Sand, found in all running Streams, I shall not insist on, because some imagine it to be only washt off from the Land and Shores the River passes over, and perhaps much of it may: But yet that Sand may be made of clear Water, this second Argument will manifest, and that is this:

*Crystals are formed by crystallizing out of a liquid or fluid.*

[p.295] That 'tis a usual Experiment in the making of Salt in the Salterns, by the boyling up, or evaporating away the fresher part of the Sea-water, to collect great quantities of Sand at each corner of the Boyler; which, after it has been well washt with fresh Water, is, in all particulars, a perfect Sand; and yet the Water is so order'd before it is put into the Boyler, that nothing of Sand or Dregs can enter with it, the Brine being first suffer'd to stand a good while and settle in a very large Fat, so that all the Sand and Dregs may sink to the bottom; after which, the clearer Water at the top is drawn off, and suffer'd to run into the Boyler. 'Tis not impossible, perhaps, but that Substance which made this Sand, might be dissolved in Water, and afterwards by evaporation coagulated; which, if so, makes not at all against, but rather argues strongly for my fourth Proposition.

*Experiment evaporating seawater. The substance of the evaporate, therefore, was in solution in the water.*

But that the other Solution is something more probable, namely, That 'tis made out of the very Substance of the Water itself, this third Argument will make probable; and that is, that any Water of what kind soever, tho' never so clear and insipid, may, by frequent Distillations, be all of it perfectly transmitted into a white insipid Calx not again dissolvable in Water, and in nothing differing from the Substance of Stone; this I have been assured by an eminent Physician, who has divers times made tryal of it with the same success. If therefore the whole Body of any Water may, by so easy an Operation in so very short a time, be transmuted into a stony Substance, what may not Nature do that can take her own time, and knows best how to make use of her own Principles?

*Or else the solution is the substance in its liquid form.*

But 4*thy*. we have many Instances by which we are assured that Nature realy does change Water into Stone, both by forming in a little time considerable Stones out of the distilling Drops of Water soaking through the Roofs of Caves and subterraneous Vaults, of which we have very many Instances here in *England*; as to name one for all at the Peak in *Derbyshire*, the pendulous Cones of this petrify'd Substance directly point at, and oftentimes meet and rest on the rising Spires, generated by the drops of Water trickling through the Roof, as I mention'd before.

*Formation of calcareous stalactites and stalagmites in caverns.*

And 5*thy*. there are divers other Waters which we need not seek after in Caves that have a petrifying vertue, and incrustate all the Chanel they pass through, and the Substances soak'd them with Stone; these are so common almost in all places, that I need not instance in any; only I cannot pass by one, which is taken notice of by *Kircher* in his *Mundus Subterraneus*, being Observations made by himself, and it has in it two Circumstances very considerable; the first is, That vegetables should grow so plentifully in a very hot Water. The second, that only such Herbs as grew in it, and not such as were Steeped in it, will perfectly, after drying, be turned into Stone, of which I shall afterwards have occasion to make more use. I shall give the History in his own Words, as they are set down in the 7th Paragraph of the 2d Sect. of the 5th Book of his *Mundus Subterraneus*, *Hæc* (says he) *experientia didici in Itinere meo Hetrusco, in quo prope Roncolanum senensis territorii Oppidum* (a Town near *Siena* in *Tuscany*) *duos fontes calidos observavi, quorum aqua per Canales ad molares Rotas vertendas ducebatur. In bisce canalibus cyperus, junci, ranunculus similesq; herbæ tanta adolescebant fæcunditate, ut quotaunis eas, ne aquæ motum interturbarent, extirpare oporteret. Extirpatas vero projectasq; in vicinum locum herbas omnes in Lapidem conversas non sine admiratione spectavi. Cujus rei causam cum a molitoribus quærerem. Responderunt aquas istiusmodi hujus virtutis esse, ut quæcuncq; inter canales, aut ipsa*

*Hooke cites the Jesuit Kircher's* Mundus Subterraneus *in which Kircher describes observations he made at two hotsprings and which Hooke considers proof of the existence of petrifying waters. Remember, by the word "waters" Hooke does not mean only $H_2O$, but all kinds of liquids and fluid substances, especially hot ones.*

*aqua excreverint herbæ mox ac extirpatæ fuerint, Lapidescant; quæcunq; vero extra-aquam in campis patentibus excreverint herbæ, ist as extirpatas nunquam Lapidescere.* I pass by his Reasons and Explications, because I think them very little to the Purpose: But the Observations themselves are very considerable, and serve for the explaining of several Phenomena I have observ'd in petrify'd Bodies, as I shall indeavour hereafter to shew, as in Corals, both Red, White, and the several Rarities of them, in Coral-[p.296] lines also, and petrify'd Mushromes, of each of which I have examined a very great variety. But this only by the by.

6*thly*. Therefore 'tis observable, that these petrifying Waters are for the most part very clear and limpid; so that to the Sight 'tis not distinguishable from other Water: But only by the Effects, and therefore by the newly mention'd Observations of *Kircher*, we find that Vegetables, that upon dry-ing turn'd into Stone; whilst green and growing flourished and spread faster than others; so that the petrifying Substance past through the finest and closest Pores of the living Vegetables, and therefore must certainly be very intimately mixt with the Water that could not be separated by so fine and curious Strainers.

But 7*thly*. To confirm this Proposition yet further, there are found in several parts of the Earth, such Waters will be intirely converted into Stone. Of this kind there are several Histories in the newly-mention'd Book, which I pass over, and shall only take notice of one for all, and that is in an Account sent to the *Roman* Coledge of Jesuits from the Masters, Surveyors and Clerks of the *Hungarian* Mines, in Answer to some Queries propounded to them. Page 183. of *Kircher's Mundus Subterraneus*, to the Query concerning the Properties and Metallick Experiments about Meneral Waters, they answer, That *Datur in fodinis aquæ genus quod in Figuram saccaro haud absimilem degenerat, viz. in Lapillos albas.*

*Hooke quotes from various ancient texts to corroborate his hypothesis that sediments can be consolidated into stone with petrifying waters and that minerals precipitate from fluids.*

And again, Page 185. of the fame, from another Prefect of the Emperial Mines in *Hungary* in answer to the same Query, we have this Account. *Reperitur quoq; aqua quædam alba quæ in Lapidem durum abit. Si vero hæc aqua ante suam coagulationem mineram cupream transiverit, tunc generatur ex ea lapis qui Malochites vocatur, quando vero aqua illa perfluit cupream mineram continentem argentum fiet ex ea pulcher lapis ceruleussimilis Turcoidi. Hæc aqua autem nullibi frequentius reperitur quam in mineris Lapidibus siliceis copiosis, & cuprum cum argento continentibus.* Whence I am apt to think, and I have many Observations and Arguments to confirm my Conjecture,

That 8*thly*, All kinds of *Talk* and *Spar*, most *Ores* and *Marchasites, Alumen Plumeum, & Asbestus; Fluores, Crystalls,* Cornish-*Diamonds, Amethysts* and divers other figured Mineral Bodies, may be generated from their Crystalization, or Coagulation, out of some Mineral Waters.

*Many kinds of crystal minerals are generated from some fluid substance.*

And to make it yet more probable, I could in the 9th place add divers Experiments, by which several of these Concretes may be in a short time made artificially by several Chymical Operations, which would very much illustrate the former Doctrin. But I hope what I have mention'd may suffice to make the fourth Proposition probable, that Waters of divers kinds may be turned in time to Stone, without being reducible again to Water by any Art yet commonly known, which being granted, my

*"Waters" of various kinds, therefore, may in time be turned into stone which cannot again return to its liquid state again by the known technology of the time.*

Fifth Proposition will follow of consequence, *viv.* That divers other fluid Substances, have, after long continuance of rest, settled and congealed

*Arguments for the 5th Proposition.*

into much more hard and permanent Substances: For if Water it self may be so changed and metamorphosed, which seems the farthest removed from the nature of a solid Body, certainly those which are nearer to that Nature, and are mixt with such Waters, will more easily be coagulated: I shall not therefore any farther insist on the Proof of this, than only to mention two Particulars, and that because we have almost every where so many Instances and Experiments; and the first is that of *Pliny* in the 13th Chap. of the 35th Book of his Natural History, in all which Chapter he gives us divers Instances of several kinds of Earth, which, by the Sea-water and Air, converted into solid and hard Stones; his Words are these: *Verum & ipsius Terræ sunt alia segmenta. Quis enim satis miretur pessimam ejus partem ideoq; pulverem appellætam in puteolanis collibus oppone maris fluctibus, mersamq; protinus fieri lapidem unum inexpugnabilem undis, & fortiorem quotidie, utiq; si cumano misceatur Cæmento.* [p.297]*Eadem est Terræ Natura & in Cizicena Regione, sed ibi non pulvis verum ipsa Terra, qualibet magnitudine excisa & demersa in mare, lapidea extrahitur: hoc idem circa Cassandriam produnt fieri: Et in fonte Gnidio dulci intra octo menses Terram lapidescere. Ab Oropo quidem Aulidem usque quicquid Terræ attingitur mari, mutatur in Saxa, &c.* to the end of the Chapter he goes on to relate divers Places where Earths, &c. are turned into Stones. Also in the 10th Chapter of the 31st Book, speaking of the Nature and Kinds of Niter, he tells about the middle of the Chapter. *Nitrariæ egregiæ Ægyptiis nam circa Naucratim & Memphim tantum solebant esse, circa Memphim deteriores; nam & Lapidescit ibi in acervis, multiq; sunt Tumuli ea de causa Saxei, fiuntq; ex his vasa, &c.*

The Second is an Observation of my own, which I have often taken notice of, and lately examined very diligently, which will much confirm these Histories of *Pliny*, and this my present Hypothesis; and that is a Part of the Observation I have already mentioned, which I made upon the Western Shore of the Isle of *Wight*. I observed a Cliff of a pretty height, which, by the constant washing of the Water at the bottom of it, is continually, especially after Frosts and great Rains, foundering and tumbling down into the Sea underneath it. Along the Shore underneath this Cliff, are a great number of Rocks and large Stones confusedly placed, some covered, others quite out of the Water; all which Rocks I found to be compounded of Sand and Clay, and Shells, and such kind of Stones, as the Shore was covered with. Examining the Hardness of some that lay as far into the Water as the Low-Water-mark, I found them to be altogether as hard, if not much harder than *Portland* or *Purbeck*-stone: Others of them that lay not so far into the Sea, I found much softer, as having in probability not been so long exposed to the Vicissitudes of the Tides: Others of them I found so very soft, that I could easily with my Foot crush them, and make Impressions into them, and could thrust a Walking-stick I had in my Hand a great depth into them: Others that had been but newly foundered down, were yet more soft, as having been scarce wash'd by the Salt Water. All these were perfectly of the same Substance with the Cliff, from whence they had manifestly tumbled, and consisted of Layers of Shells, Sand, Clay, Gravel, Earth, &c. and from all the Circumstances I could examine, I do judge them to have been the Parts of the Neigbouring Cliff foundered down, and rowl'd and wash'd by degrees into the Sea; and, by the petrifying Power of the Salt Water, converted into perfect hard compacted Stones. I have likewise since observed the like *Phænomena* on other Shores. And I doubt not but any inquisitive Naturalist may find infinite of the like Instances all along the Coast of *England*, and other

*Much like a modern geologist on a field trip, Hooke cites his own observations on the western shore of the Isle of Wight, his birthplace. He describes the erosion of the cliff shore and redeposition of the eroded materials into "layers." He concludes that these layers are made up of the materials that had been washed down into the sea from the cliff shore "by degrees" and then compacted into stone. His use of the term "by degrees" is significant here, as it shows that he is not limiting himself to catastrophic processes, but that his term "Earthquake" can also mean diastrophic movements and embody such slow processes as erosion and redeposition and range from these to the violence of earthquakes and volcanic eruptions. For a discussion of the geology of Isle of Wight, see Chapter 2.*

Countries where there are such kind of foundering Cliffs. I shall not now mention the great Quantities of toothed Spar, which I observed to be crystallized upon the sides of these Rocks, which seem'd to have been nothing else but the meer crystallizing or shooting of some kind of Water, which was press'd or arose out of these coagulating Stones; For the History of these kinds of figured Stones belong more preperly to another Discourse; namely, of the Natural Geometrical Figures, observable in Oares, Minerals, Spars, Talk, &c. of which elsewhere.

One Instance more I cannot omit, as being the most observable of any I have yet heard of; and that is, (Dr. *Castle*'s Relation) of a certain Place at *Alpsly* in *Bedforshire*, where there is a corner of a certain Field, that doth perfectly turn Wood and divers other Substances in a very short time into Stone as hard as a Flint or Agat. A Piece of this kind I saw, affirm'd to have been there buried, which the Person that buried it had shot small Shots of Lead into; the whole Substance of the Wood, Bark and Pith, together with the Leaden Shot it self was perfectly turn'd to a Stone as hard as any Agat, and yet retain'd its perfect Shape and Form; and the Lead remain'd round, and in its place, but much harder than any Iron. Of this I am promised a Sample, but have not yet receiv'd it.

But to spend no more time on the proof of that of which we have almost every where Instances, divers of which I have already mention'd, I shall proceed to the 6th Proposition; which is, That a great Part of the Surface of the [p.298] Earth had been since the Creation transform'd, and made of another Nature: that is, many Parts which have been Sea are now Land, and others that have been Land are now Sea; many of the Mountains have been Vales, and the Vales Mountains, &c.

For the proving of which Proposition, I shall not need to produce any other Arguments, besides the repeating what I find set down by divers Natural Historians, concerning the prodigious Effects that have been produced by Earthquakes on the superficial Parts of the Earth; because they seem to me to have been the chief Efficients which have transported these petrify'd Bodies, Shells, Woods, Animal Substances, &c. and left them in some Parts of the Earth, as are no other ways likely to have been the Places wherein such Substances should be produced; they being usually either raised a great way above the level Surface of the Earth, on the Tops of high Hills, or else buried a great way beneath that Surface in the lower Valleys: For who can imagine that Oysters, Muscles, and Periwinkles, and the like Shell-fish, should ever have had their Habitation on the Tops of the Mountain *Caucasus* ? Which is by divers of our Geographers accounted as high in its perpendicular Altitude, as any Mountain in the yet known World; and yet *Olearius* affords us a very considerable History to this purpose of his own Observation, which I shall hereafter have occasion to relate, and examine more particularly. Or to come a little nearer home, who could imagine that Oysters, *Echini*, and some other Shell-fish, should heretofore have lived at the tops of the *Alps, Appennine,* and *Pyrenian* Mountains, all which abound with great store of several sorts of Shells; nay, yet nearer at the tops of some of the highest in *Cornwal* and *Devonshire*, where I have been informed by Persons whose Testimony I cannot in the least suspect, that they have taken up divers, and seen great Quantities of them? And to come yet nearer, who can imagine Oysters to have lived on the Tops of some Hills near *Banstead-Downs* in *Surry*? Where there have been time out of Mind, and are still to this day found divers Shells of Oysters, both on the uppermost Surface, and buried likewise under the Surface of the Earth, as I was lately informed by several very worthy Persons living near those Places, and as I my self had the Opportunity to observe and collect.

To proceed then to the Effects of Earthquakes, we find in Histories Four Sorts or *Genus's* to have been performed by them.

The first is the raising of the superficial Parts of the Earth above their former Level: and under this Head there are Four Species. The 1st is the raising of a considerable Part of a Country, which before lay level with the Sea, and making it lye many Feet, nay, sometimes many Fathoms above its former height. A 2d is the raising of a considerable part of the bottom of the Sea, and making it lye above the Surface of the Water, by which means divers Islands have been generated and produced. A 3d Species is the raising of very considerable Mountains out of a plain and level Country. And a 4th Species is the raising of the Parts of the Earth by the throwing on of a great Access of new Earth, and for burying the former Surface under a covering of new Earth many Fathoms thick.

*The four effects of earthquakes are (1) raising a part of the earth's surface above its former level; (2) sinking a part below its former surface; (3) transpositions of parts—i.e., jumbling of the parts of the Earth together; and (4) chemical changes such as sublimations, petrifactions, and even liquefactions. The four "species" of raising are (1) raising above the former level; (2) raising the sea bottom; (3) raising into mountains; and (4) raising by accumulation of débris.*

A second sort of Effects perform'd by Earthquakes, is the depression or sinking of the Parts of the Earth's Surface below the former Level. Under this Head are also comprized Four distinct Species, which are directly contrary to the four last named.

The *First*, is a sinking of some Part of the Surface of the Earth, lying a good way within the Land, and converting it into a Lake of an almost unmeasurable depth.

The *Second*, is the sinking of a considerable Part of the plain Land, near the Sea, below its former Level, and so suffering the Sea to come in and overflow it, being laid lower than the Surface of the next adjoining Sea.

*The four "species" of sinking are (1) sinking within a continent to form a lake; (2) sinking of the land on the margin of the sea so that it is inundated by the sea; (3) sinking of the bottom of the sea; and (4) sinking by "throwing away" or washing away of the surface.*

A *Third*, is the sinking of the Parts of the bottom of the Sea much lower, and creating therein vast *Vorages* and *Abysses*.

[p.299] A *Fourth*, is the making bare, or uncovering of divers Parts of the Earth, which were before a good way below the Surface; and this either by suddenly throwing away these upper Parts by some subterraneous Motion, or else by washing them away by some kind of Eruption of Waters from unusual Places, vomited out by some Earthquake.

A Third sort of Effects produced by Earthquakes, are the Subversions, Conversions, and Transpositions of the Parts of the Earth.

A Fourth sort of *Effects,* are *Liquefaction, Baking, Calcining, Petrifaction, Transformation, Sublimation, Distillation,* &c.

The First therefore of the Effects of Earthquakes, which I but now named, was, that divers Parts of the Surface of the Earth which lay before, either below or level with the Sea, have been raised a good height above that Level by Earthquakes. Of this *Pliny* gives us several Instances in the 85th Chapter of the 2d Book of his Natural History, *Eadem nascentium Causa terrarum est, cum idem ille Spiritus attollendo potens solo non valuit erumpere. Nascuntur enim nec fluminum tantum invectu sicut Echinades Insulæ ab Acheloo amne congestæ; majorq; pars Ægypti a Nilo,*

*Corroboration of the first "species" of raising—i.e., to a level above sealevel. Hooke often quotes classical writings as corroboration of his own theses.*

*in quam a Pharo insula noctis & Diei cursum fuisse Homero credimus: Sed & Recessu Maris sicut eidem de circeiis. Quod accidisse et in Ambraciæ portu decem Millium passuum intervallo, & Atheniensium quinq; Millium ad Piræum memoratur: Et Ephesi ubi quondam ædem Dianæ alluebat. Herotodo quidem si credimus, mare fuit supra Memphin usq; ad Æthiopum montes. Itemq; a planis Arabiæ. Mare et circa Ilium et tota Teuthrania quaq; campos instulerit.*

*Meander*, and *Sandys* also, in his Travels thro' *Italy*, and the Parts of the *Levant*, gives this Instance, *pag.* 277, speaking of the new Mountain, which was produced in the Kingdom of *Naples*, in the year 1538. *The Lake* Lucrinus, says he, *extended formerly to* Avernus, *and so unto* Gaurus, *two other Lakes; but is now no other than a little sedgy Plash, choaked up by the horrible and astonishing Eruption of the new Mountain, whereof, as oft as I think, I am apt to credit whatsoever is wonderful. For who in* Italy, says he, *knows not, or who elsewhere will believe, that a Mountain should arise partly out of a Lake, and partly out of the Sea in one Day and a Night, to such a height, as to contend in Altitude with the high Mountains adjoining.*

*In the Year of our Lord* 1538 *on the 29th of* September, *when for certain Days foregoing, the Country thereabouts was so vext with perpetual Earthquakes, as no one House was left so intire, as not to expect immediate Ruine, after that the Sea had retired* 200 *Paces from the Shore, leaving abundance of Fish with Springs of Fresh Water rising at the bottom, this Mountain visibly ascended about the second Hour of the Night,* and so forwards. And again, *pag.* 281, speaking of the same Place, he says, *The Sea was accustomed, when urged with Storms, to flow in thro' the Lake,* Lucrinus *driving Fishes in with it; but now not only that Passage, but a Part of* Avernus *it self is choaked by the Mountain.* In which Histories I take notice only of these two Particulars at present. First, That that Part of the Land which lyes between *Lucrinus* and the Sea, that was oft-times before overflowed by the Sea, since this Earthquake, has been so far raised, as that now such Effects are no longer to be found. To confirm the rising of which the more, the other Circumstance of the Sea's departing from the Shore 200 Paces does much contribute. But not to insist on this, Mr. *Childry* in his *Britannia Baconica*, a Book very useful in its kind, being a Collection of All the Natural History of the Islands of *Great Britain*, to be met with in *Cambden*, or *Speed*, and some other Historians, together with such of his own as he had opportunity to observe, relates to us many considerable Passages to this purpose. In his History of *Norfolk*, he saith, That near St. *Benet's* in the *Holm*, are perfect Cockles and Periwinkles sometimes digg'd up out of the Earth, which makes some think it was formerly overflow'd by the Sea. The Fenny Grounds also of *Lincolnshire* and *Cheshire*, seem to have proceeded from the rising of the Ground; and those in *Anglesy*, where lopp'd Trees are now dug up with the perfect Strokes of the Ax remaining on them seem to have [p.300] been first sunk under Water, then overturn'd and buried in their own Earth, and afterwards the whole Earth seems to have been raised again to its former height. Of the raising of the Surface of the Earth by the overflowings and stopping of Rivers and Waters, I shall afterwards speak.

*Linschoten* gives us a Relation of the like Effects of an Earthquake that hapned in the *Terceras*. The Relation, as I find it epitomiz'd by *Purchas* in the 1677 Page of the 4th Part of his *Pilgrims*, is this: In *July, Anno*

---

*While a great deal of the descriptions he cites from old texts and writings of travelers are often exaggerations and anecdotal in nature, he believed they nevertheless provide some corroboration for his general propositions.*

*The abundance of these accounts by various authors concerning different parts of the world shows how widely read Hooke was and how he positively enjoyed his facility in the classical languages.*

*Unlike many of his contemporaries as well as later writers, Hooke was always meticulous in giving credit for his sources by naming the authors. The same cannot be said for either Newton in the 17th century or Hutton in the 18th century. Both Newton and Hutton took ideas from Hooke without crediting him. See text on the Hooke-Newton controversy and Chapter 8, "Final Assessment."*

1591. "there happen'd an Earthquake in the Island of St. *Michael*, which lyeth from *Tercera* South-East about 28 Miles, an Island 20 Miles long, and full of Towns, which continued from *July* 26 to *Aug.* 12. in which time none durst stay within his House, but fled into the Fields, fasting and praying with great Sorrow, for that many of their Houses fell down, and a Town, called *Villa Franca*, was almost razed to the Ground, all the Cloysters and Houses shaken to the Earth, and therein People slain. *The Land in some Places rose up*, and the Clifts removed from one Place to another, and some Hills were defaced and made even with the Ground. The Earthquake was so strong, that the Ships that lay in the Road, and in the Sea, shaked as if the World would have turn'd round. There sprang also a Fountain out of the Earth, from whence for the space of four Days there flow'd a most clear Water, and after that it ceased. At the same time they heard such Thunder and Noise under the Earth, as if all the Devils had been assembled together at that Place, wherewith many dy'd for fear. The Island of *Tercera* shook four times together, so that it seem'd to turn about; but there happen'd no other Misfortune unto it. Earthquakes are common in those Islands: For about 20 Years past there happen'd another Earthquake, when a high Hill that lyeth by the same Town *Villa Franca* fell half down, and covered all the Town with Earth, and killed many Men." I have transcribed here once for all the whole Relation, because there are many other considerable Circumstances in it besides the rising of the Earth, which I shall have occasion to refer to, under others of the Heads or Propositions to be proved, and therefore shall not need repetition. Two other Relations I find collected by *Purchas*, confirming this and several of the other Propositions: The one is that of *Dithmar Blefken*'s, in his History of *Island*, Page 648 of the 3d Part of his *Pilgrims*. "On the 29th of *November* about Midnight, in the Sea, there appear'd a Flame near *Hecla*, which gave Light to the whole Island: An hour after the whole Island trembled, as it would have been moved out of the Place: After the Earthquake follow'd a horrible Crack, that if all warlike Ordnance had been discharg'd it had been nothing to this Terror. It was known afterwards that *the Sea went back two Leagues in that Place, and remain'd dry.*"

*More histories relating incidents of earthquakes, receding of the sea, rising of the ground, etc.*

A Second History *Purchas* has collected out of the History of *Joseph Acosta* of the *West Indies*, Page 940 of the 3d Part: omitting for the present divers other Circumstances he takes notice of, I shall only mention that of the receding of the Sea. "Upon the Coast of *Chile*, (says he) I remember not well in what Year, there was so terrible an Earthquake, as it overturn'd whole Mountains, and thereby stopt the Course of Rivers, which it converted into Lakes: It beat down Towns, and slew a great number of People, causing the Sea to leave her Place some Leagues, so as the Ships remain'd on dry Ground far from the ordinary Road, &c." An Example somewhat like this happen'd lately in the *East-Indies*, as I was inform'd by a Letter sent thence to Mr. *D.* on *London-Bridge*. The thing in short was this: At a Place, about 7 Days Journey from *Ducca*, the Earth trembled about 32 Days; and the Sequel was, that it raised the bottom of a Lake, so as to drive out all the Water and Fish upon the Land, so that a Place which was formerly a Lake is now dry Ground. This was written from *Ballasore, Jan.* 6, 1665. The Words of the Letter I shall give afterwards.

The second Species of Effects of Earthquakes, is the raising of a considerable Part of the bottom of the Sea, and making it lye above the Surface of the Water, by which means divers Islands have been generated. Of this *Pliny*, in the 86th and 87th Chap. of the 2d Book of his Nat. Hist.

*Corroboration of the 2nd "species" of raising: i.e., the raising of a part of the sea bottom above sealevel.*

gives us several Instances. *Nascuntur*, says he, *& alio modo Terræ*, (having in the prece-[p.301]ding Chapter spoken of the Shore's rising above the Water, or the Water's deceding from the Shore, *ac repente in alto mari emergunt, veluti paria secum fasciente Natura, quæque hauserit hiatus alio loco reddente. Claræ jam pridem Insulæ Delos & Rhodos memoria produntur enatæ. Postea minores, ultra Melon Anaphe,* (of which *Strabo* makes mention in his Tenth Book.) *Inter Lemnum & Hellespontum Nea. Inter Lebedum & Teon, Alone: inter Cycladas, Olympiadis* cxxxv *ann.* 4to *Thera & Therasia. Inter easdem post ann.* cxxx *Hiera: & ab ea duobus Stadiis post ann.* cx *in Nostro ævo Thia.* Two of which Histories are also confirm'd by *Seneca*, in the Sixth Book of his Natural Questions and twenty first Chapter, where explicating the effects of Earthquakes by the commixture of Fire and Water, he says, *Theren & Therasiam & banc nostræ ætatis insulam, spectantibus nobis in Ægeo mari enatam quis Dubitat quin in lucem Spiritus vexerit. Sandis* speaking of the *Jolian* Islands, saith, "Of those there were only Seven, now there are Eleven in Number, which heretofore all flamed, now only *Vulcano* and *Strombylo*, two of that Number do burn. *Vulcano* is said to have first appear'd above Water about the time that *Scipio Africanus* died. But we have much later Instances to confirm this our Assertion: for about twenty-eight Years since, an Island was made among the *Azores* by an Eruption of Fire; of which divers have related the Story. But *Kircher* in his *Mundus Subterraneus*, from the Relation of the Jesuits, has added the most particular one. Having spoken of the exceeding height of the Pike of *Teneriff* in the *Canaries*, and of the Eruptions of Fire in it, and the hot Springs found about it, he adds, that in the Azores also there are found places having almost the same Proprieties. The *Pico de Fayal de Santo Gregorio*, being almost of equal hight, and St. *Michael*'s Island having heretofore had several Vulcans, and having been troubled with many Earthquakes, and very notably about thirty eight Years since, wherein all the Island was so terribly shaken, that the utter Ruin and Submersion of the whole was feared. The History of which, in short, is this; That "*June* 26. 1638. the whole Island began to be shaken with Earthquakes for eight days, so that the Inhabitants left Cities, Castles and Houses, and dwelt in the Fields, but especially those of a Place call'd *Vargen*, where the Motion was more violent. After which Earthquake, this Prodigy followed; At a place of the Sea, where Fishermen us'd to fish in Summer, because of the great abundance of Fish there caught, call'd *La Femera*, about 6 Miles from *Pico Delle Carmerine*, upon the first *Sunday* in *July*, a subterraneous Fire, notwithstanding the weight and depth of the Sea in that Place, which was 120 Foot, as the Fishermen had often before that found by sounding, and the multitude of Waters which one would have thought sufficient to have quenched the Fire: A subterraneous Fire, I say, broke out with a most unexpressible violence, carrying up into the Clouds with it Water, Sand, Earth, Stones, and other vast great bulks of Bodies; which to the sad Spectators, at a distance, appear'd like Flocks of Wool or Cotton, and falling back on the Surface of the Water look'd like Froth. The Space of this Eruption was about as big as a Space of Land, that might well be sown by two Bushels of Grain. By great Providence the Wind blew from the Land; otherwise the whole Island would, in all probability, have perished by the merciless Rage of these devouring Flames, such vast bulks of Stone were thrown up into the Air, about the height to seeming of three Pikes Lengths, that one would rather think them Mountains than Rocks. And which added further Horror to this dreadful Sight, was, that these Mountains returning again, often met with others ascending or being thrown up, and were thereby dasht into a 1000

*The verisimilitude of these narrations, that some islands are born from the bottom of the sea, is strengthened by the telling of the same incidents by different authors, and Hooke cites many of these from his extensive reading.*

*The Seed Certification Department at Oregon State University tells me that the "space" of this land is about one acre. It takes 120 pounds of wheat or 64 pounds of oats to sow an acre. Both come to approximately two bushels.*

Pieces; divers of which Pieces being afterwards taken up and bruised, easily turn'd into a black shining Sand. Out of the great multitude and variety of these vast rejected Bodies, and the immense heaps of Rocks and Stones, after a while was form'd a new Island out of the main Ocean, which at first was not above 5 Furlongs over; but after a while, by daily accesses of new Matter, it increased after 14 Days to an Island of 5 Miles over. From this Eruption, so great a quantity of Fish was destroy'd and thrown upon the next adjoining Island, that 8 of the biggest *Indian* Galeons would not be sufficient to contain them; which the Inhabitants fearing, lest the Stink of them [p.302] might create a Plague, for 18 Miles round collected and buried in deep Pits. The Stink of the Brimstone was plainly smelt at 24 Miles distance." Thus far he. But we have one Instance more of the Generation of an Island out of the bottom of the Sea, by an Eruption; which because it happen'd very lately, namely in 1650, and near an Island in the *Archipelago*, which *Pliny* relates to have been heretofore after the same manner produced, I shall in short relate, as it is more largely recorded by *Kircher*, in his *Mundus Subterraneus*, from the Mouth of Father *Franciscus Riccardus*, a Jesuit, who was at the same time in the adjoining Island, and was an Eye-witness of all the *Phænomena*.

*A furlong is 220 yards or one-eighth of a mile, so the island grew from five-eighth mile in length to over 5 miles in 14 days.*

"From the 24th *September* to the 9th of *October*, 1650, the Island of *Santerinum*, formerly call'd by *Pliny Thera*, was dreadfully shaken with Earthquakes, so that the Inhabitants expected nothing but utter ruine; and were yet more amazed by a horrid Eruption of Fire out of the bottom of the Sea, about 4 Miles to the Eastward of the Island: Before which the Water of the Place was rais'd above 30 Cubits perpendicularly, (I suppose he means as to appearance from the Island, otherwise 'tis but very little) which Wave spreading it self round every way, overturn'd everything it met, destroying Ships and Galleys in the Harbour of *Candie*, which was fourscore Miles distant. The Eruption fill'd the Air with Ashes and horrible sulphureous Stinks, and dreadful Lightnings and Thunders succeeded. All things in the Island were covered with a yellow sulphureous Crust, and the People almost blinded as well as choak'd. Multitudes of Pumice, and other Stones were thrown up, and carried as far as *Constantinople*, and to Places at a very great distance. The Force of this Eruption was greatest the two first Months, when all the Neighbouring Sea seem'd to boil, and the *Vulcan* continually vomited up Fire-balls. Upon the turning of the Wind, great Mischief was done in the Island of *Santerinum*, many Beasts and Birds were kill'd: And on the 29th of *October*, and 4th of *November*, about 50 Men were kill'd by it. The other four Months it lasted, tho' much abated of its former Fierceness, yet it still cast up Stone, and seem'd to endeavour the making of a New Island; which though it do not yet perfectly appear above Water, yet 'tis cover'd but 8 Foot by the Water; and the bubbling of the Water seems to speak another Eruption, that may in time finish Nature's Birth." And in the Year he writ this, which he says was 1656, there was an extraordinary boiling of the Sea, and an Eruption of Smoke. And though our Natural Historians have been very scarce in the World, and consequently such Histories are very few; yet there has been no Age wherein such Historians have liv'd, but has afforded them an Example of such Effects of Earthquakes. And I doubt not, but had the World been always furnisht with such Historians as had been inquisitive and knowing, we should have found not only *Thera* or *Santerinum*, and *Volcano* and *Delos*, and that in the *Azores*, and one lately in the *Canaries*, but a very great part of the Islands of the whole World to have been rais'd out of the Sea, or separated from the Land by Earthquakes: for which Opinion I shall

*A cubit is between 17 and 20 inches. so 30 cubits would be approximately 5 feet. Hooke considers this amount slight for the water to rise but supposes Pliny meant as it appeared from the island.*
*fourscore = 80*

afterwards relate several Observations both of my own and others, which seem to afford probable Arguments.

But to proceed to the third Kind or Species of Effects produced by Earthquakes, which is the raising very considerable Mountains out of the Plains. Of this I shall add a few Instances; but none more notable, than that of the new Mountain near *Naples* of which I said somewhat before out of *Sandys*'s Travels. In the Year 1538. *Septemb.* 29. this Mountain visibly ascended about the 2d hour of the Night, with a hideous roaring, horribly vomiting Stones, and such store of Cinders, as overwhelm'd all the Buildings thereabout, and the salubrious Baths of *Tripergula*, for so many Ages celebrated, consuming all the Vines to Ashes, and killing Birds and Beasts, and frighting away all the Inhabitants, who fled naked and defiled through the dark: And has advanced its top a Mile above the Basis: the Stones of it are so light and pory, that they will not sink when thrown in the Sea. This new Mountain, when new rais'd, had a number of Issues, at some of them smoking, and sometimes flam-[p.303]ing; at others disgorging Rivulets of hot Water, keeping within a terrible rumbling; and many perished that ventured to descend into the hollowness above. But that hollow at the top is at present an Orchard, and the Mountain throughout bereft of its Terrors. "It is reported, saith *Childrey*, that in a Parish by the Sea-side, not far from *Axbridge* in *Somersetshire*, within these 50 Years, a Parcel of Land swell'd up like a Hill; but on a sudden clave asunder, and fell down into the Earth, and in the place of it remains a great Pool." Our English Chronicles say, at *Oxenhal*, in the Bishoprick of *Durham*, on *Christmas* Day 1679, the Ground heav'd up aloft like a Tower, and continued all that day immoveable, till Evening, and then fell with a horrible noise, sinking into the Earth, and leaving three deep Pits, call'd Hellkettles. *Varenius* tells us of a new Mountain likewise raised in *Java*, in the Year 1586, with the like Effects of those I formerly named of the new Mountain; first shaking the Earth, then heaving up and throwing up into the Air the upper Parts of the Earth, afterwards the Rock and inner Parts, then fiery Coals and Cinders, overwhelming the circumjacent Fields and Towns, and killing above 10000 Men, and burning what was not overwhelmed. I have not time to reckon up the multitude of Instances I have met with in Authors; such as *Ætna* in *Sicily*, *Vesuvius* in *Italy*, one in *Croatia*, near the City *Valonia*, the *Pike* in *Tenarif*, and the *Pike* in the *Azores*, *Hecla*, *Helga*, and another in *Island*: The Mount *Gonnapi* in one of the Islands of *Banda*, which made an horrid Eruption at the same time with that in *Java*: The Mount *Balavane* in *Sumatra*: Others in the *Molucca* Islands, in *China*, *Japan*, and the *Philippines* , and in some of the *Maurician* Islands, and several other Parts of the *East Indies*. In the *West Indies* also we have multitudes of Examples, several in *Nicaragua*, and all along the Ledge of Mountains in *Peru* and *Chile*, and in *New Spain* and *Mexico* : In the Islands of *Papoys*, discover'd by *Le Mair*, joining to the South Continent in *Mar Del Zur*: All which are as so many shining Torches to direct us in the search after this Truth. There are many other Instances of Mountains, that have but lately as it were left to burn, and are cover'd with Wood and grown fruitful. So the new Mountain I formerly mention'd, has an Orchard growing where the Fire at first flamed. Another in the Island *Quimeda*, near the River *Plat* in *Brasill*: The Islands also of St. *Helena*, and *Ascension*, discovered by the great plenty of Cinders, and the Fashions of the Hills to have formerly contained *Vulcanoes*, and probably were at first made by some subterraneous Eruption, as indeed most of those Islands in the main Ocean; such as the *Canaries*, and the

*Corroboration of the 3rd species of raising: raising mountains out of plains.*

*This passage on the lightness and porosity of the stones thrown out must be one of the earliest describing the floating property of pumice.*

*This date of 1679 is wrong, as this series of discourse "Ended Sept. 15, 1668." This event happened in 1179, as Hooke relates later.*

*Tenarif = Tenerife, an extinct volcano in the Canary Islands*

*Azores*, and the *East Indian,* and the *Cariby* Islands and divers others seem to have been. A Passage, to make this Assertion somewhat more probable, I have met with in *Linschoten*'s Description of the Island of *Tercera*, which as *Purchas* has epitomized I have here added. Pag. 1670. of the 4th Part of his Pilgrims (he saith, speaking of the Island of *Tercera*) "The Land is very high, and as it seemeth hollow; for that as they pass over an Hill of Stone, the Ground soundeth under them as if it were a Cellar. So that it seems in divers Places to have holes under the Earth, whereby it is much subject to Earthquakes, as also all the other Islands are; for there it is a common thing: and all those Islands, for the most part, have had Mines of Brimstone; for that in many Places of *Tercera* and St. *Michael*, the Smoke and Savour of Brimstone doth still issue out of the Ground, and the Country round about is all singed and burnt. Also there are Places wherein there are Wells, the Water whereof is so hot that it will boil an Egg, as if it were over a Fire." Besides which, the shape of the Hills, and several other Circumstances mention'd in *Linschoten*, do make it probable that those have been all *Vulcano's*.

*It is clear from this passage that Hooke includes volcanic eruptions as part of his generic term of "earthquake."*

But to proceed to the Fourth Species of Effects of Earthquakes under this Head; and that is, the raising of the Parts of the Earth by the throwing on a great access of new Earth: Of this I have already given many Instances in the newly mentioned Histories of Eruptions, where I mentioned the overwhelming of Fields, Towns, and Woods, and the like, by Materials thrown out by these Eruptions. I shall only add one Instance or two more to confirm [p.304] this Head, and then proceed. The first is that mentioned by *Olaus Wormius*, in the 5th Chapter of the 1st Section of the 1st Book of his *Musæum*, wherein he gives an Account of an extraordinary Earthquake in *Iceland,* which fill'd the Air with Dust, Earth, and Cinders, and overwhelmed Towns, Fields, and even Ships a good way distant on the Sea; and which sent forth its Fumes with such violence and Plenty, as covered all the Decks and Sails of Ships lying on the Coast of *Norway*, some hundred Leagues distant. His Words are Page the 18th thus, *Alterum portentosæ Terræ genus, &c.* And to make this of *Wormius* the more probable, I have now by me a Paper of Dust, which was rained out of the Air upon a Ship lying at *Algiers* upon the Coast of *Barbary*, upon a great Eruption of *Vesuvius* in the Year 16--. The Relation of which, as I received it together with the Paper of Dust from that eminent Virtuoso, *John Evelyn*, Esq; I shall here annex. But which is beyond all, is the late Eruption of *Mongibell* or *Ætna*.

*Corroboration of the 4th species of raising: The four "species" of raising, therefore, are (1) raising to a level above sealevel; (2) raising from the bottom of the sea; (3) raising out of a plain; and (4) raising by throwing and heaping of earth to a place.*

*John Evelyn was indeed another of the 17th century virtuosi, but his story does not seem to be "annexed" here by Hooke. He gives several examples of massive ashfalls from volcanic eruptions.*

And to confirm this Proposition yet further, I cannot pass by a very remarkable Rain of Earth and Ashes, that happen'd in *Peru, Anno* 1600, mentioned by *Garcilasso De la Vega*, one of the Off-spring of the *Incas* of *Peru*, in his History of *America.* The Epitome of which by *Purchas*, is this, pag. 1476 of the 4th Part of his *Pilgrims.* "I might add, says he, the great Earthquakes, An. 1600, in *Peru* at *Arequepa*, the raining of Sand, as also of Ashes, about 20 days from a *Vulcan* breaking forth: The Ashes falling in Places above a Yard thick, in some Places more than two, and where least above a quarter of a Yard, which buried the Corn-grounds of Maize and Wheat, and the Boughs of Trees were broken and fruitless, and the Cattel great and small dy'd for want of Pasture. For the Sand which rained covered the Fields 30 Leagues one way, and above 40 Leagues another way, round about *Arequepa*, they found their Kine dead by 500 together in several Herds, and whole Flocks of Sheep, and Herds of Goats

*"kine" = the archaic word for cows or cattle*

and Swine buried. Houses fell with the weight of the Sand; others cost much Industry to save them; mighty Thunders and Lightning were heard and seen 30 Leagues about *Arequepa*. It was so dark whilst those Showers lasted, that at mid-day they burned Candles to see to do business.---" I could add divers other Instances to confirm this Proposition; but these may at present suffice.

But this is but one way by which divers things have been buried: there is another way which I can only at present mention, and must refer the Probation and Prosecution to some other occasion; and that is, that very many of the lower superficial Parts of the Earth, have been and continually are covered and buried by the access of Matter, tumbled and washed down by Excesses of Wind and Rain, and by the continual sweepings of Rivers and Streams of Water. Under this Head, I shall shew several Places and Countries in the World, that are nothing else but the Productions of these Causes. To this purpose, *Peter de la Valle* gives some Observations which he made in *Egypt*, in the 11th Letter dated from *Grand Caire, Jan.* 25. 1616. "Of the former seven Mouths of *Nile* (says he) there are only four left, and of those but two Navigable; the rest are either fill'd, or run no more, or are small Streams not taken notice of, or only Torrents in the time of great Rains; but I could learn nothing of them, because the great Expence of the Ancients for cleansing the Ditches, has been intermitted for several hundreds of Years." He is likewise of Opinion with *Herodotus*, that the *Delta*, and all the Lower *Egypt*, where the *Greeks* navigated in his time, was in the first Ages of the World made by the Sand and Mud of *Nile*.

*Here again Hooke includes the process of river erosion and piling up of sediments at the mouth of rivers as part of the process of moving of earth from one place to another. The cyclic process of erosion and deposition is elaborated more fully later in these lectures*

All which Histories and Particulars do manifestly enough evince, that there have been in very many Parts of the World considerable Mutations of the superficial Parts, since the beginning; and that therefore those Places where these figured petrify'd Bodies are found; though they now seem never so much foreign, and differing from the likely native Places of such animated Bodies, may notwithstanding heretofore have been in such another kind of condition, as was most sutable to the breeding and nourishing of them: Which I shall yet further manifest, by comparing the other Effects produced by Earthquakes; such as the sinking, and burying, and transposing, and overturning of the superficial Parts of the Earth.

*Hooke may seem to be belaboring the point that the surface of the earth has undergone great changes, but when one considers the prevailing thought of the time (which emphasized the permanence of terrestrial features at least since Noah's flood), one can understand why Hooke seems to be showering his audience with one example after another.*

[p.305]Another Sort of Effects, is the sinking of the superficial Parts of the Earth, and placing them below their former Position, both in respect of some Parts newly raised, and in respect of some other adjacent Parts not displaced. And this seems to be caus'd by the subsiding or sinking of those Parts into such Caverns, as by the strength of the Eruption passing below before it breaks out are made underneath: For so great is the Violence of these subterraneous Fires, that nothing almost is able to resist their Power of expanding; but spreading themselves, and rushing that way which is most easy, they carry along before them Earth, Sand, and Rocks, and Mountains, and whatever lies in their way, and raise the superficial Parts of the Earth whilst they pass underneath. And if the Parts of the Earth underneath are so loose or obnoxious to the Force of the Fire, as to be dislodged, unless the remaining Parts are very strong, and constitute a very firm Stony Arch, the Earth does easily tumble into the Holes and Hollows made by the Fire. Now it cannot be imagin'd but that all those vast Congeries of Earth, which I have already mention'd to have been thrown up, and to create new Islands and new Mountains, and the like, must leave vast Caverns below them, to be fill'd either with the Parts of the Earth that

*After expounding on the different ways of raising of the land, Hooke now turns his attention to the 2nd kind of effects of earthquakes, the various ways land can sink.*

hang immediately over them, or with the Sea, or other subterraneous Waters, if the Roofs of these Cavities be strong enough to sustain the Earth above them from sinking. And some such Power as these subterraneous Fires, seems to me to have been the Cause of the strange Positions and Intermixture of the Veins of Ores and Minerals in the Bowels of the Mountains, where, for the most part, they are now found; and even of bringing those Substances so near the Surface of the Earth, which, from the Consideration of very many Circumstances, seem to me to be naturally situated at a much greater Depth below within the Bowels of this Globe. And hence may be rendred a Reason of the Figures of these Minerals, and other Substances mix'd with them, and of the compounding and blending of several of those Substances together, whereby some of them are very strangely united and alter'd. But this I mention only by the Bye, and shall not insist on it, belonging more properly to another Head. To proceed then under this General Head, are comprised several Kinds of Effects, differing only according to the Parts of the Earth they have been wrought upon.

*As an aside, Hooke expresses his theory of the origin of ores and minerals.*

The first is, The sinking of several Inland Parts, which were before eminent, and laying them much lower into Vales. Sometimes, the sinking of a Part of the Earth to a very great Depth, and leaving behind, instead of a firm Ground, a Lake of Salt or Sea-water. Of these we have several Instances in Natural Historians. And, to pass by many others, I shall only mention such as have lately happen'd. Of this kind Mr. *Childrey*, in his *Britannia Baconica*, has collected several Instances; two out of our English Chronicle. His Relations are these, *Pag.* 62. "*August* the 4th, 1585. after a very violent Storm of Thunder and Rain, at *Nottingham* in *Kent*, Eight miles from *London*, the Ground suddenly began to sink; and Three great Elms growing upon it, were carried so deep into the Earth, that no Part of them could any more be seen. The Hole left (saith the Story) is in Compass 80 Yards about, and a Line of 50 Fathoms plummed into it finds no Bottom." Also,

*Corroboration of the first species of sinking: the sinking of parts that once were prominent into valleys, sometimes leaving behind a lake.*

"*Dec.* 18, 1596. a Mile and half from *Westram*, Southward (which is not many Miles from *Nottingham*) a Part of an Hedge of Ashes, 12 Perches long, were sunk 6 Foot and a half deep; the next morning 15 Foot more; the third morning 80 Foot more at least, and so daily. (And presently after, he says) Moreover, in one Part of the Plain Field, there is a great Hole made by sinking of the Earth, to the Depth of 30 Foot at least, being in Breadth in some Places 2 Perches over, and in Length 5 or 6 Perches. There are sundry other Sinkings in divers other Places, one of 60 Foot, another of 47, and another of 34 Foot; by means of which Confusion it is come to pass, that where the highest Hills were, there be the lowest Dales, and the lowest Dales are become the highest Grounds, &c."

*A perch is a measure for stonework equivalent to 1 rod, which is 5.5 yards or 16.5 feet. 12 perches would be about 200 feet long.*
*2 perches is approximately 33 feet.*
*5 perches is approximately 82 feet.*

And again, *Pag.* 131. he gives an Instance, upon his own Knowledge, much to the same purpose, which lately happen'd; namely,"*July* the 8th 1657[p306] about 3 of the Clock, in the Parish of *Bickly*, was heard a very great Noise like Thunder afar off; which was much wonder'd at, because the Sky was clear, and no Appearance of a Cloud. Shortly after (saith the Author of this Relation) a Neighbour came to me, and told me, I should see a very strange thing if I would go with him. So coming into a Field, called the *Lay-field*, we found a very great Bank of Earth, which had many tall Oaks growing on it, quite sunk into the Ground Trees and all. At first we durst not go near it, because the Earth, for near 20 Yards about, was

*More examples of sinking.*

exceedingly much rent, and seem'd ready to fall: But since that time, my self and some others have ventured to see the bottom, I mean to go to the Brink, so as to discern the visible Bottom, which is Water, and conceived to be about 30 Yards from us; under which is sunk all the Earth about it, for 16 Yards round at least, 3 tall Oaks, a very tall Awber, and certain other small Trees, and not a Sprig of them to be seen above Water. 4 or 5 Oaks more are expected to fall every moment, and a great Quantity of Land is like to fall, indeed never ceasing more or less; and when any considerable Clod falls, it is much like the Report of a Cannon. We can discern the Ground hollow above the Water a great Depth; but how far hollow or how deep, is not to be found out by Man. Some of the Water, (as I have been told) drawn out of this Pit with a Bucket, was found to be as salt as Sea-water, &c."

*In this narration of sinking, seawater is found to be at the bottom of the hollow.*

A considerable Circumstance also to confirm this Proposition, is a Passage in that History I have mention'd out of *Linschoten*, of the Island of *Tercera*; where he says [*and some Hills were defaced, and made even with the Ground.*]

*Kircher* in the Preface to his *Mundus Subterraneus*, Chap.2. tells us a very remarkable History of the sinking of a Town, and the Land about it, and the Generation of a Lake instead of it. *Contigit* (says he) *hac eadem hora res æterna ac immortali Memoria digna, subversio videlicet celeberrimi oppidi quod Sanctam Euphemiam dicunt; erat hoc in extrema Sinus ora situm sub equitum Melitensium Jurisdictione. Cum itaq; ad Lopicium ex vehementi Terræ subsultatione veluti exanimes in terra prostrati tandem subsidente Naturæ paroxysmo, oculis in circum jacentia Loca conjectis, ingenti nebula, paulo ante memoratum oppidum circumdatum vidissemus; ter sane post Meridiem, hora tertia præsertim Cælo sereno mira & insolita nobis videbatur. Dissipata vero paulatim nebula, oppidum quæsivimus sed non invenimus. Mirum Dictu, Lacu putidissimo in ejus Locum enato. Quæsivimus Homines qui de insolito rei eventu nonnihil certi nobis enarrare possent, sed formidabilis casus tantæq; stragis nuncium non reperimus, &c.---- Nos itineri insistentes Nicastrum, Amanteam, Paulam, Belviderium transeuntes nil aliud ad 200 Millia passuum nisi cadavera Urbium, castallorum, strages horrendas reperimus, Hominibus per apertos campos palantibus & prætimore veluti exarescentibus.* That is, "At this very time happened a thing worthy never to be forgotten, *viz.* the Subversion of the most famous Town, call'd St. *Euphemia*: 'twas situated at the side of the Bay under the Jurisdiction of the Knights of *Malta*. When therefore we had come to *Lopiz*, almost dead from the vehement shaking of the Earth, and lying prostrate on the Ground, at last the *Paroxysm* of Nature remitting, casting our Eyes towards the Neighbouring Places, we saw the forementioned Town encompassed with a great, wonderful, and unusual Cloud, which was seen by us three times, especially at Three-a-clock in the Afternoon, the Heavens being clear. This Cloud being, by degrees, dissipated, we look'd for the Town, but found it not, a stinking Lake (to our wonder) appearing in the Place of it. We sought for some Person or other, to give us some certain Account of this unusual Event; but could not find one to tell any News of this dreadful Accident and great Destruction, &c. We prosecuting our Journey, passing by *Nicastrum, Amantea, Paula,* and *Belvedere,* found nothing for 200 Miles, but the remaining Carcasses of Cities and Castles, and horrid Destructions; the Men lying in the open Fields, and, as it were, dead and withered through Fear and Terror."

*More examples from Kircher's* Mundus Subterraneus.

*Hooke translates this passage from Kircher describing the sinking of the town of St. Euphemia.*

To this purpose, give me leave to adjoin an Extract of a Letter, sent from *Balasore* in the *East Indies, Jan.* 6. 1665. "The same Star appeared in our Horizon, about the same time 'twas seen with you. The Effects in part [p.307] have already been felt here by unseasonable Weather, great Mortalities amongst the Natives, *English*, and others. We have had several Earthquakes unusual here, which with hideous Noises, have in several Places broke out and swallow'd up Houses and Towns. But about 7 Days Journey from *Ducca*, where were at that time 3 or 4 *Dutch*, they and the Natives relate, That in the Market-Place the Earth trembled about 32 Days and Nights, without Intermission. At the latter end, in the Market-place, the Ground turn'd round as Dust in a Whirlwind, and so continued several Days and Nights, and swallow'd up several Men who were Spectators, who sunk and turned round with the Earth, as in a Quagmire. At last, the Earth worked and cast up a great Fish bigger than hath been seen in this Country, which the People caught: But the Conclusion of all was, that the Earth sunk with 300 Houses, and all the Men, where now appears a large Lake some Fathoms deep. About a Mile from this Town was a Lake full of Fish, which in these 32 Days of the Earthquake cast up all her Fish on dry Land, where might have been gather'd many, which had run out of the Water upon dry Land, and there died: But when the other great Lake appeared, this former dried up, and is now firm Land."

*Quotes from a letter from the East Indies relating another example of sinking of the earth and leaving behind a lake. The year 1665 here would really be 1666, as the old calendar ran from March to March.*

To the same purpose also we have several other Instances, some later and some nearer home. "Near *Darlington* (says *Childrey*, in his *Britannia Baconica*, speaking of the Rarities of the Bishoprick of *Durham*) are three Pits, whose Waters are warm (hot, says *Cambden*) wonderful deep, call'd Hell-Kettles. These are thought to come of an Earthquake, that happen'd *Anno* 1179. For on *Christmas* Day, says our Chronicles, at *Oxenhall*, which is this Place, the Ground heaved up aloft like a Tower, and so continued all that Day, as it were immovable, till Evening, and then fell in with a very horrible Noise, and the Earth swallow'd it up, and made in the same Place 3 deep Pits." The same in the Section of *Brecknock*, says, "Two Miles east from *Brecknock*, is a Meer, called *Llinsavathan*, which (as the People dwelling there, say,) was once a City; but the City was swallowed up by an Earthquake, and this Water or Lake succeeded in the Place: The Lake is encompassed with high steep Hills, &c.---

*Examples from Childrey's Britannia Baconica.*

"Near *Falkirk*, saith *Lithgow*, remains the Ruines and Marks of a Town, &c. swallowed up into the Earth by an Earthquake, and the void Place is fill'd with Water.----*Pliny* also, in the 88th Chap. of his 2d Book of Nat. Hist. records a like Instance. *Mox & in his Montem Epopon cum repente flamma ex eo emicuisset campestri æquatum planitie. In eadem & oppidum haustum profundo alioq; motu Terræ Stagnum emersisse. Et alio provolutis Montibus insulam extitisse Prochytam, &c.* "Presently the Mountain *Epopon* (when suddenly a Flame had shon out of it) was levelled with the Plain; and in the same Plain a Town was swallow'd up into the Deep, and by another Motion of the Earth became a Lake. And in another Place, the Mountain being trumbled down, the Island *Prochyta* arose, &c."

*Examples from Lithgow and Pliny.*

The Dead Sea also in *Palestine*, was the Production of a most terrible Earthquake, and a Fire sent from Heaven: For, methinks, the Relation of the sad Catastrophe of those Four Cities, *Sodom, Gomorrha, Zeboim* and *Adma*, mentioned in Scripture, seem somewhat like that I have newly related out of *Kircher* of St. *Euphemia*. There are a multitude of other Instances which I could bring on this Head, of the sinking of Mountains

*The Dead Sea also was the result of a terrible earthquake. Here Hooke explains the fate of such sinful biblical cities as Sodom and Gomorrha by natural processes.*

and Hills into Plains, and all these into Lakes: Of which *Pliny* gives several Instances, in the 90, 91, and 92 Chap. of his Second Book. The *Pico* in the *Moluccas*, accounted of equal Height with that of *Tenariff*, was by a late Earthquake quite swallow'd into the Earth, and left a Lake in its Place. *Vesuvius* and *Strongylus*, are by late Earthquakes reduced to almost half their former Height. Many of those vast Mountains of the *Andes* in *Chile*, were by an Earthquake, *An.* 1646. quite swallow'd up and lost, as *Kircher* relates. I could add many Histories of the fatal Catastrophe's of many Towns, and other Places of Note; but these, I hope, may suffice to shew this kind also of Mutation in the superficial Parts of the Earth, to be effected by Earthquakes.

[p.308] Nor does Earthquakes only sink Mountains and Inland Parts; but such Parts also as are near to, equal with, and under the Surface of the Sea. Of this we have Instances near home, of *Winchelsea* and of the *Goodwin-Lands*, and of the Towns in *Freezland*, that have been about 400 Years since swallow'd up by the Sea; and nothing but some Towers, and the *Goodwin-Sands*, are now to be found of them. The like happen'd to several Parts of *Scotland,* as *Hector Boethius* relates. *Linschoten*, in his History of the *West-Indies*, relates among many other Histories of the Effects of Earthquakes, this considerable Passage. "Since, in the Year 1586. in the Month of *July*, fell another Earthquake in the City of *Kings*, the which, as the Vice-Roy did write, had run 170 Leagues along the Coast, and athwart in the *Sierra* 50 Leagues. It ruin'd a great Part of the City. It caus'd the like Trouble and Motion of the Sea, as it had done at *Chile*, which happen'd presently after the Earthquake; so as they might see the Sea to fly furiously out of her Bounds, and to run near 2 Leagues into the Land, rising above 14 Fathom. It cover'd all the Plain, so as the Ditches were filled and Pieces of Wood that were here, swam in the Water." There are multitudes of Instances of the like Effects in several other Parts of the World, which have been wrought by Earthquakes, which may be found in Natural Historians; which, for Brevity-sake, I omit, they serving only to prove a Proposition, which, I suppose, will be granted by any that have either seen or heard of the Effects of Earthquakes.

Now, though I find a general Deficiency in Natural Historians, of Instances to prove that the submarine Parts have likewise suffer'd the like Effects of sinking, they lying out of view, and so cannot without some Trouble and Diligence be observed; yet if we consider from how great a Depth these Eruptions proceed, and how little Distinction they make between Mountains and Plains, as to the weight of removing, we may easily believe, that the Bottom of the Sea is as subject to these Mutations, as the Parts of the Land. And since, by the former Relations, we have many Instances of the raising of the Bottom of the Sea, 'tis very probable that what Quantity of Matter is thrown to and raised in one Place, is sunk and falls into that Cavity left by another. An Island cannot be raised in one Place, without leaving an Abyss in another. And I do not doubt, but there have been as many Earthquakes in the Parts of the Earth under the Ocean, as there have been in the Parts of the Dry Land: But being, for the most part, till of late unfrequented by Mankind, and even now but very thinly, 'tis almost a 1000 to 1, that what happen are never seen, and a 100 to 1, if they have been seen, whether they be recorded: For how few Writers are there of Natural History? There is somewhat of Probability in the Story related by *Plato*, in his *Timæus*, of the Island *Atlantis* in the *Atlantick* Ocean, which he says was swallow'd up by an Earthquake into the Sea. And 'tis not unlikely, but that most of those Islands that are now appearing, have been either thrown up out of the Sea by Eruptions, such as the *Canaries, Azores,* St. *Helena,* &c. which the Form of them, and the Vulcanes in them, and the Cinders and Pumice-stones found about them, and the frequent Earthquakes they are troubled with, and the remaining Hills of extinguish'd Vulcanes, do all strongly argue for: Or else, that they are some of them at least some Relicts of that Great Island which is now not to be found; and yet we have no Records hereof. That there is as great Inequality in the Depth of the Sea, as there is in the Height of the Land, the Observations of Seamen, experimented by their Sounding Lines, do sufficiently inform us: For Hills, we have deep Holes; and for Mountains and Pikes, Abysses and Malstroons: And that these must have in all Ages been filling with Parts of the Earth, tumbled by the Motion of the Waters, and rowling to the lowest Place, is very probable; and so they would in time have been fill'd up, had not Earthquakes, by their Eruptions and Tumblings, created new Irregularities. And therefore that there are still such Places, is an Argument, that there have been of later Ages Earthquakes in some of them. Of these I shall mention one or two Instances, which I meet with in Voyages, and Relations of Travellers.

*Hooke's perceptive conclusion: that there is as great relief on the sea bottom as on land.*

[p.309] In the Relation of the Circum-navigation of Sir *Francis Drake,* speaking of the Straights of *Magellane,* he says, Pag. 35, "They saw an Island with a very high *Vulcano;*" and the next Page, he says, "They had need to have carry'd nothing but Anchors and Cables, to find Ground, the Sea was so very deep:" Which Depth is explain'd more express, Pag. 42. where 'tis said, "Being driven from our first Place of anchoring, so unmeasurable was the Depth, that 500 Fathoms would fetch no Ground." And in Page 99. of the same Relation, the Author tells, how their Ship struck upon a Rock, which Page 102. he says, at low Water was but 6 Foot under Water, and just by it no Bottom to be found, by reason of the great Depth.

*Quoting Sir Francis Drake regarding the Straits of Magellane and the great relief there.*

Mr. *Ricaut*, in a Letter of his to the Royal Society, dated from *Constantinople, Nov.* 1667, says, "That the Water runs out of the *Euxine* Sea into the *Propontis* with a wonderful swiftness, which is more wonderful in regard of the depth of the *Bosphorus* being in the Channel fifty or fifty five Fathom Water, and along the Land in most places the Ships may lye on the Shore with their Heads, and yet have twenty Fathom Water at their Sterns."

*Quoting Mr. Ricaut concerning the depth of the Bosphorus.*

Besides these effects of raising and sinking the parts of the Earth, there is a third sort, which is the transposing, converting, subverting and jumbling of the parts of the Earth together; overthrowing Mountains, and turning them upsidedown, throwing the parts of the Earth from one place to another, burying the superficial parts, and raising the Subterraneous. Of these kinds of changes there are many instances in the former Relations I have mention'd, as particularly that of *Linschoten* of the Earthquake in the *Terceras,* and that of *Josephus Acosta,* of the Earthquake upon the Coast of *Chile.* And there are a multitude of others I could here set down, but I shall only mention some of them. "Soon after, (says *Josephus Acosta,* in the same place I mentioned before) which was in the Year 1582, happened that Earthquake of *Arequipa,* which in a manner overthrew the whole City. And a little before in the same place, he tells of a terrible Earthquake in *Guatimala,* in the Year 1586, which overthrew almost all the City, and that the Vulcan for above six Months together continually vomited a Flood of Fire from the top of it. And a little after, the same Author, in the same

*The 3rd kind of Effects of Earthquakes: the jumbling of the parts of the Earth overthrowing mountains and raising the subterraneous parts; in his words, "the Subversions, Conversions, and Transpositions of the Parts of the Earth."*

place, says, "In the Year of our Lord 1581, in *Cugiano*, a City of *Peru*, otherwise call'd the *Pear*, there happen'd a strange accident touching this Subject; a Village call'd *Angoango* (where many *Indians* dwelt that were Socerers and Idolaters) fell sudenly to ruine, so as a great part thereof was raised up and carried away, and many of the *Indians* smothered; and that which seems incredible (yet testified by Men of Credit) the Earth that was ruined and so beaten down, did run and slide upon the Land for the space of a League and a half, as it had been Water or Wax melted, so as it stopt and fill'd up a Lake, and remain'd so spread all over the whole Country."

*This is possibly a description of the process known as liquefaction in which, as a result of shaking, each grain of sediment loses contact with each other grain and the whole mass behaves like a liquid.*

Nor are there wanting Examples of this kind even in this Island. Mr. *Childrey* in his *Britannia Baconica* has collected several out of *Cambden*; as that in *Herefordshire*, "Where, in the Year 1571, *Marcley* Hill in the East part of the Shire, with a roaring noise, remov'd itself from the place where it stood, and for three Days together travell'd from its old Seat. It began first to take its Journey *Feb.* 17. being *Saturday* at six of the Clock at Night, and by seven the next Morning it had gone forty Paces, carrying with it Sheep in their Cotes, Hedge-Rows, and Trees, whereof some were overturn'd, and some that stood upon the Plain, are firmly growing upon the Hill; those that were East were turned West, and those in the West were set in the East; in this remove it overthrew *Kinaston* Chappel, and turn'd two High-ways near a hundred Yards from their old Paths: The Ground that they remov'd was about twenty six Acres, which opening itself with Rocks and all bore the Earth before it for four hundred Yards space, without any stay, leaving Pasturage in places of the Tillage, and the Tillage overspread with Pasturage. Lastly, overwhelming its lower parts, it mounted to a Hill of twelve Fathoms high, and there rested after three Days travel." [p.310]

*Quoting Childrey regarding the displacement of a hill in Herefordshire in 1571.*

"At *Hermitage* in *Dorsetshire*, says *Stow* in his Summary, *January* the third 1582, a piece of Ground of three Acres remov'd from its old place, and was carried over another Close where Alders and Willows grew, the space of forty Rods or Pearches, and stopt up the high-Way that led to *Cerne*, a Market-Town, and yet the Hedges that it was inclosed with enclose it still, and the Trees stand bolt upright, and the place where this Ground was is left like a great Pit." And tis not a little observable, that at the same time that these changes happened in *America*, the like also happened in *England*, of which I shall hereafter give divers other Instances, and shall also deduce Corrolarys, that may otherwise seem very strange, and yet I question not to prove the truth of them. *Maximus* (says *Pliny*, Cap. 48. Lib. 2. Hist. Nat.) *Terra memoria mortalium extitit motus Tiberii Cæsaris principatu. XII. urbibus Asiæ una nocte prostratis.* "The greatest Earthquake that ever happen'd in the Memory of Man was in the Reign of *Tiberius Cæsar*, twelve Cities of *Asia* being thrown down by it in one Night." And again, (Cap. 83. ibid.) *Factum est semel* (says he) *quod equidem in Hetruscæ disciplinæ voluminibus inveni, ingens terrarum portentum L. Martio, Sex. Julio Coss. in Agro Mutinensi namq; montes duo inter se concurrerunt, crepitu maximo assultantes recedentesq; inter eos flamma fumoq; in cœlum exeunte inter diu, Spectante evia Æmilia Magna equitum Romanorum familiarumq; & viatorum multitudine: Eo concursu villæ omnes elisæ, animalia permulta quæ intra fuerant exanimata sunt, anno ante Sociale bellum. Quod haud scio an funestius ipsi terræ Italiæ fuerit quam Civilia. Non minus mirum ostentum & nostra cognovit ætas. Anno Neronis Principis Supremo, sicut in rebus ejus exposuimus, pratis oleisq; intercedente via publica in contrarias sedes transgressis, in Agro Marrucino*

*Quoting Stow about the displacement of a 3-acre piece of ground for 40 rods or perches.*

*Prædiis Vectii Marcelli Equitis Romani res Neronis Procurantis*. Thus English'd. "There happen'd once (which I found in the Books of the *Tuscane* Learning) within the Territories of *Modena, L. Martius* and *S. Julius*, being Consuls a great wonder of the Earth; for two Hills encountred each other charging one another with a great crash, and retiring again, a great Flame and Smoak in the Day-time issuing out from between them to the Sky, while a great many of the *Roman* Knights, their Friends and Travellers beheld it from the *Æmilian* Road. With this conflict and meeting together, all the Country Houses were dasht to pieces, many Animals that were between them perish'd. This happen'd before the *Social* War. I know not whether it were not more pernicious to *Italy* than the Civil-Wars. No less a wonder was that in our Age, in the last Year of *Nero* (as we have shewn in his Acts) when Meadows and Olive-Trees (the publick Road lying between them) went into the contrary places, in the *Marrucine* Territory, in the Lands of *Vectius Marcellus*, a *Roman* Knight, Procurator under *Nero*."

*Translation of Pliny passage quoted.*

There are many the like Instances to be met with in Authors, of the placing Parts perpendicular or inclining, which were before horizontal; so the turning of other parts upside downwards, of throwing parts from place to place; of stopping the Passage of Rivers, and turning them another way; of swallowing some Rivers, and of producing others a new; of changing Countries from Barren to Fruitful, and from Fruitful to Barren; of making islands join to the Continent, and separating parts of the Continent into Islands. There are other Relations that mention the vast spaces of Ground that have been all at once shaken and overturned, some of five Hundred Miles in length, and a hundred and fifty in bredth. Of the communication of Vulcanes (which are as it were the Nostrills or constant Breathing places of these Monsters) tho' plac'd at a very great distance one from another by Subterraneous Caverns. Other Relations furnish us with Instances of the Substances they vomit out; such as Pumice Stones, and several other sorts of calcin'd and melted Stones, and Rocks, Ashes, Minerals, hot water, Sulphur, Flame, Smoak, and various other Substances.

In others we find instances of Liquefactions, Vitrifications, Calcinations, Sublimations, Distillations, Petrifactions, Transformations, Suffocations and Infective or deadly Steams destroying all things near them, which possibly may be one cause of the scarcity of Relations where 'tis probable there have been so very many effects wrought in the World of this kind. But these I shall [p.311] not insist upon, having I fear too long digress'd on this part to shew the variety of effects produced by Earthquakes.

*The 4th kind of Effects of Earthquakes: Liquefactions, etc., which he does not elaborate here.*
*Hooke himself realizes that perhaps he has been a little excessive in relating such recorded and hearsay instances of "earthquakes" in his attempt to entertain and persuade his audience.*

There is only one thing more that I think pertinent to our present purpose, and that is the universality of this active Principle: There is no Country almost in the World but has been sometimes or other shaken by Earthquakes, that has not suffered some, if not most parts of these Effects. *Seneca* says in the Preface to the 6th Book of his Natural Questions. *Omnia ejusdem sortis sunt, et si nondum mota tamen mobilia; erramus enim, si ullam terrarum partem, exceptam immunemq; ab hoc periculo credimus, omnes sub eadem jacent lege, nihil, ita ut immobile esset, Natura concepit: Alia temporibus aliis cadunt; & quem-admodum in urbibus magnis nunc hæc domus nunc illa suspenditur, ita in hoc orbe Terrarum, nunc hæc pars facit vitium nunc illa. Tyrus aliquando infamis ruinis fuit. Asta duodecim Urbes simul perdidit. Anno priore Achaiam & Macedoniam quæcunque est ista vis mali quæ incurrit, nunc Campaniam*

*The universality of earthquakes, quoting Seneca.*

*læsit: Circuit fatum, & siquid diu præteriit, repetit. Quædam rarius, solicitat, sæpius quædam. Nihil immune esse & innoxium sinit. Non homines tantum, qui brevis & caduca resnascimur; Urbes oræque terrarum & Litora & ipsum mare in servitutem fati venit. Quo ergo nobis permansura promittimus bona fortunæ, & fælicitatem (cujus ex omnibus rebus humanis velocissima est levitas) habituram in aliquo pondus & moram credimus? Perpetua sibi omnia promittentibus in mentem non venit: Id ipsum supra quod stamus stabile non esse. Neque enim Campaniæ istud aut Achaiæ, sed omnis soli vitium est, male cohærere & ex causis plurimis resolvi; & summa manere partibus ruere.* Which I English thus. "All things are subject to the same chance; tho' they are not yet moved, they are movable; for we err, if we believe any part of the Earth excused and free from this hazzard; all are subject to the same Law; nothing is made by Nature so fixt as to be unmoveable; some sink at one time, some at another: And as in great Cities, now this House, now that House hangs tottering on Props; so on the great Face of the Earth, now this part fails, now that: *Tyre* formerly was remarkable for its Destruction: *Asia* lost at once Twelve Cities. Whatever the Power may be, the former Year *Achaia* and *Macedonia* felt it now *Campania* : Fate goes round, and repeates what it had long before acted: It brings some things often on the Stage, some seldom; but suffers nothing absolutely free and untouch'd. Not we Men only are brought forth short Liv'd, frail Beings: Cities, Countries, Shores, nay the Sea itself are the Slaves of Fate. Why therefore do we flatter our selves that the gifts of Fortune will stick by us, or that Happiness will observe any Rule or Measure, Happiness the most fleeting of all humane Things? They that promise to themselves all things fixt, surely never think that the very Ground we stand on is it self unfixt. Nor was that the frailty only of *Campania* or *Achia*, 'tis the same in all Soils and Countries, to be loosely join'd and compacted, but easily and by many ways dissolved; the whole remains while each part changes and sinks into Ruine and Alteration."

*Hooke's translation of the Seneca passage: Nothing is permanent in this world. All things change and sink "into Ruine and Alteration." Hooke himself ascribes to this principle of nature.*

*Considering the prevailing religious thought of 17th-century England, that the Earth is essentially the same as was left after Noah's flood, Hooke is indeed against the traditional belief in his accepting the "unfixity" of the world.*

Thus we see all Countries in the World are subject to these Convulsions, but those most of all that are most Mountainous: Such are usually all the Sea Coasts, therefore *Pliny* says, That the *Alps*, and *Appenine* Mountains have very often been troubled with Earthquakes. *Maritima autem maxime quatiuntur* (says he) *nec montosa tali malo carent. Exploratum est mihis Alpes Appenninumq; sæpius tremuisse.* Martine places are most shaken, nor do the Mountainous escape, for I have often found the *Alpes* and *Apennines* tremble.

*Mountainous places on sea coasts are more subject to earthquakes, although mountain chains like the Alps and Appenines (according to Pliny) also suffer tremors. "Martine" here = "maritime."*

For most probably those that are most Mountainous, are most Cavernous underneath them; to countenance which Opinion, I remember to have taken notice in certain very high Cliffs towards the Sea side, where the Hills seemed, as it were, cleft asunder, the one half having been probably foundred and tumbled down into the Sea, and the other half, as it were remaining, that at the bottom, near the Water, for almost the whole length, there were very many large Caverns, which, by several Circumstances, seem'd to be made before the access of the Sea thereunto, and not by the washing and beating of the Waves against the bottom of these Cliffs; for I observ'd in many of them, that the Plates or Layers, as I may so call those parts between the Clefts in [p312] Rocks, and Cliffs to lean contrary ways, and to meet, as it were at the top like the Roof of a House, and others of them in other forms, as if they had been Caverns left between many vast Rocks tumbled confusedly one upon another. And indeed I cannot imagine, but that under these Mountains, Islands, Cliffs or Lands, that have been much rais'd above their former level, there must be left vast Caverns, whence all that Matter was thrown, where probably may be the Seat or Place of the Generation of those prodigious Powers. But this only by the Bye; for I intend not here to examine the causes of their beginnings, force, and powerful Effects, nor of their remaining, ceasing, renewing, or the like. It being sufficient, for my present purpose, to shew, That they have been certainly observ'd to produce those extraordinary Effects from what Cause soever they proceed. That they have been heretofore in many places where they have now ceas'd for many Ages; and that they have lately happen'd in places, where we have no History that does assure us they have been heretofore. That they have turn'd Plains into Mountains, and Mountains into Plains; Seas into Land, and Land into Seas; made Rivers where there were none before, and swallowed up others that formerly were; made and destroy'd Lakes, made Peninsulas Islands, and Islands Peninsulas; vomited up Islands in some places, and swallowed them down in others; overturn'd, tumbl'd and thrown from place to place Cities, Woods, Hills, &c. cover'd, burnt, wasted and chang'd the superficial Parts in others; and many the like strange Effects, which, since the Creation of the World, have wrought many very great changes on the superficial Parts of the Earth, and have been the great Instruments or Causes of placing Shells, Bones, Plants, Fishes, and the like, in those places, where, with much astonishment, we find them.

*Hooke feels instinctively that when mountains are "thrown" up, there must be a deficiency of mass within the roots of them.*

*This description of a chaotic and unpredictable world must have been rather unsettling to some of Hooke's listeners.*

Concerning the Vicissitudes that places are subject to, in relation to Earthquakes, I find a memorable Passage sent by *Paul Ricaut* Esquire, now Consule of *Smyrna*, Dated *November* 23. 1667. "*Constantinople*, says he, is not now so subject to Earthquakes as reported in former times, there having not happen'd in the last seven Years, in which I have been an Inhabitant there, above one of which I have been sensible; but within these twenty Days in *Smyrna* fell out an Earthquake which dangerously shook all the Buildings, but did little or no harm; the Ships in the Road, and others at an Anchor, about three Leagues from hence, were sensible of it. It is reported that this City hath been already seven times devoured by Earthquakes, and it is prophesied, that it shall be so again so soon as the Houses reach the old Castle upon the top of the Hill, on the side of which remains the Ruins of the old City and the Tomb of St. *Polycarpus*, St. *John*'s Disciple, still preserv'd by the *Greeks* in great Veneration."

*A 1667 letter from Paul Ricaut, consul of Smyrna, describing an earthquake there where the destructive power of earthquakes has been felt many times.*

Another Cause there is which has been also a very great Instrument in the promoting the alterations on the Surface of the Earth, and that is the motion of the Water; whether caus'd 1*st*. By its Descent from some higher place, such as Rivers and Streams, caus'd by the immediate falls of Rain, or Snow, or by the melting of Snow from the sides of Hills. Or, 2*dly*. By the natural Motions of the Sea, such as are the Tides and Currents. Or, 3*dly*. By the accidental motions of it caus'd by Winds and Storms. Of each of these we have very many Instances in Natural Historians, and were they silent, the constant Effects, would daily speak as much. The former Principle seems to be that which generates Hills, and Holes, Cliffs, and Caverns, and all manner of Asperity and irregularity in the Surface of the Earth; and this is that which indeavours to reduce them back again to their pristine Regularity, by washing down the tops of Hills, and filling up the bottoms of Pits, which is indeed consonant to all the other methods of Nature, in working with contrary Principles of Heat and Cold, Driness, and Moisture, Light and Darkness, &c. by which there is, as it were, a

*Hooke now moves on to other causes that make changes on the surface of the earth besides the relatively violent effects of earthquakes per se—i.e., the "constant" effects of moving water that tend to modify those features created by "the former principle," earthquakes.*

*Hooke's concept of denudation: The irregularities in the terrestrial surface produced by earthquakes are countered by the forces of erosion and deposition which tend to reduce the unevenness to their original regularity.*

continual circulation. Water is rais'd in Vapours into the Air by one Quality and precipated down in drops by another, the Rivers run into the Sea, and the Sea again supplies them. In the circular Motion of all the Planets, there is a direct Motion which makes them indeavour to recede from the Sun or Center, [p.313] and a magnetick or attractive Power that keeps them from receding. Generation creates and Death destroys; Winter reduces what Summer produces: The Night refreshes what the Day has scorcht, and the Day cherishes what the Night benumb'd. The Air impregnates the Ground in one place, and is impregnated by it in another. All things almost circulate and have their Vicissitudes. We have multitudes of instances of the wasting of the tops of Hills, and of the filling or increasing of the Plains or lower Grounds, of Rivers continually carrying along with them great quantities of Sand, Mud, or other Substances from higher to lower places. Of the Seas washing Cliffs away and wasting the Shores: Of Land Floods carrying away with them all things that stand in their way, and covering those Lands with Mud which they overflow, levelling Ridges and filling Ditches. Tides and Currents in the Sea act in all probability what Floods and Rivers do at Land; and Storms effect that on the Sea Coasts, that great Land Floods do on the Banks of Rivers. *Ægypt* as lying very low and yearly overflow'd, is inlarg'd by the sediment of the *Nile*; especially towards that part where the *Nile* falls into the *Mediterranean*. The Gulph of *Venice* is almost choak'd with the Sand of the *Po*. The Mouth of the *Thames* is grown very shallow by the continual supply of Sand brought down with the Stream. Most part of the Cliffs that Wall in this Island do Yearly founder and tumble into the Sea. By these means many parts are covered and rais'd by Mud and Sand that lye almost level with the Water, and others are discover'd and laid open that for many Ages have been hid.

Of this kind the Royal Society received a memorable Account from the Learned Dr. *Brown* concerning a petrified Bone of a prodigious bigness, discover'd by the falling of some Cliffs; the words of the Relation are these. "This Bone (which he presented the Royal Society, and is now in the Repository) was found last Year 1666. on the Sea Shore, not far from *Winterton* in *Norfolk*; it was found near the Clift after two great Floods, some thousand Loads of Earth being broken down by the rage of the Sea, as it often happens upon this Coast, where the Cliffs consist not of Rock but of Earth. That it came not out of the Sea may be conjectur'd because it was found near the Cliff, and by the colour of it, for if out of the Sea it would have been whiter. Upon the same Coast, but as I take it, nearer *Hasborough*, divers great Bones are said to have been found, and I have seen a lower Jaw containing Teeth of a prodigious bigness and somewhat petrified. All that have been found on this Coast have been found after the falling of some Cliff, where the outward Crust is fallen off, it clearly resembles the Bones of Whales and great Cetaceous Animals, comparing it with the Scull and Bones of a Whale which was cast upon the Coast near *Wells*, and which I have by me, the weight whereof is 55 Pounds." Thus far he on this Subject. To this may be added the *Chartham* News, or the discovery of Riverhorse, or the *Hippopotamus* Teeth printed in the *Philos. Transactions*. N. 272. p. 882.

Nor are these Changes now only, but they have in all probability been of as long standing as the World. So 'tis probable there may have been several vicissitudes of changes wrought upon the same part of the Earth; it may have been of an exact spherical Form, with the rest of the Earths or

*Precipated = precipitated*

*Hooke's cyclical concept of nature is clearly and poetically expressed here, and it extends to the entire known universe. He also expresses here "Newton's" 3rd law: For every action there is a reaction. His understanding of the centripetal force, "the attractive Power" that keeps planets from receding, was fundamental to the expression of the Law of Gravitation. Newton had no idea of this force until Hooke informed him of it.*

*The words "almost circulate" are important as they signify a not completely closed cycle which then could never progress toward an end.*

*Growth of the Nile Delta from Nile sediments, as well as that of the Gulf of Venice from sediments of the Po River and the filling up of the mouth of the Thames. At the same time, "cliffs that wall in this Island do Yearly founder and tumble into the Sea." Hooke's own intimate knowledge of these processes on the Isle of Wight, his birthplace, give him proof.*

*This passage again demonstrates Hooke's assertion that changes have always been and are always taking place, that there have been several "vicissitudes" of them.*

Planets, at the Creation of the World, before the eternal Command of the Almighty, that the Waters under the Heaven should go to their place, which before cover'd the Earth, so as that it was αορατος και ακατασκευαστος και σκοτος επανω του αβυσσου και πνευμα δεου επεφερετο επανω του υδατος, invisible and incompleated, and the Darkness of the Deep was over it (being all over cover'd with a very thick shell of Water which environ'd it on every side, it being then in all probability created of an exact Spherical Figure, and so the Waters being of themselves lighter than the Earth, must equally spread themselves over the whole Surface of the Earth) and where the Breath of the Lord moved above or upon the Surface of these Waters. It may, I say, in probability have been then a part of the exact Spherical Surface of the Earth, and upon the command that the Waters under the Air or Atmosphere (which seems to be denoted by ζτερεωμα or Firmament; for the Hebrew Word signifies an Expansum) [p.314] should be gathered together into one place, and that the dry Land should appear. It may have been by that extraordinary Earthquake (whereby the Hills and Land were rais'd in one place, and the Pits or deeper places, whether the Water was to recede and be gathered together to constitute the Sea were sunk in another) rais'd perhaps to lye on the top of a Hill or in a Plain, or sunk into the bottom of the Sea, and by the washing of Waters in motion, either carried to a lower place to cover some part of the Vale, or else be cover'd by adventitious Earth, brought down upon it from some higher place; which kind of alterations were certainly very great by the Flood of *Noah*, and several other Floods we find recorded in Heathen Writers. If at least there were not somewhat of an Earthquake which might again sink those Parts which had been formerly raised to make the dry Land appear, and raise the bottom of the Sea, which had been sunk for the gathering together of the Waters (which Opinion *Seneca* ascribes to *Fabianus*) *Ergo* (says he) *cum affuerit illa necessitas temporis multa simul fata causas movent nec sine concussione Mundi tanta Mutatio est ut quidam putant inter quos Fabianus est.* His description of the Manner and Effects of a Flood, is fine and very suting to my present Hypothesis. This Part being thus covered with other Earth, perhaps in the bottom of the Sea, may by some subsequent Earthquakes, have since been thrown up to the top of a Hill, where those parts with which it was by the former means covered, may in tract of time by the fall and washing of Waters, be again uncovered and laid open to the Air, and all those Substances which had been buried for so many Ages before, and which the devouring Teeth of Time had not consumed, may be then exposed to the Light of the Day.

There are yet two other Causes of the mutation of the superficial Parts of the Earth, which have wrought many great changes in the World, and those are either the Sea's overflowing of a Country or Place, when forced on it with some violent Storms or Hurricans of Wind, or from the over-flowing of Rivers from great falls of Rain, or from something stopping their Course, of these we have many Instances in Voyages, and we have very often times here at *London* felt the effects of the Wind driving in the Tide with such great force, as that it has oft times overflow'd the Banks, fill'd the Streets and Cellers to the no small damage of the Inhabitants. "At *Chatmoss* in *Lankashire* (saith *Childrey*) is a low mossy Ground very large, a great part of which (saith *Cambden*) not long ago, the Brooks swelling high carry'd quite away with them, whereby the Rivers were corrupted, and a number of fresh Fish perished. In which place now lies a low Vale watered with a little Brook, where Trees have been digg'd up

*The change in Hooke's style between what he has expressed up to this point and here is startling. Before this, there was no mention of the Almighty; his entire discourse up to now has been extraordinarily modern in approach and thinking. Possibly he was criticized and is now trying his best to explain his thesis in the context of the religious beliefs of his audience. But it is clear, if one keeps reading rather than looks for isolated sentences, he much prefers natural explanations than supernatural ones. Accordingly, he refers to the act of God that caused the waters to be gathered in one place as "that extraordinary Earthquake."*

*Here he insists that there have been other floods besides Noah's.*

*Here Hooke goes back to Seneca rather than the Bible for corroboration and gives a totally modern view of the cyclic process of sedimentation, burial, raising, erosion, and exposure to the light of day.*

*Other causes of change besides earthquakes in all their varieties are the action of water or of wind; i.e., the sea's overflowing in a violent storm or flooding by rivers from great rainfall or clogging of the streambed.*

lying along, which are suppos'd by some to have come thus. The Channel of the Brooks being not scower'd, the Brooks have risen, and made all the Land moorish that lay lower than others, whereby the Roots of the Trees being loosned by reason of the bogginess of the Ground, or by the Water finding a passage under Ground, the Trees have either by their own weight, or by some Storm, been blown down, and so sunk into that soft Earth and been swallowed up: For 'tis observable, that Trees are no where digg'd out of the Earth but where the Earth is boggy; and even upon Hills such moorish and moist Grounds are commonly found, the Wood of these Trees burning very bright like Touchwood (which perhaps is by reason of the bituminous Earth in which they have been so long) so as some think them to be Fir-Trees. Such mighty Trees are often found in *Holland*, which are thought to be undermin'd by the Waves working into the Shore, or by Winds driven forwards and brought to those lower places where they settled and sunk. *Brit. Bac.* Page 167, 168."

"The Sea (as is said before) has eaten a great part of the Land away of these Western Shires. There are on the Shore of this Shire (*Cumberland*) Trees discovered by the Winds sometimes at low Water which are else covered over with Sand; and it is reported by the People dwelling thereabouts, that they dig up Trees without Boughs out of the Gound in the [p.315] places of the Shire. *Child.* p. 171. Many Trees are found and digged out of the Earth of the Isle of *Man. Ide.* p. 178."

*Several proofs from histories are given of the destruction of land by the sea.*

"In divers places of the Low Grounds and Champian Fields of *Anglesy*, the Inhabitants every Day find and dig out of the Earth the Bodies of huge Trees with their Roots, and Fir-Trees of a wonderful bigness and length. Page 150."

"At the time when *Henry* II. made his abode in *Ireland* were extraordinary violent and lasting Storms of Wind and Weather, so that the Sandy Shore on the Coasts of the *Pembrockshire*, was laid bare to the very hard Ground, which had lain hid for many Ages, and by further search the People found great Trunks of Trees, which when they had digged up, they were apparently Lopped, so that one might see the Strokes of the Ax upon them, as if they had been given but the Day before; the Earth look'd very black, and the Wood of these Trunks was altogether like Ebony. At the first discovery made by these Storms, the Trees we speak of lay so thick, that the whole Shore seem'd nothing but a lopped Grove. Whence may be gather'd, saith *Childrey*, that the Sea hath overflow'd much Land on this Coast, as it has indeed many Countries bordering upon the Sea, which is to be imputed to the ignorance of the *Britans* and other barbarous Nations, which understood not those ways to repress the fury of the Sea which we now do. p. 142. 143."

"In the low Places on the South side of *Cheshire*, by the River *Wever*, trees are oft times found by digging under Ground, which People think have lain buried there ever since *Noah*'s Flood. p. 129." "St. *Bennets* in the *Holme* hath such fenny and rotten Ground, that (saith *Cambden*) if a Man cut up the Roots or Strings of Trees it flotes on the Water. Hereabout also are Cockles and Periwinkles sometimes digged up out of the Earth, which makes some think that it was formerly overflowed by the Sea." Divers of these Effects do seem to be caused by Inundations of the Sea, tho' there are others of them that do rather seem ascribable to Earthquakes, than to Inundations caus'd by Storms; for that Earthquakes

*Burial of trees, therefore, can be caused both by inundations of the sea and by actual earthquakes.*

have produced such Effects as the burying of Trees and Plants, divers of the formerly mention'd Histories do sufficiently manifest.

The *Lignum Fossile* which is found in *Italy*, of which we have a good account given by *Francesco Stelluti* (tho' by that Author it be supposed to be generated out of the subterraneous Parts of the place where it has been found, yet) from many remarkable circumstances in this History, it seems very probable to me to have been first buried by some Earthquakes, and afterwards to be variously metamorphosed and changed by the Symptoms which usually follow Earthquakes, and which this place is much vexed with, as is indeed almost all the Country of *Italy*, to wit, the emitting of hot Steams and Smoaks proceeding from subterraneous Fires, which do their often shift their places, burn the parts of some of those Trunks into black and brittle Coles; melt a kind of Ore into the Pores of others; petrify the Substance of another sort; bake the Dirt and Clayish Substances which have soaked into the Pores of a fourth sort into a kind of Brick; rot the Parts of others, and convert them into a kind of Dirt or Muddy Earth; and so act variously and produce differing Effects upon those buried Substances, according to the Nature of the Earths, Minerals, Waters, Salts, Heats, Smoaks, Steams, and other active Instruments casually applied to the parts of the Buried Trunks, by the confusion of the Earthquakes, and by immediate application, and long continuance and digestion, as I may call it, in this Laboratory of Nature, transformed into other Substances, and exhibit all those admirable Phænomena mentioned by that Author, whereby the bury'd Bodies are transformed. Nor is it so much to be wondered at, that such Substances as Vegetables (which being exposed to the Vicissitudes of the Air and Water, are quickly corrupted and consumed, and many of them much sooner if buried in the Earth) should after so many Ages perhaps, remain intire, and rather more substantial sound and permanent than if they were newly cut down. Since if we consider the Nature of the decaying and corruption in all kind of Animal and Vegetable [p.316] Substances, we shall find that the chief cause of it seems to be from the Action of the fluid Parts upon the solid for the dissolving of them: and wheresoever the Internal Fluid is either first changed or altered by the mixture of some other heterogenous Substance, so as to loose that dissolving property as by the Intermixture of Salt, Spirit of Wine, &c. or by incorporating with it and hardning it into a solid Substance, as in Petrifactions, &c. Or, secondly, exhaled by a gradual and gentle degree of heat, and so the solid Parts only left alone, and kept either dry, or fill'd with a fluid of an heterogeneous Nature, such as unctuous and spicy Juices with watery Substances. Or Thirdly, Congealed and hardened either by cold or the peculiar Nature of the Juice itself; such is freezing and the hardning of Coralline Plants, or Submarine Vegetables, Horns, Gums, Bones, Hair, Feathers, &c. wheresoever, I say, Bodies are by these means put into such a Constitution, that the Parts act not upon one another, and continue in that state by being preserved by adventitious Moisture or soft'ning by homogeneous Fluids, they are, as it were, perpetual, unless by extraordinary Heat, many of those otherwise solid and unactive Substances are made fluid by such active Disolvents; and unless they be immersed in such Liquors or Menstruums as do of themselves dissolve and work on them; we shall not, I say, wonder at the lastingness of these buried Substances, if we consider also the various Juices with which several parts of the Earth are Furnish'd, Unctuous, Watery, Styptick, Saline, Petrifactive, Corrosive, and what not. There are some Juices of the Earth which do, as it were, perpetuate them by turning them into Stone. Others

*The various ways substances are transformed—usually in the presence of a fluid acting as solvent.*

*With the aid of heat or cold.*

do so deeply pierce and intimately mix with their parts, that they wholly, as it were, change the Nature of those Substances, and destroy that property of Congruity which all Bodies generated in the Air and Water seem to have, which are very apt to be dissolved and corrupted by innate aerial and aqueous Substances. Such are all kinds almost of oleaginous and sulphureous Substances, and divers saline and mineral Juices. Others indeed do not preserve the very Substance of those Vegetables, but insinuating into the Pores, and there, as it were, fixing, they retain and perpetuate the Shape and Figure, but corrupt and dissolve the interpos'd part of the Vegetable; of all which kinds I have seen some Specimina, as I have also of divers other Substances Pickled, Dried, Candyed, Conserv'd, preserv'd, or Mummify'd by Nature: where therefore the Substances have happen'd to be bury'd with preservative Juices, they have withstood the injury of Time; but where those Juices have been wanting, there we find no Footsteps of these Monuments of Antiquity.

*Substances thus can be petrified. Or they may be entirely destroyed.*

*Or transformed totally except for the shape of the substance.*

But to return to what I was prosecuting, another cause which may make alterations on the Surface of the Earth, is any violent motions of the Air, whereby the parts of the Earth, in dry Weather, are transported from place to place in the Form of Dust. Of this kind Travellers tell us very strange Stories of the removal of the Sands in the Deserts of *Arabia*, and other Deserts in *Africa*; and we have some instances of it here in *England*, to wit, in *Norfolk* and *Devonshire* (in the former of which there are often found natural Mummies which have been buried alive by those removing Sands, and by their driness preserv'd) But these greater and more suddain removals of Sand and Dust are not so universal, and therefore not so much to my present purpose; tho' possibly they may have been more frequent heretofore, which the Layers of Sands to be found in digging Pitts and Wells seem to hint: But that which is most universal, is very slow, and almost inperceptible, and that is the removing of the Dust from the higher Parts, and settling in the lower by the Wind or motion of the Air. This tho' its effects be almost insensible, yet being constant, must needs, in length of time, much promote the levelling and smoothing of the Surface of the Earth.

*Besides the action of water, another cause for changing the surface of the earth is the violent motion of air or wind which can transport dust.*

*The catastrophic, sudden, wind-caused sand transports are rare. What is universal is the "very slow and almost imperceptible" sand erosion and deposition. This passage again demonstrates that Hooke was not just a "catastrophist."*

I might name also another cause of the transposition of the superficial Parts of the Earth; and that is from the gradual subsiding or sinking into the Earth of the more heavy, and the Ebullition or respective rising of the more light Parts upwards. Hence we may observe, that many old and vast Buildings [p.317] and Towers have sunk into the Earth. And the like we judge of those vast Stones in *Salisbury* Plain, and we find constantly almost in all Stone Monuments placed in Church-yards, and in all old Churches unless placed on a very high place, and founded on some Rock. The Cause may possibly have great Influence where the Earth is very soft, spungy, or boggy; and possibly many of those Trees which are found in boggy Grounds, may have been buried, by having been either fell'd, or blown down by Wind, or wash'd down by some Inundation well impregnated with mineral Juices, and so made heavier than the subjacent Earth, and swallowed into it. Several of the former Relations do indeed pretty well agree with this Hypothesis; and I am very apt to think that where the Surface of the Earth has not been much alter'd since the Creation, if any such there be, if it were search'd into it would be found that the lightest Parts, lye next the Surface, and so heavier in lower Parts, which makes me imagine that the natural place of Minerals is very deep under the Surface of the Earth, and (possibly) to be found under every step of Ground, were search made under it to a sufficient depth; and that the reason why we

*The natural tendency of heavier parts to sink and of lighter parts to rise to the surface. Although some of his examples are those of settling only, Hooke is perceptive to include also the rise of less dense matter through the surface layers of the Earth— the rise of molten material toward the surface being a very essential process in modern geology.*

*The denser matter lies deeper; lighter material is found nearer the surface. Minerals therefore are formed deep in the earth and are found sometimes near the surface only because they have been "thrown up" by some "Subterraneous Eruption," which creates mountains.*

find it sometimes near the Surface of the Earth, as in Mountains, is not because it was there generated, but because it has been by some former Subterraneous Eruption (by which those Hills and Mountains have been made) thrown up towards the Surface of the Earth. And as Gold is the heaviest, so it is the scarcest of all Mettals. And I do not at all question but that there may be other Bodies or Mettals as much heavier than Gold, as Gold is then common Earth. To make these Conjectures the more probable, see what Sir *Philiberto Vernatti* writes from *Batavia* in the *East-Indies*, in answer to some Queries sent him by the Royal Society. "I have often (says he) felt Earthquakes here, but they do not continue long. In the Year 1656, or 57, (I do not remember well the time) *Batavia* was covered in one Afternoon about two of the Clock, with a black Dust, which being gathered together, was so ponderous, that it exceeded the weight in Gold. It is here thought that it came out of a Hill that burneth in *Sumatra* near *Endrapeor.*"

These fiery Eruptions in all probability come from a very great depth and with a great violence; and possible even that golden Powder that is sometimes thrown up may have somewhat conduced to the cause of the violence of it. We know not what Method Nature may have to prepare an *Aurum Fulminans* of her own, great quantities of which, being any ways heated and so fired, may have produced the Powder. However, whether so or not, it were very well worth trial to examine, whether the Flower that may be catch'd in a Glass Body, upon fulminating a quantity of such Powder gradually by small parcels, would, by being ordered as common Gold, make again an *Aurum Fulminans*: Or whether this Fulmination, which is a kind of Inflaming of the Body of the Gold, does not make some very considerable alteration in the Nature and Texture of it. Since we find that kind of Operation, to wit, inflaming or burning does considerably alter the Texture of all other Bodies so wrought on. This only by the way.

*Aurum Fulminans = explosive gold*

But to proceed to the last Argument to confirm the 6th Proposition I at first undertook to prove, namely, that very many parts of the Surface of the Earth (not now to take notice of others) have been transform'd transpos'd and many ways alter'd since the first Creation of it. And that which to me seems the strongest and most cogent Argument of all is this, That at the tops of some of the highest Hills, and in the bottom of some of the deepest Mines, in the midst of Mountains and Quarries of Stone, &c. divers Bodies have been and daily are found, that if we thoroughly examine we shall find to be real shells of Fishes, which for these following Reasons we conclude to have been at first generated by the Plastick faculty of the Soul or Life-principle of some animal, and not from the imaginary influence of the Stars, or from any Plastick faculty inherent in the Earth itself so form'd; the stress of which Argument lies in these Particulars.

*Hooke returns to his original point, his 6th Proposition: Fossil seashells, being the remains of real animals and not the result of plastic vertue of the soil or influence of the stars, found on the tops of mountains and in the deepest mines are the most "cogent" argument that the terrestrial surface has undergone great changes since Creation.*

[p.318] First, That the Bodies there found have exactly the Form and Matter, that is, are of the same kind of Substance for all its sensible Properties, and have the same External and Internal Figure or Shape with the Shells of Animals.

Next, That it is contrary to all the other acts of Nature, that does nothing in vain, but always aims at an end, to make two Bodies exactly of the same Substance and Figure, and one of them to be wholly useless, or at least without any design that we can with any plausibility imagine. The Shells of Animals, to our Reason, manifestly appear to be done with the greatest Councel and Design, and with the most excellent contrivance, both for the

*Hooke reiterates the several arguments of the organic origin of fossils—an indication that the idea is still not accepted by many of his colleagues in the Royal Society.*

Convenience and Ornament of that Animal to which it belongs, that the particular Structure and Fabrick of that Animal was capable of: Whereas these if they were not the Shells of Fishes, will be nothing but the sportings of Nature, as some do finely fancy, or the effects of Nature idely mocking herself, which seems contrary to her Gravity. But this perhaps may not seem so cogent, tho', if it be thoroughly consider'd, there is much weight in it.

Next therefore, Wherever Nature does work by peculiar Forms and Substances, we find that she always joins the Body so fram'd with some other peculiar Substance. Thus the Shells of Animals, whilst they are forming are join'd with the Flesh of the Animal to which they belong. Peculiar Flowers, Leaves, and Fruit are appropriated to peculiar Roots, whereas these on the contrary are found mixt with all kind of Substances, in Stones of all kinds, in all kinds of Earth, sometimes expos'd to the open Air without any coherence to any thing. This is at least an Argument that they were not generated in that posture they are found; that very probably they have been heretofore distinct and disunited from the Bodies with which they are now mixt, and that they were not formed out of these very Stones or Earth, as some imagine, but deriv'd their Beings from some preceding Principle.

Fourthly, Wherever else Nature works by peculiar Forms, we find her always to compleat that form, and not break off abruptly. But these Shells that are found in the middle of Stones are most of them broken, very few compleat, nay, I have seen many bruised and flaw'd, and the parts at a pretty distance one from another, which is an Argument that they were not generated in the place where they are found, and in that posture, but that they have been sometimes distinct and distant from those Substances, and then only placed, broken and disfigured by chance, but had a preceeding and more noble Principle to which they ow'd their Form, and by some hand of Providence, were cast into such places where they were filled with such Substances, as in tract of Time have condensed and hardened into Stone: This, I think, any impartial Examiner of these Bodies will easily grant to be very probable, especially if he take notice of the Circumstances I have already mention'd. Now, if it be granted, that there have been preceding Moulds, and that these curiously figured Stones do not owe their form to a Plastick or forming Principle inherent in their Substances; why might not these be supposed Shells, as well as other Bodies of the same Shape and Substance, generated none knows how, nor can imagine for what.

*Arguments against the notion that fossils are "sportings" or tricks of nature.*

Further, if these be the apish Tricks of Nature, Why does it not imitate several other of its own Works? Why do we not dig out of Mines everlasting Vegetables, as Grass for instance, or Roses of the same Substance, Figure, Colour, Smell? &c. Were it not that the Shells of Fishes are made of a kind of stony Substance which is not apt to corrupt and decay. Whereas, Plants and other animal Substances, even Bones, Horns, Teeth and Claws are more liable to the universal Menstruum of Time. 'Tis probable therefore, that the fixedness of their Substance has preserved them in their pristine Form, and not that a new plastick Principle has newly generated [p.319] them. Besides why should we not then doubt of all the Shells taken up by the Sea-shore, or out of the Sea (if they had none when we found them) whether they ever had any Fish in them or not? Why should we not here conceipt also a plastick Faculty distinct from that of the Life-principle of some Animal; is it because this is more like a

*It is the hardness of the shells that preserved them, not some "plastick Principle" that generated them.*

Shell than the other? That I am sure it cannot be. Is it because 'tis more obvious how a Shell should be placed there? If so, 'twould be as good Reason to doubt if an Anchor should be found at the top of a Hill, as the Poet affirms, or an Urn or Coins buried under Ground, or in the bottom of a Mine, whether it were ever an Anchor, or an Urne, or a coined Face, or made by the plastick Faculty of the Earth, than which what could be more absurd: And those Persons that will needs be so over confident of their Omniscience of all that has been done in the World, or that could be, may, if they will vouchsafe, suffer themselves to be asked a Question, Who inform'd them? Who told them where *England* was before the Flood; nay, even where it was before the *Roman* Conquest, for about four or five thousand Years, and perhaps much longer; much more where did they ever read or hear of what *Changes* and *Transpositions* there have been of the parts of it before that? What History informs us of the burying of those Trees in *Cheshire* and *Anglesy*? Who can tell when *Tenariff* was made? And yet we find that most judicious Men that have been there and well considered the form and posture of it, conclude it to have been at first that way produced. But I suppose the most confident will quickly upon examination, find that there is a defect of Natural History, if therefore we are left to conjecture, then that must certainly be the best that is backed with most Reason, that Clay, and Sand, and common Shells can be changed and incorporated together into Stones very hard. I have already given many instances, and can produce hundreds of others, but that I think it needless, that several parts of the bottom of the Sea have been thrown up into Islands and Mountains. I have also given divers Instances, and those some of them within the Memory of Man, where 'tis not in the least to be doubted but that there may be found some Ages hence several Shells at the tops of those Hills there generated; and as little, that if Quarries of Stone should be hereafter digged in those places, there would be found Shells incorporated with them; and were they not beholding to this inquisitive and learned Age for the History of that Eruption, they might as much wonder how these Shells should come there, and ascribe them to a plastick Faculty, or some imaginary Influence, as plausibly as some now do. I have also shewed, that Water and divers other fluid Substances, may be, in tract of Time, converted into Stone and stony Substances; and so such Liquors penetrating the Pores of these Shells, and especially if they be assisted by the benumming steams that sometimes issue from Subterraneous Eruptions, may very much contribute to the preservation of those Shells from Corruption and crumbling to Dust under the crushing Foot of Time. Besides, that the Shells themselves are so near the Nature and Substance of Stone, that they are little subject to the injuries of the Air or Weather; so that these small pyramidal Houses of Shell-Fishes seem not less lasting Monuments than those vast piles of Stones erected by the antient Inhabitants of *Egypt*, which outvye all the more curious Fabricks of *Grecian* and *Roman* Architecture both for their Antiquity and present Continuance. Nor do they exceed the Works of Architects for lasting only, but for Ornament, for Strength, and for Convenience.

Now if all these Bodies have been really such Shells of Fishes as they most resemble, and that these are found at the tops of the most considerable Mountains in the World as *Caucasus*, the *Alps*, the *Andes*, the *Appennine*, and *Pyrrenean* Mountains, to omit other Hills nearer and of less note, and that tis not very probable that they were carried thither by Mens Hands, or by the Deluge of *Noah*, or by any other more probable way than that of Earthquakes; 'tis a very cogent Argument that the superficial Parts of the

---

*Hooke's arguments against the doubters of the organic origin of fossils are eloquent and worth reading if only to admire his command of the English language.*

*The phrase "perhaps much longer" in reference to 4 or 5 thousand years before the Roman conquest is significant in showing that Hooke believed the age of the earth to be much longer than the biblical account. The Roman conquest around the beginning of the first millennium A.D. plus 5,000 years before that would already make the beginning earlier than the 4004 B.C. pronouncement of Archbishop Ussher (1581–1656).*

Earth have been very much chang'd since the beginning, that the tops of Mountains have been under the Water, and consequently also, that divers parts of the bottom of the Sea have been heretofore Mountains: For tho' I [p.320] confess I have but few Instances to prove it, besides that of *Plato*'s *Atlantis*, and some others that I have already mention'd; yet 'tis very probable, that whensoever an Earthquake raises up a great part of the Earth in one place, it suffers another to sink in another place; for Gravity is a Principle that will not long suffer a space to remain unfill'd under so vast a pile of Earth as a Mountain, unless the Substances, so thrown up, be of very hard, close and vast Stones that may, as it were, vault it: In which cases 'tis very probable (and *Kircher* and divers other Authors that write of Mines and Quarries, gives us many instances to confirm it) that these Cracks and Cliffs so left, are fill'd up with such Petrifying or Mineral Waters as do make great varieties of Stones, Marbles, Sparrs, Caulks, and Ores, and so there is made a transposition as well as a transformation. Which supposition (by the way) I think will furnish a very probable Reason of the shape of the Veins and Cracks of speckled Marbles and other Stones, of the form also of the Veins of Ores, Stones, Clays, &c. of the Earth, and of their so mixing together; of the lying of Mettals in Mountains and other Mines, &c. but of these only here by the Bye, because I refer what I have to say of that to another Subject, *Viz.* A History of the Forms and Proprieties of Minerals and Metals. To proceed then.

*Here Hooke is groping almost toward the concept of isostasy. He also instinctively realizes the necessity for a gravity deficiency in the roots of mountains as represented by a vaulting of the space inside mountains, i.e., containing a great cavity. But as he is so wont to do, his attention is diverted to mineralization and the filling of cracks.*

The Seventh Proposition that I undertook to make probable, was, that 'tis very probable that divers of these Transpositions and Metamorphoses have been wrought even here in *England*: Many of its Hills have probably been heretofore under the Sea, and divers other parts that were heretofore high Land and Hills, have since been covered with the Sea. Of the latter of these I have given many Instances already, and that which makes the first probable, is the great quantities of Shells that are found in the most Inland Parts of this Island; in the Hills, in the Plains, in the bottoms of Mines and in the middle of Mountains and Quarries of Stones. Of this kind are those Shells, which any inquisitive Man may find great quantities of in *Portland*-stone, *Purbeck*-stone, *Burford*-stone, *Northhamptonshire*-stone, out of which I have often pick'd Muscles, Cockles, Periwinkles, Oysters, Scallop, and divers other Shells that are buried in the very Body and Substance of the Stone; and indeed they may be found of some kind or other in almost all kinds of Stone. That the *Kainsham* Snail Stones, and those found in several other parts of *England,* have been the Shells of Fishes, I hope the Arguments I have already urged may suffice to evince. As also, that those Helmet Stones (of which sort I my self have found in many places of *England*, and others have furnish'd me with many more found in other parts of it) are nothing but the fillings of the Shells of a sort of Echini or Egg Fishes.

*Arguments for the 7th Proposition: Changes of land and sea areas have occurred in England.*

Now 'tis not probable that other Mens Hands, or the general Deluge which lasted but a little while, should bring them there; nor can I imagine any more likely and sufficient way than an Earthquake, which might heretofore raise all these Islands of Great *Britain* and *Ireland* out of the Sea, as it did heretofore, of which I have already mention'd the Histories; or as it lately did that Island in the *Canarys* and *Azores*, in the sight of divers who are yet alive to testifie the Truth and Manner of it: And possibly *England* and *Ireland* might be rais'd by the same Earthquake, by which the *Atlantis*, if we will believe *Plato*, was sunk. And I doubt not but any inquisitive

*Here Hooke denies that Noah's Flood, the "general Deluge," could have brought the fossils where they are found, because the flood did not last long enough. An earthquake, therefore, must have been the cause, as well as of raising islands, including the British Isles, out of the sea.*

Man that has opportunity of traveling and examining several of the Mountainous Parts and Cliffs, and of the Mines, Quarries, and other subterraneous Parts of *England*, will meet with a great many other Arguments to confirm this Supposition, besides those I have already alledg'd: But those I hope may suffice for the present to excite Men to this Curiosity, which was the chief reason of this present exercise. And this makes way for the Eighth Proposition, which is

Eighthly, That most of these Mountains and Inland places whereon these kind of Petrify'd Bodies and Shells are found at present or have been heretofore, were formerly under the Water, and that either by the descending of the Waters to another part of the Earth by the alteration of the Center of Gravity of [p.321] the whole bulk, or rather by the Eruption of some kind of Subterraneous Fires or Earthquakes, great quantities of Earth have been deserted by the Water and laid bare and dry. That divers places have been thus raised by Earthquakes has been already proved from many Histories; and then why may not all of them have the same Original, especially since there is no other more probable Cause that we know of, that should convey and place those Shells on the tops of Mountains? That they really are Shells, and have been the receptacle of Fishes, I hope the Arguments I have already alledg'd may suffice to persuade: If then they have been Shells, and have been there placed, why should we not conclude that That part hath been under the Water with as much reason as seeing Towers, &c. under the Water near ------------we do that those parts have been heretofore above Water, which Histories inform us of, or as we might have done if we had had none even from what the thing itself speaks. I think we may with as much reason doubt if an Urn should be digg'd up full with old Coins, stamped with the same impression, made of the same Substance and Magnitude of those used by the ancient *Romans*, or any other Nation, of which we have good History; First, Whether ever those Coins were made by Mens Hands, or by a plastick Faculty of Nature; for it is certainly no more difficult a task for Nature to imitate the one than the other. And, Secondly, Whether ever that Urn were made and those Coins were put into it and shaped by Mens Hands, or that they were shap'd and thrown into it meerly by Nature; perhaps those suppositions might not be impossible, but sure all Men will judge them very improbable: And I think the Case in this particular I am speaking of very much the same. First, That there is much greater reason to imagine the Shells so found to have been the *Exuviæ* of some living Creature, and next, that they have been placed there where they are found when that part was under Water, and that part to have since been rais'd up to that height above the Sea by some preceding Earthquake. There is no Coin can so well inform an Antiquary that there has been such or such a place subject to such a Prince, as these will certify a Natural Antiquary, that such and such places have been under the Water, that there have been such kind of Animals, that there have been such and such preceding Alterations and Changes of the superficial Parts of the Earth: And methinks Providence does seem to have design'd these permanent shapes, as Monuments and Records to instruct succeeding Ages of what past in preceding. And these written in a more legible Character than the Hieroglyphicks of ancient *Egyptians*, and on more lasting Monuments than those of their vast Pyramids and Obelisks. And I find that those that have well consider'd and study'd all the remarkable Circumstances to be met with at *Teneriffe* and *Fayale*, do no more doubt that those vast Pikes have been raised up by the Eruption of Fire out of their tops, than others that have survey'd the Pyramids of *Egypt*, or the

*Arguments for the 8th Proposition: Most mountains and inland places where shells are found were formerly under water.*

*Hooke's analogy with ancient coins gives some idea of how his mind worked: coins are a record of history just as these fossils tell us something about what occurred in nature in the past.*

Stones on *Salisbury* Plain do doubt that they have been the effects of Man's Labours. And they do it with as much reason; for all Conclusions that are not immediately grounded on Sense, or the result of it are but Hypothetical and from a Similitude; for since it has been heretofore and lately seen, that such Eruptions have produc'd such kind of Hills and Islands, and that the tops of these Hills do as yet burn, and that there are all about the sides of them huge Stones and Rocks, and even Mountains lying in Postures as if they had been tumbled down from the top; 'tis a rational Conclusion to say, that 'tis very probable these have had the same Original with those.

But as to those vast tracts of Ground that lye very far from the Sea, it may perhaps to some seem not impossible, that the Center of Gravity or Method of the attraction of the Globe of the Earth may change and shift places, and if so, then certainly all the fluid parts of the Earth will conform thereto, and then 'twill follow that one part will be cover'd and overflow'd by the Sea that was before dry, and another part be discover'd and laid dry that was before overwhelm'd. Now, tho' this Conjecture may at first reading seem a little extravagant, yet if we consider, that as great alterations have been really observ'd, we may a little moderate a two severe Cen-[p.322]sure; That the Magnetical Poles and Meridians of the Earth have been alter'd, and that they do at this present continue to do so is granted almost by all, and confirmed by a multitude of Observations made in divers Parts of the World, and by collecting and comparing the Observations I have met with: I suppose the Pole of the Magnetism to be at a certain distance from the Pole of its daily Motion, and that it does move round that Pole at a certain distance in a certain number of Years, and that it does annually proceed in this Circle some parts of a Degree: So that whereas the Magnetical Pole was formerly North-East of *Russia*, it is now grown North-West of it, and a little to the Westward even of *England*'s Meridian. Monsieur *Petit* Engenier to the *French* King, is of Opinion, That the Pole of the diurnal motion of the Earth alters, but I confess I cannot in the least assent to it from any of those Arguments that he alledges, but I do rather think that divers of them do make against his Hypothesis; yet 'tis not impossible but that a very great Earthquake altering the Center of Gravity, may also alter the Pole of Rotation; for we find by experience, that if any thing be laid upon one side of a large *Lignum Vitæ* Ball suspended by a String, and that Ball be turned round upon the String, it shall not turn exactly about the Point by which 'tis suspended, but about some other Point. Besides this, we know that the direction of these Poles, as to the Heavens, doth vary, for whereas, it pointed at a part of the Heavens many degrees distant from the Star in the top of the tail of the little Bear, now it points almost directly towards it. Besides this, we find that the Points of the Intersection of the Æquinoctial and Eclipitck varies, and possibly even the motions of all may vary. A diurnal Revolution of the Earth may perhaps have been made in a much shorter time than now; possibly there may have been the same alterations in the Annual, and then a Year, or a Day at the beginning of the World would not be of so long a duration as now when those motions are grown slower; for if the motions of the Heavens by analogous to the motion of a Wheel or Top, as I think I can by very many Arguments make probable, then if the Earth were (as it were) at first set up or put into a rapid circular Motion, like that of a Top, 'tis probable that the fluid Medium in which it moves, may after a thousand Revolutions, a little retard and slaken that motion, and if so, then a longer space of time will pass while it makes its Revolution now than it did at first.

---

*Hooke's idea of changes in the Earth's center of gravity as a cause of changes in land and sea areas is a bold one that received almost no approval from his contemporaries. See Chapter 5.*

*A great earthquake could alter the center of gravity of the earth and therefore alter the axis of rotation.*

*Precession of the equinoxes.*

*The diurnal motion of the Earth, like that of a top, may have slowed down so that a day at the beginning of the world would have been shorter than now. This idea is in direct contradiction to Newton's explanation in a letter he wrote to Thomas Burnet in which he tried to accommodate the Creation into six days on the grounds that a day used to be longer. See Chapter 3.*

Hence possibly the long Lives of the Posterity of *Adam* before the Flood, might be of no greater duration then Mens Lives are ordinarily now; for though perhaps they might number more Revolutions of the Sun, or more Years than we can now, yet our few Years may comprehend as great a space of time; this perhaps might deserve to be inquired into had we a certain measure of time, such as some would have a standing Pendulum of a certain length; but since we are upon suspecting, we may even doubt whether the power of Gravity itself may not alter in time; we find that the Poles of the Loadstone may be changed, that it does take up more at one time than another; that its virtue may be wholly destroyed by Fire, and some other ways; and besides that, one of these changes is really wrought in the Earth, and therefore 'tis not impossible but that even the attractive Power of the Earth (tho' I confess I think it quite differing from that of the Loadstone) may be intended or remitted; if so then the Pendulum will be no certain Standard for the examination of the length of Time by; for the more the gravitating Power is increas'd, the quicker will be the Vibrations of any Pendulum and the more weak it is, the slower are the Vibrations: But this Digression only by the bye. To return then, I say, tho' somewhat may be said for this Supposition I have started, yet I confess I do more incline to believe that what Mutations there have been of the Superficial Parts, have been rather caus'd by Earthquakes and Eruptions, which ushers in my Ninth Proposition. Namely,

[p.323] That it seems very probable, that the tops of the most considerable Mountains of the World have been under Water, and that most probably they seem to have been raised to that height by some Eruption; So that those prodigious Piles of Mountains are nothing but the effects of some great Earthquakes. This the Poets seem'd to vail under the feign'd Story of the Giants, those Earth-born Brothers waging War with the Gods, where they are said to heap up Mountains upon Mountains, *Ossa* and *Olympus* upon *Peleon*, and to cast up huge Stones and Fire at Heaven, but that at last overcome by *Jove* with his Thunder, they were buried under Mountains, the chiefest of them, namely, *Typhæus* under *Sicily*, according to *Ovid Metamorph.* Lib. 5.

*Vasta Giganteis imjecta est Insula membris*
*Trinacris,* &c.------

Thus English'd by *Sandys.*

Trinacria *was on wicked* Typhon *thrown,*
*Who underneath the Islands weight doth groan;*
*That durst attempt the Empire of the Skies:*
*Oft he attempteth, but in vain to rise.*
Ausonian Pelorus *his right Hand*
*Down weighs;* Pachyne *on the left doth stand;*
*His Legs are under* Lilybæus *spread;*
*And Ætna's bases charge his horid Head:*
*Where, lying on his Back, his Jaws expire*
*Thick Clouds of Dust, and vomit flakes of Fire.*
*Oft times he struggles with the weight below,*
*And Towns and Mountains labours to overthrow.*
*Earthquakes therewith: The King of Shadows dreads*
*For fear the Ground should split above their Heads,* &c.

And that nothing else but an Earthquake is understood by that Gigantomachia of the Poets seems yet plainer from what *Virgil* in the Third Book

---

*Hooke speculates that the old patriarchs who were described in the Bible as having lived very long lives before the flood were not any older than people today. There were simply more revolutions of the Earth around the Sun and therefore more years counted, those years being much shorter than they are today. Also typically Hooke: As long as we're doubting, we can also suspect that the force of gravity changes and the magnetic poles could also change.*

*Arguments for the 9th Proposition: The tops of the highest mountains of the world have been under water.*

*Hooke was quite interested in the verisimilitude of legends not only because they contained descriptions of natural phenomena couched in mystical language but because he positively enjoyed the classics, having been educated in them as well as were others in the Royal Society.*

of his *Æneis*, speaks in his Description of the Shores of *Sicily*.

> *Portus ab accessu ventorum immotus, & ingens*
> *Ipse, sed horrificis juxta tonat Ætna Ruinis,*
> *Interdumq; Atram prorumpit ad Æthera nubem*
> *Turbine fumantem piceo & candente favilla,*
> *Attollitq; globos flammarum & sidera Lambit*
> *Interdum scopulos Avulsaq; viscera montis*
> *Eiicit eructans, liquefactaq; saxa sub auras*
> *Cum gemitu glomerat fundoque exæstuat imo.*
> *Fama est Enceladi semiustum fulmine Corpus*
> *Urgeri mole hac, Ingentemq; insuper Ætnam*
> *Impositam, ruptis flammam experare caminis:*
> *Et fessum quoties motat Latus, intremere omnem*
> *Murmure Trinacriam & cæco Subtexere fumo.*

Thus English'd by *Ogilby*.

> *The Port was great and calm with shelt'ring Shores,*
> *But near from horrid Ruins Ætna roars;*
> *There in black Whirlwinds pitchy Clouds aspire,*
> *With sparkling Cinders mixt with blazing Fire,*
> *And Globes of Flame high as the Stars are born;*
> *Out are the Mountains Marble Entrails torn,*
> *Then upwards vomited, and melted Stones*
> *Belcht from his Stomach, hot with horrid Groans.*
> *Enceladus with Thunder struck, they tell,*
> *Under the weight of this huge Burthen fell,*
> *Above him was the mighty Ætna laid,*
> *Who now breaths Fire, through broken Trunks convey'd*
> *And as he weary turns, a Thunder crack*
> *Sicilia shakes, and Heav'n is hung with black.*

And as the Poets above-mention'd had particular Stories and Giants for *Sicily* and *Ætna*, so had they also for other Vulcans, and from the frequency of them in former Ages, about *Greece* and the other Parts of the Mediterranean, *Sophocles* calls them ο γηγενης στρατος γιγαντων the Earth-born Army of the Giants; and that nothing but Earthquakes were deciphered by these Giants may be further collected from the place where they were said to be bred, namely, the *Phlegrean* Fields in *Campania*, a part of which is now called the Court of *Vulcan*, a place that is the vent of many Subterraneous Fires. "'Tis (says *Sandys*) a naked Level, in form Oval 1246 Foot long, and 1000 broad, environ'd with high cliffy Hills that fume on each side and have their [p.324] sulphureous Savour transported by the Winds to places far distant; you would think the hungry Fire had made this Valley with continual feeding, which breaks out in a number of places. Here the Fire and Water make a horrible rumbling, conjoining together as if one were Fuel to the other, here and there bubling up, as if in a Cauldron over a Furnace, and spouting aloft into the Air at such time as the Sea is inrag'd with Tempests, &c." Besides, how well do their Actions agree with the Effects of Earthquakes, for they are said to throw up burning Trees against Heaven, and huge Rocks and vast Hills, which falling into the Sea became Islands, and lighting on the Land became Mountains. Nor does the manner of their Generation speak less, for they are said to be generated by the Blood of Heaven falling down on the Earth, that is, by the heat or influence of the Celestial Bodies operating within the Bowels of the Earth, and brought forth of her Womb in revenge to the Gods, or that they break forth with such horror and violence as if they threatned the Heavens. And he that shall read the Description of the most notable of them *Typheus*, and compare with the natural Description of an Earthquake, will easily explicate the several parts of the Poets mystical Descriptions.

This Theory which I have endeavoured hitherto to evince, tho' indeed it be very hard positively to prove, we being, as I instanced before, very deficient in Natural History, yet if we consider what has been already said, and compare it with the late Observations of divers Travailers over them, we may find it altogether more than probable. I have been inform'd by several worthy Persons, that there are great store of Shells found at the tops of the *Alps, Appenine* and *Pyrenean* Hills, which are by much the highest of *Europe*. And I have now by me several of those Shells which have been dug out of them and brought into *England*. If therefore these have been real Oyster-Shells and Scallop-Shells as upon viewing the Substance and Make of them, I see not the least cause to doubt; and that there are great quantities of them to be found in divers Parts which lye buried in the Cliffs and incorporated with the Stones; and if that these Mountains have been infested with Earthquakes both formerly and lately, as we have several Histories that testifie; and if that other Eruptions and Earthquakes have raised Mountains even out of the bottom of the Sea, and that the power of included Fire is sufficient to move and raise even a whole Country all at once for some hundreds of Miles, as Historians assure us: If to this we add the universal silence in History of any part of *Europe*, or any of other certain places of the World before the Flood, or indeed for almost two hundred Years after the Flood, I think there will be much less scruple to grant it probable that the *Alps*, and divers other high Mountains, on whose tops are found such numbers and varieties of Sea-Shells, may have been heretofore raised up from under the Sea, and now sustain'd by the sinking of other Parts into the places from whence they were raised. This the very form of them will also very much argue for; for I have been inform'd by several that took diligent notice of it, that the parts are continually tumbling down from the higher parts to the lower, and that some of them do seem to overhang very strangely, which cannot in any probability be imagin'd to be the form of the first Creation, it being contrary to that implanted Power of Gravity, whereby all the parts of it are held together and equally drawn towards the Center of it, and so all the parts of it ought to have been placed in their natural position which must have constituted an exact Sphere, the heaviest lowest, and lighter at the top and the Water must have covered the whole Surface of the Earth, which seems to have been indeed their first position, according to the Description of *Moses* in *Genesis*, besides all those Hills that have been made by subterraneous Eruptions are of the like Structure; such as the *Pike* of *Tenariffe*, the *Pike* of *Fayale*, the new Mountain in *Italy*, *Ætna* and *Vesuvius*, all which seem to have been made up of great Stones thrown up out of the Mouth of their several Vulcans, many of which lie in such tottering postures that oft times they tumble down to the bottom, and make great destruction of the parts beneath; of this we have lately had several memorable Examples. To mention only two or three, we are inform'd by Historians, that among the [325] *Alps* in the *Grisons* Country, a Town named *Plura*, seated in a Plain at the Foot of the *Alps* near the River *Maira*, and continuing by estimation at the time of its fatal Catastrophy, at least fifteen hundred People was, by the falling down of a great part of a huge high Mountain that hung over the said place upon the twenty sixth of *August* 1617, together with the Inhabitants, in a moment crusht and buried deep in the Earth, and that there

*Shells have been found at the tops of the highest mountains in Europe—e.g., the Alps, Appenines and Pyrenees.*

*The arrangement of the Ur-earth, because of gravity, has the heaviest parts nearest the center and lighter layers toward the top and the water covering the surface This description happens not to contradict Moses, and Hooke is not above mentioning it. But mostly, his mention of the Bible is very sparse in this first series of lectures on earthquakes.*

is nothing now left in the place thereof but a vast abyss or bottomless Gulf. And we are now newly inform'd by Letters brought out of *Italy* that a great part of the City of *Ragusa* has been this Year destroyed by the like falling down of some part of the Mountain above it.

The Tenth and last Conjecture which I shall at present mention (as reserving some others which will seem at first sight much more strange and extravagant, till I can by a sufficient number of Observations make them more plausible) is, that it seems not improbable but that the greatest part of the inequality of the Earths Surface may have proceeded from the subversions and overturnings of some preceeding Earthquakes.

*Arguments for the 10th Proposition: The greatest part of the inequality of Earth's surface is the result of earthquakes.*

And for making this Conjecture probable, I might repeat all the Arguments I have already urged to make probable the Generations of Islands, Mountains, Abysses, &c. but that I suppose will be needless, they having been so lately mention'd. I could also instance in a multitude of other smaller effects produced by Earthquakes, of making the Surface of the Earth irregular; but those are so numerous, and so very well known in those places where Earthquakes are more frequent, as in *Italy, Turky,* the East and West *Indies, &c.* that I shall not insist on them. To this I might add the universality of Earthquakes, there being no part of the World of which we have any good account, but we find to have been some time or other shaken by Earthquakes; and 'tis very probable had we receiv'd any certain account of the State and Constitution, and being of the Earth in its Infancy (as I may say) or first Being of the Earth after its Creation, when 'tis not improbable but the parts of it that lay uppermost and next the other were more fluid and soft, we might have had a thousand other observables. Of which I shall say more hereafter when I mention some other Conjectures.

*Hooke's use of the word "conjectures" at the end of this paragraph shows that he is quite aware himself that he is speculating, keeping his mind open to all possibilities. Giving free rein to his imagination but within the confines of reason and evidence, both factual and hearsay, is what makes these discourses so fascinating.*

Thus much only I shall add at present, that from what I have instanced about Petrifactions and the hardning of several Substances, it seems very probable, that in the beginning the Earth consisted for the most part of fluid Substances, which by degrees have setled, congealed, and concreted, and turn'd into Stones, Minerals, Mettals, Clays, Earth, &c. And that in process of time the parts of it have by degrees concreted and lost their Fluidity, and that the Earth itself doth wax old almost in the same manner as Animals and Vegetables do; that is, that the moisture of it doth by degrees decay and wast either into Air, and from thence into the Æther; or else by degrees the Parts communicating their motion to the Fluid ether either grow moveless and hard, almost in the same manner as we find the Bodies of Animals and Vegetables when they grow old in their several proportinate times, all the Parts tend and end in solidity and fixtness, the Gelly becomes Gristles, and the Gristles a Bone, and the Bone at length a Stone, the Skin from smooth and soft grows rough and hard, the motions grow slow, and the moveable Parts and Joints grow stiff, and all the Juices decay and are deficient. The same thing happens in Trees and other Vegetables. If therefore the Parts of the Earth have formerly, in all probability, been softer, how much more powerful might Earthquakes be then in breaking, raising, overturning, and otherwise changing the superficial Parts of the Earth? Besides, 'tis not unlikely but Earthquakes might then be much more frequent before the Fuels of those subterraneous Fires were much spent. That the Parts of the Earth do continually grow harder and fixt and concrete into Stone, I think no one will deny that has consider'd the Constitution of Mountains, the Layers and Veins of them,

*Here Hooke mentions his idea to which he returns later: That the earth waxes old in spite of the cyclic processes that take place.*

*The Earth was formerly more fluid and soft.*

*Earthquakes were stronger and more frequent before the subterraneous "Fires" were much used up.*

the Substances mixt with them, the Layers of the several Earths, Sands, Clays, Stones, Minerals, &c. that are met with in diging Mines and Wells. The Nature of Petrifying Waters, the shapes of Crystals, Ores, Talks, Sparrs, and most kind of precious Stones, Marbles, Flint, [p.326] Chalk, and the like, every of which are by their forms sufficiently discover'd to have been formerly fluid Bodies, and whilst fluid, shaped into those forms: One or two undeniable Instances I shall add of the fluidity of Flints, and that shall be that I have now by me, a Flint that has so perfectly filled the Shell of an *Ecknius*, and inclosed it also that it has received all the impressions of the cracks of the Shells both on the Concave and Convex Part thereof, and has exactly filled all the Holes and Pores thereof, and has so perfectly received all the shape thereof as if it were nothing but Plaister of *Paris* tempered, Wax, or Sulphur that had been melted and cast on it; notwithstanding which it is a Flint so hard as to cut Glass very readily, and is of a very singular and uniform Texture; to this I might add many others of the like kind, which have the impressions of these and other Shells, and yet are some Marble, some Pebbles, some Agats, some Marchasites, some Ores, some Crystals, &c. Some Flints I have marked with impressions as exactly as if they had been soft Wax stamp'd with a Seal.

Further, That the Subterraneous Fuels do also waste and decay, is as evident from the extinction and ceasing of several Vulcans that have heretofore raged; which Considerations may afford us sufficient Arguments to believe that Earthquakes have heretofore, not only been much more frequent and universal, but much more powerful. If to this we do add what I formerly mention'd, that there seems to be no other more probable and intelligible Cause in Nature of the inequality of the Earths Surface, the natural Principle of Gravity reducing the Parts of it as near to an exact spherical Figure as their Solidity and forc'd Postures will permit, and consequently (as I mention'd before) the natural form produc'd by Gravity would be a multitude of Spherical Shells concreted of the several Substances of which it consists, incompassing each other, not unlike the Orbits or Shells (for we have no proper name for that kind of hollow Spherical Figure) of an Onion, or as the *Ptolemaiick* Astronomers do fancy the solid Orbs of the Heavens, ranged every one in its distinct Order according to its Density and Gravity; that is, that which hath been heaviest would have approach'd nearest the Center, or at least nearest to that part which is attractive and the cause of Gravitation, if such a Body there be in the middle of the Earth, and the next lighter in the second place, and so on to the third, fourth, fifth, &c. according to several degrees of Gravity and Density, they would have taken their several Quarters, and so Water would always have covered the Face of the Earth, and the lightest Liquor would always have been at the top, and the Air above that, and Æther above that; and as in Fluids so also in Solids, the Shells of Gold would have been the lowest of any Body we yet know, that of Quicksilver next, that of Lead next, and so the rest in their order, which seems also really to have been the form of the Earth, till disturbed by Earthquakes, which I conceive to be the reason of the scarcity of those heavy Bodies of Metal near the Surface, and of the greatest scarcity of Gold which is the most heavy, and that it is not to be found but in such places where there have in probability been great subversions by Earthquakes, as in Mountains, or in Rivers running out of Mountains, or in Earth washt and tumbled down from Mountains, and such like places, as by many Circumstances may be guest to have been formerly deeper under the Surface of the Earth.

*The replacement of a fossil shell by flint that perfectly fills the shell, as well as enclosing it so that it received all the impressions of the shell shows that the flinty material must once have been fluid to have penetrated all the cracks and crevices so exactly.*

*Hooke speculates about the core of the Earth.*

*The various layers arranged by density are only disturbed by earthquakes and that is the reason the heaviest metals like gold are rare on the Earth's surface.*

There is yet one Argument more that to me seems very good, and that is fetcht from no less distance than the Moon and the Sun by the help of Telescopes. These Bodies, as I have formerly hinted in the latter end of my Micrography, seem to have the same Principle of Gravity as the Earth, which, as I have there argued, seems probable from their Spherical Figure in general, and the several inequalities in particular, visible by the help of Telescopes on the Surface of the Moon, and the several Smoaks, and Clouds, and Spots that appear on the Surface of the Sun; and as they have that Principle in common with the Earth, so it seems to me that they are not free from the like motions with those of an Earthquake: For as to the Moon 'tis easily to be perceiv'd through a Telescope, that the whole Surface of it is covered over with a multitude of small Pits or Cavities which are incompassed round with a kind of protuberant Brim, much like the Cavities [p.327] or small Pits, which are left in a Pot of Alabaster Dust boyled dry by the Vapours which break out of the Body of it by the heat of the Fire; and all the inequalities that appear on the Surface of that Body, seem, by their form, to have been caus'd by an Eruption of the Moon, somewhat Analogous to our Earthquakes; all those Pits in the Moon being much like the Caldera or Vent at the top of Vulcans here on the Earth, or like those little Pits left at the top or surface of the Alabaster Dust by the natural subsiding of that Dust in the place where the Vapours generated within the Body of it break out. I need not, I think, spend time in urging Arguments to prove the sufficient powerfulness of the Cause to produce Effects as great as any I have ascribed to it, as being able to raise as great and high Mountains as those of the *Alps, Andes, Caucasus, Montes Lunæ*, &c. especially since even of late we are often informed of as great effects elsewhere, and even of the shaking and moving those vast Mountains by our latter and more debilitated Earthquakes, tho' those Mountains are now in probability much more compacted and tenacious by the since acquired Petrifaction, than they were before their first accumulation; and tho' 'tis not unlikely but the Fuel or Cause of the Subterraneous Fire may be much wasted and spent by preceding Conflagrations; Yet possibly there may be yet left in other Parts sufficient Mines to produce very great effects if they shall by any accident take Fire; and 'tis not impossible but that there may be some Causes that generate and renew the Fuel, as there are others that spend and consume it.

*Hooke describes the craters of the moon and likens them to pits produced by bubbling hot alabaster, therefore similar to eruptions. Earlier, in 1665, Hooke also performed impact experiments by throwing bullets onto wet pipe clay (*Micrographia*, p. 244–245) Also see Drake and Komar (1984).*
*Here is another example that Hooke's "Earthquakes" also include volcanic eruptions.*

*The imaginative Hooke, with no knowledge of radioactivity, speculates on the energy of the Earth, the subterranean heat: Could it be generated as well as spent?*

From all which Propositions, if at least they are true, will follow many others meer Corollaries which may be deduced from them.

First, That there may have been in preceding Ages, whole Countries either swallowed up into the Earth, or sunk so low as to be drown'd by the coming in of the Sea, or divers other ways quite destroyed; as *Plato's Atlantis*, &c.

Secondly, That there, many have been as many Countries new made and produced by being raised from under the Water, or from the inward or hidden Parts of the Body of the Earth, as *England*.

Thirdly, That there may have been divers Species of things wholly destroyed and annihilated, and divers others changed and varied, for since we find that there are some kinds of Animals and Vegetables peculiar to certain places, and not to be found elsewhere; if such a place have been swallowed up, 'tis not improbable but that those Animal Beings may have been destroyed with them; and this may be true both of aerial and aquatick

*Corollaries deduced from his Propositions:*

*1st corollary: Whole countries or regions have been destroyed.*

*2nd corollary: Whole regions have been raised from under water or from the Earth's depths.*

*Arguments for Proposition no. 11, i.e., extinction of species and generation of new ones, have become 3rd and 4th corollaries.*
*3rd corollary: Whole species have been destroyed, i.e., become extinct, or changed to adapt to a new environment.*

Animals: For those animated Bodies, whether Vegetables or Animals which were naturally nourished or refresh'd by the Air would be destroy'd by the Water. And this I imagine to be the reason why we now find the Shells of divers Fishes Petrify'd in Stone, of which we have now none of the same kind. As divers of those Snake or Snail Stones, as they call them, whereof great varieties are found about *England*, and some in *Portland*, dug out of the very midst of the Quary of a prodigious bigness, one of which I have weighing near          Pound weight, being in Diameter about          Inches, which I obtain'd from the Honourable *Henry Howard* of *Norfolk*: We have Stories that there have been Giants in former Ages of the World, and 'tis not impossible but that such there may have been, and that they may have been all destroyed, both they and their Country by an Earthquake, and the Poets seem to hint as much by their *Gigantomachia*.

*Reason for extinction: destruction of environment.*

*These dimensions were left blank by Hooke, but typically large fossil ammonites weigh 50–100 pounds and are more than a foot in diameter.*

Fourthly, That there may have been divers new varieties generated of the same Species, and that by the change of the Soil on which it was produced; for snce we find that the alteration of the Climate, Soil and Nourishment doth often produce a very great alteration in those Bodies that suffer it; 'tis not to be doubted but that alterations also of this Nature may cause a very great change in the shape, and other accidents of an animated Body. And this I imagine to be the reason of that great variety of Creatures that do properly belong [p.328] to one Species; as for instance, in Dogs, Sheep, Goats, Deer, Hawks, Pigeons, &c. for since it is found that they generate upon each other, and that variety of Climate and Nourishment doth vary several accidents in their shape, if these or any other animated Body be thus transplanted, 'tis not unlikely but that the like variation may follow; and hence I suppose 'tis that I find divers kinds of Petrify'd Shells, of which kind we have none now naturally produced; of this sort are many of those Helmet Stones which have been made by the Petrifactions of Substances in the Shells of several sorts of *Echini*, whose sorts have been destroyed by the alteration of the Nature of that part of the Sea where they were produced; and hence 'tis we find scarce any Shell-Fish in our *English* Sea that has a Shell like those sorts of *Nautili*; from whence our *Keinsham* and other sorts of Snake-Stones are produced.

*4th corollary: New varieties of the same species have been generated because of a change in the environment.*

Fifthly, 'Tis not impossible but that there may have been a preceding learned Age wherein possibly as many things may have been known as are now, and perhaps many more, all the Arts cultivated and brought to the greatest Perfection, Mathematicks, Mechanicks, Literature, Musick, Opticks, &c. reduced to their highest pitch, and all those annihilated, destroyed and lost by succeeding Devastations. Atomical Philosophy seems to have been better understood in some preceding time, as also the Astronomy evinc'd by *Copernicus*, the *Ægyptian*, and *Chinese* Histories tell us of many thousand Years more than ever we in Europe heard of by our Writings, if their Chronology may be granted, which indeed there is great reason to question.

*5th corollary. There might have been great civilizations in the past that have been totally destroyed by huge disasters. Although questioning the validity of Egyptian and Chinese chronologies, which are many thousands of years longer than ever heard of in European writings, Hooke evokes their histories to support this corollary. One has the impression, therefore, that it was the European or biblical chronology, that Hooke was really questioning.*

Sixthly, 'Tis not impossible but that this may have been the cause of a total Deluge, which may have caused a destruction of all things then living in the Air: For if Earthquakes can raise the Surface of the Earth in one place and sink it in another so as to make it uneven and rugged with Hills and Pits, it may on the contrary level those Mountains again, and fill those Pits, and reduce the Body of the Earth to its primitive roundness, and then the Waters must necessarily cover all the Face of the Earth as well as it did in the beginning of the World, and by this means not only a learned Age

*6th corollary: A global disaster like the Deluge, therefore, could have occurred that destroyed all living things save those survived in Noah's Ark—i.e. " 'tis not impossible."*

may be wholly annihilated, and no relicks of it left, but also a great number of the Species of Animals and Plants. And 'tis not improbable but in the Flood of *Noah*, the Omnipotent might make use of this means to produce that great effect which destroyed all Flesh, and every living thing, save what was saved alive in the Ark.

Seventhly, 'Tis not impossible but that some of these great alterations may have alter'd also the magnetical Directions of the Earth; so that what is now under the Pole or Æquator, or any other Degree of Latitude may have formerly been under another; for since 'tis probable that divers of these parts that have such a Quality may have been transpos'd, 'tis not unlikely but that the magnetick Axis of the whole may be alter'd by it, after the same manner as we may find by experiment on a Loadstone, that the breaking off and transposing the parts of it, do cause a variation of the magnetick Axis.

I could proceed to set down a great many other Corollarys that would naturally follow from these Principles if certainly proved. But this Essay I intended only as a hint or memorandum to such Gentlemen as travel or any other inquisitive Persons, who for the future may have better opportunities of making Observations of this kind, that they may be hereby excited, or at least intreated to take notice of such Phænomena as may clear this Inquiry tho' never so seemingly mean and trivial, since it seems not improbable but that they may discover more of the preceding duration and alterations of the World than any other Observations whatsoever, and that thence may flow such instructions as may be of some of the most considerable uses to humane Life and Society, to which end all our Philosophical Studies and Inquiries tend. Ended *Sep.* 15.1668.

*One needs to read the entire* Discourse of Earthquakes *to conclude whether Hooke was confined by the biblical chronology. It is clear that he was not, although he acknowledged that the flood was possible.*

*7th corollary. Great changes to the surface of the globe might also cause a variation in the magnetic axis of the Earth.*

**No. 2.** [p. 365]*Of the standing of the* Mercury *in the Tube to the height of 75 Inches, read May 28, 1684.*

That *Theories* are not altogether useless, we may perceive by the happy invention of the ingenious *Galileo*, and the addition of the acute *Torricellius*, which two completed the Experiment of the *Æthereal Vacuum*; the further Improvement and Observation of which hath produc'd the *Barometer*, now useful for predicting the variation of the Weather, and the *Pneumatick* Engine much more prolifick of discoveries; the causes of most of whose Phænomena are sufficiently obvious, and certainly known to be the Gravitation and Spring of that part of the *Atmosphere*, which is call'd *Air*, and agreed to by the most accurate of the Modern Philosophers. But Mr. *Hugens* [sic] about twenty Years since having tried Mr. *Boyle*'s Experiment of making Water descend in a Tube, the Orifice of which was inclosed in an exhausted Receiver, found that if the Water were first well freed from the Air that is usually latent in it, and then inclosed, the Water would not descend in the Pipe, tho' the pressure of the Air were wholly taken off; this occasion'd the trial to be made here with *Quicksilver* instead of Water, and by many Experiments it was at last found by Mr. *Boyle* the Lord *Broucher* [sic], and several others, that the *Quicksilver* also when the Tube was very well freed from the latent Particles of Air, would not part from the top of a Tube, tho' it were twice as high as the usual height the *Quicksilver* used to stand at; and tho' there were no more pressure upon the stagnant Vessel than was usual, and that the bottom of the Tube were as open and free for the *Mercury* to run out, as was usual for Experiments of the Mercurial Standard. This seem'd at first to overthrow the Theory of the Gravity of the Air, and was made use of by some *Antagonists* to that purpose, but with little reason; for it was observ'd, that in the making this Experiment if a little Jog were given to the Tube in which the *Mercury* thus remain'd suspended, the *Mercury* would immediately leave the top and fall down in the Tube to the usual height of about thirty Inches, and there exhibit all the same Phænomena as the common Mercurial Standard or Torricellian Experiment had been observ'd to do. However it could not but affect the inquisitive after the causes of things, with a disire of satisfying themselves with some probable Conjectures at the causes of this so strange an Effect, some supposing one thing, some another; what my Conjectures were, and still are, I shall in brief declare and leave them to be consider'd by such as have better Abilities, and shall please to trouble themselves with such inquiries.

Since I first made the Experiment, I saw an absolute necessity of a pressing Fluid very much more subtile than the Air, and yet consisting of parts of a determinate bulk, which would easily strain through and pervade the Pores of Glass, Water and other Bodies impervious to the Air, but could be kept out by the nearer Conjunction of some of the constituent parts of those Bodies which constitute Pores of a much less Magnitude or Capacity, which fluid I suppos'd might be somewhat of the Nature of the second Element of *Descartes*, tho' for many Reasons drawn from Experiments, I suppos'd it to have many differing Proprieties from those which he ascribes to his, and I saw also a necessity of supposing a third Element consisting of a matter yet more subtile and fluid, [p.366] as he supposes, and more then that, of several other fluid Matters, some more subtile than others, each of which have their proprieties distinct, and are the causes of this or that Phænomenon in the World, of which there hath as yet been no intelligible reason given of their Power and Original, as I

---

*This lecture about pressure on Mercury, although not properly a subject on earthquakes, nevertheless contains some interesting ideas. Waller states that this lecture was first read to Lord Brouncker, the first president of the Royal Society, in the 1670s and with some small changes in the first page read again to the Royal Society in 1684.*

*Hugens = Huygens*

*Broucher = Brouncker*

may hereafter shew in the Explication of some of them. And I do believe, from that little insight I have had of the Operations of Nature, that all the sensible part of the World is almost infinitely the least part of the Body thereof, and but, as it were, the *Cuticula*, or outward Filme of things; whereas that which fills up and compleats the space incompass'd by that Filme consists of a multitude of insensible Bodies, each of them as distinct in their Natures and Operations, as Air and Quicksilver, or any other two sensible Bodies we can name; for so many, so curious, and so minute are the insensible workings of Nature, that without supposing some such Instruments as these we shall quickly find a *non plus* in the explication of any one appearance in Nature. But for the finding out the Number and Nature, of these Elements we can proceed but by slow and single steps, and 'tis not to be expected but from a long and close prosecution of Nature as we see that the pressure of the Air was not detected till *Galileo* and *Torricellius* happily light upon that Notion, and this second Element was not experimentally manifested till the makng of this Experiment of the *Mercury*'s standing much above the height discover'd by *Torricellius*. The matter on which the Loadstone works will perhaps be found another, and that which causeth Gravity a Fourth. But this by the bye.

My Notion and Explication of this Phænomenon is this, That there is another fluid Body; this Fluid is the Menstruum or Liquor into which the Air is dissolv'd like a Tincture of Cocheneel into Water, which, as I have explain'd in the 15th Observation Microscopical, *Page* 96, and 97, doth penetrate the Pores of the Glass Water, and several other terraqueous Bodies (possibly all such as are transparent, for that I have not a sufficient supellex as yet to determine positively; nor is it material for the Explication of this Phænomenon, as other more curious and critical Experiments shall be found out, it will be time enough to determine it.)

Next that this fluid, as all other sensible fluids we meet with, hath a greater Congruity or Incongruity to this or that Body it is contiguous to, and therefore doth more readily join to this Homogeneous than to that Heterogeneous Body, whether Solid or Fluid, and doth more easily penetrate the small Pores of the Homogeneous, and not without some difficulty the Pores of the Heterogeneous Bodies. And in short, that this Congruity or Incongruity of it to other Bodies doth make it perform the same kinds of Effects with those we find perform'd by sensible Fluids and Solids, such as I have explicated in my sixth Mycroscopical Observation.

Thirdly, That this Fluid doth not at all penetrate the Body of *Quicksilver*, tho' *Quicksilver* may be penetrated by a great number of other more subtile Fluids, such as those which cause Gravity, Magnetism, Fluidity, &c. if at least it shall be found necessary by future Experiments to ascribe those three Properties to more than one fluid.

Fourthly, That several other Liquours whose greater Pores are penetrated by this Fluid, may yet be sustain'd by it above the level equivalent to twenty nine Inches of *Quicksilver*, so long as the Pores are not so far seperated as to admit the parts of this Fluid between them where they are more neerly contiguous, and have some more subtile fluid Body only between them. Of this kind Water well purged of Air may be one, as the Experiment of the not subsiding of Water purg'd of Air doth manifest.

Fifthly, That this Fluid hath a pressure every way analogous to the pressure of the Air, and that this pressure is much greater than that of the Air.

*Here Hooke explains his concept of Congruity and Incongruity, an idea he had discussed much earlier and also in* Micrographia *under Observ. no. 6, and refers to materials at hand in a chemistry laboratory situation: Some fluids have more of an affinity to penetrate the small pores of some bodies than others (i.e., at the atomic level), as, for example, some solids are more easily dissolved in some liquids. But Westfall extrapolates this idea out of context to apply it to Hooke's gravitational ideas in an attempt to discredit the significance of Hooke's contribution in this matter. See Chapter 1, note 21, and Chapter 4.*

Sixthly, That there is no need (for the explicating any Experiment I have yet heard of) of supposing it to have a springy Nature like that of the Air, since all the Phænomena may be solved without it.

For the more intelligible Explication of this Solution, I shall indeavour to shew an Experiment very much like it, in sensible Bodies. I took then a small Glass Cane *a, b, c,* open at both ends, then having procur'd a long small Glass [p.367] Pipe in a Lamp almost as small as a Hair, I brake it into a great many short ones and made of them a Stopple by binding them together Fagot-wise with thread, and melting Wax or Cement about them, so as that none of their Perforations were stopt, I put them into the end *a, b,* for a Stopple, then I had another Cane of Glass big enough to contain the former wholly as Def. which was fill'd with Water; then the first Tube with the open end downwards was immers'd into the said Tube Def. till the Water had fill'd the whole, and the ends of the small Pipes *iiiii*, then gently raising up the Tube *a, b, c,* out of the Water, I found I could raise it so high as that the Water in the Tube *a, b, c,* did stand above the Surface of *g, h,* the Water in the Glass Def. some Inches. Wherein 'tis observable, that tho' all the ends of the Pipes *iiii* were pervious to the Air, yet by reason of a great Congruity of the Water to Glass than of Air, the Air was not able to force its way thorough without the help of the Gravity of the Cylinder of Water *a, g, h, b*; the same Experiment I tried also with *Quicksilver,* by making the Stopple *k,* of Brass, and instead of the small Pipes caus'd to be drill'd, thro' the same a great number of small holes, then by the help of *Aqua Fortis* I caus'd all those holes to be whited with *Quicksilver* then holding my Finger against those holes, and filling it with *Mercury,* and stopping the other end, and immerging it under other *Mercury* in a Dish, by degrees I rais'd the same, and found that the Air would not force its way in at the above-said drill'd holes, till the end *A, b,* was rais'd above the level of the *Mercury* in the Dish some Inches.

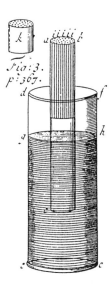

From both which Experiments 'twill not be difficult to understand my explication of this Phænomenon of the extraordinary hight of the *Mercury* in a Tube, well fill'd, and perfectly cleans'd of Air: For if we suppose in the former Experiment, that the Ambient Air doth represent the Ambient Fluid, whose pressure we do suppose, and that the Perforations of the small Pipes do represent the imaginary Pores of Glass, and that the Water with which it is fill'd doth represent the *Quicksilver* in the new Experiment, and that we suppose that *Quicksilver* hath a greater Congruity to Glass than the other, and that consequently it keeps the other from getting a Body within the hollow of the Tube by stopping it at its first entry, 'twill be easie to imagine how, tho' the Glass can be supposed all over Porous, through which the Æther can pass, the Ambient new Fluid can by its more free pressure on the Surface of the bottom, keep the *Mercury* suspended forty five Inches above the former Standard of thirty Inches.

Nor doth the second Experiment explicate it less naturally; for the Air represents the Æther, or what other name soever it be call'd by: The *Quicksilver* in the one represents the *Quicksilver* in the other. Immerse that Dish in a Bucket of Water, and you shall find that the top of the Tube will be rais'd considerably higher above the Surface of the *Mercury* in the Dish before the *Mercury* will leave the top; Then the Water, under which the *Mercury* is immers'd, will represent the Air or Atmosphere, and the holes in the Brass Stopper, the Pores of the Glass, the additional raising of the height of the *Mercury* after 'tis put under Water will shew how part of the seventy five Inches is ascribable to the pressure of the Air, and the other height will shew how another part of it is ascribable to the pressure of the Æther. I think I need not explain it farther, only 'tis observable, that tho' the Air finds a difficulty to make its first entry into the small holes of the Glass Pipes, or of the Brass Stopper, yet after it hath got through, and that there is Air within the Tube as well as without, it very readily and freely maintains its Passage, and the same *Phænomenon* also happens in the *Quicksilver* Experiment, for as soon as ever the *Mercury* begins to seperate from the top of the Tube, and the Æther hath a Body within the Tube, it readily falls down to the height supported by the pressure of the Air. The Reason of the two preceding Experiments, to wit, of the suspension of Water in the Tube whose end is stopped with a bundle of small Glass Pipes, as also of the suspension of the *Mercury* in the Tube whose end is stopp'd with the perforated peice of Brass, will be, I think, sufficiently manifest to him that shall thoroughly consider the Nature of the Congruity and Incongruity of Bodies to one another; somewhat of my Thoughts concerning the same I have formerly deliver'd in the [p.368] sixth Observation of my Micrography, which was indeed but a cursory Meditation for the solving of the Phænomenon then mention'd; but whosoever shall thoroughly examine the Nature and Power of it, will, I doubt not, find it much more universal. To me indeed it seems to be not only the cause of this extraordinary *Phænomenon*, but of the Conglobation and Tenacity of most Liquors of the Tenacity Springyness, Sonorousness, Malleability, &c. of all solid and hard Bodies: But of this elsewhere when I have occasion to examine what is the cause of Congruity itself, which I do not suppose a first Principle, but rather of a second, third, or fourth Rank, which being more universal, must be ascended to by degrees, after the Synthetick method. To proceed then, I did heretofore propound in the twentieth Page of my Micrography as a thing worthy trial to examine what Power was requisite to force a Liquor through holes of several bignesses made in a Heterogeneous solid, and fill'd with some Liquor Homogeneous to that solid; for were that accurately done I judge this Experiment of the extraordinary height of the *Mercury* above the usual Standard would give us a demonstration of the bigness of the Pores of Glass; for since we find that a hole of ------of an Inch will make the *Mercury* stand suspended one Inch in height a hole of -----of an Inch will make it stand suspended two Inches; a hole of -----of an Inch will make it stand three Inches, it will follow that a hole of ----- of an Inch will make it stand forty five Inches, and a hole of -----of an Inch will make it stand a hundred Inches, which minds me of several other Experiments worth trial, for determining this controversy; such as these;

First, Whether some Glasses are not more porous than others, and consequently whether the *Mercury* will not stand to a much greater height in Tubes made of Glass of a more opacous or more refracting Substance than in Tubes of a more transparent or less refracting Substance.

Secondly, Whether in a Tube made of Lead very intire from holes and perfectly cleans'd of Air and rubb'd with *Mercury* that doth every where stick to the same; if the said Tube be fill'd with very well cleans'd *Mercury*, the *Quicksilver* will not stand suspended to a much greater height than it doth in Tubes of Glass, for if Lead, Silver, &c. be impervious to this fluid Substance that so freely penetrates Glass, it seems not improbable, but that the *Mercury* may stand suspended to a very much great height, and if so it will be a certain way of finding out the force or pressure of this fluid; from the determination of which will follow probably the reason of the Strength,

Weight, Sonorousness and Springyness of Metals. And I am the more inclin'd to believe that this Experiment will succeed, because I judge that the same fluid that conveieth height, is the cause of this Phænomenon; and whatsoever Body is perfectly impervious to Light, is also impervious to this fluid. But herein I would be understood not to mean such Bodies, as by the thickness of their bulk and some degree of opacousness, do intercept the direct passage of this fluid Matter, and so by consequence cause a kind of opacousness, as a thick Body of Red and Blew, &c. Glass which notwithstanding are not perfectly opacous Bodies, because, when made very thin, they are transparent of a Red or Blew Colour: For such Bodies, tho' they may intercept the direct passage of the Light, yet may they admit the fluid freely to pass through their winding Pores, and so may not perhaps keep the *Mercury* suspended much higher than a Tube of Crystal-glass; whereas I am very apt to think, that if there could be a Tube made of a Substance perfectly impervious to this fluid Matter, the *Mercury* may possibly remain suspended as many Feet as it doth now Inches; but this trial will more fully inform. Now that a Body may be pervious to some Liquors and yet not pervious to Light, is evident by the Experiment of forcing *Mercury* through the Pores of Wood, for if you take a Pipe of Beech, Elm, Oak, Firr, Ash, or the like, of four, five, six, eight, or ten Foot long, and stopping one end thereof, you erect it with the open upwards and fill it with *Quicksilver*, you shall find that the *Quicksilver* will as freely and plentifully pass through the Microscopical or Imperceptible Pores of the same, almost as it will be strain'd through the [p.369] Pores of Cloath, Linnen or Leather, and will thereby so fill the Pores of the Wood as to make it feel almost as heavy as Lead; by this way I have been able to force *Mercury* into the Pores of Charcole, and divers other Vegetable Substances, whereby the Pores of the same are made very conspicuous, by placing small peices of those Substances at the bottom of a Glass Tube of four, six, eight, or ten Foot long, and filling the Tube with *Mercury* over them, for those and most other Vegetable Substances, will, by the pressure of such a Cylinder of *Mercury*, be fill'd with *Quicksilver*, and thereby plainly discover the Shape and Texture of their Pores. I have not had the opportunity to try Bones, Horns, Teeth, Hair, Quills, and the like animal Substances this way, tho' it seems to me very probable, that their Pores may be discover'd this way, at least by lengthing the Cylinder, and making the pressure yet greater, or by a condensing Engine: Nay, I am inclin'd to believe, that *Mercury* may be forc'd even through the Pores of Glass itself if the Cylinder pressing be sufficiently lengthned; for by this Experiment of the suspension of *Mercury* at seventy five Inches high, it seems that *Mercury* has a greater Congruity to Glass then the pressing Fluid or Æther hath to the same, and therefore 'tis not improbable but that a force as great in proportion to the bulk of *Mercury*, as the force of the Æther is to the bulk of the Æther, may force it through the Pores of Glass, that it may be subtile enough to do it, seems probable from this, that it doth so readily penetrate the Pores of Gold, Tin, Lead, Silver, &c. those Bodies with whom it hath a perfect Congruity even with meer apposition and contiguity; and therefore 'tis not improbable but that a degree of force may make it penetrate the Pores of Glass, which in probability are much greater than those of the congruous Metals, especially since we find it can be forc'd into the Pores of Wood, Cork, Pith, Coles, &c. so as to drive out the Juices contain'd in them; whereas those Juices having a great Congruity, do penetrate them by meer apposition. Now that this penetrancy of *Mercury* into Glass is not meerly conjectural, I shall shew you by taking notice of certain Spots or Stains which I have found in polish'd Looking-glass-Plates after they have remain'd a long while foil'd, and then being unfoil'd, for I have very plainly seen with a Microscope that there hath been in the place, where spots appear, an infinite number of exceeding minute Parts of *Mercury* which seem to be gotten into the very Pores of Glass, and can by no kind of rubbing be fetcht out without wearing away so much of the very Substance of the Glass itself: What therefore is thus done accidentally by duration, might in probability be much better done by pressure, if we were able to make it considerable enough, as by letting down Glass in *Mercury* to a very great depth under Water, where that can be done; or to a considerable depth under the pressure of *Mercury*. It may possibly be Objected, that if *Mercury* hath a greater Congruity to Glass than this suppos'd Fluid, why doth not the *Mercury* without much force penetrate the Pores of Glass at first, and so running through it, make it appear opacous. To which I Answer, That tho' I suppose *Mercury* to have a greater Congruity to Glass than this Subtile Fluid, yet that it hath not a perfect Congruity, but rather an Incongruity in respect of other Fluids that are more Congruous, as Air. Nor hath that a perfect Congruity, but rather an Incongruity in respect of Water; for there may be infinite degrees of Congruity, as Water salt hath more Congruity to Glass than Water fresh, Waters than Vinous Spirits, Vinous Spirits more than Oils, Oils more than Airs and Fumes, and they more than *Mercury*, and *Mercury* than this fluid Æther, or what other Name so ever we call it by; and in every one of these degrees of Congruity or Incongruity there may be a multitude of other Subdivisions; as for instance, under the first Head there may be a very great variety; I know some, acid Liquours that will of themselves, without any force, penetrate the Pores of Glass so as to dissolve it into a Powder, whereas others will not at all penetrate or dissolve it by any means I have yet found: But this part of Congruity and Incongruity by which solid Bodies become dissoluble by Fluids, and whereby Fluids readily penetrate each other, and unite with one part of a Fluid, and separate or precipitate another, belonging to another Subject, I shall pre-[p.370] termit at present, and only take notice of some things that may be pertinent to the Inquiry under Consideration; and those may be these;

First, That there is no difficulty at all in admitting, that within the same Liquor, which to the sight appears uniform, there may be a great variety of Fluids of differing penetrancy, for we find in *Aqua Regis* for instance, that there are the *Sal Armoniack* parts that help to penetrate Gold; the *Nitrous* that penetrate Silver, Copper, &c. the *Flegme* that will penetrate neither: There have been few Experiments made of the penetrating of one Fluid by an other, beside that I formerly shew'd of Water and Oil of *Vitriol*, 'tis a copious Head, and contains much of information; Copper and Tin melted are an example of it. From which Observation we may without difficulty suppose the Air (as it is commonly taken) to be a Body consisting of a great variety of Fluids, of which this Æther we suppose may be one; and possibly the principal which takes up the greatest space, and whose Effects are the suspension of *Mercury* above the height suspended by the pressure of the Air and the like; tho' yet I suppose it not the subtilest, there being many Experiments that do seem to require a much more subtile and penetrant Fluid, of which more elsewhere. The Elastical Part of the Air that causes the Phænomena of Springing a Second, the Steams of Bodies a Third, the Nitrous part a Fourth, each of which have several degrees of penetrancy, and may possibly be several distinct Fluids, tho' when blended altogether they make that compound Body, which we call the Element of

Air. Now as the Air consists of a variety of Fluids, so 'tis not unlikely but that each of these may differ in their proportionate quantity, and in their respective Gravity; so that if we should take the whole bulk of the Air or Atmosphere we might possibly find it made up of divers Fluids, as of the Fluids *A, B, C, D, E, F, G, H,* &c. and each of these of differing proprieties, both as to penetrations, Quantities and Gravity, Congruity, and the like; and that That part of it which is next the Earth might be a compound of the Fluids *A, B, C, D,* and *G, H,* extending to a certain number of Yards above the level of the Sea, the next part of the Atmosphere immediately above it may consist of *B, C, D, E,* and *G, H,* and having nothing of *A,* or *F*. The third Region may consist of *C, D, E, F, G, H,* and have nothing of *A,* or *B*: And this seems probable, First, because we find that there are several distinct Surfaces of the Air, upon which the several Regions of the Clouds seem to swim like Froth upon the Surface of Water; for 'tis obvious to any that shall observe it, to see the under Surface of the Clouds smooth and level, and the upper in confus'd heaps, and further, that all the under Surfaces of Clouds appearing at the same time lye as near as one can judge by the Eye in the very same Level. Next that the make of the Clouds in a higher Region are quite differing. Thirdly, that the parts of the Air in several heights from the Earth, have differing proprieties, as it hath been found in very deep Wells, that the lower twenty Fathoms were all possess'd by a Damp, or an Air in which no Fire would burn, or Animal live. We are inform'd also, that the Air at the top of some exceeding high Mountains is of such a Nature as will not serve for Respiration. Possibly the presence of *A,* in the lowest, may be the cause of the first Effects and the want of *B,* at the tops of Mountains may be there the cause of those other Effects.

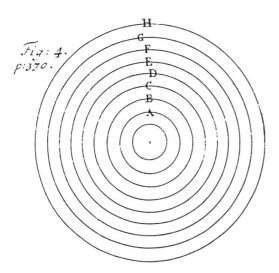

**No. 3.** *This series of lectures was read between December 8, 1686 and January 19, 1687.*

[p.329] I am not, I suppose, the only Person who hath heard some Persons (with what Reason I dispute not) ask what the *Royal Society hath done for so many Years* as they have met. And other Persons as confidently affirm that they have done *just nothing*. Nor am I ignorant that the same Reflections have been made upon me in particular with more severe Aggravations. As to what concerns my self I shall not now spend time in answering, designing to do it by another way. But as to what concerns this Honourable Society, I conceive it might be a satisfactory Answer to assure such Objectours that this *Society* have been imployed in collecting such *Observations*, and making such *Experiments* and *Trials* as being fitly apply'd and judiciously made use of, will very much tend to the advancement of *Natural Knowledge*: And tho' the things so collected may of themselves seem but like a rude heap of unpolish'd and unshap'd Materials, yet for the most part they are so qualified as that they may be fit for the beginning, at least of a solid, firm and lasting Structure of Philosophy.

But because some of those may doubt whether really there be any such Collection made, and more of the practicableness of making such a Use and Application of them, and will not acquiesce and be satisfied with the effects hereof that future times may produce, but are desirous to see some *Specimen* of what may be hoped for, by seeing the Ground designed and set out, the Foundation laid, and the Workmen beginning to raise the Walls, and make use of the Materials that are said to be got in readiness for such a Fabrick.

I conceive, it may not be altogether unseasonable this following Year nor improper for this *Honourable Society's Care* to make some attempt of that kind by shewing some *Specimen* of such a Structure raised from Observations and Collections of their own, that it may appear that they have not disquieted themselves in vain, in heaping up such a Treasure which they know not who shall enjoy or make use of; that is, to see whether any of these things they have been collecting, will afford sufficient Evidence to ground a deduction of a higher Nature upon, such as is more obscure to be seen, or more difficult to be ascertained of, to see whether, when a weight comes to be laid open the Stones or the Bricks, the natural Observations, or the Artificial Trials and Experiments, they will not crush under it, and fail of sufficient Solidity; and if they do, whether such may not be fit for other places, and whether it will not be necessary to seek out for some others that may be more firm and solid, and such as are of a closer and better concocted qualification, which may be more powerful to sustaine a higher Superstructure, and a greater weight of Argument to be laid upon them.

In order to this it is necessary (for the Architects at least) to know, 1st. What are the particular qualifications necessary for the several Materials of their designed Fabrick. 2dly. The Methods by which these qualifications may be examin'd. 3dly. The place where, and the means how Materials, so qualified may be proved, without which præmeditated Design, Knowledge, [p.330] and Care; a Collection, tho' very great, made at a venture must needs contain abundance of such as will be of little use for the end aimed at, and not only so, but will also prove a great Obstruction and Confusion in finding out such as are proper, and in separating the Good from the Bad.

*And before the publication of Newton's* Principia.

*The meetings of the Royal Society were mercilessly satirized in some contemporary plays and poems by various authors. Hooke's work with the microscope was singled out for ridicule in a play by Shadwell. See Chapter 1.*

*Since Hooke's vast experience in rebuilding London after the Great Fire, he seems to enjoy using architectural metaphors for the advancement of knowledge.*

The Structure aimed at, is a true and certain knowledge of the Works of Nature, and this is designed to be attained as fast as may be, and to be perfected as far as may be; or the end of the Inquisition is the promotion and increasing of Natural knowledge.

The methods of attaining this end may be two, either the Analytick or the Synthetick. The first is the proceeding from the Causes to the Effects. The second from the Effects to the Causes: The former is the more difficult, and supposes the thing to be already done and known, which is the thing sought and to be found out; this begins from the highest, most general and universal Principles or Causes of Things, and branches itself out into the more particular and subordinate.

*Hooke's methodology is clearly stated in this lecture. For a discussion on this subject, see Chapter 4.*

The second is the more proper for experimental Inquiry, which from a true information of the Effect by a due process, finds not the immediate Cause thereof, and so proceeds gradually to higher and more remote Causes and Powers effective, founding its Steps upon the lowest and more immediate Conclusions.

An Inquisition by the former Method is resembled fitly enough by that Example of an Architect, who hath a full comprehension of what he designs to do and acts accordingly: But the latter is more properly resembled to that of a Husbandman or Gardener, who prepares his Ground and sows his Seed, and diligently cherishes the growing Vegetable, supplying it continually with fitting Moisture, Food, Shelter, &c. observing and cherishing its continual Progression, till it comes to its perfect Ripeness and Maturity, and yields him the Fruit of his Labour. Nor is it to be expected that a Production of such Perfection as this is designed, should in an instant be brought to its compleat Ripeness and Perfection; but as all the Works of Nature if it be naturally proceeded with, it must have its due time to acquire its due form and full maturity, by gradual Growth and a natural Progression; not but that the other method is also of excellent and necessary use, and will very often facilitate and hasten the progress to Perfection. An Instance of which kind I designed some Years since to have given this Honourable Society in some of my Lectures upon the motions and influences of the Cælestial Bodies, if it had been then thought fit; but I understand the same thing will now be shortly done by Mr. *Newton* in a Treatise of his now in the Press: But that will not be the only Instance of that kind which I design here to produce, for that I have divers Instances of the like Nature wherein from an *Hypothesis* being supposed or a premeditated Design, all the *Phænomena* of the Subject will be *a Priori* foretold, and the Effects naturally follow as proceeding from a Cause so and so qualified and limited. And in truth the *Synthetick* way by Experiments Observations, &c. will be very slow if it be not often assisted by the *Analytick* which proves of excellent use, even tho' it proceed by a false position: for that the discovery of a *Negative* is one way of restraining and limiting an *Affirmative*.

*Here Hooke alludes to his priority in the matter of gravitational laws, referring to Newton's upcoming treatise as "the same thing" he had already lectured to the Royal Society.*

*Here he demonstrates his advanced view of the philosophy of science: The value of a negative discovery is its limiting effect on a positive one.*

But not to spend more time at present upon the more particular explications of These Methods, which would make of it self a very large Discourse, I shall proceed to the Subject which I began to discourse upon the last meeting, premising only in general what I think necessary thereunto, an Explication of what I understand by natural Knowledge, or the Knowledge of Nature.

By *Knowledge* then in the highest *Idea* of it, I understand a certainty of information of the Mind and Understanding founded upon true and undeniable Evidence.

[p.331] True and undeniable Evidence is afforded either immediately by *Sense* without Fallacy, or mediately by a true Ratiocination from such Sense.

I call that *Sense* without Fallacy, where the *fallacies* of Sense being detected and known, the Evidence produced thereby is examined and found to be free and clear of all such *Fallacies*.

I call that true *Ratiocination* from such Sense, where being sure of the Premises, the Conclusion necessarily follows from them; which is the method of Reasoning made use of in *Geomotry* [sic], and by which we arrive at as great a certainty of things unseen as seen. Thus *Ovid* describes the method of *Pythagoras*, in scaling the Heavens.

> _____*Isq; licet cæli Regione Remotus*
> *Mente Deos adiit, & quæ natura negabat*
> *Visibus humanis, oculis ea pectoris hausit.*

Now tho' in Physical Inquiries, by reason of the abstruseness of Causes, and the limited Power of the Senses we cannot thus reason, and without many Inductions from a multitude of Particulars come to raise exact Definitions of things and general Propositions; yet by comparing of varieties of such Inductions we may arrive to so great an assurance and limitation of Propositions as will at least by sufficient to ground Conjectures upon, which may serve for making *Hypotheses* fit to be enquired into by the *Analytick* method, and thence to find out what other Experiments or Observations are necessary to be procured for the further progress in the *Synthetick* which will question-less so far inform us of the general and universal progress of the Operations of Nature, that nothing but what is really the truth shall be proposed but the absurdity and insufficiency thereof will presently be detected and proved.

So that tho' possibly we may not be able to produce a *Positive* Proof, yet we may attain to that of a *Negative*, which in many cases is as cogent and undeniable, and none but a willful or senseless Person will refuse his assent unto it. Thus much I thought was necessary to premise in order to what I have further to propound to the Consideration of this noble Society upon the Subject I have discoursed of formerly only by way of Queries; it being my aim at present to see what *Positive* or *Negative* certainty at least may be attained concerning the same, either from the information of Sense freed from Fallacies or from the superstructures of Reasoning.

I propounded then two Hypotheses for the solution of the *Phænomena*, observ'd in *Petrified Bodies*, of the resemblance of Animal and Vegetable Bodies; such as the Shells and Bodies of Fishes, the Bones, Teeth, Hornes, &c. of Fish, and also of Terrestrial Animals, the Wood, Leaves, Bark, Roots, and Fruits of Plants and Vegetables; which resembling Bodies are found to be of variety of Substances, sometimes of Stone, as Flint, Marble, Black, White, Gray, and of various other Colours, of Free-stone, *Portland*-stone, Chalk, and an infinite variety of other Stone, some harder, some softer: Of various kinds of Clay, Earth, Sand, &c they have been found near the tops of the highest Hills, and the bottom of the lowest

*Hooke uses the reasoning process he applied in the question of the origin of fossils to illustrate his method in the quest for knowledge.*

Wells and Mines, in the middle of the solid Quarries of Stone and other Minerals, and those the most remote, or at least very far distant from the Sea. Some of these have the perfect representation of the Figure of such Creatures and other Substances as are now well known, others of such as have some analogy and likeness to them, yet different from what are known of those Species to which they seem to belong, either in Magnitude, Figure, Colour, &c. yet retain such characteristick marks as seem to indicate them to belong to this or that Species of Animals or Vegetables. Some of these are nothing but perfect Stones of several kinds, others are inclosed with a Substance seeming to be the same, with that of the very Animal or Vegetable they resemble. Add to this, that in as many varieties of places there have been found others of these Figured Bo-[p.332]dies, which have as to Sense the very same Substance and Figures with those of the corresponding Species of Animals or Vegetables, and do not seem to be at all of a petrify'd and stony Substance, but rather of an Animal or Vegetable; of every of which Particulars I have viewd and examined Instances. And if it were not for detaining you too long, could have here produced a more particular Account and Description. But they being so generally known at least so universally almost spread over the whole Earth, so that no Country almost but doth afford them, I thought it would be needless.

The Question now is how we shall come to a certainty of Knowledge concerning them, by which we may be able to understand what they really are: And Secondly, What was the Cause of them. Thirdly, How they came to be disposed, placed, or made in those parts where they are, or have been found. Fourthly, These Discoveries being made to satisfaction, of what use or benefit will it be to Mankind, or how shall we be the wiser, or how will this Knowledge be an improvement of Natural Knowledge? Which is the aim of this Society.

*Why study fossils? To gain knowledge—which is the aim of the Royal Society, lampooned by some in the 17th century.*

For Answer then to these several Queries I shall propound these following Considerations.

*First*, As to the way of knowing of what Substance they are, I conceive there can be no better way than what we generally use when we inquire into the true Nature or Substance of any other Body when it is delivered to us to be examined or denominated; for instance, if a peice of Metal be delivered to be examined, it will first be viewed to see what Metal it resembles in Colour and Consistence, or in such other obvious and sensible Qualifications as may enable one to judge or guess what kind of Metal it is; as suppose it resembles Gold for closeness and ponderousness, these give intimation enough of examining it yet a little more curiously, since all is not Gold that glitters, and it may be that some Counterfeit in those more obvious Qualifications has mimicked that noble Metal, to do this, it is tried further by being put into good *Aqua Fortis*, to see whether that will dissolve it, for if it doth, it cannot be Gold. Secondly, The Specifick Gravity thereof is more curiously and exactly found by the help of Scales and Weights, by which it is weighed in and out of Water, because if thereby it be found to be lighter than Gold it cannot be Gold. Thirdly, By Hammering and a Gold Beater, 'tis found to be Malleable, and by looking through the Leaf,, 'tis found of a transparent Greeness and reflects a true Golden Colour, then 'tis tried by copelling and found fixed in the Fire, then it tarnishes not in the Air, and Amalgams Readily with *Mercury*, then its Solution in AR tinges, the Skin and Nails red, and a further trial shews it

*One can apply certain tests to determine the nature of fossils, just as gold can be ascertained to be gold if subjected to certain tests, because counterfeits cannot withstand them.*

to tinge Glass of a Ruby Colour. Now if it bear all these several ways of Examination, and answers to the properties of Gold, it may safely be concluded to be true Gold, and whoever shall deny it to be such must be looked upon as one that doth it without Reason, unless he can produce a further Criterion by which it shall be found to be very differing from it. Now, tho' this Metal were found at the bottom of a Mine a hundred Foot under Ground, or at the top of a Hill a hundred Foot above the Level of the Plain; or in the Pores of a growing Vegetable; or in the Tooth or Thigh of an Animal, tho' possibly it may be difficult to assign the Reason or Cause how it came to be placed there; yet the Examiner hath the Evidence of Sense to assure him that this Metal is Gold, and he ought to conclude and acquiesce in it, that such it is; otherwise there can be nothing at all known that it is this or that Body, and then there is an end of all further Inquiry or Experiment.

Now though there may not be known so many various ways of examining every other kind of Body, as, by reason of the value of it, there have been found out for Gold; yet in many Bodies at least a much less number will serve the turn to give assurance, that the Body so examined is of this or that kind and in many the very outward form as visible to the naked Eye, but especially if the inward visible appearance of the Substance be joined with it, will [p.333] be assurance sufficient to force an Assent or Conclusion of what kind the Body is that is so examined, and it ought not to be denied to be such without as evident a manifestation to the contrary.

This Discourse I have been the larger in premising, because, till it be agreed what is sufficient evidence to prove a natural Body of this or that kind there can be nothing done. If Experimental Philosophy, and if Poofs of this kind will not suffice, I cannot expect that all that I shall bring to prove the *Hypothesis* will be of any validity. That then which I shall indeavour to prove is;

*"Poofs" = Proofs*

*A recapitulation of his hypotheses concerning fossils.*

First, That there have been, and daily are found, the real Shells of Fishes in such parts of the Earth as are much above the present Level of the Sea, and others buried at a very great depth under the Surface of the Earth, where notwithstanding, 'tis evident that they were not there placed by any humane Power or Design.

Secondly, That many of these Shells are of a form differing from any of those Shells of that Species to which they seem to belong, which are commonly known at this Day.

Thirdly, That there are others of them which to all appearance are of the very same Species now known and to be found living.

Fourthly, That there are many of these Shells which are and have been in process of time, fill'd within and inclosed without with divers sorts of Earth, such as, Clay, Lome, Sea Sand, and divers other kinds of Sand, Mud, Chalk, &c.

Fifthly, That those filling and enclosing Bodies have been, and are by degrees, in process of time, petrify'd and hardn'd into Stones of differing Natures, Hardness and Contextures retaining the Impression, Form, or Signature of those Shells, so inclosed.

Sixthly, That those Shells so filled and inclosed, as above said, are according to the differing Nature of the Petrifactive Liquor or Juice:

sometimes also Petrify'd retaining still the same Figure they were of when the Petrifactive Juice began to operate upon them.

Seventhly, That others of them remain yet perfect Shells without Petrifaction or Alteration, when as yet both the Substance that fills them, and that which encloses them is converted into Stone of differing Natures and Textures.

Eighthly, That many of these Shells are either by length of Time, or by the Nature of the Petrifactive Juice perfectly rotted and decayed so as to be easily frangible between one's Finger and Thumb into a very fine Powder and yet serve to give a perfect Mould or Shape to the inclosed and inclosing-Stone.

Ninthly, That in many cases the Shell is not only Petrify'd by the said Juice, but sometimes after the impression hath been made as aforesaid, the Shell hath been clearly dissolved and carried away from the inclosing and inclosed Substances, leaving only the Space empty where the said Shell hath been placed.

Tenthly, That it sometimes happens that the Substance that filled the Shell hath been Petrify'd, and after the Shell hath been rotted away, the Petrify'd Body that was inclosed and had received that Impression, hath been afterwards inclosed with a Substance which hath afterwards Petrify'd and so inclosed it in Stone.

Eleventhly, That these kinds of Shells or the Petrify'd Substances that have been formed by them, have been for all Antiquity, and are at this Day to be found in most parts of the known World.

Twelvethly, That they are most of them differing from one another, tho' all referable to some Species of Shell-Fishes now to be found; yet in many particulars each of them also differing from them; that is, those that are found in one Country or Region, are very differing from those of another Country or Region, and that not only as to the Nature of the Petrify'd Substance inclosed and inclosing, but also as to the Figure and Make of the Shells themselves; and many of those also differing from the shape even of [p.334] those Shell-Fishes which are now to be found in such parts of the Seas which are situated near to the places that they are found in.

Concurring to this Head I shall prove that Shell-Fishes of the same Species in differing Countries now to be found, have many differences one from another, as much as any one of those Petrify'd Bodies have from any of the present Shells.

I have in my former Discourses explain'd the end and aim of these my Inquisitions, namely, to make some Use and Application of several Observations and Experiments that have been Collected in order to deduce some Doctrine from them, which may serve to direct such further Inquisitions as shall be necessary for the perfecting of the same, or at least to find whether such are wanting, and of what Kind and Nature they are.
The Doctrine aimed at, is, the Cause and Reason of the present Figure, Shape and Constitution of the Surface of this Body of the Earth, whether Sea or Land, as we now find it presented unto us under various and very irregular Forms and Fashions and constituted of very differing Substances.

*Hooke reiterates the purpose and aim of his discourses on "Earthquakes"—i.e., to find the causes of the irregularities of the surface of the Earth.*

Now, because when we look into Natural Histories of past Times, we find very few, if any, Footsteps of what alterations or transactions of this Nature have been performed, we must be fain to make use of other helps than what Natural Historians will furnish us with, to make out an account of the History thereof: Nor are there any Monuments or Medals with Literal, Graphical, or Hieroglyphical Inscriptions that will help us out in this our Inquiry, by which the Writers of Civil Histories have of late Years been much assisted from the great curiosity of modern Travellers and Collectors of such Curiosities.

*Writers of civil histories have the advantage of written documents, monuments, inscriptions, etc.*

The great transactions of the Alterations, Formations, or Dispositions of the Superficial Parts of the Earth into that Constitution and Shape which we now find them to have, preceded the Invention of Writing, and what was preserved till the times of that Invention were more dark and confused, that they seem to be altogether Romantick, Fabulous, and Fictious, and cannot be much relied on or heeded, and at best will only afford us occasions of Conjecture.

*Natural historians deal with changes and developments that preceded the invention of writing. Stories handed down about the formation of the Earth before that invention, therefore, are deficient and not reliable. Hooke must include in this category, therefore, the biblical account.*

For Proof then of the first Proposition that, &c. I could produce a multitude of Authorities fetched out of printed Authors, and as many others that I have received from the Relations of very credible Persons that have found them themselves, but it would be too long, I shall therefore only name one who was formerly a worthy Member of this Society, and well known to divers here present, and that was Dr. *Peter Ball*, he passing over the *Alps* to go into *Italy* by a narrow Passage, where there was on the one hand a prodigious high Cliff above him, and on the other hand, as prodigious deep Precipice below him, observed in the Cliff a Layer of Sea Sand and Shells for a very great length buried under that high Mountain above; he had the curiosity to take up and bring home with him into *England* divers of them which he dug out of the said Layer of Sand which he shewed me, and I found them to be true Oyster-shells, not Petrified but remaining perfect Shells, one of which he gave me, and had divers others which he kept; he observed also, that there were divers other Substances among the Sand as if it had been upon the Sea-shoar. To this I shall add an Observation of my own nearer Home, which others possibly may have the opportunity of seeing, and that was at the West end of the Isle of *Wight*, in a Cliff lying within the *Needles* almost opposite to *Hurst-Castle*, it is an Earthy sort of Cliff made up of several sorts of Layers, of Clays, Sands, Gravels and Loames one upon the other. Somewhat above the middle of this Cliff, which I judge in some parts may be about two Foot high, I found one of the said Layers to be of a perfect Sea Sand filled with a great variety of Shells, such as Oysters, Limpits, and several sorts of Periwinkles, of which kind I dug out many and brought them with me, and found them to be of the same kind with those which were very plentifully to be found upon the Shore beneath, now cast out of the Sea. [p.335] This Layer is extended along this Cliff I conceive near half a Mile, and may be about sixty Foot or more above the high Water mark.

*Examples of the occurrence of sea sand and shells high above the present sealevel can be found in the Alps and also nearer Hooke's home in the western part of the Isle of Wight.*

*The layer of sea sand and shells exposed in the cliff on the Isle of Wight and within the rocks known as the Needles is more than 60 feet above the high water mark. See illustration of the Needles, Chapter 2*

Another Instance I observed nearer this place, and that was in St *James's Fields*, where St.*James's* Square is now built, in which place when they were making Bricks of the Brick Earth there dug, they had sunk several Wells, which I judge might be near twenty Foot in depth, to procure Water for that purpose; going down into several of those, I found, at the bottom, a Layer of perfect Sea Sand, with variety of Shells, and several Bones, and other Substances, of which kind I dug out enough to fill a small Box and

*A layer of sea sand and shells and also fossil bones occur in wells dug in St. James's Square and Park.*

shewed them to Mr. *Boyl*, and also to this *Society*. And I was informed also, that the same kind of Substances were found in digging of a Snow Well in St. *James's Park*; and I doubt not but whoever shall sink any where thereabout to that depth will find the same things. Now whoever will consider the Positions and Circumstances of the said places, will easily grant, I suppose, that they could not be there placed by the Industry of *Man*, but must be ascribed to some other cause to be fetched from *Nature*.

As to the second Head, That, *&c.* I shall produce several Oyster and Cockle Shells which have been and are to be found in many Parts of *England*, which in many particulars of their shape, do differ from those of the Oysters and Cockles now to be found; yet upon examination of them they may be found to be true and perfect Shells by all sensible Qualities, except only their exact shape, and therefore I conceive that to be sufficient Evidence to prove them to be really such, because it is all the Evidence the Matter is capable of. If in digging a Mine, or the like, an artificial Coin or Urne, or the like Substance be found, no one scruples to affirm it to be of this or that Metal or Earth he finds them by trial to be of: Nor that they are *Roman, Saxon, Norman*, or the like, according to the Relievo, Impression, Characters, or Form they find them of. Now these Shells and other Bodies are the Medals, Urnes, or Monuments of Nature, whose Relievoes, Impressions, Characters, Forms, Substances, *&c.* are much more plain and discoverable to any unbiassed Person, and therefore he has no reason to scruple his assent: nor to desist from making his Observations to correct his natural Chronology, and to conjecture how, and when, and upon what occasion they came to be placed in those Repositories. These are the greatest and most lasting Monuments of Antiquity, which, in all probability, will far antidate all the most ancient Monuments of the World, even the very Pyramids, Obelisks, Mummys, Hieroglyphicks, and Coins, and will afford more information in Natural History, than those other put altogether will in Civil. Nor will there be wanting *Media* or *Criteria* of Chronology, which may give us some account even of the time when, as I shall afterwards mention.

*Hooke hints that fossils can provide a chronology, giving us some account of when they were formed.*

As to the Proof of the third Proposition, *viz.* That, *&c.*, All those Instances I have named are of sufficient evidence, for that those which I found in both those places I mentioned were of the same kinds with those that are now to be found near those places, as whoever shall examine will find.

And the fourth will also from the same, and hundreds of others be as evident, and therefore I suppose none will scruple to assent to this Proposition, *viz.* That, *&c.* Page 333, especially if the truth of the former be granted, which I conceive cannot be denied.

For Proof of the fifth Proposition, namely, That, *&c.* The place I mentioned before near the *Needles* in the Isle of *Wight* afforded a most evident and convincing one as could well be desired, which was from the following Observation. I took notice that the aforesaid Earthy Cliff did founder down and fall upon the Sea-shoar underneath, which was smooth and Sandy, and bare at low Water so as to be walked on, but at high Water a great part of it was covered by the Sea. I observed several great lumps of the said Founderings lying below, some whereof, which lay next the Cliff, tho' they were somewhat harden'd together more than they were above in

*See illustration in Chapter 2 of such a lump that broke from the cliff in one night in Freshwater Bay, near Hooke's birthplace in the town of Freshwater.*

the Cliff, were yet not [p.336] hard enough to be accounted Stone; others of them that lay further into the Sea were yet more hard, and some of the furthest I could not come at for the Water, were as hard I conceived as *Purbeck* Paving (which is taken up from the Shore of *Purbeck*, lying just opposite to it on the West side of this Channel or Passage) divers of these Stones I observ'd to be made up of the peices of Earth that had foundred down from the Cliff, which I was assured of by carefully observing and finding divers of them to consist of the several Layers, and in the same order as I saw them in the Cliff; among the rest I found divers that had the Layer of Sea Sand and Shells which I had observed in the Cliff inserted in the Stone with the adjoyning Layers all petrify'd together into a hard Stone. Here I found multitudes of the said Shells I before mention'd to have observ'd in the Cliff, mix'd loosely with a Sea Sand; now together with the said Sand both fill'd, inclos'd, and petrify'd altogether, and I broke off many peices of the said Rocks, where I found the said Petrifactions, and found them much like other Stones I had seen from other Inland Quarries here of *England*, wherein I had observ'd also such kind of petrify'd Shells, tho' how they came there to be so Petrify'd I could not be so well inform'd. For that which I conceive was the cause of this suddain Petrifaction (for I conceive those that I examin'd had not been Stones for very many Years, which I judged by their distance from the present Cliff, and from the quantity thereof, which *Communibus Annis* did founder down) was that close by this Cliff, there is a vitriolate or aluminous Spring or Rill which runs into the Sea, where formerly those Salts have been made of it by boyling, but has been now omitted for many Years. These saline Springs or Rills I conjecture mixing with the Sea-Water, may be the cause of the said Petrifaction, and the want of it is the cause why other founderings in other parts of the said Cliff are not at all so Petrify'd. Now from the assur'd Observation of these Petrifactions, I cannot but judge that the truth of this Proposition will most evidently appear, and needs no other to confirm it. However I doubt not but that any one who should there lay a part of the said Cliff shaped and marked as he pleas'd for his own assurance, would find the same very hard Petrify'd in two or three Years, which may not be unworthy of farther Inquiry and Trial for such as have opportunity.

*Hooke speculates that in order for the loose sediment to harden into stone, there needs be the presence of a cementing agent, which he conjectures may come from the saline springs. This cement he refers to as Petrifactive Liquors. These liquors can "insinuate" themselves into the smallest "pores" of materials. This explanation is reminiscent of the lit-par-lit replacement process.*

As to the Proof of the sixth Proposition, it will not be difficult, the preceding being once granted for that there may be thousands of Instances of that Nature found in the Stones dug out of divers of our *English* Quarries; some of which Stones are found full of such Petrifactions.

Upon this occasion I think it not improper to mention an Observation which I have often taken notice of, which is of the Flints which are generally found intermix'd with Chalk in Quarries of that kind of Stone. I have observ'd then that these Flints are nothing else but the Body of the Chalk united together, and, as it were, first diluted by a petrifick Juice, and by that dissolved into it, and so make a uniform close Body which by degrees doth all petrify and harden together into that solid hard Body of the Flint. This I found by taking notice of the Nature of those Flints when broken, and how the Grain, Colour and Hardness of them was situated especially towards the edges; for there where the Juice seem'd to be almost spent, the Flinty Body appears of a midling Nature between Chalk and Flint, besides I have observ'd sometimes other Bodies inclosed, and sometimes lumps of Chalk also, toward which the Limb-parts of the Flint were colour'd and terminated just as towards the incompassing Chalk. And from the curious and sharp running and mouldings which I have observed in Flints, I conceive that the first Liquid Substance of it was altogether as fluid as Water, tho' it were imbu'd with a Saline Sulphureous or other petrifactive Tincture. These Indications may be much more plainly manifested by such a peice of Flint than 'tis possible for any one to describe by words; and therefore I shall omit the farther mention of them till I can meet with a Flint to shew them. I mention this here only to shew that the petrifactive Juice is often found to insinuate itself into the closest Pores of Body, by reason of its great fluidity which inables it oft to petrify even the very Bodies and Substances of the Shells themselves.

[p.337] But tho' some of these Petrifactive Liquors be thus fluid, yet they are not all; and thence it comes that many Shells remain unpetrify'd, tho' the Substances that fill'd them and enclos'd them be so wrote upon, which was the seventh Proposition I undertook to prove. This I can make evident by divers of the Petrifactions that are kept in this Repository, and by thousands of others which I have seen: And any one that will but diligently examine them will find the very Shells themselves preserved Shells, tho' inclosed in the middle of a Stone, as of *Portland, Purbeck*, and divers other Inland Stones here dug in *England*. And I am promised to have sent me a flake of a Stone which is very hard, which notwithstanding is all over full of Shells. I sah Shells, for that I cannot call them any thing else, since to all sensible trials they are so, both as to Figure and Substance.

*Here Hooke repeatedly refers to the repository of the Royal Society where presumably there must have been a collection of fossils and other specimens. Even though the society moved from Gresham College after Hooke died and Newton became president, Newton must at least bear partial blame that none of this repository survived. Also, all the hundreds if not thousands of fine instruments made by Hooke for Royal Society experiments are also lost.*

As to the Proof of the Eighth, I cannot produce a more pregnant one than the *Echinus* or Helmet-Stone, found by Mr. *William Ball* upon the Shore of *Devonshire*, near *Exeter*, which he presented to the Society, and I suppose may yet be seen in the Repository, for by that alone it will plainly appear, that there had been formerly a Shell that had caused both the formations of the containing and contained Flint, there being just the due thickness of such a Shell vacant between them, but there may be hundreds of others produced of the like kinds if it were needful.

*Although the original shell is gone, it has left both a cast and a mold.*

A Proof of the Ninth and Tenth, *viz.* I think the large *Cornu Ammonis* may afford; for here it seems plain, that a great part of the Shell was wasted away before the perfect Petrifaction of both the inclosing and inclosed Stone, tho' part of the Shell be yet remaining sticking between them.

*Some of the shell of this large ammonite is still present in this specimen.*

Tis a hard matter to make a positive Proof of the Eleventh, *viz.* Because of the infinity of them that would be necessary, yet I think it would not be difficult to bring credible Testimonies enough to supply one for each Country, and that I suppose may suffice to make it probable that they may be found in all others, since, as I shall afterwards prove, they have been produced all by the same cause.

*Fossils are found in every country.*

As the the Twelfth Proposition which I undertook to prove, *viz.* That most of those Shells or other Substances found as above, whether Petrify'd or not Petrify'd, are in the first place differing from one another in many particulars both of Figure and Substance, tho' yet they retain such particular Characteristicks as are sufficient to denote and show to what Species they belong, either of Vegetables or Animals, whether of Fishes or terrestrial Creatures, such as are now to be met with alive; that is, not only that such as are thus found in one Country, are differing from those which are found in another: And in Petrify'd ones this is not only remarkable in the Substance inclosing and inclosed, but also in the magnitude, Figure and

Make of the things themselves; and in the second place many of them do considerably differ from the shape of those Shell-Fishes, and other Substances which are now to be found alive in such parts of the Seas as are nearest situated to the places where these Fossil or Land Shells are now to be found. For the proof of which I have no better means than to have recourse to the Substances themselves, which have been so found, of which there is an excellent Collection in the Repository of this Society, though I have also seen divers other instances in other Collections and Observations which I have elsewhere met with, which I cannot now produce. Yet one Instance for all I suppose may be this great Voluta which I have here produced, that was taken out of a Quarry in *Portland* (and I believe that those two other great ones in the Repository which I begg'd of the late Duke of *Norfolk* for the Repository are of the same kind and from the same place) for by these I think it plainly enough appears, that they are very differing from all the other Substances or particular Petrifactions that are in the Repository, both in Magnitude, Colour, Shape and Substance including and included, and even in the very Substance of that which I call, and shall prove the Shell; and not only do they thus differ from the Petri-[p.338]-factions Fossile or Land found Shells, but they differ also from all the known sorts of Shells of that Species of Fishes, to which I would refer them, which are now to be found any where near that place alive, nay, in any part of the World that I yet knew of; notwithstanding all which, they do retain, I conceive, certain Characteristicks of their Form, which show them to have belong'd to that Species of Shell-Fishes which are call'd *Nautili*. These *Nautili* are describ'd by *Gesner, Aldrovand, Johnston,* and others, where you have their Names and a Picture or two of the Shells, and some Stones also tending to a Description of the Creature and two Species of them; but he that shall think to find any such Characteristicks by reading their Descriptions and seeing their Pictures of them, will be much mistaken. And indeed it is not only in the description of this Species of Shells and Fishes, that a very great Defect or Imperfection may be found among Natural Historians, but in the Description of most other things; so that without inspection of the things themselves, a Man is but a very little wiser or more instructed by the History, Picture, and Relations concerning Natural Bodys; for the Observations for the most part are so superficial, and the Descriptions so ambiguous, that they create a very imperfect Idea of the true Nature and Characteristick of the thing described, and such as will be but of very little use without an ocular Inspection and a manual handling, and other sensible examinations of the very things themselves; for there are so many considerable Instances that may by that means be taken notice of, which may be useful to this or that purpose for which they may be instructive, that 'tis almost impossible for any one Examiner or Describer to take notice of them, or so much as to have any imagination of them. It were therefore much to be wisht for and indeavoured that there might be made and kept in some Repository as full and compleat a Collection of all varieties of Natural Bodies as could be obtain'd, where an Inquirer might be able to have recourse, where he might peruse, and turn over, and spell, and read the Book of Nature, and observe the *Orthography, Etymologia, Syntaxis,* and *Prosodia* of Natures Grammar, and by which, as with a *Dictionary*, he might readily turn to and find the true Figure, Composition, Derivation and Use of the Characters, Words, Phrases and Sentences of Nature written with indelible, and most exact, and most expressive Letters, without which Books it will be very difficult to be thoroughly a *Literatus* in the Language and Sense of Nature. The use of such a Collection is not for Divertisement, and Wonder, and Gazing, as 'tis for the most part

*Some shells found as fossils differ from any living shellfish known.*

*The great Voluta which Hooke acquired for the Royal Society repository from the duke of Norfolk must be a giant ammonite from the Portland quarry. Portland stone is an Upper Jurassic, very dense, limestone. Hooke recognizes that many of these animals are not found alive anywhere today, although some seem to be related to the nautilus.*

*Hooke launches into an eloquent plea for the amassing of a fossil collection—not for wonderment, but for study by scholars.*

thought and esteemed, and like Pictures for Children to admire and be pleased with, but for the most serious and diligent study of the most able Proficient in Natural Philosophy. And upon this occasion tho' it be a digression, I could heartily wish that a Collection were made in this Repository of as many varieties as could be procured of these kinds of Fossile-Shells and Petrifactions, which would be no very difficult matter to be done if any one made it his care: For *England* alone would afford some hundreds of varieties, some Petrify'd, some not. There are few Quarries of Stone here in *England* I believe, but if they were look'd into some kind or other of these Petrifactions might be found in them: I have observ'd them in Marbles almost of all varieties of Colours, as Black, White, Red, and otherwise Speckled: I have seen them in great varieties of Flints and Pebbles, in various sorts of hard Stones, as *Purbeck, Portland, Yorkshire, Kentish, Northamptonshire,* &c. I have seen many of them of Coperose or Vitriol Stone, or *Pyrites*, and *John Bauhine*, and others have described many of them of that Nature. Others of these are found above Ground, and others also under Ground very deep sometimes unpetrify'd and remaining perfect Shells, Bones, Woods, Roots, &c. and have been found by several sorts of trials to be truly so, not only in External Figure, but also in the Internal and Substantial Parts of them; so that in truth there is no manner of Reason to doubt them to be of those very Substances they so perfectly and fully resemble.

*As many varieties of fossils as could be procured should be in the collection. There can be no doubt then of their organic origin. New specimens can also be identified by comparison with the collection.*

But if yet there should be some one that should make a doubt of their identity or sameness with such Substances as they seem to resemble, I would willingly know what kind of Proof will satisfie such his doubt, and by what Indications or Characteristicks he will know a Shell of an unknown Species (for [p.339] such may be shewn him) when it shall be presented to him, or a peice of Wood of some strange Tree brought from an unknown place; if he will say by the relation of the bringer, that I conceive is not becoming a good Naturalist; and so one might have been impos'd on by the Relation of the incombustible Linnen which was here examin'd; but if he will say by its Properties, which he finds the same with that of Shells, or Vegetables, or other resembling Substances, then I answer, that the same will in these be manifestly shewn. Now, the more of these certain Characteristicks of the several Species of Bodies there are known, the greater certainties and assurances will be afforded by the artificial and strict Examination of them. As for instance, the knowing the Existence and Form of the microscopical Pores of Wood, is a better Characteristick to know that a Substance is Wood than the outward Figure and Appearance thereof, which may be artificially or accidentally imitated, by which means I found that a peice of Lignum Fossile sent from *Italy* by Cavalier *Pozzo* to Sir *Beorge Eut.* and by him supposed to be only Earth shaped into that form and not to be real Wood, as *Stelluti* also indeavours to prove. By the examination, I say, and discovery of the microscopical Pores thereof with a Magnifying-Glass to be like those of Firr, I produced a better Argument that it was really Firr than any *Franciseo Stelluti* has argued to prove it Earth. Another was, that it burnt as Wood, and made Coles like those of Wood, with microscopical Pores; had I had enough of it I could have examin'd it by Distillation, and varius other Chymical Probations; for the more of Testimonies and Confessions are fetch'd from these Examinations and Wracking, the greater will be the Evidence of the true Nature of those Substances so examin'd, tho' oftimes the Evidence afforded by some one, may be sufficient clearness to save all further Enquires: Such as these the Lord *Verulam* call'd *Experimenta Crucis*, which serve to direct the Inquierer

*Lord Verulam = Francis Bacon*

to proceed the right way in making his judgment. These are such marks as I call Characteristicks, which expresly determine and limit the Nature and Species of the Body under Consideration. For Instance, I conceive that all those Petrify'd Substances which are call'd Snake-Stones in *English*, from some resemblance imagin'd of a Snake coyled up; and in Latin (*Cornu Ammonis*) or Sand Horns possibly from their being found in those Sandy Deserts.

These Petrifactions, I say, I conceive to be nothing else but the Petrifactions of several sorts of Substances that the Shells of some sorts of *Nautili* happened to be mix'd with, whilst those Substances were yet very soft and Liquid, and before they came to be hardned into Stone by the Petrifactive Agent. This Conception I grounded upon these Characteristicks, which in examining a great many of them I have found. First, That in very many of them I have manifestly seen the real moulding Shell there preserved, together with the moulded Substance.

Next I conceive, that this Shell did belong to the Species of the *Nautili*, or sailing Fish, from these Characteristicks. First, That the Shell is of a true Conical Figure from the *Base* to the *Apex*. Secondly, That this Cone is turned into a *Voluta* or Spiral Cone, so that the *Azis* thereof doth perfectly lye in the same Plaine. Thirdly, That this Spiral being a true proportional Spiral, is continually at certain distances intercepted by Diaphragmes; so that those Diaphragmes being taken as Bases of several Cones, the Cones shall be found to diminish in a series Geometrically Proportional. Fourthly, That every one of these Diaphragmes is perforated with a hole similar and proportional also according to a Geometrical Series.

*"Azis" = axis*

To these I might add other accidental Proprieties of the flating, crenating, depressing, ridging, stringing, and the like, ornamenting, as it were, of the outward sides of this voluted conical Body, and the undulation and foliation, as I may call it, of the Diaphragme, and the Fringing and Ruffling thereof; all which are found of great variety in this or that Subalternate Species, as is also the Section of the Base, or that of the Diaphragme; but these are not to be looked upon as Characteristicks or Differences to denominate a new Species.

[p.340] And here by the bye I cannot but take notice of the imperfect and inaccurate Description of this so curious a Fish as the *Nautilus* must needs be, if one may guess at the curiosity thereof from those descriptions, which I find in *Johnston* out of *Aristotle, Pliny, Bellonius, Piso, Cardan, Fauconerius*, and others, and from the curious make of the Shell, for by all those descriptions I cannot imagine any one can get any tolerable Idea or Notion, what the make of so wonderful a Fish must be that has such an admirable quality as to buoy himself as *Pliny* says, *ex alto mari* from the bottom of the Sea, and make himself to swim and sail upon the top of the Water, and at pleasure, or for fear presently to sink himself down again to the bottom. This will appear so much the more wonderful to one that shall consider the great pressure of the Water at the bottom of the Sea, and in how differing a stae of compression this Animal must be at those two places, and by what power it becometh able to make itself so light at the bottom to rise and seem half out of the Water, and yet presently so heavy as to sink down to the bottom, and this without Finns or Tail to move itself. Now as this Property is peculiar to this Fish only, so is the make

*Hooke digresses to marvel at the characteristics of the nautilus and its ability to go from the bottom of the sea to the surface and back again with equal ease.*

of the Shell differing from all the Species of Nature besides, and as I conceive is the Engine by which he performeth this admirable Exploit; for the whole Shell is divided into a multitude of Cells or Cabins separated and distinguished one from another by several Diaphragmes or Partitions without any other perforation, save one small one, through which passeth a small Pipe, which I take to be the Gut of the Animal; this Gut doth not fill a two hundred part of the Cavity through which it passeth, and the remaining part must either be filled with Air or Water. Now if it be filled with Water, as probably 'tis, when he sinketh himself to the bottom, 'tis prety hard to conceive how he filleth it with Air under so great a pressure and at such a distance from it as to buoy himself up, unless it be caus'd with such a fermentation of the Excrements of the Gut, or other Juices of the Body as doth produce an artificial Air, which serves for that purpose; which seems to me to be the true Cause, especially since I find *Gulielmus Piso* to add this Remark to his History and Description of it. *Cum damno meo Plinii Discriptionem verissimam esse compertus sum namdum talem pisciculum* (speaking of the *Nautilus* of *China*) *in mari captum imprudentius manibus meis contrectassem, tantus ardor manum invasit, tanquam si aqua ferventi suffusa esset, & nisi apposito statim allio conraso cum aqua mihi ipse subvenissem, procul dubio præ dolore in febrim incidissem: Unde ego ipsum piscem de Holothuriorum esse genere contenderim, ut quæ omnia in maria fluctuantia, eam aerem calorem attrectantibus inurunt quod & fallacissimi omnium mortalium* Chinenses *noverunt, qui illa Orjza miscent, ut liquorem suum Destillatitium (quem Arac hos hic vocamus) tanto callidius reddant, pernicioso invento, quod hinc miseri nostri Socii navales, sanguinis sputam, phthisin, marasmum deniq & ipsam tandem mortem incurrant.* By whilch it plainly appears, that the Juices or Excrements of this Fish are of a strange fermenting or burning Nature which may be the cause of so singular and wonderful an Indowment, which whether it be so or not, I could heartily wish that some Person curious in Anatomy that has the opportunity of meeting with them alive would give us a more accurate Description of its external and internal Formations and Qualifications.

But to leave this Digression, which I have the longer insisted upon to shew the great imperfections of the Descriptions of the Species of Nature and their Qualifications and of the varieties of them (for that I have seen two Species of this sort not described or mentioned in any Author) and of how great use a good Collection and Description of them would be, as particularly concerning this very Fish I shall have occasion shortly to mention. To leave, I say, this Digression, we may from this perceive how little able we are from the want of this Knowledge and Collection, to conclude, that because we do not already know a Fish or Shell exactly of the shape of this or that *Cornu Ammonis*, therefore that it could never have been any such Shell, since it then cannot presently be proved that there is at present, or ever was any such Fish in being, which some possibly too confident of their Omnisciency may Object, because they know none such themselves, or have read of them; and therefore that there is more reason that such Arguments as are drawn from the examina-[p.341]tions of the Substances, and the Characteristicks of the Form should be of sufficient evidence to evince that these Bodies that have these Qualifications could not be formed but for such purposes, as those Animals which we are informed of, we know have all parts fitted for each singular and surprizing use designed; for it is certain that Nature doth nothing *frustra*, but manifestly with an admirable and wise design, the truth of which Maxim

*Just because we do not know the existence of a living ammonite does not mean that it has never existed. Hooke, therefore, understood the concept of extinction and was not bothered by religion in this respect. A religious-minded person in the 17th century, on the other hand, would not be able to accept extinction. Accordingly, some, like John Ray, advocated that the unknown species might exist somewhere in the world.*

will more and more evidently appear, the more the Works thereof are curiously examined and searched into; and no unprejudiced person that thoroughly examins them can fail of being convinc'd of the Truth and Certainty thereof, there being such a Harmony, Consent and Uniformity, as I may so speak, in all its Operations, and a gradual transition from one to another, that it is evident that all these kinds of Petrifactions have been moulded by some Animal or Vegetable Substance, as by Shells, Bones, Teeth, Fruits, Woods, &c. and that many of them are the Substances themselves, yet unaltered.

Now this being proved or granted, which I conceive the inspection and examination of the things themselves will most powerfully effect; it must follow as a Consequence of that Phænomenon, that all parts almost of the present Earth extant and appearing above the Sea, have been for some considerable time under it, and covered therewith. Since I conceive there is scarce any Country in the World where these Monuments of Antiquity, these Medals of Nature, or these Sea Marks and Evidences are not to be found either above, or at some depth under Ground, and some not very deep; particular Testimonies of which Truth I have collected many out of the few Natural Historians I have had the opportunity to peruse since I have had this Notion; and I doubt not but that abundantly more may be collected even out of Books. But inquisitive Naturalists, if it were made an Head of Inquiry, would questionless meet with multitudes of other Instances almost every where not as yet handed by any Historian, of which truth I have been assur'd by many Testimonies from other Persons; but of this I have spoken already sufficiently.

*All parts of the Earth must have been at one time or another under the sea; the evidence is in the fossils which Hooke refers to as "medals of Nature."*

From the comparing of which Evidences with several other pertinent Circumstances that may be observ'd may be deduced Conclusions very instructive as to the preceding and subsequent State also of this World. *Nam Res accendunt lumina Rebus*, and the understanding the History of the Course and Progress of Nature preceding will afford sufficient information of the Method of proceding, which in most things we may find to be very constant, uniform and regular. By such means we have arriv'd to the present Knowledge of Cælestial Motions, and by the like, to that we have of the Motions of the Seas and Winds, and tho' none of these are yet come to their highest perfection, yet Inquiry, and Ratiocination, and Comparison will carry us much further towards that end, which the comparison of the present state thereof with what it was two or three hundred Years since, will give us good reason to hope.

*Hooke is convinced that nature is knowable.*

*The present is the key to the past.*

It remains then to inquire by what means these prominent Parts of the Earth which at present are dry Land, came to be so, since by these Testimonies it is, I conceive, evident that they have been for some time under the Water.

And here in the first place I think it will be evident, that it could not be from the Flood of *Noah*, since the duration of that which was but about two hundred Natural Days, or half an Year could not afford time enough for the production and perfection of so many and so great and full grown Shells, as these which are so found do testify; besides the quantity and thickness of the Beds of Sand with which they are many times found mixed, do argue that there must needs be a much longer time of the Seas Residence above the same, than so short a space can afford.

*Noah's Flood is rejected as the cause of the phenomenon that the present dry land had been for some time under the sea. Hooke believes that the thickness of the sediment in which fossil shells are found is evidence that the land was under water for a much longer time than the 200 days of Noah's Flood.*

Nor could they proceed from a gradual swelling of the Earth, from a Subterraneous fermentation, which, by degrees should raise the parts of the Sea above the Surface thereof; since if it had been that way, these Shells would have been found only at the top of the Earth or very near it, and not buried at [p.342] so great a depth under it as the Instances I mentioned of the Layer of Shells in the *Alps* buried under so vast a Mountain, and that near the *Needles* in the *Isle of Wight* found in the middle of an Hill, could not rationally be so caused.

*Neither could the terrestrial features be due to a swelling of the Earth, as then the shells would be at or near the surface and not buried in layers under mountainous regions, as in the Alps, or found in the middle of a cliff, as at the Isle of Wight.*

Nor could it proceed wholly from a washing of the Water from off the Face of those parts of the Earth, for the same Reason, for how should the Mountain come to be placed on the top of them.

Now, if after all these topicks of Proofs, there shall yet remain some who will not allow any of them to have been Shells, because they are found in the middle of Stone; I have, as a suppliment, added my Observation of the Place where, and the Manner how they may be observ'd to be so inclosed into the Body of a solid Stone, namely, at a place near the *Needles*, at the West end of the *Isle of Wight*.

*Hooke again uses the area near his home as evidence that these fossils represent real shells even though they were found in solid stone.*

With such now as shall not think all, or any of these convincing Arguments to prove them Shells, I cannot, I confess, conceive what kind of Arguments will prevail, since these sensible Marks are, in all other things, the Characteristicks and Proofs by which to determine of their Nature and Relation, and why they should not be allow'd to be so in this particular Case I cannot well conceive.

The great scruples I find are these; First, That they know not how they could come to be placed where they are and have been found; some Conjectures at which I shall after shew.

*The objections of the skeptics are (1) they don't know how the fossils came to be where they are found, and (2) many of the shellfishes found (e.g., the ammonites) are much bigger than the live animals found on the coast of Portland.*

And, Secondly, That many, nay most, of them are of somewhat a differing Shape, and of a much great Magnitude than are the Shell-Fishes of the like Animals to be found upon the Coast of *Portland*, or near the places where they have been found; and indeed against this my Hypothesis or Assertion I find none more pressingly urged than this, that there is not one to be found either in the Seas near those Parts where such are found, nor in any part of the known World, any such Animals or Vegetables as those which are supposed to have afforded the Substances of some of them, or the Moulds of some other; and particularly it has very much been urged upon the Consideration of the Petrifaction or *Cornu Ammonis* taken out of the Quarry of Stone in the Isle *Portland*, whether it could be reasonably supposed that ever there were in the World a Species of the *Nautilus* of this shape, and of so vast a bigness, of which it is supposed the World has not afforded an equal in a living Species. And I perceive that the very supposition is looked upon as very extravagant and ridiculous. However, it may be possibly worthy some Mens Considerations to inquire, First, Whether there may not yet be found in the World many Species of Shell-Fish they have not hitherto heard of, or seen in the Writings of Natural Historians, or in relations of Voyages, or by their own Experience.

Secondly, Whether the exceeding greatness of this Shell be a sufficient Argument to conclude it ridiculous to suppose, that there could be a living Fish that might fill so great a Shell, since I shew'd the last Day out of *Maudelslo* and *Olearius*'s Travels, an instance of Oysters found in *Java*, that seem'd much to exceed this Magnitude: And possibly some here present may have seen, as well as my self, the great pair of Shells in the *Musæum Harveanum* before the Fire in 1666. And that the Shells of a *Pinna Marina* are now to be seen in this Repository, which exceed the common bigness of a Muscle as much as this *Cornu Ammonis* doth the smaller sorts of *Nautili*, and varieth also as much from them in Shape: And that hotter Countries, such as are in the *Torrid Zone*, produce Turtles or Sea Tortoises, abundantly more exceeding the smaller sorts of these of colder Regions, of which there are Testimonies enough to be had both from Natural Historians and Travellers, which it were necessary I could produce.

[p.343] But because it may be upon this Head further Objected, That all those extraordinary great Species are the productions of the *Torrid Zone*, or the hotter Climates, and not of the colder, and such as lie so far remov'd towards the Poles as *Portland* or *England* do, about which there are now no living Fishes to be found that any wise come near to that Magnitude, but are of much smaller size and of different shapes.

Therefore before the Opinion be wholly rejected, I would desire them to consider, whether it may not have been possible, that this very Land of *England* and *Portland*, did, at a certain time for some Ages past, lie within the *Torrid Zone*; and whilst it there resided, or during its Journying or Passage through it, whether it might not be covered with the Sea to a certain height above the tops of the highest Mountains. And further, how deep this may have lain below the Surface of the Sea, when it might have been in that Passage, and how long time it may have spent in such a state, and how long since it may have been emerged. Such as are better versed in ancient Historians than I ever have been or hope to be, may possibly resolve some of these Doubts, or at least may prove the impossibility thereof, which may save further trouble of inquiry: But if after inquiry it should be found that Natural History is defective in that particular, then I will indeavour to see what Helps and Histories will be pertinent towards the determination of these Queries.

*These are bold thoughts indeed: that England could have been in the Torrid zone in the past, been covered with the sea, and "journeyed" to its present position with respect to the system of longitudes and latitudes. These ideas are related to Hooke's concept of polar wandering. See Chapter 5.*

And in order to determine the Possibility or Impossibiblity of this Matter, I could wish it were well considered further, whether the Superficies of this Ocean be equally distant from a Central Point in the Bowels of the Earth, and whether any other perpendiculars to the Surface thereof, besides those of every single Parallel, and its Poles, do tend to any other Point of its Axis; and if there should be found more than one Point, then what are the limiting or terminating Points of a Line of such Points; that is, at what distance they must be from one another, or from a Central Point? This I mention'd in two of my preceding Lectures, the one read about ten or twelve Years since, and in the other about two Years since; in both which I indeavour'd to shew that the form of the Earth was probably somewhat flatter towards the Poles than towards the Equinoctial, since which I have met with some Observations that do seem to make a probability in my Conjecture and Hypothesis.

The *Antipodes* were once thought a Chimera, length of time hath made that notion more reconcileable to Sense and Reason; these may possibly at first hearing appear much more extravagant, and Time that brings all to Light, may possibly evidence them to be nothing but *Chimæra*'s; I will not prejudge, or pre-possess, but leave them to their Fortune. However it

*Chapter 5, "Polar Wandering on an Oblate Spheroid Earth," has a detailed discussion of this and following paragraphs. Note that Hooke is making it clear here that he had treated this subject some 10 or 12 years before the date of this lecture, or around 1674. By 1674, Newton had not been convinced of the oblate spheroid shape of the Earth.*

were desirable by the Experience and Inquiry of a short time to dispatch and hasten the Growth and Ripenings of the Productions of Nature, since the Experience and Duration of a Man, whether he looks forward or backward, is very short in comparison of what seems requisite for this Determination; his Sight is weak and dim, his Power and Reach much shorter, yet may it be worth considering (tho' he cannot lengthen or prolong his limited time either past or to come), whether by Telescopes or Microscopes he may not see some hundreds of Years backwards and forward, and distinguish by such Microscopes and Telescopes Events so far distant both before and behind himself in time, as if close by, and now present? And whether by Instruments he may not extend his Power, and reach things far above his Head, and far beneath his Feet, in the highest parts of the Heavens, and the lowest parts of the Earth; for could he perform things of this Nature and Quality as they ought to be, he would lengthen his Life and increase the injoyments thereof by a multiply'd and condens'd knowledge of times past, and of times also yet to come.

*Hooke predicts the power of technology and instrumentation in extending human knowledge.*

But before we come to this last Expedient, I could wish we had a good Account and Collection of what Histories pertinent to this, or any other Natural Inquiry are to be found in Printed or Written Authors, which I conceive is yet a *Desideratum*; and that this is possible to be so I shall mention one Observation, tho' not pertinent to this present Enquiry, yet to another which I have read formerly before this Society, *Viz.* about the Chinese Character and about the Chinese Printing. Inquiring then about *Tartary* [p.344] and *China*, upon occasion of the Discourse that was here lately made, I found that in *Purchas* his Pilgrims is a part of the Works of *Roger Bacon* publish'd, whereby I find that he so long since knew they had a way of Printing, and had a better account of their Character than any one, or all we have since that time. *Sciendum quod a principio Cataiæ Magnæ Nigræ usq; ad sinem orientis Sunt principaliter Idololatræ sed mixti Sunt inter eos Saraceni & Tartari & Nestoriani, qui sunt Christiani imperfecti, habentes Patriarcham suum in oriente.* This *Cataia magna nigra* is one of the North Provinces of *China*, and the *Patriarcha* is the *Lamos* mention'd in the Voyage of *Verbiest*.

*An interesting digression: Hooke was fascinated with the Chinese language. He made a study of Chinese characters and could pronounce and reproduce in writing many of them. One tract is published in the* Philosophical Transactions, *correctly showing characters running from the top of the page to the bottom and the columns of characters reading from right to left. Hooke provided an English phoneticization of the characters, showing he could also pronounce them (in the Cantonese dialect).*

*Jugres qui habitant in terra ubi Impertor moratur,----Sunt optimi Scriptores, unde Tartari acceperant Litteras eorum & illi Sunt magni Scriptores Tartarorum & Scribunt a sursum in deorsum & a Sinistra in dextram, multiplicant Lineas & legunt. Zebeth Scribunt sicut nos & habent figuras Similes Nostris. Tanguæ Scribant a Dextra in Sinistram sicut Arabes & multiplicant Lineas ascendendo. Cataii orientales Scribunt cnmpunctorio, quo pingunt Pictores, & faciunt in una figura plures literas comprehendentes unam Dictionem, & ex hoc veniunt Characteres qui habent multas Literas simul. Unde veri Characteres & Philosophici sunt, compositi ex literis & habent sensum Dictionum.* Thus much concerning the Character, where I shall note only by the bye, that both the *Jugres* and *Cataians*, those of *Tebet* and *Tangut*, may be said to write all the same way with us, for that they differ only in the Position of the Page as to the Eye when read or writ. Next, as to the use of Printing, he says in the same Page, speaking of the Money of the *Cataians. Istorum Cataiarum moneta vulgaris est charta de bombasio in qua imprimunt quasdam Lineas.* This I suppose he took in part out of the Voyage of *Gulielmus de Rubriques*, a *French* Frier, who wrote an account of his Travels into those Eastern Parts to the King of *France*, and for divers Reasons I believe it to be a very true Relation, for I find in the thirty sixth Chapter of his Book as follows.

"The common Money of *Cataia* is Paper made of Bombast the length of an Hand, upon which they imprint Lines, like the Seals of *Mangu*, they write with a Pensil wherewith Painters Paint, and in one Figure they make many Letters comprehending one word. The People of *Thebet* write as we do, and they have Characters very like ours. They of *Tangu* write from the right Hand unto the Left, as the *Arabians*, and multipy the Lines ascending upwards. *Jugur*, as aforesaid from above downwards." This is very much the same with *Roger Bacon*, whereby we had above four hundred Years since a hint of the Chinese Printing; as also that the Chinese Characters were compounded of certain Elements, which expressed both a literal and philosophical Word. I have one Observation more to add before I leave this Digression, and that is in answer to another Objection which was made against my Conjecture of the deducing the Name of *Cornu Ammonis*, or Sand Horns from a probability that they might possibly be found in those Sandy Deserts of *Pentapolitana* in *Africa*, now call'd *Barca*, which lieth West of *Egypt*, between that and *Africa Minor*, almost opposite to the *Morea* of *Greece*, a large and barren sandy Desert, troublesome to be travailed in, by reason of the instability of the Footing, and for that the Sand is thrown to and fro by the Wind, in the midst of which stood the Temple of *Jupiter Ammon* whose Effigies was adorn'd with Horns supposed to be Rams Horns, but I conjectur'd they might possibly be the resemblance of those petrify'd *Nautili*, found in that Sand. To this Conjecture I have only this to add, First, That *Lucan* in the describing this Idol, calls him *Corniger*, which seems to argue, that the Statue had Horns. But which seems more to agree with my Conjecture, is what is related of the form of this Idol by *Curtius*, that it was without the form of any Creature, but like a round Boss or Navel, (*Umbilicus* is the word) beset with Jewels; this was carried in Procession by the Priests in a guilded Ship hung with Bells on both sides, &c. by which it should seem that the very Idol itself was nothing but such a *Nautilus* Petrify'd, as I have produced, beset round with Jewels for ornament, and carry'd in a Ship possibly as a Hieroglyphick, to signifie the manner of some eminent Deliverance of that Country from a former Flood, or the use of Ships in that place, whilst an Island and that Desert was cover'd with Water. But this is only Conjectural, which I submit to further examination.

*Hooke's speculation on the origin of the name* Cornu Ammonis, *the Horn of Ammon, from which the name ammonite is derived.*

[p.345] But to leave this Digression and proceed. I say, it were very desirable in order to the solution of this and divers other Inquiries in Natural Philosophy; that we had a Collection of such Observations as are to be found, already made and recorded in Natural Histories, to see what Light such Histories would afford, which may be perform'd by the joint Labour of many Persons who would peruse and collect such Matters; but possibly it may be believ'd that little can be found pertinent to this Inquiry, as indeed I fear there will be no great matter; yet *Pliny* in the tenth Chapter of his thirty sixth Book takes notice of a matter which is not altogether impertinent, affirming, that an Obelisk set up by *Augustus* for shewing the length of the Day, was found after some time to go false.

But upon this I build no great matter, and I fear the ancient Observations will in general help us no great matter, though they may give us cause of suspicion, as particularly concerning the Latitudes of Places, of which Mr. *Vernon* takes notice that the present Latitude of *Athens* is near two degrees differing from that assign'd it by *Ptolomy*, which is remarkable, it being of a Place so eminently known in former Ages. But upon neither of these can much be built as to the accuracy of determining such a motion; tho' they may serve well enough for hints for Inquiry farther conerning them. Monsieur *Pettit* has also written a Treatise to prove that the Latitude of *Paris* is differing from what it was formerly. *Scaliger* also had a notion of some such matter, but I cannot tell what he would have, nor do I believe he well knew himself. The place is quoted in *Chilmedes* English Edition of *Hues de Globis*; others also have mention'd it, but none have determin'd it or brought it to a certainty. I did therefore upon this occasion, where I am discoursing concerning the general Form and the proprieties or Motions of this great Body of the Earth, think fit to insert it as a thing worthy of determination; since 'tis not improbable but that there may be some such motion of the Earths *Axis* as may alter both the Latitudes of Places, and also the position of the Meridional Line. And that this may not seem so absurd, we may consider the alteration of the *Axis* of the Earth in respect of the fixt Stars long since discover'd, and the variation of the magnetical *Axis* discover'd first about fifty Years since by some of the Professors of this Colledge.

*Hooke is again introducing the idea that the Earth's axis may have changed with respect to the surface of the Earth, polarwandering, and hence various places could have had different coordinates in the past. One should not think this idea absurd, because if the Earth's axis could alter with respect to the fixed stars, as is known, why should it not also alter with respect to the surface of the Earth itself? He therefore clearly makes the distinction between these two types of axial shifts, which has not been understood by others. See Chapter 5 for a detailed discussion.*

But now the Question is how these general Queries can be determin'd, that is, First, *Whether there be any alteration of the gravitating Center of the Earth.*

Secondly, *Whether the Body of the Earth be of a true Spherical or Oval Figure, and thence whether it hath one or infinite Centers of Gravitation.*

*Oblate spheroidal shape of the Earth.*

Thirdly, *Whether the Axis of its Rotation do change its Situation or Position in respect of the Parts of the Earth;* and thence, *Whether the Latitudes and Meridional Lines of places do differ in process of time*, and if so in the

*Polar wandering.*

Fourth place to determine *What is the particular motion that causeth it, and by what steps it hath devolved for the time past, and will proceed for the time to come.*

**No. 4.** *Concerning the Figure of the Earth and variations of the Earth's axis. This lecture was read January 26, 1687.*

[p.346] My First Proposition then is this, That we should suppose First, That this Globe or Ball of the Earth was carried round the Sun in the plain of the Ecliptick, making an entire Revolution in that Plain once in a twelve month, and thereby making the Sun to appear to pass continually in the Ecliptick Line, as *Pythagoras, Aristarchus Samius, Copernicus,* &c, have supposed.

*The first four points are a reiteration of what was already known by astronomers of the day.*

Secondly, That this Globe or Ball whilst it maketh one such Revolution, is likewise whirled round three hundred sixty and five times, and about 1/4 upon an *Axis*, or imaginary Line passing through, or near the Center thereof, which *Axis*, is all the while kept in an Inclination to the said Plain of 23-1/2.

Thirdly, That this *Axis* doth continually keep a Parallelism to itself very near; all which *Axes* at present respect a Point in the Heavens, not far distant from the last Star of the Tail of the little Bear call'd the *Pole-star*, but heretofore 'twas at a greater distance from it.

Fourthly, That this *Axis* doth, in process of time, vary its respect to that Star or Point of the Heavens, and by degrees proceed nearer towards it, not directly, but in a Circle parallel to the Ecliptick, or whose Center is the Pole of the Ecliptick. Thus far I take the same with the Hypothesis of *Copernicus* and his Followers. But

Fifthly, I suppose yet further, that the *Axis* of the *Diurnal Rotation* of the Earth had also had a progressive motion, and hath, in process of time, been chang'd in position within the Body of the Earth, and consequently that the Poler points upon the Surface of the Earth, and have alter'd their Situation; so that the present Polar Points have formerly been distant from those Poles that were then; and consequently that those former Polar Points are now remov'd to a certain distance from the present, and move in Circles about the present.

*This fifth point is original with Hooke—i.e., variations of the Earth's axis with respect to the surface of the Earth, or polar wandering. See Chapter 5 for a detailed discussion of this important concept.*

Sixthly, I suppose that the Form of the Surface of the Water at least, is, and hath been, ever since the duration of the Earth, of an *Oval Form*, whose longest Diameters lye in the Plain of the Equinoctial, and whose shortest is the *Axis* itself of the said Rotation.

*The shape of the Earth is described clearly here as an oblate spheroid although the term "oblate" was not known at that time.*

Seventhly, As a Consequent of this I suppose the Center of *Gravity* of the Earth to be drawn out into a Line into the *Axis* thereof, and consequently into infinite Centers, there being one for every Parallel Line upon the Surface of the Earth, and that no Perpendiculars but those of the Poles and Æquinoctial, respect or tend directly to the Central Point, but that all the Perpendiculars from the other Parallels respect certain Points in the opposite Parts of the *Axis* which are so much the further remov'd from the Center by [p.347] how much the nearer the Parallels approach the Polar Points; which Points of Gravitations and Position of Perpendiculars in respect of the *Axis*, may be determin'd both *a Priori* by Theory, and also *a Posteriori* by Experiments or Observations.

*Because of the shape of the Earth, only those lines perpendicular to the surface from the poles and the Equator end at the center of the Earth. All other lines intersect the axis within a limit corresponding to the degree of oblateness.*

Eighthly, As a Consequent of these, I suppose, that in process of time there will be caused an alteration of the gravitating Power and Tendency of the Parts of the Earth, both Solid and Fluid, and that according as the Positions of them are alter'd in respect of the Polar Points, either present Precedent or Subsequent, there will be caused in the

Ninth Place, and indeavour of sliding, subsiding, sinking and changing of the Internal Parts of the Earth, as well as External, tho' the latter will be more powerful, as being more affected by the Rotation thereof; and this may cause in the

Tenth Place, an alteration in the Magnetical Power and Vertue of the Body of the Earth, especially of such Parts as are more loose and of a more fluid Nature. And

In the Eleventh Place, may be a cause also of some of those *Tremores Terræ*, or Earthquakes, which have in all Ages been in the Earth, tho' we have no Histories or Records that have preserved the Memory of them, but only such Signs and Monuments as they have left by the unequal ragged and torn Face of the Surface of the Land and Bodies that are discovered; which proves that they had some time an other Position than they are found to have at the present.

These two last notwithstanding I do not suppose the only causes of these Effects of Earthquakes, no nor the Principal, but only as concurring and adjuvant Causes which may have their Effects in some measure, but how far and how powerful they may be supposed, will be proper to be resolved under the Heads of Magnetisms and Earthquakes, and more especially under that of the Air. The same Principles or Suppositions will also produce in the

Twelfth Place; a more than ordinary swelling or rising of the Sea in those Parts which are near the Æquinoctial, and a sinking and receeding of the Sea from those which are near the Poles; so that as any Parts do increase in their Latitudes, so will the Sea grow shallower, and as their Latitudes decrease, so must the Sea swell and grow high; by which means many submarine Regions must become dry Land, and many other Lands will be overflown by the Sea, and these variations being slow, and by degrees will leave very lasting Remarks of such States and Positions, in the superficial Substances of the Earth.

And hence also will follow in the Thirteenth Place, a great alteration and variety of the Productions of those Parts which are thus alter'd in their Position, whether they are parts of the Sea or parts of the Land; for as there seems to be somewhat which is peculiar to this or that Soyl or Spot of Land whereby this or that Animal or Vegetable doth grow and thrive and increase both in Quantity and Quality, and the contrary: So is there also somewhat in the Climate and Position to the Sun and Heavens, which doth as powerfully at least, if not much more, affect the Productions, Propagations, *&c.* of Plants and Animals. And as 'tis a known Observation, that in the same Country, this or that Field, or Soil is more effective for this or that use; so 'tis as well known that the transplanting of animate Subjects to differing Climates, tho' the Soil seems of the same Nature, doth as effectually co-operate in the changing or alteration of them. And hereby a fruitful Land may be turned into Barrenness, and be made unfit for Productions as well as Barren and Useless may be made Fruitful; for that the Temper and Constitution of a Soil may be such as to be fit for many purposes in some Climates, which in others is fit for nothing.

*Changes in the center of gravity and in the terrestrial magnetism then could cause the "sliding, subsiding, sinking," etc., of both the internal and the external parts of the Earth, although the external shifts would be more powerful.*

*Such changes then would cause earthquakes and other effects associated with earthquakes, such as volcanic eruptions.*

*Because of the equatorial bulge, the sea would rise near the Equator and recede from the poles; thus exchanges of land and sea areas could take place as polar wandering proceeds.*

*The environment then would also change, and certain animal and plant life would flourish or not as they adapt to these changing environments.*

[p.348] From hence also will follow in the Fourteenth Place, That many places which by degrees are made Submarine, will be cover'd with various Coats or Layers of Earth; so that the former Surface of it, when Land will not only be drown'd with Water, but buried under Earth; for that, as the parts of the Land, are continually washed down, and by the Rivers carried into the Sea, and there deposited in the Submarine Regions, so much more powerfully and plentifully are the higher parts of the Submarine Regions by Tides, Currents, and other Agitations of the Water, removed and transported into the lower, partly by sinking out of the muddy Water, but principally by tumbling and rowling down from the higher, which sorts of covering or burying Earth must be posited in certain Layers or Stratifications of divers kinds of Substances, according to the nature of those which are this or that way brought thither, and there deposited. Hence also it will follow, that the Earth itself doth, as it were, wash and smooth its own Face, and by degrees to remove all the Warts, Furrows, Wrinckles and Holes of her Skin, which Age and Distempers have produced.

*Many changes in the surface of the land occur "by degrees" as with the wearing down of high places by water erosion and the piling up of "layers or stratifications" of sediment. These changes, therefore, are not catastrophic in the sense that changes are wrought by earthquakes. Another example that Hooke was not solely a catastrophist.*

And hence in the Fifteenth Place will follow, That such Regions as have for a time been Submarine, and produced Substances of Animals or Vegetables proper for them, when they come to be dry Land and to lye above the Waters, must produce Animals and Vegetables proper and peculiar to that Soil, Element and Climate they are then furnish'd with; preserving in the mean time the Characteristicks and Marks of the former Qualifications, when in another Condition.

*As the exchange of land and sea areas take place, changes in animal and plant life occur, but "characteristics and marks" of the former condition are preserved. Hooke is convinced, therefore, that we can read past history in the rocks!*

But some possibly may be ready to say before a thorough examination, that this is only a supposition, and that there are no such Phænomena as here are put for the Supposition: Others, that 'tis foolish to make an Hypothesis for the solving of any one Phænomenon. Others may possibly demand how comes this to be now discover'd, which none hath hitherto known? Or how is this to be proved? By what History? By what Signs and Tokens? I must leave every one to his own freedom to judge as he sees cause, and censure as he pleases; however, I conceive it ραον μωμασθαι η μεμεισθαι easy to play the Momus or the Mimick. *Sed siquis quid rectius istus noscat, candidus impertiat.* But if he know better let him not, hold his Tongue but tell us. I shall not impose on any; I propound it only as a Hypothesis, and have shewed what will be the Consequences of it, whether there be Phænomena answerable to be observ'd let it be examin'd; and let there be produc'd another Hypothesis that will solve the various Phænomena that are to be every where met with better; for that I have no farther design in propounding it than to have it strictly examin'd, and in order thereunto to have such Observations made and taken notice of for the future as may ascertain the Truth whether for or against it.

*Hooke is aware that his ideas here are rather revolutionary, but he defends his presenting them as a hypothesis and welcomes other ideas. It's "easy to play the Momus or the Mimick" may refer to the fact that others have tried to plagiarize these ideas as well as others.*

Yet give me leave to add a word or two, before I wholly leave it to its Fortune.

First then, I say, That what is here supposed is not impossible. First, 'Tis not impossible from the Natural History now to be met with the things supposed; for that all things may be the same as they now appear, and yet this may be true; for no one Phænomenon, that I can think of, is contradicted by it, either fetch'd out of ancient Histories, or yet Collected by present Observation. As there are no Observations of Latitudes, or fixed, accurate meridian Lines, or Eclipses for the Oval Shadow of the Earth, or Mensuration of Degrees to find their difference in differing Latitudes. Nor Secondly, Is it impossible from the Nature of the things supposed, for that there is as yet no certain Cause assigned, why the Earth doth move upon the *Axis*, it now doth, and not upon another, nor why it should always continue and remain the same without change, contrary to all other motions in Nature. Nor is it impossible because not discover'd before, which yet is more than can be positively proved; for if so, then would Magnetical Motions fall under the same Censure, as also, Optick-glasses, Guns, Printing, and other new dis-[p.349]coveries. And by the same Argument the Motion of the Sun, and *Jupiter* upon their Axes, the Reality and Revolution of the Satellites of *Jupiter* and *Saturn*, the Ring of *Saturn* and the Belts of *Jupiter*, and the like might be condemned.

*In an attempt to forestall criticism, perhaps, Hooke defends his hypotheses as not only not impossible, but probable, on the grounds that his observations support them.*

*It was Hooke himself who first discovered the giant spot on Jupiter and the fact that it moved from left to right across the planet, demonstrating that Jupiter rotated on its axis.*

Secondly, I say for it, that 'tis no more folly to invent new Hypotheses to solve Phænomena in the Earth, than it was in *Pythagoras, Ptolomy, Copernicus, Ticho, Kepler,* and others in the Heavens; for that each of them conceiv'd by such Hypotheses to solve the Phænomena more agreeably to the other appearances of Nature; whereas yet no one of them has hit the right I conceive, and I shall, I hope, in due time demonstrate.

But in the Third Place, for Affirmative, I say, 'tis not only possible, but probable, and altogether consonant and agreeable to the rest of the Works of Nature, and even to the very Constitution and Phænomena to be observ'd upon the Earth itself.

And First for the *Oval Figure of the Sea and Body of the Earth in some measure.* If the gravitating Power of the Earth be every where equal, as I know no reason to suppose the contrary, then must this Power be compounded with a contrary indeavour of heavy Bodies to recede from the Axis of its Motion, if it be supposed to be mov'd with a diurnal Revolution upon its Axis, and consequently a part of the gravity of such Bodies towards the Center must be taken off by this *Conatus*, which is every where oblique, but only under the Æquinoctial, which must therefore most diminish its Gravitation, and consequently the gravity will act the most freely and powerfully under the Poles, and the more powerfully the nearer the Bodies are plac'd to those Poles; and that Phænomena do answer to this Theory, has been verify'd, first by Mr. *Hally* at St. *Helena*, and since by the *French* in *Cayen*, and now lately in *Siam*, in all which places it is affirmed, that 'twas necessary to shorten the Pendulum to make it keep its due Time.

*The oblate spheroid figure of the Earth has been demonstrated by expeditions on which the pendulum had to be shortened nearer the Equator as clocks lost time there.*

In the Second Place for the *Variation of the Axis of Rotation* in the Body of the Earth. I say it is consonant to all the other motions of Nature: For first it is found that the *Axes* of the *Ellipses* of the Planets do vary a little, I say a little (tho' Mr *Street* only will have them not to vary at all) because all Astronomers have hitherto affirmed, that they do, and from my own Mathematical Hypothesis I collect the same, tho' it be but a little, yet it is somewhat, since there is some impediment in the Medium. Next there is also a motion in the Nodes, all which are very eminent in the Moon. And again, the direction of the *Axis* in the Earth is varied as to its respect to the Heavens, which the precession of the Æquinoxes do manifest. Nay yet further, the *Axis* of the Magnetical Motions which is within the very Body of the Earth, and seems even to go through its very Center, hath, about fifty Years since, been prov'd to vary also somewhat analogous to this which I have supposed, whereby both the Magnetical Latitudes, and Magnetical Meridians have most certainly been varied; which seems abundantly more difficult to be granted than this which I propound, did not certain Observations both here at home and all over the World confirm the truth of Matter of Fact, and that because this doth seem to prove a motion of a Magnetical Core or Magnetical Globe of the Earth, within this outward earthy and watery Shell; whereas this which I suppose is nothing but a progression of the *Axis* of Rotation, which may be caus'd by the visible accidental Mutations of the outward and superficial Parts, as well as by other unknown alterations which may succeed within the Bowels of the Earth. So that 'tis very probable that there is some such motion of the siad *Axis*, since we are certain both of outward and inward changes.

It only remains then Positively and Experimentally, or Historically to prove the Reality thereof. Now the motion of the Mutation thereof being but slow, as I conceive, and the Observations of the Antients Recorded in Histories necessary for this purpose, being so unaccurate and uncertain for such a determination as this, I fear they cannot be rely'd upon; but whatever shall be alledg'd as a proof of this Theory, will be attributed to a fault in the Antient Observation, as that *Ptolomy* puts the Latitude of *London* 52 10 and the longest Day 17 Hours. Nor will I insist on the Latitude of *Athens* [p.350] found by Mr. *Vernon*, to differ near a Degree: Nor on the Latitude of the *Herculean Streights*, which varies as much from the present, as that of *London*, tho' all these were remarkable places, as was also *Constantinople*; but rather rely upon Observations to be made for the future; the way of performing which I shall treat of hereafter, whereby I shall shew, how, in a short time, the same thing may be determin'd as well as by so long a time.

*It is known that the terrestrial magnetic poles have shifted, a fact which seems to Hooke should be more difficult for acceptance than his own thesis of wandering of the geographical pole.*

*It has been suggested that Hooke started to lecture on the classics [see later discourses] in order to look for evidence of changes in coordinates over time to support his idea of polar wandering, but it is clear from this passage that Hooke quite distrusts the accuracy of such records in ancient writings.*

**No. 5.** *Concerning the same subject, i.e. the Figure of the Earth and the alteration of the Earth's Rotational Axis. Read February 2, 1687.*

[p.350] What I propounded the last Day by way of an Hypothesis, may possibly be look'd upon not only as very extravagant, but very improbable; from the last of which I hope I did then clear it; and as to its extravagancy, I hope I may be able to shew, that there have been suppositions altogether as extravagant, which yet have not only been made, but accepted and imbraced, and for many Ages as stifly defended as the most probable. My Instance shall be in the *Ptolomaick* Hypothesis of the Heavens, which, that you may the better judge of, I have here a Book to shew the whole Design and Intrigue of it, in which the same and all its parts are most curiously delineated, whereby all the Wheel-work may be at once discover'd; and if it be desir'd to be made in Clock-work, I have another Author that shall give the bigness of the Wheels, and the number of the Teeth and Pinions necessary to accomplish the same in Clock-work: And yet when all is done, there will want as many more to make out all the irregularities of appearances exact; the reason of which proceeded from one false Principle, that one Body was capable of no more than one simple motion, whereas in truth there is no body mov'd but is capable of, nay, actually mov'd by thousands.

But it may possibly be said that this *Hypothesis* was the Product of an Age not so inquisitive and able to judge as the present, which will hardly be impos'd on with such improbabilities; nor was all this clutter thought necessary at first, but the maintainers of that Opinion, to make out the appearances, as well as they could, have since found it necessary to help out the first Invention by additional Expedients; and if these were sufficient, I conceive it might yet be an acceptable *Hypothesis*, tho' we have no *Medium* to prove that there is any such thing in Nature as a *Solid Orbe*, or a moving *Genius*.

The like favour I hope may be allow'd to what I propound, if upon due examination the *Phænomena* are answerable to what the *Hypothesis* does hint.

Now what would be consequential to what I have propounded, I shew'd the last Day; it only now remains to examine whether Phænomena do answer.

First then to determine whether the Figure of the Sea from North to South be Oval, swelling towards the Æquinoctial and depress'd towards the Poles, it will be necessary to make some few Trials, Observations and Experiments.

And First for Experiments that may be made here. Let a Bowl or Bubble of Glass be made and melted in a Lamp, and when so melted let it be blown into a hollow Ball or Bubble, which will naturally form and Shape itself into a round and spherical Body, especially if the Substance be of an equal thickness and equal heat, which let be examin'd; then let the same be melted again as before, and as it is blowing, let it be mov'd round upon the Pipe, by [p.351] which it is blown, by a pretty quick Circular Motion, and you will find that instead of the Spherical Figure it will receive an Oval one, such as I suppose the Surface of the Sea to have. This Experiment I shall by and by shew here (which was accordingly done).

*For detailed comments on the contents of this lecture, refer to Chapter 5.*

*Hooke's demonstration of the oblate spheroid shape of the Earth—i.e., with equatorial bulge and polar flattening. The shape is called a prolate spheroid by Hooke, using the longer axis in the equatorial plane as reference rather than the shorter rotational axis. The word "oblate" was not known then.*

Now in this Experiment here are evidently two kinds of Powers that cooperate in the production of this Form: The first is that of the Congruity of the Matter, which, as I have many Years since in a small Treatise, Printed in the Year 1660, proved, doth shape the Glass into a true Spherical Figure, and so maketh every part to indeavour towards the Center of the whole. The next is that of the vertiginous Motion, which giveth to every part, an indeavour to recede from the *Axis* of the vertiginous Motion; this driveth the shape of the whole into that Oval Form it receiveth and retaineth.

*The attribute of congruity of matter allows the glass to form into a spherical figure, while the twirling action, or "vertiginous Motion," drives the shape into the oval form.*

The same Experiment may be much better made at the Glass-house, where a greater quantity of Glass may be melted, and that more equally and a quicker Motion may be given, which will make the Experiment the more sensible, the Glass retaining its melted heat much longer. Besides, it may be there tried with a solid lump of Glass which will receive the same Figure from a vertiginous Motion about the Puntilion. And again, to make the Glass Oval the other way, the same is whirled round with a motion wherein the Puntilion is made the Radius of the vertiginous Motion.

A second Experiment to shew that the Water doth naturally recede from the Poles towards the Æquinoctial is this. Take a round Dish of Water, and let it be set upon a Stand where it may be gently mov'd round upon an Axis passing through the Center of the Dish perpendicularly; first observe the Surface of the Water when it stands still without motion, there you find it smooth and horizontal; then move the stand gently round by degrees, till you find the Water begins to receive the motion of the Dish; then examine the Surface thereof and you will perceive the Water to sink in the middle, and to recede and swell towards the Circumference of the Dish: And the better to satisfie you I have prepared the Experiment which I will by and by shew. The Experiments are plain and common, yet I humbly conceive not less instructive to the present Controversy, than the most pompous and more chargeable Experiments.

*It is interesting to note that Alfred Wegener of continental drift fame repeated this same experiment. See Chapter 5.*

This last Experiment doth hint, that the Convexity of the Sea near the Poles of the Earth must necessarily be much flatter than elsewhere, and not only less Spherical than the rest of the Sea, but possibly plain, nay, beyond a plain, possibly Concave, for that the Water cannot but have or receive from the vertiginous Motion, an endeavour to recede from the Center of that Motion, and the Gravity of the Earth working there more powerfully and freely. But this only by the bye. But which seems more material, I conceive that a Degree of Latitude, if there measured would be very much longer than a Degree of Latitude under the Æquinoctial, of which I shall speak more by and by.

*Hooke's argument: If his Earth's figure is correct, a degree of latitude near the poles would measure longer than one near the equator, as of course it does.*

In the next place then we are to consider what other Observations and Trials will serve to the direct and positive proof of this *Hypothesis, That the Figure of the Earth is that of a prolated Sphæroide, not of an oblong Sphæroide, nor of a Sphære.* And those may be ranged under two Heads, First, Such as are consequential Proofs drawn from the similitude in Nature's Operations, on other Bodies similarily affected. And Secondly, Those which more immediately and positively prove the Effects thereof upon the very Body of the Earth itself.

*If one considers the axis of rotation as reference, the figure of the Earth would be an oblated spheroid. Hooke obviously is thinking in terms of the long axis of his "prolated spheroid" to be in the equatorial plane. Also, as noted already, the word "oblate" was then unknown.*

The first sort of Observations are to be fetch'd from the Cælestial Bodies, such as we are assur'd by Observation have a vertiginous Motion about their Axis, as Hypothetically only we suppose the Earth to have; such are the Body of the Sun Primarily and Principally, which was discover'd by *Galileo*, and prov'd and perfected by *Scheiner*; next the Body of *Jupiter*, which was first found to move about its own Axis, in the Year 1664, and which has since been perfected by *Cassini*. Now, if by exactly examining the true Diameters of the Sun when we are in the plain of its Æquinoctial (which is in the beginning of *June* and of *December*), if I say by Trial, we find that the Diameter *per Axin* of the Sun is shorter than the Diameter of its Æquator, then there will be a [p.352] further probability that the like may be in the Earth if it be so mov'd, as is now generally supposed: The like trial may be made of the like Axis of *Jupiter* though the Trials will be therein more difficult, as being much less sensible, from the smallness of the Difference; however 'tis worth examining, as it will be to examine also the Diameters of *Mercury* and *Venus* when they pass under the Sun, tho' we are not yet assur'd of their vertiginous Motion, and if Mons. *Gallets* Observation may be credited, such a *Phænomenon* was taken notice of by him in the late transit of *Mercury Sub Sole*, as appears by his account of the Passage of *Mercury Sub Sole,* Printed in a Treatise by itself, and in the Journal *des Scavans*. Now, if this Observation do answer in the Diameters of the Sun, it will afford us also a further information of the Nature of that Glorious Body, and will, I conceive, prove it to be of a fluid and yielding Substance, especially the shining and superficial parts thereof. Trial also may be made of the like Diameters of the *Moon*, tho' her vertiginous Motion in comparison of her bulk, be the slowest of all we yet know as turning round on her Axis but once in a Month. The like may be made of the Body of *Saturn*, when the Ring is so posited as that the Diameters that lye in the longer and shorter Diameters of the Ring may be plainly discover'd; what the reason of that Ring may be I shall discourse of elsewhere. These I suppose will be the easiest and soonest made, and if judiciously and accurately perform'd, with a due regard of Refraction, and the true position of the Axis, will give a great probability or improbability to this supposal, but still I confess it will afford no more than a probable Argument either for or against it: However, that probability being very great, and the trial not very difficult; it will be well to make the Observations, especially those of the Suns Diameter, with all imaginable accurateness, which may be done to a very great one, if there be fit Instruments and sufficient Care used therein, so as very many times to out strip all that I have hitherto met with of that kind, the whole method of which will be too long and tedious now to explain; however, if I can procure Assistance, I resolve to try it this following *June*, which is much the best time of the whole Year to avoid the inconveniency of Refractions, and the true Phænomena thereof I will produce here, without being biassed for this Hypothesis, for which I have no further concern than as it shall be found agreable to the truth of Appearances. Now, tho' I confess also, that I cannot expect that the difference of the longer Diameter in the Sun from the shorter will be very much in regard of the very strong power of Gravity in that Glorious Body whereby it is able to detain all the planetary Bodies in their Orbs from running from him, and even that of *Saturn* so vastly remov'd; yet when I compare that with the Magnitude of its Body, and the time of its Rotation, I am apt to think that accurate Trials may discover some sensible difference, which I must leave to Trial.

*Hooke's plans to measure the differences between the diameters of the longer and shorter axes of the Sun and planets, assuming that planets like Mercury and Venus rotate on their axes in a "vertiginous Motion."*

The second sort of Observations or Trials necessary to prove this Hypothesis, which are direct and positive, and may be truly call'd *Experimenta Crucis*, according to the Lord *Verulam* are principally two,

*Lord Verulam = Francis Bacon*

which are sufficient to prove it thoroughly, tho' the other should fail; the first is to procure an exact trial to be made of the Time that a Pendulum Clock will keep under or near the Æquinoctial, which is adjusted exactly to the time by the Sun or Stars in a much greater Latitude; or the trial of such a Clock in two places very much differing in Latitude after the Clock hath been exactly adjusted in time, to one of those places; because such a difference if it be found and determin'd will be of sufficiency to determine the proportional co-operation of these two Powers. As for instance, this may be sufficiently examin'd by a Clock adjusted in *England*, and tried in the *Barbadoes*; if Care and Accurateness be used in both these places, which I conceive might be easily procur'd by the Favour and Assistance of this Honourable Society. The second which is a much more difficult Experiment, but yet much more positive and convincing than any other, is the measuring of the quantity of a Degree of Latitude upon the Earth, in two places very much differing in Latitude; the one as near as might be towards the Pole, as upon the Ice in the *Finniek Gulf*, as Monsieur *Thevenot* proposeth, which might be procur'd by Mr. *Hevelius* at *Dantzick* or Dr. *Rudbeck* at *Stockholm* in *Sweedeland*, who might do it himself or [p.353] procure it to be done at the North end of that Gulf, which would be yet better, and by some Persons in *Jamaica*, or other parts nearer the *Æquator*. These last trials, if accurately made, would be undeniable Proofs of this supposition, if it should be certainly found that a Degree in the more Northern Countries were more large than a Degree in the more Southern Climate, and the Experiment with the Pendulum Clocks would likewise more exactly adjust the true Gravity of the Earth consider'd simply without the composition of the vertiginous Motion. And thus much for the first part of the Hypothesis, that the Figure of the Water above the Earth is that of a prolated Sphæroeid whose shortest Diameter is that of the Axis of its Rotation.

Next for the examination of the second Part thereof (namely, whether the Axis of its Rotation hath and doth continually by a slow progression, vary its Position with respect to the Parts of the Earth; and if so, how much, and which way, which must vary both the Meridian Lines of Places, and also their particular Latitudes) it had been very desireable, if from some Monuments or Records of Antiquity, somewhat could have been discover'd of certainty and exactness, that by comparing that or them with accurate Observations now made, or to be made, somewhat of certainty of information could have been procur'd: But I fear we shall find them all insufficient in accurateness to be any ways relied upon; however, if there can be found any thing certain and accurately done, either as to the fixing of a Meridian Line on some Building or Structure now in being, or to the positive or certain Latitude of any known place, tho' possibly those Observations or Contructions were made without any Regard or Notion of such an Hypothesis, yet some of them compared with the present state of things might give much Light to this Inquiry. Upon this account I perus'd Mr. *Graves* his Description of the great Pyramid in *Ægypt*, that being Fabl'd to have been built for an Astronomical Observation, as Mr. *Graves* also takes notice. I perus'd his Book I say, hoping I should have found, among many other curious Observations he there gives us concerning them, some Observations perfectly made, to find whether it stands East, West, North and South, or whether it varies from that respect of its sides to any other part or quarter of the World, as likewise how much, and which way they now stand; but to my wonder, he being Astronomical Professor, I do not find that he had any regard at all to the same, but seems to be wholly taken up with one Inquiry, which was about the measure or bigness of the whole and its parts, and the other matters mention'd are only by the bye and accidental, which shews how useful Theories may be for the future to such as shall make Observations; nay, tho' they should not be true, for that it will hint many Inquiries to be taken notice of which would otherwise be not thought of at all, or at least but little regarded, and but superficially and negligently taken notice of. I find indeed, that he mentions the South and North sides thereof, but not as if he had taken any notice whether they were exactly facing the South or North, which he might easily have done. Nor do I find that he hath taken the exact Latitude of them, which methinks had been very proper to have been retain'd upon Record with their other Description. [Here by the bye because it agrees with a former Conjecture, I here proposed, concerning those stupendious Works, namely, that the Core of them was probably some natural Rock cut and shaped fit to be cased or cover'd with another sort of Stone, which was at that time much contradicted, by Affirmations, that the whole Country and Place of their Station was nothing but Sand. Give me leave to take notice that Mr. *Graves* does affirm, That the great Pyramid is founded upon a natural Rock which riseth above the rest of the Sand, and that the Rooms about the second Pyramid are hewen and shapen out of the natural Rock; and I doubt not but that if they were all examin'd, they would be found to be so and nothing else, which would much alleviate the stupendious Labour and Work of Men that must otherwise have been supposed to be made use of; but this only by the bye.] To proceed then where I left, I say that I conceive it were very desirable for the future, that those I have mention'd, and several other particular Observations, were purposely [p.354] made for that such would give a great light to judge and make a true valuation of the State and Nature of places and things, which in most Descriptions we find altogether wanting. As among many other things I could hint, I should be very glad to find such a Description of the Nature of the Sand of those Parts as would inform me whether it have not been all a Sea-Sand: I say, not only of this Country of *Ægypt,* which is so exceeding plain, and so exceeding Sandy, with many cragged Rocks rising out of it; but of *Arabia Deserta,* and *Arabia Petræa.* and all the parts near the *Tigris* and *Euphrates,* and all the parts on this side of *Egypt,* as the Region of *Barca* and *Pentapolitana,* and many other which are said to be all smooth and cover'd with Sand; for Observations designedly made, would easily discover whether such Sands had been owing to the Sea, or to some other Cause, which, by some curious Observations I have met with in the Travels of *Peter de la Valle* and *Bellonius,* and others, I judge they have. I shall here present you with one of them. *Pietro della Valle parte terza Lettera* 11da *d'Aleppo* Aug. 5. 1625. *Vidi per terra molte Conchiglie marine, lustre dentro comme Madre Perle, parte intere, e parte spezzate, che in Luogo tanto lontano dal mare mi marvigliai come potessero trovarsi, vidi anco sparsi per tutto molti pezzi di Bitume, che in quell terreus sulmastro, e che in qualche tempo dell anno per allagarsi d' acqua si genera, del quali ne presi e tengo mostra appresso di me.* NB. This Place is betwixt *Bassora* and *Aleppo,* in the Deserts of *Arabia,* fourteen Days Journey from the Sea.

In *English* thus, "I Saw on the Ground many Sea-Shells shining within like Mother of Pearl, some whole, some broken, I much wonder'd how they could be found in a place so far distant from the Sea; I saw also scattered every where many bits of Bitumen, which in this salt Earth and Soil is generated and rises upon the Water at some times of the Year, of

*To Hooke, theories were important and they afforded future inquiry. He was perhaps aware that many of his ideas are before their time.*

*Hooke, who loved to digress, now does so to discuss how the pyramids were constructed.*

*Hooke is convinced that given a complete description of sands from any desert or any other place, he could discern whether they were marine sands or not. This idea that one can discover the provenance of sand grains is a distinctly modern geological concept. As this passage indicates, Hooke would find the presence of fossils a useful tool.*

*Translation of the above passage.*

which I took some, and keep the Specimens by me to shew." Moreeover, I hoped to have found something remarkable to my purpose in the Voyages of Sir *George Wheeler*, where he hath describ'd *Greece* and *Athens* in particular, and all the remarkable places about it, which are Places the best described of any thing of Antiquity, and more especially in his Description of the Temple of the Eight Winds, which is said by *Vitruvius* to be given to the City of *Athens* by *Andronicus Cyrrhastes*, and is remaining intire to this Day, all except the Vane or Weather-Cock at the top. I expected, I say, I should have met with some very exact and curious Observations, which methinks the very design of the place should have hinted, of the true Position of it as to those eight parts or *Plagæ mundi*; but I find nothing more to this purpose but that each Wind answer'd exactly to the compass, in the mean time not telling what was the variation of that compass at that time or place; however, he doth shew that the Position and Latitudes of places do much differ from what they had been described to us, but then how far we may relie upon antient Observations, will be a further doubt.

*One has the impression that Hooke would have loved to travel to all kinds of foreign lands to make observations, but he had to rely on reports of travelers, as Hooke himself was extremely occupied with his various responsibilities.*

I should be glad that such as are better read in ancient Records would for the future at least take notice of any Observations they meet with which may afford some light to this Inquiry; and so for that Matter I must there leave it; for tho' I could accumulate many Observations which do seem to make for it, yet the uncertainty and unaccurateness of the Observations of the Ancients in this particular make me omit them.

*Again he reiterates that ancient records are unreliable as to accuracy.*

And so I am reduced at last to such Observations as have been made in latter times, and with more accurateness and diligence, and with better Instruments, and to what may be purposely made with Instruments a hundred times more exact, and with designed and pertinent Observations for this very end; and such Observations will be principally of two kinds, First, Such as examine and state the exact Position of the Meridian Line of places even to a single second, or to a greater accurateness if required. And Secondly, Such as examine and state the true Latitude for that from some few such Observations accurately made, as they ought, more may be proved by seven Years Observations than by seven hunder'd Years Observation of the Atients, nay tho' they were again multiplied by seven. But of this shall discourse in my next.

*The need for accurate measurements of longitude and latitude for which, in the 17th century, there was great interest.*

**No. 6.** *This lecture was read February 9, 1687, a continuation of the subjects discussed the previous week*

[p.355] I Hoped I had by my Discourse at the last meeting evidenced the first part of my Position which I deduced as a Corollary from the diurnal Motion of the Earth, namely, that such a motion must cause a recession of the Sea from the Polar parts towards the Æquinoctial, which must necessarily make the Surface thereof of a prolated Sphæroidical Figure. But I perceive some notwithstanding the Experiment, which shewed of the recess of the Water from the Center, do yet doubt of the Consequences thence deduced with reference to the Earth, and seem'd not to be satisfied that the two Methods which I propounded for the examination and determination thereof were sufficient.

Now that I might not leave any rub behind which might be a stumbling Block at the entrance, I have now prepared a short demonstration of the necessity and infallible certainty thereof, as it is a deduction from an Hypothesis which is now by most Philosophers and Astronomers granted, namely, the diurnal Motion of the Body of the Earth upon its Axis.

In order to which Demonstration I must premise this principle of Motion, That *every Body that hath received, or is moved with any degree of motion if it receives no other motion from any other Body whatsoever, will constantly persevere or continue moved with the same velocity in the streight Line of its tendency infinitely produced.* The reason of which is this, that no Reason can be assigned why its Motion should cease where there can be no impediment. Nor is there any reason why it should deflect to any side out of its direct way, since from the supposition there can be no new motion added to it from any other Body. Now this being a Principle will not admit of any other Demonstration than that of Induction from particular Observations in Natural Motions, by which all such Principles are made; for whosoever shall strictly and accurately examine and analyse all local Motions, will find hundreds of instances that after a due analysis is made do sufficiently evidence the universality and certainty of this Principle in all local Motions.

*Hooke's expression of Newton's 1st Law of Motion is independently conceived.*

From which Principle it will follow, that any Body moved Circularly with any degree of velocity (whilst some way continu'd to move about that Center) will at the instant that containing Power is remov'd, proceed to move directly forward in the straight Line of its tendency, which straight Line is tangent to that Circle in which it aquired, or had its imprest velocity; for the conteining Power, which by a continual atraction or otherwise towards the Center, kept it in that Circulation, ceasing, and no other Body whatsoever impressing any new motion upon it (as is supposed in the first Proposition) the Body must continue to move in the streight Line of its Direction without any Deflection, Retardation, or Acceleration.

*The "containing Power" = centripetal force, a concept that Newton had not conceived and that Hooke communicated to Newton in 1679, allowed the latter to understand the physics of celestial motions and develop his laws of gravitation. See Chapter 3.*

From hence it will follow, that the farther it is moved in that Line, the more and more will it recede from that Center of Motion to which it was detained, and that for a short time with Spaces in a duplicate proportion of the times it spendeth, or of the Spaces it passes in that tangent Line, namely, in the proportion of the smaller Secants. This, as shewn by *Gelileo* [sic] and others, I pass over without farther proof.

*Gelileo = Galileo.*

[p.356] From hence it will follow, that in all Circular Motions that make their Revolutions in equal times about the same Center, but in Circles of

258  DISCOURSE OF EARTHQUAKES

differing Radii, the recess in equal times will always be in the same proportion as the Radii of those Circles, or as the Tangents or Secants of the same Angle at the Center; this will be plain by the Scheme, where *a* represents the Center of the Motion, *eg, di, cl,* &c. Similar Arches of different Circles on the same Center *a*, the Bodies placed in *b, c, d, e,* are put to pass their respective Arches *bn, cl, di, eg,* all in the same time; now the Tangents *ef, dh, ck, bm,* being in the same proportion with their respective Radii, and their respective Secants, their respective receding from the Center *a*, will be in proportion to their Radii.

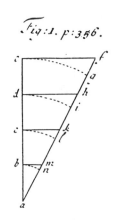

Hence it follows, that the recess of the Parts of the Earth from the Axis of the diurnal Rotation will be in the same proportion as the Sines complement of the Latitude of those places, which recess is no where directly from the Center of Gravity, but under the Æquinoctial it being every where perpendicular to the Axis of Rotation.

Now the simple Gravity of the Earth as a Globous Body at rest can be no other than to the Center of that Globe, it being consider'd only as a Globe without any Circular Motion, as I shall prove when I speak of Gravity. And this Gravity every way equal, it will thence necessarily follow, that by the composition of those two Powers acting on Bodies, there will necessarily follow these Consequences, First, That every Meridian Line upon the Surface of the Sea, is of an Elliptical Figure, whose shortest Diameter is in the Pole, and whose longest is in the Plain of the Æquinoctial. Secondly, That the Gravitation of the Earth, as moved on an Axis, is in every Latitude different, the least under the Æquator and the greatest under the Poles.

Thirdly, That the Perpendiculars or Lines of Gravity or Descent do no where, except under the Poles and Æquinoctial respect the Center of the Earth; but other Centers in the Axis of its Rotation, let *Abc,* represent a quarter of the terrestrial Globe Orthographically projected upon the plain of a Meridian, where let *a,* represent the Center, *b,* the Pole, *ab,* the Axis, *ac,* the Æquinoctial, let *αε, βg γi* represent the Radii of certain parallels of Latitude, whose Rotation about the Axis *ab,* gives each of them a proportion of velocity corresponding to their length or distance from their Axis of Motion *ab,* that is in proportion to the Sine complement of the Latitude of the place or parallel. Let *c, g, b, r, n, y,* represent a very thin Superficies of the Globe of the Earth or Sea; let *ac, ae, ag, ci, ab,* represent the natural Lines or Rays of Gravity tending to the Center of the Earth all of equal length and equal power as to Gravity. The parts then in the Figure being understood, I proceed to the Exposition of the Doctrine, let *g,* then represent a Body somewhere placed upon the Superficies of the Earth; I say, this Body will be affected or moved with a double Power: First, By a Power gravitating towards the Center *a,* which is the same where ever the Body be placed; this gives it a power of descending from *g,* to *n,* in a certain space of time. Secondly, by a levitating power in the Line *βg,* whereby in the same space of time it would ascend from the Center of its Motion *β,* from *g,* to *h.* Now draw, *no,* parallel and equal to, *gh,* and draw, *og,* and *oh.* Now because in both these Motions the acceleration is in duplicate proportion of the times it spendeth in passing them, it follows, that the Motion composed of both those Motions shall be made in a straight Line, namely, in the Diagonal Line *go,* for *g,* being by Gravity carried to *n,* and by Levity, as aforesaid, removed from *n,* to *o,* the place of the Body *g,* at the end of that time, shall be found *o.* The same

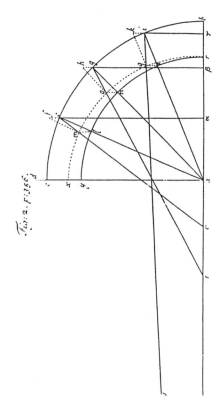

Demonstration will serve for *c, e, i,* and *b, Mutatis Mutandis*; whence it follows necessarily, that the Lines of Descent of such a Body are not to the Center of the Earth at *a,* but to some other point of the Axis of Motion, as *t, v,* &c. Secondly, The Figure of the Water will be Oval, or truly Elliptical, as *x, m, o, q, r,* because *xy, ml, on, qp,* &c. are all proportioned to their respective Radii. Thirdly, The power of this compounded Motion will affect all Bodies in differing Latitudes with differing Gravity, which were the proprieties to be proved.

[p.357] From which demonstration it plainly appears, that the Consequences I have deduc'd from the Hypothesis of the diurnal Rotation of the Earth are necessary, and cannot, according to the Laws of Motion, be otherwise than what I have deduc'd, not frivolous Suppositions taken up at random to solve one Phænomenon, but such as will give light to many other considerable effects of Nature, as I shall demonstrate in explaining several other Phænomena both of the Earth and of the Heavens.

It further appears also, that the Experiments or *Criteria* I have propounded, are both pertinent and sufficient to determine and state this Enquiry without any other, and that they are neither impossible nor very difficult to be procur'd to be try'd with accurateness enough.

Now as these may easily enough be procur'd by the mediation of this Honourable Society; so I doubt not but they may, with little more trouble, procure such Observations and Experiments to be made as would afford great Light towards the perfecting several other parts of useful Knowledge, some few of which, if judiciously and pertinently contriv'd so as to be plain and easy, would give us the determination of many old, yea, and many new Theories, possibly not hitherto thought of, some of which I shall hereafter have occasion to mention. Such Observations will be worthy the Care of this Society, and will be better than accidental and casual Trials, which, tho' surprising and pleasant, are at best but like those of the seekers of the Philosophers Stone and Perpetual Motion, who generally make trials at a venture, to see if there good Genius or Fortune will direct them to meet with what they seek; whereas indeed all Experiments ought to be directed to some end for the examination of some supposed Truth, and for that end to take notice of all such Circumstances as may give any information concerning it, whether it be for Confirmation, Confutation of such a Doctrine, and if so, the plainer and the more obvious the Experiments are, the better.

If yet there shall remain any doubt either in the deduction of this Conclusion, or the sufficiency of the Experiments to determine and state the truth thereof, I would very willingly explain any part thereof.

The next part of my Hypothesis is, that by many Observations I conceive that there may be in the Rotation of the Body of the Earth, a change of the Axis of that Rotation, by a certain slow Progressive Motion thereof, whereby the Poles of the said Motion appear to be in superficial parts of the Earth, which heretofore were at some distance from the then polar Points or Parts. I have waved all the Observations that I have hitherto met with in Histories which might seem to favour this Hypothesis, as having found them irregular and unaccurate enough in Observations of this kind, and have put the whole stress of its Proof, or rather Examination, upon trials to be made for the future. But because this Motion, if any, seems to

*Hooke explains further his hypothesis that the rotational axis has shifted with respect to the surface of the Earth—so that the poles have not been always where they are today. This idea of rotational axis shift was entirely original with Hooke. For further details, see Chapter 5.*

be very small and slow, and therefore since the Age of Man, which is very short comparatively, seems insufficient for such a purpose; I have therefore indeavour'd to carry Mahomet to the Mountain, since I cannot bring the Mountain to Mahomet, and that is by contriving such ways as may perform that in a short time, which, by the Methods of the Antients, could not be perform'd in less than some Ages. This Contrivance consists only in the exactness of Instruments, and the accurateness of making Observations; for if for instance we are not sure of the truth of the Latitudes of places recorded by *Ptolomy* and *Strabo* to a degree or two, as I can shew hundreds of places that differ more than that from the Truth; and if any new Method we may be able to make Observations either of the Latitude of a Place, or of the true Meridian of any such place to a single second Minute, than we may by such a means arrive to a certainty in a three thousand six hundred part of the time that could be arriv'd at by such Observations as theirs are; wherein their Defects lay whether in their Instruments, or their way of using them, or their negligence in computing, or the want of our present arithmeticial Art, and of proper and accurate Tables, or in the Doctrine and Practise of plain and Sphærical Trigonometry, I do not inquire; but certain it is, we have at present a great advantage of them in all these particulars, but above all, most eminently by the Knowledge and Use of Optick Glasses, especially as they are [p.358] applied to Mathematical Instruments, for by them only we are truly made Gygantick, and our Eye from the little Ball of less than an Inch in Diameter is grown to be of fifty, sixty, nay a hundred Foot and more in Diameter, and may be made able to do some thousands of times more than what our bare Eye alone without the use of such helps can perform; and therefore tho' *Hevelius* might have some reason to be uneasy, and so to rail at me for asserting of this Truth to the World after he had publish'd his *Machina Cælestis* to shew he had made use of the best Instruments in the World for his Observation; yet why Dr. *Wallis* and his Adherents, some of which have made of the very Contrivances which I Publish'd, should with so much Gall write against me for it, I cannot but wonder: But I doubt not but to prove to all the World in my own Vindication, that neither the one nor the other had any reason but ill Will for what they did, and at the same time to prove the truth of every particular which I have asserted in that Book to any that will believe his own Eyes; but not to trouble you any further with this Controversy at present, designing suddainly to publish my Answers to them where they may be seen more at large, I shall proceed to the Methods of making Observations both of the Meridian Line of any place, and also of its true Latitude in respect of the Heavens to the accurateness of a single Second.

And here only I have one or two Postulata to premise, which I suppose every one that hears it will readily grant; it is no more but these; First, That it is possible, nay, practicable, to find a Point below perpendicular, to a Point above, tho' the distance between them be a hundred Foot, and to be certain of the truth thereof to the exactness of a Second Minute.

Secondly, That the Refraction of the Air at sixty Degrees Altitude above the Horizon does not at vary the Azymuth of a Body a single Second Minute.

A Third Truth I will put by way of a Postalutum [sic], that 'tis possible, nay, easily practicable to distinguish the parts of a far remov'd Object by the help of Telescopes long enough, tho' they really appear to the Object Glass of that Telescope less than that of a single Second Minute.

*Hooke rambles on about the accuracy and superiority of the instruments and mathematics of his time over those in the days of Ptolemy.*

*Hooke and Hevelius participated in a controversy regarding the use of telescopic sights for astronomical observations. Hooke was one of the earliest users of telescopic sights, while Hevelius stubbornly relied on his own eyesight. See Chapter 1. John Wallis, mathematician, engaged Hooke in a dispute over the axial shift idea. For this history, see Chapter 5.*

These I conceive so easy and certain, that I have put them as *Postulata*; but yet if any doubt of their certainty, I do undertake to prove the Truth of either of them both by Experiment or Demonstration, which of the two shall be judg'd most convincing.

The next thing to be shewn is how to order a Telescope so that it may be made a sight, that the true Line in which the Object appears may be certainly determin'd, and this, be the Telescope sixty or a hundred Foot or more in length, and how to make by this, an Instrument as large as the said Telescope is a sight.

For the performance of these Qualifications there will be no greater difficulty, than the making a Tube for such a Telescope, or, if that be thought too much, it may be done by two small Scaffold Poles joyned together in the middle with convenient Lines to keep them streight, or if this be still thought too difficult, it may be done by fitting the Object Glass in one Cell, and the Eye Glass and Thread-sights in another, with Lines strained between them to keep them directly parallel to each other; but the best way is by a Telescope Tube of a due length and bigness for the Object-glass made use of.

Supposing then a Telescope of sixty, eighty, or a hundred Foot in length thus fitted with a Tube, to find the true Line of Direction, I fasten to the Cell that holds the Object Glass a Needle with the point outwards against the middle of the Glass. And the like I also fix in a small sliding Plate, that lieth upon the Cell, that holdeth the Thread-sight together with the Eye-glass: This Plate, by a very fine Screw, I can cause to flip out or in at pleasure, till it be adjusted to a Line from the point of the Needle fasten'd to the Cell of the Object-glass: To adjust this Telescope then for a sight, I direct it to some very remote Object in the Horizon, and fix a Pin or Wire just touching the point of the Needle at the Object-glass; then having found and remarked some convenient Point of the Object in the Horizon, I move the Tube till the Thread of the sight exactly lie upon it; then inverting or turning the Tube, making the under side upermost, and the uper side undermost, the right the [p.359] the left and the left the right; I cause the point of the Object-glass Needle, as also, that of the Cell of the Threads to touch the same Points as before inversion, then looking at the point of the Horzontal Object, I see whether the afore-said Line of it do cover the same part of the Object as it did before inversion. If it happen so to do, then I am certain that the Telescope is already adjusted; but if it do not, it will be adjusted by moving the sliding Plate with the Needle at the Eye-cell: When thus adjusted, these two Needles points become the Indexes to my Instrument, for exactly taking the visual Line of the Object, I observe to as great accurateness as is desired. Having thus prepar'd the sight for my Instrument, I make choice of some Tower of a convenient height for the resting the end of the Tube that holdeth the Object-glass, and order it so that the Needles point may touch a fixed point upon the same; then I make a Board below upon the Ground lying Horizontal, whereupon the other end of the Tube may be slid Horizontal and easily remov'd at pleasure. The Object I make choice of is the Pole-star, or the Star in the tail of the lesser bear. I by this means observe its most Eastern and Western Excursion the same Night; or if it happen that one is in the Night and other in the Day, by means of this Telescope I can plainly see it, tho' the sun shine. Now the Needles point at the Object-glass touching in both Excursions, and the Needles point at the Eye-sight shewing the two Azymuthes of the said

*Preparation of his telescopic sight. This passage shows Hooke's considerable experience in constructing telescopes and making astronomical observations.*

Excursions, wherein the Refractions can have no effect to make the Ray bent, it will be easy enough accurately to divide the space between the two Excursions into half, and as easy to fine the Point below perpendicular under the Point above marked by the Needles Point of the Object-glass, or which for this purpose will be better to find a point above upon some building of equal height with that of the Object-glass, by which two Points I direct my Telescope to the Horizon either Northwards or Southwards, and find what Objects lies directly in the Meridian Line, which I diligently note and draw the Landscape of that part of the Horizon which appears through the Glass when so posited with the very point of the same cover'd by the Thread-sight; which done, I continue the said Landscape by the help of the Telescope till I bring in some remarkable known Object, by means of which I shall be able a Year or two after to find the same again, when the same trial is again to be repeated with the same care. In order to determine this Question, whether the Meridian Line upon the Surface of the Earth do change; by which means if it be alter'd but a Second or two, I shall be able to distinguish it in the Horizontal Landscape. Now tho' this Experiment upon the whole Matter may seem troublesome and difficult to be perform'd duely as it ought, yet if we consider the Importance thereof in this Matter, and how much can be done by the Care of one Man in a short space, which by the Method of the Antients was to not be expected from the performances of any one or many under the expectation of some Ages, I conceive the Experiment may be look'd upon in the whole as compendious, cheap and easy; there being nothing therein so difficult but that two Men may every day, for some days together repeat the Observations and Trials after the apparatus is made ready and put in order, and need not spend above four Hours in twenty four to make them sufficiently accurate.

Nor will it be very difficult in this City to find a convenient Building or Tower for the resting the end of the Telescope of a hundred Foot long if it be made use of; or of finding a good prospect of a far distant Meridional Object in the Horizon, whether towards the North or South, they being both or either equally sufficient for this Observation: And if a fifty or sixty Foot Telescope be made use of, which will be able to perform the Observation accurately enough, with a little more Care and Circumspection, tho' with less Labour, and Pains, and Charge, there are Houses enough to be found of sufficient height.

Now this Experiment I conceive sufficient to perform what is design'd, or to be expected from it, as to this Inquiry, and all things consider'd, I conceive the best, tho' I could produce some others if there be occasion; and further, I conceive the same to be free from all material Objections: As [p.360] First, If it be Objected, that the Refraction of the Air doth make the Cælestial Objects to appear out of their true places. I say, that in this Experiment it can have no effect, because the Azymuth and Circle of Position only are sought and those the Refraction of the Air alters not; for the Star being only to be observ'd when it is either ascending in its most Eastern Azymuth, or descending in its most Western Azymuth, the effect of any would be the same, since they are so found and observ'd at the same Altitude both ascending and descending; and tho' the Refraction should raise them, and the whole Circle, and the said Star to a sensible higher Position than the the Truth, yet the Points to be observ'd being both of equal Altitude, the effect will be the same, which will no way disturb this Observation.

Next if it be Objected, that this Star doth alter its distance from the Pole-star every year, and that will make the Excursions less in the succeeding Year. I say, as to this Inquiry, it would have no effect, tho' it should alter ten times more, because the middle between the Excursions in the same day, is that which is sought in this Experiment.

If the parallax of the Earths Orb be Objected, which is the most material, I Answer, That the succeeding Experiments are to be try'd again when the Earth is in the same part of the Ecliptick, which will fully answer any Scruple thence.

And upon the whole, I cannot think of any other; but if any of this Honourable Society can think of any Material, I would desire to be inform'd of them, that I may think of some means of remedying them; or if they think of any other more convenient and certain, that I may put them in practise. Some other of my own I shall propound the next Day and leave them to the Judgment of the Society to chuse the most fit.

This, I hope, May save the Labour of searching into Records of Antiquity, of all which if I may be allow'd to judge by those I have met with, I believe they will at best afford us but uncertain and unaccurate Observations, and I do very much doubt whether ever there were above two thousand Years since, any Meridian truly set to the certainty of less than one Degree; so that tho' we had found by the great Pyramid, that there was either some considerable variety from the present Meridian, or that it were now in the very Meridian Line, the Conclusions drawn from either of them would have been but conjectural, since it might have been placed true, and have since varied, as it is found, or it might have been placed wrong, and since have move'd to a Truth, or the contrary.

*Testing his hypothesis with accurate yearly observations using his precise instruments, Hooke believes, would be much more useful than relying on ancient records.*

Whereas, since by this Experiment, we may be able to find the Meridian true to the three thousand six hundred part of a Degree, and these Observations may be made by one and the same Man, and with the same Instrument in the same place, and at the same time of the Year, and of the Day; I conceive that one Years Observation will more ascertain us in this particular, than if we had Records of Observations made, as those I have met with 3600 Years since, which is the Expedient I have thought of for redeeming or expanding the power of the short Line of the Life of a Man.

**No. 7.** *Other Methods of determining the same Question, read Feb. 23, 1687, although Waller dates it the previous week, February 16.*

[p.360] In order to determine whether the Meridional Line of Places did alter, I did in my last Discourse wave all antient Observations, as fearing there might be wanting in them that certainty and accurateness of Observations that might be sufficient to assure us of the Matter of Fact, and that might be convincing as to the Reality or Nullity of such an alteration, as I suppos'd.

But because possibly there may be some Observations of a latter date which have been here made within our present reach, which are of more accurateness, I would propound it a thing not unfit to be examin'd, whether the Position of several of the most eminent Cathedrals built by the *Gothick* Architecture, wherein great regard, if not Religion, seems to have been had of the Position of them, according to four quarters of the Horizon, *viz.* E, W, N, and S. And to this end, because nearest this place, I could wish [p.361] it were tried at *Westminster Abby*, which is intirely built after that Mode; whether that be truly so plac'd that the four Ends or Fronts thereof do exactly fact those Quarters, and if not, which way the variation may be, and how much it really is at this present. The same Observation may be procur'd to be made at several other Cathedrals, as at *Salisbury, Winchester, Chichester, York,* &c. where there are such Buildings, which will be with no great difficulty procur'd by the Mediation of this Society.

*To find out whether positions of meridians have changed, Hooke proposes measuring the orientation of great gothic cathedrals which were known to have been placed exactly according to the points of the compass.*

Among other places worth examining, I could wish that the great Dialstone in the Privy Garden at *White-Hall*, were one, for that I conceive there was very great Care and Accurateness used in the placing thereof; and tho' this may seem, if compar'd with others of a very short continuance since its first placing, yet it may be with probability enough suppos'd to have been so much more accurately plac'd, that That alone may possibly make it preferable to any other whether Ancient or Modern.

Now because the ways publish'd for finding the true Meridian Line have really much of difficulty in them, and require both a great Apparatus and a considerable time to make the Observations necessary for this purpose, without which the Informations and Examinations will be very unaccurate and scarce to be rely'd upon, therefore I have contriv'd an Instrument by which, in a few Minutes of Time, the exact Meridian Line, at any place, may be easily and with accurateness enough, that is, to ten Seconds, if need require; and this free from Exceptions of Refraction, Declination, &c. by which the true Position of any Building, Monument, &c. may be presently discover'd and computed.

The first Instrument from this purpose, is a Telescope of what length shall be thought convenient to be easily used and manag'd; as suppose one of six, twelve, or fifteen Foot, this must in the first place be fitted with Eye-sights, plac'd upon a thin piece of Looking-glass Plate, on which must be drawn with a very fine Diamant, such Lines and Circles as I shall direct, the Center of all which Circles is to represent the true Polar Point in the Heavens, at the time of the Observation of which more by and by.

*Hooke would supervise the construction of a telescope capable of exact sightings along a direction.*

This Sight-glass being fix'd in the Tube, the next thing to be done is to fix two pieces of Brass, or some other convenient Metal, which may have each of them a small hole to hold a small clew of Silk fit to bear a

Plumbet or such other Instrument as I shall direct; these holes must be so plac'd as that an imaginary Line drawn over the ends of them may be exactly parallel to the axis of the Telescope which passeth through the Center of the Sight-glass.

Thirdly, Into these holes must be fitted small Silken Lines with Plumbets hanging at them, which two Plumb-Lines will (when the Axis of the Telescope passing through the Center of the Sight-plate is directed to the Polar Point) hang in the plain of the Meridian.

Fourthly, The Axis of the Telescope may be easily directed to the Polar Point, by bringing three or more Stars of the *English* Rose into their proper Circles and there fixing it.

*The axis of the telescope can point directly at the Polar star by positioning it with reference to the stars in the constellation known as the English Rose, discovered and named by Hooke.*

Fifthly, The *English* Rose is a Constellation in the Heavens discoverable only by a Telescope, consisting of six Stars in the Rose itself, and several other in the Leaves and Branches, one of these is in the Center of the Rose, and five in the five green Leaves of the Knob: This I have somewhere describ'd about ten Years since, but have mislaid them at present; the way of finding them I then shew'd to Sir *Chr. Wren*, and some others of this Society at the time when my Instrument was fixed for that purpose.

The Instrument or Telescope being fix'd in this Position, the two Plumb-holes represent the true Axis of the Earth, and accordingly will serve to determine both the plain of the Meridian, and also the inclination of the Axis to the plain of the Horizon; so that by the same Observation both the Meridian Line may be determin'd, and also the elevation of the Polar Point, [p.362] which may be various ways most exactly measur'd and determin'd. Now this is a second way of determining the true Meridian Line to what accurateness shall be desir'd, for that the length of the Telescope is not limited, but may be us'd of what length soever may be made, tho' it may be three hundred or four hundred Foot, for that the Object-glass may be fix'd at the top of some Tower or Steeple, and the Sights and Eye-glass at the Ground. But on this I shall not at present inlarge, because, whenever there shall be occasion of trial, I can easily direct the whole Apparatus.

*Thus positioned, by the same observation the telescope can determine both the plane of the meridian and the inclination of the axis to the horizon—i.e., the latitude (assuming, of course, one knows the time of day).*

As to the second Use thereof, which is for taking the Altitude of the Polar Point above the Horizon; this way is far beyond any I have met with, and is liable to one only Objection (as I conceive) and that is the Refraction of the Air, which elevates the same somewhat beyond its due Limits, but then if compar'd with the best yet propos'd, I conceive it to be less subject than any other, and to come nearer to a certainty and exactness: However I grant it to have that Objection good against it; but if we consider the use of this Observation as it is design'd to examine the Latitude of one and the same place after the interval of some few Years, the Objection is of no validity, for that the Refraction of the Air at the height proper for *London*, viz 51, 32, is hardly sensible; but then the difference between the Refraction of the first and second Observation is yet much less discoverable, so that for this purpose 'tis as effectual as the best.

But because some may yet further desire to free the Observation from Refraction, I have contriv'd another way, much less subject to it; which way will also find the true Meridian Line to great exactness: Not to make any long preamble to it 'tis this, Make choice of some notable fix'd Star that passeth over or near the Zenith of the place, as here, for *London* the

*A way to overcome the refraction caused by the atmosphere.*

*Lucida Draconis*, or the last Star in the tail of the great Bear; 'tis easy, by the way I have already publish'd in Print to find the Zenith, and the Meridian Line passing through it.

Having fix'd all things requisite for this, about an Hour or two before the Star comes to the Zenith or Meridian near it, find and observe exactly its bearing, which may be done with a Telescpe of fifty, sixty, or more Feet in length, then by an exact Pendulum Clock number how many Seconds or half Seconds of Time pass before it arrive at the Meridian, which note and remember, then prepare to observe the place of the same Star after so many Minutes, Seconds, and half Seconds have pass'd, after the Star hath pass'd the Meridian, note the Point also. There are then given three Points, which, with the help of Calculation, the time being taken, it will give the Latitude of the place to a great exactness; and if a Line be drawn from the most Eastern to the most Western Observation, this will give the true E, and W, line; and if the same be divided in half and through the same, and the Meridional place of the said Star, a straight Line be drawn, this will give the true Meridian Line.

So much for the methods of observing the Latitudes and Meridian Lines of places for the time to come.

But because there have been some of our later Observations of the Latitudes of some places which have been with very great Care and Accurateness made; I could likewise wish that the Latitudes of those places where they have been so made, might be a new examin'd, to see whether any considerable difference can be found which cannot well be ascrib'd to the defect of the preceding Observations. As the Latitude of *Uraniburg, London, Paris, Rome, Bolonia*, &c. tho' yet I fear we shall be apt to ascribe what difference shall be found to the faileur [sic] of the preceding Observations.

*Hooke fears that whatever difference in latitudes can be measured this way could still be attributable to the inaccuracy of the observations.*

This is all I have at present to propound concerning the external Figure of the Water and Earth. As to the motions thereof I shall propound some Conjectures after I have consider'd the Figure and Constitution of the next great Fluid in compassing the Earth, which is the Air; after which I shall propound some Conjectures at the various internal motions of those great Fluids, which concern the Currents and Tides in the one, and the Winds in the other.

**No. 8.** *Figure of the Air or Atmosphere.   Read March 9, 1687.*

[p.363] I have in my former Lectures propounded my Thoughts, and the Reasons of them concerning the Figure of the Body of the Earth, and the tendency of the Perpendicular Lines of Gravitation; as also concerning the probability of a variation of the Axis of Rotation in the Body of the Earth. I have likewise shew'd the influence of those Principles upon which I grounded those Thoughts, upon the Body of the Waters incompassing this Earthly Body.

I come in the next place to consider the Figure of the next great fluid Body, incompassing both the one and the other, and that is the Atmosphere or Mass of Air.

And here for the present I shall only consider so much of the Nature and Constitution of this Body, as seems necessary to the explication of the Figure thereof.

It is now very well known, that this Body is of such a Constitution, that a greater degree of Heat, or a lesser degree of Pressure will effect a greater degree of Expansion, that is, will cause the same parcel of Air to occupy or fill a larger space of rome; next that the same parcel of Air, when rarify'd, will weigh no more than when condens'd, tho' it fill a greater space, because the real quantity of the parts that compose this Air, are still the same, tho' there may be a greater quantity of other matter that fill the Interstitia or Spaces between them.

Thirdly, That the Atmosphere is compos'd of three kinds of Substances, one more fluid than the other, two of which, namely, the less fluid Cause considerable effects upon the subjacent parts of themselves, and upon other Aqueous and Terrestial Bodies by their Weight or Gravitation. These I only name at present, designing more fully to explain them and their Causes when I discourse of the Substance and Constitution of the Air.

These three things then for the present being taken as Suppositions, and the tendency of the Lines of Gravitation, being, as I have prov'd, to differing and various Centers, it follows that the Figure of the two lower parts of the Air must be of a prolated Sphæroidical Figure, and that much more considerably differing from that of a Sphærical Form than that of the Earth or Water.

For First, It is very evident, that the more gross Parts thereof are carry'd along with the subjacent Parts of the Earth, with an almost equal swiftness; say almost, because in the wide and open Ocean there is some kind of loss of swiftness and lagging behind, which, as I conceive, (as *Galileo* and many others have done) is the Cause and Original of the Eastwardly Winds within or near the Tropicks; from this Rotation then will follow a considerable levitation of such parts of the Air as are whirl'd round from West to East with such a Rotation; that is, those parts which are mov'd swiftest will have the greatest indeavour of Recess from the Axis of Motion; and those which are mov'd slower will have a less, as I shew'd before in my Explication of the Figure of the Water.

[p.364] Next it is evident from the springy Nature of the Air, that the less the pressing is upon the Body thereof, the more will it expand and stretch itself and possess a greater space. Now the quantity of Air towards the Æquinoctial, having a greater Levitation upwards, or less Gravitation towards the Earth, a greater quantity of the Air must go to make up the Cylinder that gives an equal Pressure, and consequently the Surface or Extent of the Air towards the Æquinoctial must upon this account be much higher than towards the Poles.

But in the Third place this Oval Figure of the Air must necessarily be increas'd by the differing Degrees of Heat and Cold; for that a greater Degree of Heat doth expand, and of Cold doth condense the Body thereof. Now it is evident that the Degrees of Heat near the Æquinoctial are very great in comparison to what they are near the Poles. And consequently, upon this account, also the Body of the Air towards the Æquator, must be very high and rarify'd, and the Body of it towards the Poles must be very low and condens'd; from which two Causes it will necessarily follow, that the Figure of the Body or Mass of Air, incompassing the Body of the Earth and Water, must be of a prolated Sphæroidical Figure, much more prolated towards the Æquator than that of the Water.

*An equatorial bulge is present also in the shape of the atmosphere encompassing the globe.*

From which Considerations, I conceive, some Reasons may be drawn of several Phænomena taken notice of by Travellers; such as the frequency of Foggs and Mists and various sorts of *Parhelia* and *Paraselenæ*, in and near the Polar Regions; all which argue a dense and heavy Air. And of the Hurricanes, Tornadoes and the Storm call'd the *Bulls Eye*, which descends from a great height with great precipitation into the lower Regions of the Air, and of the frequent and violent Rains in the *Torrid Zone*; all which *Phænomena* are indications of an Atmosphere much more extended upwards, and of the vaporous Parts carry'd to a much greater height than elsewhere.

*Reasons for different climates in different parts of the world.*

From these Considerations also will follow a necessary motion or tendency of the lower Parts of the Air near the Earth, from the Polar Parts towards the Æquinoctial, and consequently of the higher Parts of the Air from the Æquinoctial Parts towards the Polar, and consequently a kind of Circulation of the Body of the Air, which I conceive to be the cause of many considerable *Phænomena* of the Air, Winds and Waters, which I shall more fully explain when I come to consider the Constitution and Motion of the Body of this great Mass, whose Figure and external Form only at the present I am considering.

*Hooke is possibly the first to propose this idea of the circulation of the air, a precursor of the Hadley cells described almost 100 years later.*

Nor shall I at present explain any thing farther concerning the two more fluid Substances that help to compose or fill the space which is taken up by the Atmosphere, because my present Subject leads me only to consider that part of the Air which is call'd the Atmosphere, and to speak only of the Figure thereof, of which I have no more to add at present, but shall return to consider the Nature and Motions of each of these three great Masses, *viz.* the Earth, the Water and the Air.

First then for the Internal motions of the Earth; there are two principally taken notice of; the first is that of Gravitation, the second is that of Magnetism.

Of the first of these I have some Years since discours'd more particularly, and therefore shall omit it at present.

Of the second of these, namely, Magnetism, I shall only propound my Hypothesis now and explain it more particularly in my next Discourses.

My Hypothesis then is this, First, That all magnetical Bodies have the constituent Parts of them of equal Magnitude and equal Tone.

Secondly, That the Motion or Tone of one Magnetical Body is convey'd to that of another by means of a Dense Medium.

Thirdly, That the motion of the Dense Medium is Circular and Vibrating.

From which three Suppositions all the *Phænomena* of Magneticks will be most evidently and clearly, even *a Priori*, deduc'd.

**No. 9.** *Confirmation of former hypothesis by histories. Read December 7, 1687.*

[p.371] I Have indeavour'd to discover and prove the true Figure of this Body of the Earth upon which we Inhabit, and likewise to give some Conjectures concerning the *Form* and *Shape* of the *Superficial* Parts thereof. This I have done in order to comply as near as I could with a Natural Method of Natural History: This great Body being the Mother of all Terrestial Productions, which make up the greatest part of *Natural History*; and the Foundation, as it were, upon which, not only all that History, but all the other Parts and Superstructures almost do rest; for from the Productions of this we take our Principles, we raise our Axioms and Maxims, we form our Similitudes, we make our Observations, Experiments and Trials, and by Analogy from Comparison and Similitude we deduce our Conclusions. I thought it therefore not improper, since Natural History will carry us into forein Parts of the World, very far remov'd from this our Country of the Earth, to be first of all a little acquainted, at least, with what we have at home, that thereby we may the better be able to observe and judge of what those far remote Parts may present us with, whether they be like our own or not, in what they agree, and in what they differ, that these we know at home may be the Standards and Touch-stones of all the rest we meet withal Abroad.

In prosecution of this Method, I began first to shew what seem'd to me to be the most likely Figure of the whole Body, which I shew'd for several Reasons seem'd to be of a prolated Sphæroidical Figure, not of a perfectly Globular, as most Authors suppose and affirm, much less of an Oblong Oval, as the ingenious Author of the sacred Theory of the Earth, and some others, have indeavour'd to make probable.

From this I deduc'd the prolated Sphæroidical Figure of the Waters also, and more eminently of the Air or Atmosphere, and from that deduc'd these Conclusions, That the Lines of Gravitation or Perpendicularity did not tend to one single point, as all hitherto have asserted, but to infinite points in the middle parts of the Axis.

And that a Degree, or a 360th part of the *Æquinoctial* did not agree exactly with any one Degree in a *Meridian*, and thence that the Magnitudes

*Starting with this discourse, Hooke indulges in a favorite sport—interpreting fables by ancient writers that in his opinion really relate histories, couched in mythological terms, of terrestrial changes that occurred in the ancient world. As such, Hooke believes, they are worth reading.*

*Hooke recapitulates what he had covered concerning the figure of the Earth and axial shift with respect to the surface.*

*The "ingenious Author" is Bishop Thomas Burnet, who wrote and published the Latin edition of the* Sacred Theory of the Earth *in 1681.*

of the respective Parallels were not to be estimated as if the Body were truly Globular.

From this I deduc'd a necessity of a differing Gravitation of the same Body in differing parts of the Earth, and thence a necessity of a differing length of Pendulums to measure by their Vibrations the same quantity of time, by which the universal Standard of Measure, by some suppos'd from the length of a Pendulum, became questionable and dubious.

I have likewise shewn what Observations of Celestial Bodies were likely to be assistant to the perfecting and confirming of these Matters, at least of discovering the Truth whether really so or not.

I have also inquir'd concerning the fixedness and instability of the Terrestial Axis, and shewn some Arguments to induce us to believe that it may have and suffer a mutation, and not be always fixt in the same parts of the Earth, and by what methods that may be ascertain'd in a short time with more exactness [p.372] than many Ages of Observations made with less accurateness would have done.

And from thence I have deduc'd what would be some of the necessary Consequences of such a mutation; such as the differing Latitude of places in differing Ages. The differing Azymuth of Places as to one anothers Position; the differing Altitudes of Places with respect to the Superficies of the Sea, as the Emerging of some places from below that Surface and the sinking under, and the being overwhelm'd by that Surface in others, and consequently of changing the Nature, Soil, Climate, &c. of the superficial Parts of the Surface; to which, as I conceiv'd, some alterations might be ascribed.

But Lastly, I shew'd that the ruggedness and inequalities of Hills and Dales, Mountains and Lakes, and also the alterations of these superficial Parts of the Earth, as to the seeming Irregularities thereof at present, seem'd to me to be most probably ascribable to another Cause, which was Earthquakes and Subterraneous Eruptions of Fire. That there had been many such alterations I indeavour'd to prove from the almost universal Disposition of those curious Medals of former Ages now found in the petrify'd Monuments of the parts of several both Terrestial and Aquatick Animals and Vegetables, but especially by those Productions of the watery Element found in places now far remov'd from the Sea, and far above its Level; of which I have produc'd several Instances, some of which, and those very considerable, were procur'd by the inquisitiveness of a Person here present.

*Earthquakes and subterraneous eruptions are the causes of the irregularities of the Earth's surface, and the proof is in the occurrence of fossils that indicate exchanges of land and sea areas.*

I have made some Excursions out of this Method; as First, in order to answer the Doubts and Scruples of some, and the Obloquies of some other Persons, who, I hope, are now, or will be somewhat better satisfy'd, which I wish all might be, for I have no desire to impose Conjectures and Inquiries as Demonstrations, but only to shew what Arguments have inclin'd me to be of these Opinions, which, whether sufficient, I must leave to their better Judgments and Examinations, hoping at least that no preposession will hinder them from examining them with Candor and Indifference, as I Indeavour to do in all my Inquiries. Next by some Experiments made for the clearing some accidental Discourses at the meeting, as those about the best ways of communicating force at a

*It is clear that Hooke's ideas had critics.*

distance, and of making a Pendulum to observe by Trials the Velocities of the parts of Pendulous Vibration, and to make a Pendulum that shall, without Clock-work, continue moving twelve Hours or longer. And Thirdly, By accidental Observations made of the growth of Trees and some others; nor will it, I hope, be taken amiss that I indeavour to produce such Arguments as occur to me, that seem to favour these Conjectures, tho' possibly much better may be shewn by others eithers for or against them; however give me leave to alledge what I can to answer such as I conceive are not sufficiently cogent Arguments against what I have supposed.

One of the most considerable Objections I have yet heard, is, that History has not furnish'd us with Relations of any such considerable changes as I suppos'd to have happen'd in former Ages of the World; I do confess our Natural History as to these and many other matters of the first Ages is very thin and barren, but yet I conceive not wholly devoyd of Instances, nay, possibly if they be look'd into with a little more attention than hitherto has been used, they may be found to contain many more than has of late Ages been imagin'd. Some things of this kind, I fancy I have detected, of which I shall produce some, together with some Remarks upon them, which I have added, they are, I conceive, related as true Histories; but whether so or not I must leave others to judge who are better Antiquaries and Criticks.

*One of the objections to Hooke's ideas is that there is no written history to document what he has espoused as to the cause of the irregular surface of the Earth.*

What Learning and Accounts of Ancient Times the *Ægyptians* might have in their Histories, who are said by *Plato* in his *Timæus*, to have had accounts of great alterations in the World for nine thousand Years before *Solon*, which is now above two thousand Years since, it is very hard to guess from that short account that is there given of it; yet since of all the Records that are to be met within the Ancient Historians to this purpose, this is the most considerable, I thought it would not be improper to relate it on this occasion, by reason that tho' it should be accounted *fabulous*, as some have thought, and [p373] to be only a Fiction in *Plato* in order to lay a Scene for his Republick; yet there is so much of Probability in it (bating only his number of Years) and so much of Reason and Agreement with the State of things, that if it be not a true History, it will at least shew that *Plato* himself had, at that time, some such Notion or Imagination of the *Preceding State of the Earth*, and that he saw, or found at least, some very good Arguments for his being so; *Plato* then in his Dialogue maketh *Critias* thus speak, "Hear, O *Socrates*, a wonderful indeed, but yet a true History, which *Solon*, the wisest of the seven Wise Men, related to my Grandfather *Critias*, as the old Man hath since told me; among other things he told me of the memorable Actions of this City (*Athens*) by length of Time, and Death of many, quite obliterated. But among the rest he related one remarkable Passage, which I think now proper to acquaint you with, and it was an old History, which he being then about ninty Years old, told to me when I was about Ten, upon a solemn Day, when I, with divers other Boys, as the Custom was, were wont to recite divers Verses by Heart to see which could excel, among which were divers Verses of *Solon*: And I remember I heard my Grandfather then say, that if *Solon* had but committed to Verse, not what he did for refreshing of his Mind, but seriously, and like other Poets, the History, which he, returning out of *Egypt*, resolv'd to have written (from which, by disturbances which he met with at home, he was interrupted in perfecting) neither *Homer, Hesiod*, nor any other Poet, would have been comparable to him. This was of the greatest Affair that had been transacted by this City, of which we have no

*But he believes that some ancient accounts, as, for example, Plato's* Timæus, *could have happened. Plato tells of great alterations in the world for 9,000 years before Solon, who was more than 2,000 years before Hooke's time, or more than 11,000 years. Clearly this passage shows that, even though he added the aside in parentheses—"(bating only his number of Years),"—one has the impression that he believed the account to have "so much of Probability in it." The aside, in my opinion, was added as an afterthought to placate the literal interpreters of the Bible. See Chapter 6 for a discussion of Hooke's concept of time.*

remains at present, by reason of the length and injury of Time. The Summ of what I remember was, That *Solon* going into *Egypt* to *Saim*, at the Mouth of the *Nile*, when *Amasis* was King, was there receiv'd honourably. There he inquiring of those Priests which were most skilful in those Matters concerning the Memorials of great Antiquity, found, as he related, that neither himself nor any other Greek knew any thing of Antiquity; and when he to provoke the Priests to tell him some of their Knowledge, had, in there presence, spoken concerning the most antient Actions of the *Athenians* of *Phoroneus* and *Niobe*, and of *Pyrrha* and *Deucalion*, after the inundation of the World, and of the times when those had happen'd; one of the Seniors of the Priests cried, O *Solon, Solon,* you Greeks are all Boys, not one Old Man among you. *Solon* asking him why so? The Priest answer'd, because you have young Heads always that contain nothing of ancient History, of ancient Opinion, or of Old Mens Science, which has happen'd to you by reason that there have been already, and shall be many and various Destructions of Men: But the greatest of all will be caus'd necessarily, either by a Conflagation of Fire, or an Inundation of Water; but the lesser by innumerable other Calamities: For what you tell of *Phaeton*, the Son of the Sun to have got into his Father's Chariot, and not knowing how to Drive like his Father, had fired the Earth, and with that Flame had almost set Fire to the Heavens, tho' it may seem fabulous, yet 'tis not without its truth in some sense. For in long process of time there is a certain *permutation of the Cælestial Motions which a vast Inflammation must necessarily follow.* Whence such as inhabit high and dry Places will suffer more than such as are nearer the Sea and Rivers. Now our *Nile*, as it is in most other things very wholesome for us, so will it preserve us from such a Destruction. But when the Gods of the Waters shall wash away the Filth of the Earth by a Flood, those which feed Sheep and other Cattle at the tops of the Mountains will scape the danger; but your Cities that are situated in the Plains, by the impetuosity of such Floods will be swept into the Sea. But in our Region we have no Water descends from above, but all ours springs out of the very Bowels of the Earth; which is the reason that with us the Records and Monuments of the most antient things are safely preserved. Whence it comes to pass, that where neither too great a Storm of Rain nor any extraordinary Fire happens, tho' sometimes more, sometimes fewer, yet still some Men always escape. Now whatever we hear that is worthy notice, either acted by our selves, you, or any other Nation we keep described in our Temples: With you indeed, and other Nations things lately [p.374] done have been committed to Writing, and preserved by other Monuments. But in certain periods of Time there come from the Heavens certain Destructions which depopulate all; whence the following Generations are depriv'd both of Letters and Learning. Whence you are all again made Boys, rude, and altogether ignorant of preceding Matters. Hence 'tis that what but now you speak of, O *Solon*, differs very little from Childish Fables. First, In that you make mention but of one Inundation, whereas many preceeded. Next, That the stock of your Ancestors which was most Eminent, and of the best, you know nothing of; whence both thy self and the other *Atheneans* had your Birth, which was a small Remnant that scaped the publick Destruction: Which becomes unknown to you; for that this Remnant and their Posterity for many Years wanted the use of Letters; whereas your City before that had excelled both in the Arts of War and Peace, of which we had a full account. [So he proceeds to tell how they had Records of their own City for nine Thousand Years, and of the Laws, &c. as also, of long times for *Athens*, which I pass over, and only mention what seems to

*Solon's story as related by Critias to Socrates, recorded by Plato.*

*In the bracketed passage here, Hooke interjects the Egyptian priests' claim of records of their city for 9,000 years and also the long-time records of Athens. Clearly Hooke favored a non-biblical chronology.*

relate to Natural History. He proceeds then,] "Many wonderful Actions of your City are preserved in our Monuments; but one above the rest for Greatness and Virtue exceeds; for 'tis said, that your City resisted a numberless company of Enemies, which coming out of the Country where the *Atlantick-Sea* now is, had conquer'd almost all *Europe* and *Asia*; for at that time was that navigable Streight which is call'd that of *Hercules Pillers* which had near the Mouth, and as it were in the very entrance of it, an Island then said to be bigger than *Lybia* and *Asia*, through which was a Passage to other Neighbouring Islands, and from the Islands was a prospect to the main Lands lying near the Shoar, but the Mouth of the Streights was very narrow. This Sea was truly the Ocean, and the Land was truly a Continent. In this *Atlantick Island* was a most great and wonderful Power of Kings, who Rul'd over, not only that whole Isle and many others, but over the Greatest part of the Continent, and even over those which were near us, for they Reigned over a third part of the World, which is call'd *Lybia* even unto *Ægypt*, and over *Europe* even to the *Tyrrhene* Sea; the whole power of these collected together, invaded both ours and your Country, and even all the Lands within the *Herculean* Streights, but both your and our Country repell'd them; the manner I omit. Afterwards by a prodigious Earthquake and Inundation which happened in a Day and a Night, the Earth cleaving swallow'd up all those War-like Men, and this Island of *Atlantis* was drown'd by a vast Inundation of the Sea, by which means that Sea became unnavigable, by reason of the Mud of that sunken Island which was left." The rest I omit." Now,

*The story of the disappearance of Atlantis.*

Whether this Relation be a Fiction or Romance invented by *Plato*, or a true History, I shall not now dispute, only by all the Circumstances of *Plato*'s relating of it, I conceive he design'd to have it to be reputed a true History and not a Romance, for that his design for laying a Scheme for his imaginary Government, needed no such Fiction, and accordingly he made very little, if any, use of the Circumstances of it that relate to Natural History. However, be it what it will, it evidently shews that *Plato* did suppose and believe that there had been in many preceding Ages of the World, very great changes of the superficial Parts of the Earth by Floods, Deluges, Earthquakes, &c. for as much as he could suppose a Continent or Island as big as the third part of the known Earth, to be by one Earthquake sunk into the Sea and overwhelmed by it.

*Plato did not need to invent a fiction to lay a foundation for his imaginary government, and therefore he told it as a true story. But whether the story of Atlantis was true or not, evidently Plato believed that there had been great changes in the surface of the Earth caused by floods, earthquakes, eruptions, etc.*

I think therefore I may at least conclude, that divers of the Antients, and particularly *Plato*, had some knowledge of past Catastrophys of some parts of the World. And those to have been caus'd by Earthquakes and fiery Eruptions, such as had sunk some places into the Sea and rais'd other places out of it, of great Floods also and Inundations by Rains and Eruptions of the Sea: And that some of those had happened in *Greece*, others without the Streights Mouth, and others elsewhere, and at another opportunity I shall produce a Cloud of Witnesses to this effect which, I conceive, will put it past dispute, but because this Relation has been possibly too long, I shall only add one Re-[p.375] lation more, because it seems to relate to the remainders of the Island of *Atlantis*, and it seems to be of a later date much than the *Egyptian* Stories.

That which I mean is the History of the *Periplus* of *Hanno* the *Carthaginian*. When it was writ I know not, but sure it was very ancient, 'tis lately in the Year 1674, Publish'd by *Abrahamus Berkelius*, with some fragments of *Stephanus Byzantinus*, with the Commentaries of *Gesnerus* and *Bochart*, being but short I have put it into English.

It pleas'd the *Carthaginians* that *Hanno* should sail beyond the Columns of *Hercules* and build *Lybyphenician* Cities, he went then with sixty sail of Ships each rowed with fifty Oars, in these were transported to the number of 30 Thousand Men and Women with necessary Provision and Stores. After we had sailed two Days without the Columns; the first City we built we call'd *Thymiaterium*; under this lay a large Plain, thence carried Westward we made *Solunte* a Cape of *Lybia* cover'd with Wood, where having built a Fane to *Neptune*, we tacked about and sailed προζ ηλιον towards the South, half a Days sail into a Lake not far from the Sea; filled with many and large Canes, where were fed Elephants and various other wild Beasts; having passed this Lake in one Days sail, we built those Maritime Cities, *viz. Caricus, Gytte, Acra, Melissa* and *Aranibys*; sailing thence we arriv'd at the great River *Lixus* which falls out of *Libya*. Near this the *Nomades* (a sort of Grasiers or Cattle-herds) and *Lixitæ* feed their Cattle, with these having made Friendship we stay'd sometime. Beyond this the savage *Æthiopians* live, whose Country is full of wild Beasts, and intercepted with great Mountains from which the *Lixus* flows. Those Mountains the *Troglodita* inhabit a strong sort of People swifter in Running than Horses, as the *Lixitæ* told us. From hence we coasted two Days Southwards, and then one Day more προζ ηλιον and in a Bay found a small Island five Stadia in Compass, where we left some Planters and called it *Cerne*; this, by the Journal of our Voyage, we judged to be in the same parallel with *Carthage*, and as far without the Columns as *Carthage* was within. Hence we enter'd a great Lake, through which past a great River, which we called *Chreses*. There we found three Islands bigger than *Cerne*. From these in a Days Voyage we reach'd the inermost parts of the Lake: It was incompassed with vast Mountains, inhabited by Savages, who threw Stones at us. Thence sailing we past a large River full of Crocodiles and Hippopotams, and return'd to *Cerne*. From hence we past twelve Days by the Coast towards the South, all inhabited by *Æthiopians*, much afraid of us, and not understood by our Interpreters; the last day we discover'd great Mountains covered with Woods, which were of various Kinds and Odoriferous. Coasting round these Mountains we found an immense opening of the Sea, that side which was next the Continent was a plain Country, from whence by Night we perceived Fires from all places, some greater some lesser. Watering here we Coasted along for five Days till we came to a great Bay, which they called εσπερον κεραζ Here we found Lakes and Islands, where landing we found nothing by Day but Woods, but in the Night we saw many Fires, and heard an innumerable noise of Drums, Trumpets, Cymbals, and the like; wherefore being afrighted, and our South-sayers commanding us also to leave it, we Coasted χωραν διαπουρον θυμιαματων the burning Coast of stinking Vulcano's, from whence there run out into the Sea Rivers of Fire, and the Earth was so burning hot that our Feet could not indure it. Hence therefore we hasted and for a Days sail we saw all the Land full of Fires in the Night; but in the middle of these was one vastly bigger than the rest, so that it seem'd to touch the Stars; this, in the Daytime, we found to be a prodigious high Mountain called Θεων οχημα or Chariot of the Gods, in three Days sail more we past all the fiery Rivers, &c.

The reason why I have been so particular in translating the whole Story, is because I conceive it is an instance in History so considerable, especially as to the preceding Relation of *Plato*, that I can hardly believe there is a better Instance to be found. *Plato* tells us of the Island of the *Atlantis* that it was by an Earthquake some Thousands of Years before him sunk into

*Hooke's translation of this Greek phrase "toward the South" seems to be inaccurate. The Greek words literally mean "to the sun," that is "to the east."*

*This sentence supports the meaning of the Greek words as mentioned in the note above— i.e., two days southward and then one day more "to the east."*

*Translation: Horn of Evening*

*Hooke's translation here is less poetic. Rather than "the burning Coast of stinking Vulcano's," The Greek actually says "the burning land of incense."*

the Sea, but yet so that it left many Lakes and unnavigable Places. This gives us a [p.376] Relation of a Navigation (over the very place where the *Atlantis* was placed and sunk by the former Relation) in the times of *Philip* of *Macedon*, or sooner, as some suppose; these Navigators find the Coast of *Africa* without the Mouth of the Streights to trend Westward almost προζ εσπεραν *ad Occasum. Gesner,* in his Notes upon this place, seems a little startled, and says, *Atqui mihi videtur ambientibus Africam omnis post columnas Navigatio converti vel ad Meridiem vel ad Orientem & postremo ad aquilonem*; not thinking, I suppose, of this Supposition. He seems also to be as much to seek about the situation of *Cerne,* but at last he thinks it may be the *Maderas* (p. 85) which I conceive to have lain North-westward from it, but with divers Gulphs and Bays in which were divers great Lakes and Islands, divers Mountains likewise and some Rivers. But which is most considerable, a great part of this Island to the South was then all on Fire. Now comparing this Relation to the present State of those parts, we find all that Continent which they passed by between the Columns and *Cerne,* to be wanting, for 'twill be hard to reconcile the Relation with the present State of that Country, so in probability sunk and cover'd with the Sea; for *Cerne* by this Description, lying in the same Latitude with *Carthage*, and as far from the Pillars without as *Carthage* was within, it must have lain to the North or Northwest of the *Maderas,* from which place the Coast of the main Land seemd then to trend South for twelve Days Voyage as far as the *Canary Islands* are now found, or somewhat farther, from whence it turned away to the Eastward. About these Islands, I conceive, was the Land that was all on Fire, multitudes of which they saw in the Night, and heard the noise of the Vulcanes, and Rivers of Fire running into the Sea, and in some places found the Earth so hot as to burn their Feet. That which directs me the better in this Conjecture, is the prodigious *Vulcano* mention'd, called θεων οχημα the Chariot of the Gods, by reason of its prodigious height, seeming to touch the Stars. This, in all probability, seems to have been the same with the present *Pike* of *Tenarif* which tho' it burns not now, yet, yet there are present Evidences enough, as I have been told by those who have been at the top of it, to prove it to have formerly been a *Vulcano*. And if they had now been wanting, yet no longer since than *Sebastian Munster*'s time it was known to be so, and in his Geography he has so described it. Besides, this by late Example, as in 1639, and by a latter in *Ferro*, which I have Printed, it appears, that those Vulcanoes are not Strangers to those Parts even in this Age: But I have detained you too long with those Conjectures, yet if all Circumstances be examined in the Relation of *Plato's Atlantis,* and in that of *Hanno's Periplus* and compar'd with the present Condition of those Parts, I conceive there will appear many Reasons to make us conclude that there have been in those parts prodigious alterations somewhat like those I have supposed in my Hypothesis, which may serve as an instance of History for such Mutations. The next opportunity I shall produce many other, which, I conceive, will as plainly speak the same thing, according to the Mind and Intention of most of the Ancients, and this is to take off the odium of Novelty.

After the foregoing Passages quoted out of *Plato's Timæus* and the *Periplus* of *Hanno*, I shall adventure to present this illustrious Assembly with some of my Conjectures at the meaning of the Fables of the Poets, but first to say something as to that of *Plato* and of the *Periplus,* which last is suppos'd by several Authors to be very Ancient. From both those

*Here the Greek translation = to the west. this is supported by the latin words* ad Occasum.

*Gesner and also Hanno's Periplus seem to corroborate the story that after the sinking of Atlantis many lakes, unnavigable places, and fires resulted.*

*Possible location of the great disaster of Atlantis.*

*Hooke concludes, therefore, what he has proposed: earthquakes and volcanic eruptions having altered the surface of the Earth, is nothing new; the idea has been depicted in histories.*

Relations compar'd together, there seems at least to result a probability, that there has been some great changes of the superficial Parts of the Earth, where the now *Atlantick* Ocean without the Streights of *Gibraltar*, as they are now call'd, is; and then we have certain Histories now to prove that the main of *Africa* or *Libya* hath extended Westward beyond the *Maderas*, and Southward as far as somewhat farther than the *Canaries*. I have given the Reasons why I entertain'd those Conjectures, which I submit to the Judgment of such as are more knowing and better read in Historical Matters.

**No. 10.** *Waller labels this discourse as "part of another lecture to the same purpose," i.e., corroboration of actual natural events through mythological stories of poets. Read December 14, 1687.*

[p.394] I have, in some of my former Discourses, indeavoured to shew some Probabilities, that the Mythologick Stories of the Poets did couch under those monstrous and seemingly impossible representations of Actions performed by humane Powers, some real and actual Catastrophies that had been caused by the Body or Face of the Earth by other Natural Powers, of which the *Ægyptians, Chaldeans, Greeks*, or some other learned Nations had preserved some Histories or Traditions among the more learned part of them; which, that they might the better conceale their Knowledge, and keep it to themselves, and abscond it from the Vulgar, and such as were not initiated and admitted into their Fraternities, they had contrived and digested into fabulous Stories, which, as they might serve to amuse and awe the Vulgar by the Dæmonology they had thereupon superstructed, so they might serve to instruct and inform the Adepti, or such as were admitted to the true interpretation and understanding of what they knew, of the real History that was concealed thereby, as also of their Philosophical or Physical Hypothesis for the Explication and Solution thereof. I think it cannot be doubted that the Theogonia of *Hesiod* was of this nature, which if it was *Hesiod*'s (of which, yet I confess their are some Moderns make a doubt) it seems to have been some of the first Notions which the *Greeks* had obtained of these Matters from the *Ægyptians* or *Phœnicians*, or some other of the Eastern Nations; except we suppose that *Orpheus*, who preceded both *Hesiod* and *Homer* near five 100 Years, might in those times have known and communicated some what of what they had by the same Methods procured. The Histories of those times are very dark and uncertain, and nothing convincing can be built upon them. It will be therefore but lost Labour to indeavour to prove my Conjectures from Histories, or hints to be sought among those few Fragments which are now to be met withal among the Relicks of written Antiquities. Those, if such there were, (as being committed to small and perishable Substances) have been more easily drowned and swallowed by time, or buried and overwhelmed with the Dust of Oblivion: And the Copy or Counterfeits of some of them, which have been made by some of those we now call the Ancients (though with respect to them they are to be accounted Modern) seem to have been but very imperfect, and to have been like Structures made up and peiced of the Rubbish, Ruins and Fragments of those

*Ancient poetry telling of monstrous and fictional stories may represent actual historical events.*

*Hesiod was one of these poets who narrated actual events that the Greeks received from the Egyptians or Phœnicians couched in fictional and fantastic claims.*

*The histories of even earlier times than Hesiod and Homer are dark and uncertain.*

Antiquities which they in those times could rake together; so that though some great Buildings have been by these secundary Ancients erected; yet being made up of such Fragments or Parts of those more ancient sacred Piles by the new Disposition and Order of them they now appear a preposterous *Moles*, yet we cannot but conceive that they had some better and more certain informations of those more ancient Histories or Traditions than what we now can find; and we cannot think so mean of them as not to believe they did in some measure comprehend the Intention, Meaning, and Drift, or Design of those that preceded them; and tho' they wanted a compleat knowledge, yet from the knowledge they had of the then Ruins, they were better inabled to Judge and Conjecture concerning them, than we now can. And tho' their Conjectures might not be all right, yet we cannot but think they might be tollerably near the matter, and that they did acquaint Posterity by their Writings what those their Conjectures were. And of this Nature I take the Metamorphosis of *Ovid* to be, who, I conceive, had made it his study to inform himself as fully as he was able of what was then to be found concerning that knowledge, and out of those informations he compiled that Book which was to comprise all the Records of Antiquity concerning the Changes and Catastrophies that had happened to the Earth from the Creation unto his own time, which his four first prefactory Verses do plainly enough declare. *In Nova sert animus mutatas dicere formas, Corpora dii captis nam vos mutasti & illas, Aspirate meis, Primaq; ab Origine mundi, In mea perpetuum deducite tempora Carmen.* Which is as much as to say, My design in [p.395] this Book is to speak concerning the various alterations and transformations which the Bodies or superficial Parts of the Earth have, by the Divine Powers, undergone; for to those he doth ascribe them, *Nam vos mutastis & illas*, and therein to comprize all the knowledge I have been able to procure from the very first Creation or Original of it, even to these very times in which I live. And accordingly we find him to begin this his History, even with the beginning of the Creation of the Earth itself, and therein to have followed the Traditions, Opinions, and Doctrines of the most Ancient Sages concerning its manner of Formation out of a preceeding Chaos; which Doctrine that it was very ancient, and indeed the most ancient of all others concerning the Origination of it, I think the Learned and Ingenious Dr. *Burnet* in his *Archæologia* has sufficiently proved, and therefore I shall not need to say any thing concerning it; only I would make this one occasional Remark, That how ancient soever it was, it did not favour of an unlearned or ignorant Age or of a first beginning of real Knowledge, for that we find by *Ovid*'s Copy of it, that it contained a more refined Conception concerning the Figure, and Shape, and Properties of the Earth, than many of the Greek Philosophers (who in probability were many hundreds of Years after those first Sages) had concerning it. Some of those Greek Philosophers making the Earth to be of the form of a Drum or Cylinder, others of an infinite Column, others of a Skiff or Boat, or of a floating Island in the midst of an infinitely extended plain Ocean, and others of other extravagant Shapes; whereas we find that the Doctrine of the Chaos made it to be of a Sphærical Form, *Solidumq; coercuit orbem*, to consist of Land and Water, to have a proper Gravity that kept all its parts in that shape, or his *Tellus (Elementaq; grandia traxit & pressa estgravitate sua)* to be involved with the Air, and that again with the Æther. *Hæc Super imposuit liquidum & gravitate carentem Æthera nec quicquam terranæ fœcis habentem*, to be suspended in the Air, or Æther, or space of Heaven without being supported by any imaginary Foundation, as those Greeks fancied. *Circumfuso pendebat in aere tellus, Ponderibus librata suis*. Nay, and by

*The ancients are closer to those dark times than the present, and therefore their writings, e.g. those of Ovid, are better informed of the events of those times.*

several other Passages and Expressions of this Book, it is clear, that in those very ancient times, whenever they were, for 'tis hard certainly to limit them; the Learned Men that then lived, had arrived to a very great height of Natural Knowledge, especially of that part which concerned the Cosmography or Constitution of the Universe; and by that Expression, *Ignea convexi vis & sine pondere Cæli emicuit, summaq; Locum sibi legit in Arce.* It seems plain that they placed the Sun in the Center of the universe, and made the Earth to move about it. *Principio terram, ne non æqualis ab omni Parte foret, magni Speciem glomeravit in orbis.* But this only by the bye; for I know the common interpretation of these places, is altogether differing from what I now give, yet were it now my business, I think I can shew sufficient Reasons to persuade any unprejudiced Person that what I have given is the designed meaning of them; but I proceed to shew the general design of *Ovid* in this Book. After the Description of the formation of the Earth, he comes to describe the first times of its continuance; that is, the αδηλου, or unknown Ages of the World, of which he makes four, the Golden, Silver, Brazen and Iron, in the last of which comes in the Mythologick and Historick, for that he himself hath Mythologized also some of the Historical Times and Events. What space of Time he allows to each of these Ages it doth not so readily appear, but it is certain that the *Chaldeans, Ægyptians, Brachmans,* and some Heathen Historians have assigned spaces large enough and even beyond belief almost; and Mr. *Graves* tells us, that the Chinese do make the World 88640000 Years old. He begins the Mythologick Times with the Gygantomachia, which to me seems to be nothing else but a Description of some prodigious Earthquakes or Eruptions. And that by the Giants he plainly means nothing else but the Subterraneous Fires or Accensions which break out, and throwing up before them the Earth, seemed to threaten the very Heavens by piling Mountain upon Mountain (*Affectasse ferunt Regnum Cæleste Gygantes Altaq; congestos Struxisse ad Sidera montes*) I shewed before in the interpretation of the Rape of Proserpine, where it plainly appears what was meant by Typheus one of those Giants, who is said to lie buried under the Island of *Si-*[p.396]*cily*, and therefore shall not need to say more upon that Subject. After the breaking forth of these Subterraneous Streams and Flames, we find *Ovid* describing them to be burnt off with Lightning. *Tum Pater omnipotens misso perfregit Olympum Fulmine & excussit Subjectam Pelion Ossa. Obruta mole Sua cum corpora dira jacerent, perfusam multo Natorum Sanguine terram immaduisse ferunt, Calidumq; animasse cruorem.* This we find to be a general Concomitant or Subsequent of such Eruptions, and it were easy to produce many Examples of it in our late Eruptions; and 'tis also as usual for many of those places that have been thrown or raised up into Hills to be sunk or tumbled down again, *Excussit subjectam Pelion Ossa.* So we are told of a Hill that lately rose up by *Catanea*, which soon sunk again. [*Obruta*] by this, I think, is plainly signified the Eruption of fiery Streams or Rivers of melted Minerals out of those Orifices or fiery Vents, such as in the two last Earthquakes in *Sicily* have broke out of *Ætna*, and overflowed and burnt up and destroyed several Towns, Villages, Fields, &c. for what can better express the moving, raging and devouring Qualities of such a stream of Liquid Fire, than to call it an animated, or living scalding Gore from its red and fiery Colour, its scalding and burning Heat, its fluidity and rapid Motion, and its devouring and consuming Power; but it would be too tedious to insist on all the remarkable Circumstances and Expressions, which, I conceive, makes it plainly enough appear what was the Design and Scope of the Story; nor need I mention the Description of

*In the middle of this summary of the content of Ovid's* Metamorphosis, *Hooke interjects, without comment, the age of the world according to the Chinese, more than 88 million years. This is one of the many clues that Hooke knowingly, but secretly, felt that the Earth had to have been older than the biblical account.*

it by other Mythologers, as *Claudian, Hygynus, Antoninus, Liberalis.* Nor will it, I hope, be needful to answer any thing to those who would interpret it another way: Some making it to be only a Description of a Rebellion; others a disguising of the History of the Tower *Babylon.* I shall rather leave it to the Judgment of every one to make choice of which interpretation he shall, upon duly considering the relation, think to be most aggreeable to the whole drift of the Book. And what I now deliver I would not have to be taken otherwise than only as my Reasonings and Conjectures upon the like Considerations: For as I observed before, the Poet has so couched all his Relations and Expressions as to comprize a Physical, a Moral, and an Historical Meaning in them. And it may be so interpreted as if it were designed to describe some particular Earthquakes, or some particular Rebellion, or the general Rebellion of wicked Men against Heaven, and the Divine Powers, or the attempt of those at *Babel*, and at the same time it may also be found designedly to contain in brief the Theory of Opinions of the most antient Physiologers which they held concerning the Causes and Effects of Earthquakes upon the uper Face of the Earth; which to me, I confess, seems to be the principal aim and design of this Story of *Ovid*, as well as of the most part of the rest of the Book, which I design, God willing, to prove more expressly and particularly in a Treatise upon this Subject, so soon as I have settled some Affairs, which have hitherto hindred me from perfecting that and many other Subjects.

*Hooke has been criticized by some historians as one who started many things but finished few. It is true that he was involved in so many projects that it would have been humanly impossible to complete them all. This passage shows he was quite aware of this problem himself. But the critics should realize that the totality of what he did achieve was astronomical in human terms.*

I shall not need here to say any thing concerning the Custom of the *Greeks* in those former Ages of turning all their Histories into Mythologick Poetry; 'tis plainly enough proved by that Relation I read the last day out of *Plato's Timæus*; and it was not only used by them but by divers other Nations, as the *English* and *Germans*, as you will know. I suppose the reason was for the better fixing it into the Minds of the Youth by a kind of indelible Character, as *Plato* expresses it: Which could not be forgotten; for extravagant Marks we know are the great helps of Artificial Memory, for that they raise extraordinary Attention, and that extraordinary Attention and Wonder does stigmatise or burn in as twere indelible Ideas in the Memory. Pleasure also is another help to fix Ideas, and that Poetry and Songs contribute to, and the activity of the Spirits in Youth work the Effects more powerfully, and make them more durable. These, I imagine were the Reasons why the *Ægyptians, Greeks* and other Nations converted their true Histories into these Romantick Fables: Not that I do here undertake for the truth of History in every Fable, for I conceive that there are as various kinds of Fables as there are of Histories. Some are repeated and believed Fables which are true Histories, others are believed true, but are really Fables: Some are believed Fables [p.397] and are really so, and others are believed true and really are so. But of this fourth Head I fear is the smallest number; but we must take the best Evidence we can to confirm our Belief of those that are generally so reputed: Among which none has been more looked after of late than Medalls, Inscriptions, and real Monuments, yet remaining of the preceding Persons and Actions, these are by all looked upon as a most undeniable Proof to confirm a written History, and yet we know that many things of this kind have been counterfeited, yet that cannot be said of all: Now, if these that may be counterfeited be yet looked upon as more Authentick than Written History, then certainly these Medals, Inscriptions, or Monuments of Natures own stamping, (which I alledged to prove an Hypothesis) which 'tis impossible for Art to counterfeit, might in reason be looked upon as Proof sufficient tho' no History could be produced. If I saw a perfect Medal, tho' I could

*Again he refers to fossils as the "inscriptions" of nature itself.*

not be assertained whether it were Antique or Counterfeit, yet I could certainly conclude it had been made by Art from the sensible Characteristicks of it; now it seems very strange to me that so many evident Characteristicks as may be plainly discovered in those figured Bodies should not force an assent; but truth will in time prevail; but to give as much satisfaction as I can to all Doubts, I will pitch upon one or two of the Fables of the Metamorphosis for instances, to shew that they were designed to convey a certain History very much differing from the first appearance of the Fable. I will begin with those of *Perseus, Atlas, Andromeda* and *Medusa*, because, as I conceive, they have relation to the *Herculean Columns*, and to the *Atlantis*, or those parts of *Libya* which were near it; they are somewhat long, however I must beg your Patience to explain them a little more fully, and I will be shorter in the rest.

*Perseus* from περιζεω *circumferveo*, I take to signifie hot inflamed Air or Lightning which is the Earthy Exhalations set on fire by the Air dissolving them; he is said to be the Son of *Jove*, that is of Ætherial or Elementary Fire begotten in a shower of Gold or Fire from Heaven, that is Lightning. He carries with him the *Gorgons* Head haired with Vipers, the Picture of Lightning.

> *Viperei referens spolium mirabile Monstri*
> *Aera carpebat tenerum stridentibus alis.*

> *Bearing the spoil adorn'd with snakey Hair*
> *With clashing Wings he rends the yielding Air.*

This I take a proper Description of Thunder and Lightning, fiery Serpents representing the Emanations of Lightning, or the wrigling flashes of it darting out sometimes: 'Tis represented as held in the Hand of *Jupiter*, sometimes in the Mouth, sometimes in the Claws of his Eagle, and we shall find afterwards in the Fable, that the Actions of *Perseus* against the Sea Monster or the Flood are compared to those of *Jupiter*'s Eagle.

> *Cumq; super Libycas Victor penderet arenas*
> *Gorgonei capitis, guttæ cæcidere cruentæ,*
> *Quas humus acceptas varios Animavit in angues.*

> *And while the Victor hover'd in the Air,*
> *The drops that fell from* Gorgon*'s Bloody Hair,*
> *By Earth receiv'd, were turn'd to various Snakes.*

These are the effects of Heat in those sandy, hot, burning Countries, and I conceive this alludes to the Snake-Stones, or Thunder-bolt-stones, as well as the living Serpents; for the vertue of the *Gorgon*'s Head, which is Subterraneous Eructations or Damps, was the petrifying Quality converting all things to Stone.

> [p.398]*Inde per immensum ventis discordibus actus,*
> *Nunc huc, nunc illuc, exemplo Nubis aquosæ*
> *Fertur, & ex alto, seductas æthere longe*
> *Despectat Terras, totumq; supervolat Orbem:*
> *Ter gelidos Arctos, ter Cancri brachia vidit,*
> *Sæpe sub Occasus, sæpe est sublatus in Ortus.*

> *Thence carry'd by discordant Winds he's hurl'd,*
> *As watery Clouds through the expanded World;*
> *Now here, now there, on the far distant Plains*
> *He casts a glance, then heav'nly Arches gains;*

*Hooke's own words here prove that he did not launch into these studies to look for changes in longitude and latitude to support his polar wandering hypothesis, as Oldroyd (1989) suggested, but rather to show that fables often conveyed true histories. He would have distrusted the accuracy of ancient measurements.*

*Both the Greek and the latin words mean "boil round about."*

*In these and following paragraphs Hooke does an excellent job of interpreting the myths as cover stories for actual happenings of earthquakes and volcanic eruptions.*

> *Thrice the cold Bear, thrice the hot Crab his Eyes*
> *Survey, as oft to West or East he flies.*

This I conceive very properly apply'd to Lightning, which is now here, now there, all over the World.

> *Jamq; cadente die veritus se credere Nocti,*
> *Constitit Hesperio, regnis Atlantis in Orbe*
> *Exiguamq; petit requiem, dum Lucifer ignes*
> *Evocet Auroræ, Cursusq; Aurora diurnos.*

*The letter J was unknown in classical Latin. Italian humanists created the letter to represent the consonantal "I"; it is not in use now in classical texts.*

> *And now not trusting to approaching Night,*
> *Doth on th' Hesperian Realms of Atlas Light,*
> *And craves some Rest, 'till Lucifer displays*
> *Auroras blush, and she Apollo's Rays.*

This describes the settling of this fiery Vapour about the Westermost parts of *Africa*; where

> *Hominum cunctis ingenti corpore præstans*
> *Japitionides Atlas fuit: Ultima Tellus*
> *Rege sub hoc & Pontus erat, qui solis anhelis*
> *Æquora subdit equis, & fessos excipit axes.*

> *Gigantick Atlas Empire here possest*
> *O're Lands extended to the farthest West;*
> *Where* Titans *panting steeds his Chariot steep,*
> *And bath their fiery Fet-locks in the Deep.*

It was a Country that lay farthest Westward where the Sun seemed to set in the Sea.

> *Mille greges illi, totidemq; armenta per herbas*
> *Errabant -------*

> *A thousand Flocks, a thousand Herds there Graz'd*
> *On verdant plains ------.*

It was a delicate Country for Pasture and Cattle.

> *Et Humum vicinia nulla premebat.*

> *No Neighbouring Lands offended this.*

It was an Island not joined to any Continent.

> [p.399]*Arboreæ frondes auro Radiante nitentes*
> *Ex auro Ramos, ex auro poma ferebant.*

> *The dazling Trees there glitter in the Air,*
> *Which golden Fruit and gilded Branches bear.*

Its Rivers and Rivulets all abounded with Gold or golden Sand.

Rivers are very properly Mythologised by Trees, the greater Body of Water resembling the Trunk, the lesser Rivers the Branches, the Rivulets, Fountains, Springs and Sources, the Twigs and Leaves; and the Hills and Mountains the Fruit: For as I have already, upon another occasion, hinted, Trees receive the greatest part of the Sap from the Air and little from the Earth: And they distribute more moisture to the Earth from their Bodies by their descending Sap than they draw from it by their Vessels, and as the Sea Returns the Water it receives out of the Rivers into the Air, whence it circulates again into the Fountains and Rivulets by condensation and Rain, so doth the process of Nature also operate in the manner of returning the moisture into the Leaves, as I shall upon another occasion more particularly explain, having mentioned it only upon this occasion to shew how properly the Rivers are Mythologised by the Trees, Branches, Leaves, and Fruit.

The next Verses expressing *Perseus*'s Addresses to *Atlas* is Poetical, as also of *Atlas*'s Resentment, upon remembrance of an old Prophesy that *Parnassian Themis*; or all knowing Predestination had fore-shewn, *viz.* That time should come when an Off-spring of Celestial Fire should destroy that golden Country; for fear of which it is said,

> *Id Mentvens solidis Pomaria clauserat Atlas*
> *Mænibus, & vasto dedit servanda Draconi,*
> *Arcebatq; suis externos finibus* ------.

> *This fearing, he his Orchard had inclos'd*
> *With solid Cliffs: A Dragon too oppos'd*
> *All Entrance* ------,

This *Hesperian* Garden was incircled with high Cliffs and encompassed round by the Sea.

> *Huic quoq; vade procul, ne longe gloria rerum*
> *Quas mentiris, ait, longe tibi Jupiter absit*
> *Vimq; minis addit, manibusq; expellere tentat*
> *Cunctantem, & placidis miscentem fortia dictis.*

> *Begon, said he, for fear thy Glories prove*
> *But Counterfeit, and thou no Son of* Jove.
> *Then adds uncivil Violence to Threats;*
> *With strength the other seconds his intreats.*

This Island had not been troubled with Thunder, Lightning, Earthquakes, or Eruptions, poetically thus described; and how these came on by degrees, and Barrenness with Drouth increased, and how the Inhabitants endeavoured to prevent it by their Labours; but at last because they strove against the Course of Nature, the Poet makes *Perseus* say,

> *Accipe Munus, ait, Lævaq; a parte Medusa*
> *Ipse retro versus squallentia protulit ora.*

> *Take then, said he, they due Reward, to's view*
> *Shewing* Medusa's *Head, his own withdrew.*

[p.400] That is, the subterraneous Eruption, and therewith the petrifactive quality exerted itself, upon that Country, and as a Consequence thereof,

> *Quantus erat; Mons factus erat nam barba comæque*
> *In Sylvas abeunt, juga sunt humeriq; manusq;*
> *Quod caput ante fuit, summo est in monte Cacumen:*
> *Ossa Lapis fiunt, tum partes Atlas in omnes*
> *Crevit in immensum* -----.

> Atlas *to a Mountain, equal to the Man,*
> *Was turn'd, where Hair and Beard was, Trees began*
> *To grow, his Shoulders into ridges spread,*
> *And what was his, is now the Mountains Head:*
> *Bones turn to Stones, and vastly all increase* -----.

A prodigious Mountain is raised, and the *Hesperian* Garden or Country lost, this Mountain being the only remains thereof. Now before this Metamorphosis of the Country of *Atlas* into that Mountain, *Perseus* had destroyed the *Gorgons*, and but off the head of *Medusa*. These *Gorgons* were said to inhabit certain Islands lying near *Atlas*, they were called the *Phorcidæ*, of which there were two which were said to have but one Eye between them; possibly a *Vulcano*.

> ------*Gelido sub Atlante jacentem*
> *Esse locum solida tutum molimine molis,*
> *Cujus in introitu, geminas habitasse sorores*
> *Phorcidas, unius sortitas Luminis usum.*

> ------*Under frosty* Atlas *side*
> *There lay a Plain with Mountains fortify'd,*
> *In whose access the* Phorcidæ *did lye*
> *Two Sisters, both of them had but one Eye.*

This he takes with him; that is, I suppose there began the Earthquake or the Subterraneous Vapour kindled, and thence extended to the farthest extreamity of those Islands, possibly the *Atlantick*.

> *Id se solerti, furtim dum traditur, astu*
> *Supposita cepisse manu; perq; abdita longe,*
> *Deviaq; & Sylvis horrentia fana fragosis*
> *Gorgoneas tetigisse domos, passimq; per agros*
> *Perq; vias vidisse hominum simulacra ferarumq;*
> *In silicem ex ipsis, visa conversa medusa.*

> *How cunningly thereon his Hands he laid,*
> *As they from one another it convey'd;*
> *Then thro' blind Wasts and rocky Forrests came*
> *To* Gorgon's *House; the way unto the same*
> *Beset with forms of Men and Beasts, alone*
> *By seeing of* Medusa, *turn'd to stone.*

By which it seems to have extended a great way and to have been very Rocky, Cragged and Uninhabitable, where Men and other things had been before that time petrified. Being there arrived, he finds *Medusa* asleep; that is, I supposed, the *Vulcano* not burning: But by this new Eruption the Head of *Medusa* is taken off, and the Vapour or Eructation riseth into the Air, partly in Flames and lightning and fiery Vapours, which is *Perseus*; partly in watery Vapours and Wind, which is,

> [p.401]*Dumq; gravis somnus Colubros ipsamq; tenebat*
> *Eripuisse Caput collo; pennisq; fugacem*
> *Pegason & fratrem, matris de sanguine natos.*

> *And how her Head he from her Shoulders took,*
> *E're heavy sleep her Snakes and her forsook;*
> *Then told of* Pegasus, *and of his Brother,*
> *Sprung from the Blood of their new slaughter'd Mother.*

This may represent the mounting of fiery Eruptions, which rise as swift as *Pegasus*, and shine like his Brother, who was supposed to brandish a golden flaming Sword. But I must hasten, *Perseus* having performed those Exploits of sinking the *Atlantick* and raising Mount *Atlas*.

284  DISCOURSE OF EARTHQUAKES

*------Pennis ligat ille resumptis,*
*Parte ab utraq; pedes, teloq; accingitur unco,*
*Et liquidum motis talaribus aera findit:*
*Gentibus innumeris circumq; infraq; relictis*
*Æthiopum propulos Cephæaq; conspicit arva.*

*His Wings at's Feet, his Faulcion at his Side*
*He sprung in th' Air: Below, on either Hand,*
*Innumerable Nations left, the Land*
*Of Æthiope, and the* Cephean *Fields survey'd.*

The fiery Vapours flies over several Countries till it comes to the Country of the *Æthiopians*, and the Plains of *Cepheus* or the *Drones*. Here he finds *Andromeda* chained to a Rock expecting to be devoured by a Sea Monster. *Andromeda*'s Name and Description agrees with that of an half drowned and Rocky Country, by turns overflowed with the Sea at High Water and covered with Sand (*Jusserat Ammon*) and seems to be that part of *Africa* where *Jupiter Ammon*'s Temple was built; which since raised, is all Sandy, and therefore is called *Ammon* or Sandy. *Perseus*, its said would have thought her Marble, but that he saw the waving of her Hair by the Wind, which may signifie that some Reeds, or such Water-plants, might grow among the Sand and Rocks sometimes overflowed by the Sea. Here (to make it short) *Perseus*, or the fiery Eruption, raiseth the frontier Parts to the Sea, and repelleth the Tide of Flood from overflowing and drowning the Land. And so *Andromeda* becomes Ἀναδρομη, raised and freed from the Inundation; this is so very plainly specified in the Description, that bating a little poetical Expression about Love; which *Ovid* had well studied, having a gust for it. It seems plainly to design the History of such a Metamorphosis, and very probably of that very Country I before-mentioned; and were it worth while, I am apt to believe, that the Histories designed by all the other Fables may be discovered; for that they have all such Histories couched under them, I do no ways doubt; not only for the Arguments I at first mentioned, but for several others. Moreover I do conceive, that there is a Chronologoe of the preceding times to be discovered out of them, and that they are written, not fortuitously, but with great Care, and according to the due Order in which they happened. And tho' possibly this do not at first view so plainly appear, nor can be so undeniably demonstrated without a more perfect knowledge of the *Hieroglyphick* and *Mythologick* Characters, yet I am almost certain that some such Chronogick [sic] Account is couched in the Fable, and may, if well examined, be detected; and to conclude, I am apt to believe, that in this Mythology is contained the greatest part of the *Ægyptian* and *Grecian* History of the preceding Ages of the World; the truth of which I do not undertake to defend, we must take that as we do all other Histories, upon trust till we can have better [p.402] Proof either for or against them. And if these be so admitted, this Book will furnish a sufficient number of such Catastrophies that have happened in former Ages to make the Hypothesis, I have indeavoured to explain, to seem at least probable, if not necessary, and neither so absurd or impossible as some have asserted.

**No. 11.** *Ovid's Metamorphoses, Book 1. Read January 4, 1688.*

[p.377] *Varro* has distributed the Ages of the World into three, *viz.* the αδηλον, μυθικον, and ιστορικον of the αδηλον we know nothing from Heathen Writers; of the μυθικον we must look for an account from the Fables of the Poets, *Homer, Hesiod, Ovid, &c. Ovid,* to pass by *Hesiod* and *Homer*, is said to have imitated the Greek Poet *Parthenius*, and has left us a very large History of the changes that had anciently happened in the World, his whole Metamorphosis, being, as I take it, written for that purpose: We are extreamly obliged to *Pliny* and some few others, as all well know, for what they had collected out of others, or wrote from their own Observation and Knowledge.

*These three Greek words mean, respectively, unknown, mythic, and historical.*

Now, that *Ovid*'s Metamorphosis was penned for this end we may find by the 4 first Verses.

> *In Nova fert animus mutatas dicereformas*
> *Corpora, Dii captis (nam vos mutastis & illas)*
> *Aspirate meis, Primaq; ab origine mundi*
> *In mea perpetuum deducite tempora carmen.*
>
> *I sing of Beings in new shapes array'd,*
> *Assist ye Gods (for you the Changes made,)*
> *That from the Worlds Beginning to these Times*
> *I may comprize their Series in my Rimes.*

*Ovid's* Metamorphosis, *a prime example of real histories couched in mythological terms, popularly read in the 17th century by the intelligentsia, is one such "account of the ages and times of the duration of the earth."*

That is the time of *Augustus Cæsar* in which he Lived.

The Hypothesis in *Ovid* (for I conceive it only an Hypothesis in him) is this, that the præ-existent Matter of the World was first, a quantity of Matter without any particular form, *Rudis indigestaq; moles,* a rude disorder'd Mass, and yet it had the property in it which (when directed afterwards to some Center) was weight, which as yet he calls *Pondus iners* unactive weight. Secondly, It had in it the seminal Principles, which were afterwards to effect the Productions, these he calls *discordia semina rerum*, the jarring Seeds of things, as being then *non bene junctarum*, not well conjoined, no not to form the Sun, Moon, or the Earth, the primary or secundary Planets, *Nec cirsumfuso pendebat in aere Tellus, ponderibus librata suis*, nor did the self-poiz'd Earth encompast round hang in soft Air; these Verses do seem to glance at an *Hypothesis* I have formerly acquainted this Society with, somewhat of which Mr. *Newton* hath Printed. *Tellus, Pontus & Aer*, Earth, Water, and Air were yet all confounded with each other, like Mortar or Mud. *Instabilis Tellus innabilis unda.* The Earth unstable, Waves for Keels unfit, which it comes to attain afterwards, and remains so for sometime, till by degrees again it lost it when *Astræa* left it, which was just before the Gygantomachia; for *Astræa*, as I shall by and by make appear, is the Virgin and primitive Smoothness and Stabiliy of the superficial Parts of the Earth, from (α) the first or Primitive, as (α and ω) *Alpha* and *Omega*, and σερηα stability, *Et Virgo cæde madentes ultima Cælicolum terras Astræa reliquit.* The last of Deities from Blood polluted Earth *Astræa* flies; for like moist Pap or Mud, by degrees the watery and Aerial exhaleing, it settled into a smooth, tender, and uniform Substance, like the Youthful and Virgine Constitution, but a farther separation of the Fluid Parts makes the Earthy, Dry, Rough, Rincled and Chopt; inclining to the Countenance and Constitution of Age, and the Virgin Beauty is fled: For a while there was a jumble, *Corpore in uno,*

*Hooke is quick to point out, however, that Ovid's story is only a hypothesis.*

*Another reference to Newton publishing a hypothesis with which Hooke himself had "formerly acquainted" the Royal Society.*

*frigida pugnabant Calidis, humentia siccis, mollia cum duris*, the Cold, the Hot; the Moist, the Dry ones fight; the Soft, the Hard, all incorporated strove together, *Sine pondere habentia pondus* with weight, yet weightless, that is, they all being Bodies had a capacity of being weighty, but a gravitating or attracting Center not yet being existent, they had no actual Gravity any way; but so soon as *hanc Deus & melior litem Natura diremit*, God and the better Nature ends this War; that is, God and Nature had made the gravitating Center, presently the heavier descend towards it, the lighter rise from it

> [p.378]*Et Cœlo terras &terris abscidit Undas*
> *Et liquidum sprisso secrevit ab aere cœlum.*
>
> From Sky the Earth, thence Floods divided were,
> And liquid Æther from the thicker Air.

The Atmosphere inclosed the Ball, and was distinct from the Æther; 'tis remarkable that he makes the Water the lowest in this and the following Account.

> *Ignea convexi vis & sine pondere Cœli*
> *Emicuit.*
>
> Of the convex and weightless Heav'n the bright
> And fiery Power shin'd forth.

He seems to make it by the word *Emicuit* to be at the first encompassed with a shining Fire like a Star or Sun, for its place was *in arce*, above all; within this was Air.

> *Proximus est aer illi levitate locoque.*
>
> The next to this in weight and place is Air.

The Earth is assigned next.

> *Densior his Tellus Elementaq; grandia traxit,*
> *Et pressa est gravitate sui.*
>
> Prest by its weight Earth sinks, to which repair
> The heavier Elements.

And the Water lowest.

> ---------------*Circumfluus humor*
> *Ultima possedit solidumq; coercuit orbem.*
>
> The Floods at last sink in
> From every side, yet leave a spherick Skin.

So that it seems there was a notion that the middle part of the Ball of the Earth was filled with Water as well as the outside covered with it: To which also agrees *Des Cartes* Theory and that of the ingenious Dr. *Burnet* in his *Theoria Sacra.* Thus far, I suppose, it will easily be granted that the Poet gives us a short History of the formation of the Earth, and 'tis as plain that the twenty eight Verses following are to the same effect, wherein he describes the cutting and forming the Face of the Earth into Lakes,

*Ovid, therefore, has given a short history of the formation of the Earth in these verses.*

Seas, Rivers, Hills, Dales, &c. the dividing the whole into Zones, and assigning the use of Air, for Clouds, Rain, &c. nor has he yet Personated or Mithologized any thing, but in the twenty ninth Verse following, *viz.* The sixtieth Verse of this first Book he begins calling the Winds Brothers, *Tanta est discordia fratrum,* &c. the Sense of all the rest is plain till the eighty second Verse, where he begins to personate Actions Mythologically; for speaking of the formation of Man,

> -------- *Natus homo est,* &c.
> *Sive recens tellus seductaq; nuper ab alto*
> *Æthere cognati retinebat semina Cœli;*
> *Quam satus Japeto, mistam fluvialibus undis,*
> *Finxit in effigiem moderantum cuncta Deorum.*

[p.379]Man's Born, &c.

> -------- *Or th' Earth new gain'd*
> *From nobler Æther, some Seeds still retain'd*
> *Top Heav'n ally'd, which Earth* Prometheus *took*
> *And mixt with Waters of a living Brook*
> *Made Man like th' all-commanding Deities.*

From this place onwards he seems to Mythologize the most part of his History, of which he gives notice in the eighty sixth and eighty seventh Verses.

> *Sic modo quæ fuerat rudis & sine imagine Tellus*
> *Induit ignotas hominum conversa Figuras.*

> *So what was rude and shapeless Earth, puts on*
> *When chang'd, the unknown Character of Man.*

Hitherto hae had spoken of things as Dead and Unactive Earth, but from hence forth he will describe the Earth as changed and clothed with the various shapes of Men and Persons, and so having described the Formation or first Generation of all things Physically and plainly, he comes next to tell the Age or Ages of the World, and what Periods of Life or Being it hath had, and the States it hath been induring those several Periods.

The first Age or Childhood of the World he calls the Golden Age: Gold is soft, flexible the most ductile of Metals, it has the best Lustre, and has always had the greatest Esteem. This state of the Earth he represents to be like that of Childhood, wherein all things are gay and pleasant, all things flow plentifully and smoothly; the Skin or Shell is yet smooth, succulent and soft, moisture and heat abound; so that things sprouted forth and Flourish: There is a continued Spring, all things are Budding, Blossoming, and bearing Fruit at the same time, no need of Art as yet to help the progress of Nature forwards; or to regulate it, no one part of Nature intrenched, invaded, or hindred the free progress of another; there was plenty and enough, for all Rivers flowed with Milk and Nectar, and Honey drop'd from the Leaves of Trees.

All these Poetical Expressions, which the Author seemeth to speak, as of Men, and their Actions, and Enjoyments, I take to be significative of all acting Powers of the Earth whether Vegetative or Animal, *Per se dabat omnia tellus, Ver erat æternum. Sponte sua sine lege sidem Rectumq; colebant.* The Earth gave all things of itself, Spring was Eternal, and Justice observ'd without Law, &c.

Now, tho' all that happened in those times of the World, fell within the Age which *Varro* calls the *Adelon Tempus;* that is unknown as to the Heathen Writers, yet I look upon this Account almost as considerable, if not more, than those things which fall within the Mythologick; for I take this to be the Summe and Epitomy of the Thoughts and Theories of the most ancient and most knowing Philosophers among the *Ægyptians* and *Greeks;* and howmuchsoever there may be some who slight and neglect and villify the Knowledge, Doctrines and Theories of the Ancients, which Humor I am apt to think proceeds from their ignorance of what they were, and the difficulty of attaining the knowledge of them: Yet certainly former times wanted not for Men altogether as eminent for Knowledge, Invention, and Reasoning as any this present Age affords, if not far before them; for if we do believe a time of the Creation or Production of the Earth (as we have somewhat more of Argument to persuade us than possibly the Heathens had from the History thereof written by *Moses*) then 'tis very rational to conclude, that in the more Youthful Ages of the World, there was a much greater Perfection of the Productions of it, and that before those many and great Alterations and Catastrophies that have since happened, and before the senile Iron and decaying Ages of the same, wherein every thing by degrees grew more Stiff, [p380] Rocky, Unactive and Barren, and so a degeneration of the Productions thereby seems a necessary Consequent. In the times, I say, that preceded all or many of these, it seems very rational to conclude, that it might produce Men of much longer Life, bigger Stature, and with greater accomplishments of Mind (of all which we have very good Testimonies without the Argumentations, Histories, Traditions or Theories of the Heathen Writers) upon which account tho' this Description of the Genesis of the Earth, and the first Age of the World should be supposed to be but the Theory or Philosophy of some of the most eminent Men, as *Orpheus, Pythagoras, &c.* in Ages so much nearer to those more active Ages of the Earth, yet, upon that account, they may, I conceive, be well worth our inquiring into, to see, at least, how Consonant those things are which they thought Reason, to that of ours at this present. Some possibly may be of *Aristotle*'s Opinion that the Earth was eternal: But I am apt to think that such as are so, have not so fully consulted their own Reason and Experience, nor much troubled themselves with that Speculation. We found that the *Ægyptian* Priests by that Passage I quoted out of *Plato,* had the notion of the *Genesis* Mutations, Catastrophies, by Fire and Water, and the like of the Earth, if we will not allow them to have the History of them, or the Accounts of so many 1000 Years as *Plato* mentions. But it will by some be required perhaps, by what means can we judge of any such preceding Age? I answer, That possibly the petrified Shells that lye in the Repository, and the prodigious Bones and Teeth that have been found buried in the Earth, of which the Repository affords some instances, and more might be fetched elsewhere: These, I say, might to some unprejudiced Men prove Arguments, but for others 'tis best to let them enjoy their own Thoughts. But to return to the Subject I was indeavouring to prove, namely, That the Metamorphosis of *Ovid* was a continued account of the Ages and Times of the duration of the Earth. I say, so far as I have gone, namely, to the end of the Golden Age, none will doubt but that this was the design of it, to relate what were the most celebrated Opinions concerning its Formation and first Ages, and as I conceive more particularly that of *Pythagoras*, who had spread and left his Doctrines in *Italy* long before *Ovid*'s time.

We come next to the 313 Verse where he begins to give an account, tho' very short, of the Youthful time of the Earth, which he calls the Silver Age. *Postquam Saturno tenebrosa in Tartara misso, Sub Jove Mundus erat, Subiitq; argentea proles*. After a long time was past and buried in Obscurity, the World had got a new Face and was under the Regiment of *Jupiter*, which signified the *Æther* and *Celestial Fire*; Before this 'tis said in the Golden or Infant Age of the World, *Ver erat æternum; placidiq; repentibus auris, mulcebant Zephiri natos sine semine flores*. The Air and Earth was moist and tepid, which made a continual Spring, but now that moisture is dried up, and fervour, heat and driness is got into the Air. *Subiit argentea proles,* now *Jupiter antiqui contraxit tempora veris, perq; hiemes æstusq; & inæquales Autumnos, & brene Ver, Spatiis exegit quatuor annum*. This ingress of *Jupiter* caused those strange changes in the Air, that we in part now feel; for 'tis not immediately the heat of the Sun that makes that difference in the heat of the Air, tho' that be also a Cause. But as I shall have occasion to treat in an other place 'tis the Constitution of the Air, nor is it the oblique Radiation (as all which one consent affirme) nor the nearness to, or distance from the Sun, but it is the ingress of *Jupiter* that makes the Air susceptible of these Mutations. *Tunc primum siecis aer fervoribus ustus canduit,* &c. then entered Lightening and extraordinary Heats; and so he proceeds in the description of the other Seasons and Constitutions of Air, *Semina tum primum longis cereolia Sulcis, obrnta* [sic]*sunt,* &c. The Earth being now dried having lost much of its Infant softness and moisture, needed some helps to make the Seeds grow. After this juvenile Age was past over, then *Tertia post illem successit a henea proles, Savior Ingenio & ad horrida prompior arma, non Seclerata tamen*. All the aforesaid Qualities increased, the Earth growing drier and drier, and the Air more intemperate, but yet it produced no direful Effects of terraneous or aerial Catastrophies. But *De Duro est ultima ferro*. Now the Shell of the Earth is Petrified, and the Iron Con-[381]stitution is introduced, all its Rocks and Iron Mines. *Protinus Erupit venæ pejoris in avum omne Nefas*. Then followed all the dismal effects of Subterraneous and Superterraneous Dissentions, Conflagrations, Floods, Earthquakes, the Sea overwhelming the Lands, and the Lands getting out from under the Seas, here Islands, there Lakes, here Mountains, there Voragoes and Abysses, and multitudes of other Confusions which rased and mangled the superficial Parts of the Earth, so that no place was free from the effects of these discordant Principles. *Astrea*, as I said before, which signified the Virgin, *Juvenile* smooth, soft, and even Face and Constitution of the Earth which it first received from the gentle Influence of the Heavens, and preserved in the Infant, *Juvenile* and pretty well in the Virile or brazen Ages. Now, that the Earth was arrived to its old Age, Wrinkles, Chops, Furrows, Scarrs, and the like, had not left one spot of *Astrea* unblemish'd, then she is said to have left it. This is a short account of this Iron or old Age of the World, of which I suppose the whole following Metamorphosis is written; this in good part falling within the Mythologick History of the Poets, but the Genesis and three preceding Ages, I look upon to belong to *Varroe's* Αδηλον *tempus*, and to be the Epitome of the Theories of the most antient and most approv'd Philosophers. This I could in part prove, as I could also many other Passages of this Discourse, by Quotations out of other Authors among the Antient, and also by the consent of many more Modern Writers. But that possibly might seem too tedious, and I

*Ovid's "Silver Age" = the young Earth*

*Golden Age = Earth's infancy*

doubt not but there are others who having more applied their Studies that way will do it more fully. The first of the memorable events of the Iron or old Age of the World is described in the next following Verses.

*Iron Age = Old age of the Earth*

> *Neve foret terris securior arduus Æther,*
> *Affectasse ferunt Regnum cæleste Gigantes*
> *Altaq; congestos struxisse ad sydera montes.*

> *But least high Heav'n should unattempted rest,*
> *Aspiring Thoughts the Giants Minds possest,*
> *Mountains they rais'd 'gainst the ætherial Throne.*

Now the dismal effects of the old Age of the Earth appear, the outward Shell of the Earth being now hardned and petrified, and the Pores of Emanation stoped so that the fiery and watery Vapours and Rarefactions below the same, could not now find their usual transits; these are said to conspire against Heaven to break out of that Prison of *Tartarus*, where *Jupiter* had lately thrust down and inclosed *Saturn, Saturno tenebrosa in tartara misso*, and to force their Passage into the open Heaven, where *Jupiter* now prevails; these therefore fermenting together had raised the subterraneous Parts into many Cavities and *Cryptæ*, and therefore were said to have a thousand Hands, being so many Caverns and far extending *Cryptæ*, wherein these subterraneous Sprits[sic] convened, in which lay their strength; and because such *Cryptæ* are winding and not streight, they were called *Anguipedes* like Snakes; these at last break forth and make Mountains, lay *Pelion* upon *Ossa, Altaq, congestos struxere ad sidera montes*. Then *Jupiter* is said to have rent the Heavens with his Lightning and to have buried them at last with Mountains heaped on them; that is, these Vapours having made Eruptions and thereby carried the Earth up with them, so as to make Mountains one of the top of the other, the Vapour got into the Air where it produced hideous Lightning and so spent it self in the Air, and the Mountains being left, and the Vapours that raised them spent, *Jupiter* is said to have destroyed them and buried them under those Mountains: One of these is said to be buried under *Sicily*, and to breath through the Mountain *Ætna*. But I must not stay too long upon the particular Explication of every thing concerning it, it may be sufficient for me at present to hint the meaning in general; only 'tis to be noted, that the Blood of these produced a generation that was of the same kind; that is, that the remainders in the Earth were of the same kind.

[p.382] These remainders of the first Effects were so prodigious that they made *Jupiter* groan and grow white hot with Anger, that is, made Thunder and Lightning, and call a Council of the Gods,

> *Terrificam Capitis concussit terq; quaterq;*
> *Cæsariem, cum qua terram, Mare, sydera movit,*
> *Talibus inde modis, ora indignantia solvit.*
> *Non Ego, pro mundi Regno magis anxius illa*
> *Tempestate fui, qua centum quisq parabat*
> *Injicere anguipedum captivo brachia Cœlo.*

> *The Thund'rer oft this dreadful Tresses shakes,*
> *At which the Heaven, the Earth, and Ocean quakes,*
> *And thus he his affronted Mind exprest.*
> *Not a more anxious thought my Mind possest*

> *For the Worlds Empire, when the captive Skies*
> *With hundred Hands the Snake-feet did surprize.*

It seems this was as great a Conflagration, or Collection of subterraneous Spirits, and like to be as dreadful as the preceding, nay greater, for that was but one single Enemy, but one small part to be destroyed; but now there is an universal defection, all must be destroy'd; for speaking of the last Eruption,

> *Nam quanquam ferus hostis erat, tamen illud ab uno*
> *Corpore, & ex una pendebat Origine Bellum.*
> *Nunc mihi, qua totum Nereus circumsonat orbem,*
> *Perdendum est mortale genus, &c.*

> *For tho' the first was a fierce raging Foe,*
> *From one Original the whole did flow,*
> *And all the War depended on one Head.*
> *Now whereso'ere the silver Waves are spread,*
> *I must destroy Mankind.*

And why must all this be? Why *Jupiter* being informed of this designed Conspiracy, coming down found *Lycaon* had laid a design to destroy not only the *Semidei, Fauni, Nymphæ, Satyri,* and *Sylvni,* that were the terrestrial Deities of the Plains, Rivers, Woods and Hills; but even *Jupiter* himself, who ruled the celestial Deities, the Æther, Air and Meteors, all which he had call'd together, who

> *Confremuere omnes studiis ardentibus.*

> *A Murmur rais'd with an inflam'd desire.*

But who is this *Lycaon*? Λυηαων, as the Word signifies, is Dissolution, the general Congregation of the Sulphureous, Subterraneous Vapours being every where pent in, threaten'd a general Dissolution and Catastrophy of the whole World at once, and so would not only overturn Hills, Plains, Rivers and Woods, but set on Fire and destroy the Air; for, as in another place he expresses it,

> *Vis fera Ventorum cæcis inclusa cavernis*
> *Expirare aliqua cupiens, luctataq; frustra*
> *Liberiore frui Cælo, cum Carcere Rima.*
> *Nulla foret toto, nec pervia flatibus esset,*
> *Extentam tumefcit humum: Seu Spiritus oris*
> *Tendere Vesicam solet.-----*

> [p.383]*Winds raging force within close Caverns pent*
>   *Desirous to break out at any Vent,*
>   *Long strives in vain t'injoy a freer Field*
>   *Of Air, the well-clos'd Pris'ns no Crannys yield;*
>   *At last it stretches out Earths hide-bound Shell,*
>   *As with strong Breath blown up tight Bladders swell.*

The whole Earth was big with these collected, subterraneous fiery Spirits and watery Exhalations.

--------*Partim ferventibus artus*
*Mollit aquis, partim subjecto torruit igne.*

--------Part soft with the boyling Waters, part
He roasts with Flames beneath.

*Jupiter* therefore descending destroys him *vindice flamma*, that is, fires into Lightning such as had broken out,

*Territus ipse fugit, nactusq; silentia Ruris*
*Exululat.*

Frighted, to dark and silent Groves he flies
In these he howles aloud.

This made the subterraneous Vapours fly to other places and make a noise under Ground, and in some places where it broak out, it had

-------*Veteris vestigia formæ:*
*Canities eadem est, eadem violentia vultus,*
*Idem Oculi lucent, eadem feritatis Imago*

He still the marks of his old Form retains:
The same gray Hair, the same stern Look remains,
The same Eyes stare with wildness still the same.

The same white tops of Mountains, the same gaping devouring Mouth, the same flaming Eyes, the *Caldera* at the top yielding Fire, the same frightful and terrible Aspect, like that of a devouring Wolf; and that this is the meaning of the shape of a Wolf which *Lycaon* is said to be transformed into, is more plain by what is said in the eleventh Book, Verse 365, of *Psamathes* being turned into a Wolf, where *Antenor* is introduced telling a story to *Peleus* of a devouring Wolf destroying Men and Cattle which had come out of the Sea: It will be plain to any that shall read it, that an Earthquake is there meant by the description of the Wolf, but I must not now insist upon it.

*Hooke interprets the whole legend of Lycaon turning into a wolf as description of an earthquake.*

But to proceed, there was yet but a stop put to some small *Vulcano* or Eruption which had destroyed but some small Country *de Gente Molossa*, some of which it had overflowed with Water, and destroy'd some other parts with Fire.

---------*Sed non Domus una perire*
*Digna fuit, qua terra patet fera regnat Erynnis.*
*In facinus jurasse putes.*

Thus one House perish'd by revenging Flame
Deserv'd by all, the Furies all possess;
You'd think the World conspir'd in Wickedness.

But this was not sufficient to vent these subterraneous imprison'd Spirits; but an universal Catastrophy was necessary, because *Erynnis* ruled over the whole Globe; *Jupiter* therefore is said to have considered which way to effect it, whether by an universal Conflagration by fiery Eruptions

> [p.384]*Jamq; erat in totas sparsurus fulmina terras,*
> *Sed timuit ne forte sacer tot ab ignibus Æther*
> *Conciperet flammas, totusq; ardesceret Axis.*
>
> *And now he just was ready to let fly*
> *His Light'ning, but he fear'd the sacred Sky*
> *Should catch the Flame, and Heav'ns whole Axis blaze.*

He concludes at last to do it by an Inundation.

But I must not dwell too long upon the Explication, which with this notion will plainly appear to him that reads the Poet's Description. Next this follows the Story of *Python*, which is nothing but the Corruption and ill effects of it from the Mud and Stagnations left by the Flood, which the Sun by its Rays by degrees destroys, drying it up. And the next of *Daphne* turned into a Laurel by *Apollo*, is nothing but the pleasant verdures the Sun produced upon the Earth, inriched by the Inundation after it was dried. I could proceed, but I fear I have already wearied you with this Recital, which was only designed as a Specimen to shew what I hinted the last Day, namely, That this Mythologick History was a History of the Production, Ages, States and Changes that have formerly happened to the Earth, partly from the Theory of the best Philosophy; partly from Tradition, whether Oral or Written, and partly from undoubted History, for towards the latter end we find accounts of many things our Histories reach, as *Orpheus*, the *Trojan* War, *Pythagoras, Romulus, Rome, Numa,* and it comes down even to the Death of *Julius Cæsar*, and the Reign of *Augustus*, under whom he lived.

*Hooke's interpretation of some of the classics must have pleased his Royal Society audience, many of whom were educated in the Oxbridge classical mode.*

**No. 12.** *A lecture read Feb. 15, 1688. According to Waller, confirming what the Author had said before as to Earthquakes and their Effects, but actually given in answer to critics.*

[p.403] I need not repeat what I have formerly said as to the several curiously figured Stones found in many parts, nay, I may say, all parts of the Earth, that they are really the several Bodies they represent, or the mouldings of them Petrified, and not, as some have imagined, *a Lusus Naturæ* sporting her self in the need less formation of useless Beings.

*Lusus Naturæ = sport of nature*

I shall only add some Confirmations of the Conclusions I then deduced from an *Hypothesis*, which I took the liberty to propose. And First,

As to the Sphæriodical Figure of the Earth, and thence of the Decrease of Gravity towards the Æquator instead of the Increase, as most of the followers of *Des Cartes*, Mr. *Hobbs*, and divers others of the Modern Naturalists assert; tho' it were at first much opposed, yet I find that it is now by divers not thought so improbable but that it may be supposed; and tho' I find the consequent Supposition as yet opposed, yet I question not but in time to make that appear to be necessary also; but every thing must have its time. As to that also which I have Published, how unlikely soever it may appear, I hope [p.404] also to be able to produce very good Arguments for it, and that it was not an Hypothesis proposed at random, as some may imagine.

*Hooke returns to the subject of the figure and shape of the Earth and that gravity decreases towards the Equator, ideas which have been opposed by some.*

But because these in themselves, tho' fully proved, were not sufficient to solve all the *Phænomena* of Nature as to the Disposition of those figured Bodies, whether Shells or other Substances; therefore in the Fourth place I laid down as a Supposition, that the superficial Parts of the Earth had been very much altered by *Subterraneous Eruptions*, whereby divers Parts that had before such Eruptions or Earthquakes been under the Sea, had been raised out of it and been made Islands; and that other parts that had been dry Land had been sunk into and covered by the Sea; that Vallies had been turned to Cliffs, and Hills to Vallies or Lakes, and the like.

*He reiterates: Fossils prove that the surface of the Earth had been altered by subterraneous eruptions or earthquakes, so that land and sea areas have exchanged places.*

This was likewise opposed and thought very improbable, because for so long time as our History will reach backwards, it was affirmed there had happened no such change; and therefore, because it was supposed no such History could be produced, this also was to be rejected, and we must again have recourse to the *Lusus Naturæ* as the only expedient to give satisfaction; only some kind of Subterraneous Passages were thought of, by which Oysters and other Fish might be conveyed to the middle of the *Alps*, going along with the Stream of the Water from the Sea to supply the Springs and Fountains at the top of the Hills.

*To this idea there was also opposition.*

I confess it seemed to me a little hard, because I could not give the Pedigree of the Fish, therefore I should not be allowed to believe it a Fish, when I saw all the sensible marks of a Fish; and that, because I could not tell who it was, or upon what occasion that caused the Stones on *Salisbury Plain* to be dispersed in that irregular Regularity, that therefore I must allow them to be a *Lusus Naturæ*, or placed there by *Merlin* or, some such unknown way, and not by the Hands, Labour or Workmanship of some such Men as are now living. Nevertheless that I might, as far as I was able, satisfy these Objections also, I produced the History of *Plato* as brought out of *Ægypt* by *Solon*, concerning the Island of the *Atlantis*.

*Hooke's tone is rather sarcastic here in mimicking the attitude of opponents of his ideas. See Chapter 6 for an interpretation of this and similar passages.*

But this tho' related by *Plato*, with all the Circumstances, as if he believed it a true History, was yet supposed to be only a Fiction of *Plato* to lay the Scene of his Common-wealth, or at best a Fable of the *Ægyptian* priest to magnify the knowledge of the *Ægyptians* as to the History of preceding Ages. I confess the account of the nine thousand Years is Argument enough to make the whole History to be suspected as a Fiction; but yet till we are certain what space of Time is there signified by a Year it will be a little hard to reject the whole for that Circumstance, since most of the other Circumstances of it are more probable.

And that they were thought so by divers of the Ancients, is plain from several Testimonies that might be alledged; I shall only mention what *Strabo* says of it in his second Book, where examining whether *Eratosthenes* had duly amended the της οικουμενης πινακα of the Antients he adds, that το δε εξαιρεσθαι την γην ποτε, &c. *Eratosthenes* (says he) has done well in expounding the manner how the Surface of the Earth may be changed, by relating how the same may sometimes be raised, and sometimes be sunk by Earthquakes, and various otherways changed, as we also have in many particulars Enumerated, by which also he hath properly shown how the History, which *Plato* relates concerning the Island of *Atlantis*, as it was brought out of *Ægypt* from the Priests there, by *Solon*, may be well believed not to be a Fiction, but a true History. From which Passage it is plain, that both *Strabo* and *Eratosthenes* did look upon this History of *Plato*, or rather of the *Ægyptians*, as very probable; *Pliny* also was of the like Opinion, as appears, not only by his mentioning the History of *Plato*, but by the several other Mutations, which he relates to have been made by the means of Earthquakes.

But because the Scene of this Tragedy of *Atlantis* was placed very far backwards in times remote, and that we have not other History of this change but what *Plato* is pleased to relate, I did therefore indeavour to produce some History concerning the changes that had happened since that time, namely, within the reach of the Greek Histories, in the same place where this *Atlantis* [p.405] was said to be sunk down into the Sea. For this I produced the History of the *Periplus* of *Hanno*, the *Carthaginian*, as it is set forth by *Berkelius*; from which I collected that at the time when this Expedition was made, the place where the *Atlantis* was said to be sunk, was found to be partly Sea and partly Islands, and that the same extended, as I conceive, as far to the Westwards almost as the *Madera* Island, about which place was then found a small Island called *Cerne*; from which Coasting Southwards for twelve Days, they found the Land all on Fire, and one prodigious high Mountain flaming out at the top, called *Theon Ochema*, or the Chariot of the Gods; which, by all Circumstances, I conceived to be the same with the now Pike of *Teneriff*. Supposing which Relation true, I deduced thence that there must needs have happened great changes in those Parts between the time of this Expedition and the present; for that all those places which seem to be described by that History are not now to be found in the places where they are by that Relation placed. But this Relation was also looked upon as fabulous, because I produced no other Authority for it besides the Relation itself and the Testimony of *Pliny*. But those who are better read in Ancient History, may find that it was by most of the Ancients supposed real; and all agree that the *Phœnicians*, of whom the *Carthaginians* were a Colony, were very skillful in Astronomy, Navigation, Arithmetick and Traffick, and that they were

---

*This passage has been chosen by some people as indicating that Hooke adhered to the biblical chronology—i.e., 9,000 years is patently out of the question—too long. But if read carefully, one can conclude that he believes the "Circumstances" as probable. Also, passages elsewhere unquestionably deny the strict biblical chronology. Clearly he wanted to believe Plato's narrative as a true history and not as a fiction. To believe that Hooke adhered to the biblical chronology is to make his entire study of the classics an exercise in futility See Chapter 6.*

*Eratosthenes had amended the pictures of the world (Greek translation) of the Antients. The rest of what Eratosthenes says is translated in the text.*

*Strabo, Eratosthenes, and Pliny all believed this history of Plato—i.e., the disappearance of Atlantis—as very probable.*

*As supportive evidence Hooke had earlier produced the history of Periplus of Hanno, the Carthaginian.*

the first introducers of these among the Greeks, together with the Knowledge and Use of Letters. And from those Particulars I noted in the said Relation, it seems to me very evident, that they understood what Longitude and Latitude was, and knew how to keep account of their Course and Distance; and tho' the Interpretation thereof which I produced, be differing from all the other I have yet known: Some supposing it to relate wholly to the Coast of *Africa* as it is shaped, at present, and others in other Situations, yet whoever (taking this Notion of their skill in Astronomy and Navigation along with them) shall strictly examine the Relation itself, he will, I conceive, be persuaded to be somewhat of my Mind. *Strabo* therefore says of them. *The* Sidonians *were reported to be good Artists in various things, as it is also manifest by their Actions; as also, good Philosophers, Astronomers and Arithmeticians, and such as well knew the secret of Numbers and of Sailing in the Night also.* From which consideration I conceive, that κατ ευθν in Relation can signify nothing else but the same Parallel of Latitude, or the same straight Line with that from *Carthage* to the Mouth of the *Streights*.

> *The credentials of these ancient people are very respectable, and it follows that they should be credible.*

> *Greek translation = straightforward*

I was not then able to quote the place in *Aristotle* which relates to this discovery of the *Carthaginians*, tho' I was well assured I had met with such a Relation; 'Tis in his Book, περι θαυμασιων ακουσματων, in these Words, εν τη θαλασση τη εξω Ηρακλειων στηλων φασιν υπο καρχηδινιων υησον ευρεθηναι ερημην εχουσαν υλην τε πανταδαπιην και ποταμους πλουτους και τοις λοιποις καρποις θαυμαστνω απεχουσαν δε πλειονων ημερων δαυμαιω, &c. "In the Sea that lies without the Pillars of *Hercules*, they say, that, by the *Carthaginians*, there has been discovered an Island deserted, but abounding with variety of Woods, and rich Rivers fit for Navigation, abounding also with variety of Fruits, distant from the Continent several Days Sail. *Pomponius Mela* also mentions the extream *Atlantick* to be inhabited by a wild sort of People, which he calls *Ægypanes*, *Blemmeæ* and *Gamphasentes*, and a kind of *Satyrs*. And in his Third Book and tenth Chapter, he says, *Ultra hunc sinum mons altus ut Græci vocant* θεων οχημα *vehiculum Deorum perpetuis ignibus flagrat*, &c. *Diodorus Siculus* also mentions some such Island in his Fifth Book την λιβυην χειται μεν πελαγια νησος, &c. That is, "Against *Libya* there is situated an Island in the Ocean, considerable for Magnitude, several Days Sail distant towards the West, &c." I could cite several other Authors who mention some such Place, which may have relation to the discoveries of the *Carthaginians* in *Hanno*'s Expedition; all which do plainly make it appear, that his discoveries were towards the West, and not towards the South; and therefore it seems very probable, that at that time there were Islands both greater and smaller to the Westwards of the Streights Mouth, which are not now to be found, and consequently they must have suffered a Submersion by some intervening Catastrophies, which was the thing I indeavoured to deduce from it. Nor will it seem so unlikely if we will [p.406] but consider the alterations by Eruptions out of the Sea near the Islands of the *Canarys*, and in one of the Islands also within these few Years; the former was the Eruption out of the Sea, in the Year 1639, which *Athanatius Kircher* has given a description of, and I have received a relation of it by Word of Mouth from two Persons who were both upon the *Tenariff* Island at the same time, and had each of them often observed it tho' at a considerable distance: The latter Eruption happened within a few Years since in one of the same Islands, of which I have Printed the Relation in one of my Collections.

> *Greek translation = concerning reports of wondrous things*

> *The translation of the rest of this Greek passage as provided by Hooke includes the words "fit for Navigation" referring to the rich rivers, but this phrase is not in the Greek words.*

> *Greek translation = chariot of the gods, or in Latin,* vehiculum Deorum

> *The last word of this paragraph, "Collections"* = Philosophical Collections *published by Hooke 1679–1682 while he was secretary of the Royal Society. The* Philosophical Transactions, *which were supposed to be carried on by his co-secretary, Nehemiah Grew, had suspended publication during this period, after the death of Henry Oldenburg.*

And thus, I hope, I have given some ground to believe that the Antient Historians knew and gave some Credit also to both these Relations, namely, that of *Solon* and that of *Hanno*, which, by that Passage of *Aristotle*, appears to have been made either before or in his time.

I came in the next place to shew that the *Metamorphosis* of *Ovid* contained many Histories of great Changes and Catastrophies that had happened long before his time to the parts of the Earth; which, tho' rapped up in Mythology and Mascarade, yet those disguises being removed, it will not, I conceive, be very difficult to make appear what the true Histories are, which now pass Incognito. To this purpose I did observe that *Ovid* has in some part or other of his Fable, given Marks or Characteristicks by which it may be found what the History is which he doth there Mythologize; this he doth very often in that part which serves as a Link to join the Story into a continued Chain or in the Etymology of the Names, tho' often times also in the process of the Poem. And 'tis usual with him all along to have and mix a treble Design in each of them, namely, an Historical, a Physical, and a Moral; and this he hath done with great Judgment and Subtilty of Invention, and upon several occasions, he makes Excursions into this or that Design, and prosecuting it for a time, as if he had no respect to the other two; but yet, if well examined, it will be found, I conceive, in most that he influenceth, even there also, that design by the other two. If I had leisure to prosecute this Speculation, I conceive I could trace most of these his Designs, but it would be too long a Work; however, I hope in a short time to be able to give several Instances and Examples out of those I have more attentively examined, which may suffice to shew that there is a probability in this Conjecture how differing soever it be from other Commentators.

*He repeats again that Ovid's* Metamorphosis *contains real histories of terrestrial changes, though couched, "wrapped up," in mythology.*

But those perhaps may not be thought sufficient for the present Dispute to satisfy such as demand positive and direct Histories, and undoubted Records of such Changes, as I have supposed necessary to make out the Hypothesis of the figured Bodies, which are found to be real Shells, &c. and to have been by them disposed and situated in the places where they are now found: For such therefore I shall prepare a Cloud of Witnesses, which, unless they will deny all History, will stand the Proof.

I confess I cannot see any Circumstance in the Story of *Hanno* that should render it suspected, since 'tis granted by all, that the *Phœnicians*, of whom the *Carthaginians* were a Colony, were so early eminent in Arts, especially in that of Navigation and Traffick; so that we find *Solon* made use of them; and that *Sanchoniathon Beritius* before the time of the *Trojan* war, did write the Theology of the *Phœnicians* (as *Porphyrius* relates) in the *Phœnician* Language: And that the Philosophy of the Greeks was derived principally from them; as also Astronomy, and even the knowledge and use of Letters. For *Thales* was a *Phœnician*, and *Pherecides* who was the Master of *Pythagoras*, and the founder of the Italick Phylosophy, and co-equal with *Thales* learned it out of the occult Books of the *Phœnicians*. And from *Pythagoras* his Philosophy sprang and flowed both the *Platonick* Philosophy, and also the Philosophy of his Scholar *Aristotle*, tho' somewhat altered by the Pipes it ran through. So that tho' we have but very little of the History of those Times, yet by those few Fragments dispersed here and there, we may be sufficiently satisfied they were able, and actually did make as great Voyages and Discoveries, as that of *Hanno*; of which there are divers Relations mentioned by *Herodotus* in his Second Book.

*The Phœnicians enjoyed a very advanced state of knowledge and civilization that were passed down to the Greeks.*

**No. 13.** *This Lecture was read February 22, 1688, although Waller records February 15.*

[p.407] But tho' (notwithstanding what I have alledged) all these Histories shall be looked upon as Fictions and Romantick without any real Ground, yet what I have indeavoured to shew by Experiment and Inspection, and the deductions made therefrom, will not be found destitute of good Authority, proved from very eminent Authors both Antient and Modern, to make out the Truth and Certainty thereof. As first to prove, that those Bodies were found at the tops of Mountains, and that they were notwithstanding asserted to be Shells; we have the Testimony of *Herodotus* in his *Euterpe* or second Book, and twelveth Section, where speaking of the Country of *Ægypt*, as having been mostly raised by the Mud and Sand of the *Nile*, he says, the whole Country was of such a Soil, only the Mountain above *Memphis* was Sandy, and had *Conchilia* or Fishes Shells upon it, and abounded with Salt, so that it corrupted the *Pyramids*. Which Passage is very pertinent to my present purpose, and is also fully confirmed by *Aristotle*: For it seems all the lower *Ægypt* was a Plain, which had rose by the settlement of Mud of the *Nile*, which he says, in the space of nine hundred Years, had been raised eight Cubits, or twelve Foot; for that eight Foot rise of the *Nile*, in the time of *Myris*, overflowed all *Ægypt*, and in his time there was necessary sixteen Cubits, or twenty four Foot swelling to overflow it. So that he seems to understand, that all the lower *Ægypt* had been at first Sea, and that the *Nile*, by degrees, had filled it up to the height of the Plain, and so had covered all the bottom or Sand of the Sea, only the Mountainous part above *Memphis* was above that level, and so that kept its old bottom or covering of Sand and Shells. This seems to be the meaning of what he argues for; but yet I must needs say, that does not solve all the Difficulty; for how comes this Mountain to be so much higher than the Plain, as it was then raised by the *Nile*, and thence that Plain to be much above the Sea that was thereby excluded, unless we do suppose also that some Subterraneous Power did raise that Mountain above the level it was of when covered by the Sea, or that the Sea had sometimes been so high as to cover that Mountain, which, tho' *Herodotus* takes no notice of, yet *Aristotle* does fully solve it; but be it which way soever, 'tis Testimony enough of the matter of Fact, that *Herodotus* himself calleth them κογχυλια, and observed the exudation of the Salt. But this Place is also observable for another Passage, and that is to confirm Conjectures I formerly acquainted the Society with, concerning the *Pyramids of Ægypt*, namely, that I conceived them to be *founded on and Ashler'd, as it weere about a Core of Rock*, and by this Discourse it is plain, that the place where they stand is so qualified; for by this Description 'tis plain, that it is described as Rocky, and covered with Sea Sand; for that *Herodotus* takes notice it had both Sea-shells, and Salt mixed with it; but this here only by the bye. Before I leave *Herodotus* his Testimony, I cannot but take notice of another Passage in the same Book, in the 74th. and 75th Paragraphs; the translate runs thus, *Circa Thebas sunt Sacri Serpentes Nihil omnino hominibus noxii, pusillo corpore, binis præditi cornibus e summo vertice enatis, quos defunctos in Jovis Æde sepeliunt, huic enim Deo Sacros illos esse prædicant;* thus far is the story he is told by the *Ægyptians. Est autem Arabiæ Locus, ad Butum urbem fere positus ad quem Locum ego me contuli quod audirem volucres esse Serpentes. Eo cum perveni, Ossa Serpentum aspexi & Spinas multitudine Supra sidem ad Enarrandum quarum acervi erant magni, & his alii atq; alii minores ingenti Numero. Est autem hic Locus ubi Spinæ projectæ Jacebant hujuscemodi, ex Arctis*

*The deposition of the Nile had converted what once was sea to land, and the sea sand had been covered by the mud of the Nile.*

*But this process does not explain the presence of the mountainous part above Memphis, which then must have been raised as a result of subterraneous power.*

*Greek word = seashells*

*Testimony of Herodotus*

*montibus exporrigitur in vastam planitiem Ægyptiæ Contiguam. Fertur ex Arabia Serpentes alatos, ineunte statim vere, in Ægyptum volare, sed eis ad ingressum planities occurrentes aves Ibides, non permittere sed ipos interimere, & ob id opus Ibin magno in honore ab Ægyptiis haberi, Arabes aiunt.*

Part of this story is what he was told, part what he saw; he was told of flying Serpents, which the Bird *Ibis* met over that Valley, and so devoured them leaving only the Back-bone. I have heard many stories told of our Snake-stones or *Cornua Ammonis*, and I have seen some to confirm the story with a very formal Head carved on them; I think not long since here was one shewed in this place; I am apt to think the Spines of the Serpents *Herodotus* there found in such plenty and such variety of bignesses, were no other than those *Cornua Ammonis*, and thence, I conceive, proceeded the superstitious [p.408] Custom when they found any of these Stones or Spinæ they carried and buried them in the Temple of *Jupiter Ammon*, and it seems to me a farther confirmation of what I formerly hinted concerning the Stone adorned with Jewels, and carried in processions by the Priests of that Temple, mentioned by *Pliny* and several others. But to proceed, to this Testimony and Opinion of *Herodotus*, I shall add that of *Pythagoras*, as related by *Ovid* in his 15th Book of the Metamorphosis, *ver.* 262, and so onwards. *Vidi eqo quod fuerat quondam Solidissima tellus, esse fretum. Vidi factas ex æquore terras: Et procul a pelago conchæ Jacuere Marinæ. Et vetus inventa est in montibus anchora summis.* From which Testimonies 'tis plain that this Phænomenon of Shells was taken notice of by the Antient Historians and Philosophers: And I am apt to think that this might, in some measure, spread among them, the Notion of general Deluges that in preceding Ages had happened, as *Pythagoras* seems to hint in this place, by supposing them to happen after a certain long Revolution of time; and that *Thales,* and many others, supposed that the Principle, from whch all things sprung, was Water: And that the Passage in the Fragment of *Sanchoniathon*, where he speaks of the first original of all things, says, That in the *Phœnician* Language it was called μωτ, which possibly may be much of the signification of מנע in the Hebrew, which signifies Motion; that is, Fluidity, for so the interpretation of *Sanchoniathon* seems to make it τουτο πινεζ φασιν ιλυν οι δε υδατωδουζ μιξεωζ σηφιν και εκ ταυτηζ εγενετο τασα σπορα κτισεωζ και γενεσιζ ολων. Which some will have to be Mud, others the Corruptions of watery mixtures (as if μωτ were derived from מות, *Mors* Death or Corruption) from which sprung the Seeds of all living Creatures, and the Generation of all things. That the *Ægyptians* threw the History of the Flood so far backwards, and make it so differing from the Chronology of the Bible, I take it to be for no other Cause but to make the World believe they were preceding to all others in Antiquity of History and Chronology: To which purpose *Herodotus* tells a pleasant Relation of *Psammiticus*, that the *Ægyptians* before his time had vaunted themselves to be the first People upon the Face of the Earth, but he having a mind to be informed of this by Experiment, caused two Children to be bred up in a Desert Place by a Shepherd, so that they should not hear any Language at all spoken, to the end to see what Language they would naturally speak of themselves; from hence he supposed they would speak the first and most Natural Language. This having been done, and the Children grown two years old, the Shepherd opening the Door of the place where they were so kept and fed with Milk, they both reached out their Hands to him crying *Beccos*, which the Shepherd taking notice of acquainted *Psammiticus* with,

*It is interesting to learn the origin of the word "ammonite"—i.e., named after Jupiter Ammon. Ammon or Hammon was actually a Libyan deity but was worshiped at Rome under the name of Jupiter Ammon.*

who inquiring what that word might signify in any Language, was informed that *Beccos* signified Bread in the *Phrygian* Language; from which time *Herodotus* says, the *Ægyptians* lost their seniority, and granted the *Phrygians* to be the first and themselves the second People for Antiquity. So that tho' their account of Years may be hence supposed to be uncertain, yet their Learning and their lasting Monuments of their former greatness, namely, the *Pyramids, Obelisks, Colossi, Labyrinths*, and the like, shewed them to have been long before *Herodotus* his time very considerable for Arts and Literature. And that they had some Records of a preceding Flood, I have before mentioned, whether the same with that of *Noah*, or some more particular Flood, which those of *Deucalion* and *Ogyges* seem to have been, I leave to the learned Antiquaries to determine. I could produce several other Testimonies to shew they had the notion of a Deluge.

*The ancients therefore had the notion of a deluge, but whether this was the same as Noah's flood Hooke cannot say.*

But it seems to me very improbable, that these Shells should have been the effect of *Noah*'s Flood by reason of its short duration; which was not long enough of continuance to produce and perfect those Creatures in so short a space to the bigness and perfection they seem to have had. It must therefore have been either some particular Floods of a longer duration; or else the places where they are found, must have been some times or other the bottom of the Sea, and afterwards raised by Subterraneous Motions, Swellings, or Eruptions: Which, whither those just immediately preceded the end of the general Deluge in the time of *Noah*, and that That part which before the Flood was Land, did sink, and became covered by the Sea, and those [p.409] parts which were before under the Sea, did, by degrees, towards the determination of the Catastrophy, rise and swell up into Land, Hills and Mountains, I leave to the Learned to determine. Certain it is, that there were some very great changes of the superficial Parts of the Earth at that time; since it is said, that all the Fountains of the Deep, or Abysse, were broken up; the Scripture renders ερραγησαν πασαι αι πηγαι της αβυσσου και οι καταρρακται του ουρανου ηνεωχθησαν. There seems by the Expression to be a twofold supply of Water to cause this Flood, the one by the opening of the *Stereoma* or Firmament in the middle of the Waters, *viz.* that of the στερεωμα εν μεσω του υδατος, and of the στερεωμα του ουρανου, the gathering together of the Waters above the former στερεωμα; but υποκατω του ουρανου, these God called the Sea, and the parts of the Earth that were uncovered thereby he called dry Land. The *Stereoma* του ουρανου is assigned for the place of the Stars, Sun and Moon, &c. and has always the Epithet of του ουρανου joined with it. *Ovid* likewise who seems to allude in some measure to this History of the Creation delivered by *Moses* in the first Chapter of *Genesis*, says, *Circumfluus humor ultima possedit solidumq; coercuit orbem.* I think, were it proper to the present Subject, I could give a very plausible account concerning the manner of that Deluge, as it is expressed by *Moses*; tho' it differ from all that I have yet met with, yet I can prove it warranted both by the Text and by genuine Physical Principles; but it would be too long a digression for the present Subject, and I shall shortly have a more proper opportunity to demonstrate the inner Parts and Constitution of this Globe, my present Business being to explicate the Phænomena of the outward and superficial Parts, and to prove that the Bodies, which I have asserted to be Shells, have been so reputed by the Antient as well as the Modern Historians: Next to shew that they are, or have been, found in most Parts of the World. And if those two be proved, then will necessarily follow that there must have been some time

*Here Hooke again questions the duration of Noah's flood—as being insufficient to produce the ammonites, and therefore they must have been the result of a much longer flood or floods, or else the places where they are found were once the bottom of the sea. The latter hypothesis is, of course, what Hooke adhered to.*

or other such Catastrophies, Metamorphoses, or Mutations as must have caused those parts, which were once the bottom of the Sea, to be now, or at the time when they were so observed, to be dry Land.

*Some parts of the Earth that had once been the bottom of the sea is now dry land.*

That they were so esteemed by divers of the Antients I have in part shewn and could inlarge upon that Head, but that I would likewise shew that they have been so esteemed by the most eminent of our Modern Naturalists, and for this I could produce the Testimonies of *Georgius Agricola, Cardan, Gesner, Aldrovandus, Ferranti Imperatus, Wormius, Calceolariius, Bauhinus, Belonius, Fracastorius, Cisalpinus, Fabius Columba, Stevinus*, and a great many others yet more Modern, besides the Testimony and Opinion of divers others, who have themselves declared their Judgment by word of Mouth; but this would be too great a wasting of Time to prove that which carries in itself the true *Medium* of its Proof and Demonstration, which is by sensible examination. I shall therefore only give one Instance or two for all, and that is, First, That of *Fabius Columba*, who has writ a Treatise on purpose to evidence this Truth by many Arguments; 'tis at the end of his Treatise *De Purpura*. *Nituntur quidam* (says he) *acanis Naturæ in medium adductis, omni Responsione Seclusa, Linguas Serpentinas aut Glosso-petras; quia non Solum mari, proximis & insulis, sed etiam Longe dissitis, copiose Reperiri traduntur, ab ipsa formatrice natura, sic genitas atq; Lapideas esse: Vel qui dentes esse dicunt, non Carchariæ, Lamiæ, Malthæ aut ejusdem generis Cætaceorum, sed illis similes sponte sic ortos Quin etiam id tantum Naturam produxisse eo loci, quod ratione materiei aptum erat ad formam illam Recipiendam affirmant. Hoc Arbumento in dubium Revocare videntur, an unquam Locis illis mare fuerit,* Quod probatissimi antiquiores Philosophi & Historici affirmarunt. Nos *quidem dicimus hujusmodi concretionem* non *esse* lapideam, *ex ipso* aspectu, Effigie *rei, ac tota substantia: Ac neminem censemus tam Crassa Minerva Natum, qui statim primo* intuitu *non affirmarit* Dentes *esse* Osseos *non lapideos, Sed præter* aspectum *omnia quæ Ligneam, osseam, & Carneam Naturam habent,* Ustione in Carbonem *prius abeunt, quam in Calcem, aut cinerem. Ea vero quæ tophace vel Saxea Sunt Natura, non in Carbonem sed in Calcem abire, nisi liquuntur propter vitream aut Metallicam mixtionem. Cum igitur hi* dentes *statim* assati *transeant in* Carbonem & *tophum adhærentem minime, clarum crit osseas esse dentes non Lapilleas.* He hath many other Arguments to confirm this Truth, [p.410] which would be too long to trouble you with at this time, and I only proceeded so far that I might give an occasion to have the Experiment now tried in the presence of this Society, there being several of that kind in the Repository, and the trial being very easy will not be long in making, all things being in a readiness for it. I shall only add one more Testimony, which is of *Andreas Cisalpinus* in his first Book *de Re Metallica* and second Chapter. *In fodinis metallorum seu Marmorum* (says he) *aliorumq; Saxorum Nunquam vivens Corpus Reperitur. Etsi enim aliquando in eorum Cæsura ostrearum testæ aut cætera Conchilia Reperta sunt; hæc recedente Mari & Lapidescente Solo inibi derelicta in Lapides concreverunt. Ubiq; enim ubi nunc est arida aliquando affuisse Mare testatur Aristoteles. Hoc enim modo Censere magis Consonum est rationi, quam putare vim animalem, intra Lapides, rudimenta animalium ac plantarum Gignere ut quidam putant.* He hath not told where *Aristotle* hath maintained this Doctrine; but whosoever shall examine his Writings, shall, by many Passages in them find, that he was fully of this Judgment: And more fully in the fourteenth Chapter of the first Book of *Meteors*, where also he confirms the same Sentiments of *Herodotus*, which I have

*This thesis has been supported not only by ancients but also by naturalists of more modern times.*

*Clearly, Hooke was extremely well read in spite of his many responsibilities and activities. Unlike several of his contemporaries and others of later generations, he always provided the source of his information and gave full credit to others whether it was an idea, a passage from a book, or a fossil specimen found.*

newly quoted; and concurs likewise with the Doctrine of the alterations that are caused by slow degrees of Progress, which I have Hypothetically explained by the Oval Figure of the Earth, and the alteration of that shorter Axis to differing parts of the Earth: But this only by the bye to shew how much soever that Hypothesis were exploded by a learned Dr., by reason of the Consequences that would follow from it; yet *Aristotle* (though he hath not explained by what means and in what manner) hath asserted Mutations as great and much after the same method of Progression, as those which are alledged to be the *extravagant Consequences of that Hypothesis*, some of which I shall have occasion to mention at another time. So that to conclude for this time I hope I have shewn good grounds to evince. First, That these kind of Bodies are either Animal or Vegetable Substances. Secondly, That the places where these are found must have sometimes been covered with the Sea. Thirdly, That the general Deluge of *Noah* was not of duration enough to effect it, unless the manner of its effecting were after that which I proposed, by changing that part which was before dry Land into Sea, by sinking; and that which was Sea into dry Land, by rising underneath it. Fourthly, That the universality of the *Phænomena* over the whole Earth seem to argue for this manner. Fifthly, That there have been several particular Floods, as that of *Deucalion*, the *Atlantick*, &c. which being caused, for the most part, by Earthquakes, may have been the causes of divers particular *Phænomena*, such as the raising of some parts from under the Sea, and the sinking of others into it, or into Lakes.

*Hooke advocates changes in the earth's surface by "slow degrees of Progress"—another indication that he was not solely a catastrophist. His axial shift idea was attacked and ridiculed, in Hooke's words, "exploded by a learned Dr." This learned Dr. was Wallis. See Chapter 5 for an account of this dispute.*

*He provides here a brief summary of his hypotheses.*

**No. 14.** *Read February 29, 1688.*

[p.410] I have, in my former Lecture, proved how early and how generally the Phænomena of Shells were taken notice of by the most antient Historians and Philosophers, and I could have given many other Instances to confirm it, if it had been thought necessary. And thence, I conceived, might be continually revived the Traditions and Theories concerning preceding Floods and other Catastrophies that had happened to the Earth in Ages long preceding. But because, among the Philosophers, I only quoted that place in *Plato* about the *Atlantis*, which was thought to be a Fiction (however that shewed he had such a notion) and the Doctrine of *Pythagoras* as reported by *Ovid*, which was thought Poetical (tho' as I conceive all those Mythologies have certain Historical and real Truths thereby represented) I shall therefore add one out of *Aristotle* which I hinted the last Day out of his first Book of Meteors. "The same parts of the Earth (says he) are not always dry or moist, but they receive a change from the increase or defect of Rivers; therefore parts bounding Sea and Land change often, nor is the same part always Sea or always Land, but is changed in time, and that which was Sea is Land, and that which was Land is Sea; but this is in a long process: This arrives from interior changes of the Earth, which from a long Constitution grows old, as the Bodies of Plants and Animals, and that not singly the [p.411] Parts but the whole. It may therefore for a time be moist, and by degrees grow dry and old: This may happen both by the decreasing of Rivers and also of the Sea; but these happen not, but in a long time, in comparison of our short Life, which is the cause they are not noted, the change being so little in the space of one Life, and so several Ages pass before they are finished; whence the memory of them is lost. (He adds much more to the same effect to explain his Notion) Exemplifying his Doctrine by *Ægypt*, which (says he) has been all made by Mud of Rivers, and is observed continually to grow drier, and the Lakes filling up by degrees have been inhabited, and length of time has obliterated the memory of such changes; for all the present Mouths of *Nile*, except the *Canobic*, have been cut by Art; for old *Ægypt* was that only about *Thebes*, as *Homer* testifies, who lived not long after these Changes; for he mentions *Thebes* only as if *Memphis* had not yet then been, so he proceeds in explicating, and instances again in saying, so marshy places grow better by draining, but dry grow worse and barren, as it has happened to the Country about the *Argives*, and the *Mycenæans*, and what has happened to these parts, the same may be conceived of the whole. So many parts which have been Sea have been added to the Continent; and the contrary, those that do *respicere ad pauca*, ascribe these changes to the Heavens, but they are mistaken; but they are to be ascribed to Causes that happen after a long process of time, as that of *Deucalion*'s Flood, which happened only to *Greece* about that part which is called old *Hellas*, which is that about the present *Dodon* and *Achelous*; this happened from great abundance of Rains which are generated by the Mountains, which are by degrees changed and so produce differing effects." He exemplifies his Doctrine further by *Ægypt*, and the Country where the Oracles of *Jupiter Ammon* was, saying, "'Twas formerly Marshy, but by degrees dried and grew parched; so not only the present most famous Rivers will come in long process of time to be dried and changed, but the Sea also; and that which was Sea will be Land, and the Land will be Sea." I have here given the sum of his Doctrine which he doth much inlarge upon to explain it; but to save time, I have only abstracted the meaning, and given you the Epitomy of it that may easily enough be more fully explained or read at

*Quoting from Aristotle to support the exchange of land and sea areas idea.*

large in the fourteenth Chapter of his first Book of *Meteors*. By the whole it plainly appears, that *Aristotle* was of the Opinion that all the dry Land of the Earth had been sometimes covered with the Sea, which he seems to be informed of by the then present *Phænomena*, as he plainly expresses in his description of *Ægypt*, and of the Country about the Oracle of *Jupiter Ammon*; and 'tis not to be doubted but one of those *Phænomena* and possibly not the least considerable was that of the Sea-sand and Shells, which I shewed the last day *Herodotus* had taken notice of. I do therefore humbly conceive (tho' some possibly may think there is too much notice taken of such a trivial thing as a rotten Shell, yet) that Men do generally too much slight and pass over without regard these Records of Antiquity which Nature have left as Monuments and Hieroglyphick Characters of preceding Transactions in the like duration or Transactions of the Body of the Earth, which are infinitely more evident and certain tokens than any thing of Antiquity that can be fetched out of Coins or Medals, or any other way yet known, since the best of those ways may be counterfeited or made by Art and Design, as may also Books, Manuscripts and Inscriptions, as all the Learned are now sufficiently satisfied, has often been actually practised; but those Characters are not to be counterfeited by all the Craft in the World, nor can they be doubted to be, what they appear, by any one that will impartially examine the true appearances of them: And tho' it must be granted, that it is very difficult to read them, and to raise a *Chronology* out of them, and to state the intervals of the Times wherein such, or such Catastrophies and Mutations have happened; yet 'tis not impossible, but that, by the help of those joined to other means and assistances of Information, much may be done even in that part of Information also. And tho' possibly some may say, I have turned the World upside down for the sake of a Shell, yet, as I think, there is no one has reason for any such assertion from any action I have hitherto done; yet if by means of so slight and trivial Signs and Tokens as these are, [p.412] there can be Discoveries made and certain Conclusions drawn of infinitely more important Subjects; I hope the attempts of that kind do no ways deserve reproach, since possibly 'tis not every one that takes notice of them, nor one of a hundred that does, that will think of a reason; besides, much greater conclusions have been deduced from less evident and more inconsiderable Marks, if we respect Bulk, Magnitude, or Number, and much more weighty Consequences may, and will in time, be drawn from seemingly more trivial, and much lighter and slighter Indications, yet where the Testimonies are clear, certain and self-evident, they are not to be rejected for their bulk, tho' it be so small as no Eye or Sense can reach it unless assisted by Engines, as the Sight by a Microscope, Telescope, and the like: In how few Letters, Words, or Characters is the History of the World before *Noah's* Flood? Is it therefore not to be believed because we have not as many Volumes of its History as there are now to be found words? In how little room will the History of the Flood be contained if *Homer's* Iliads could be boxed in a Nutshell? But to leave every one to the freedom of his own Thoughts, I shall proceed to what I thought was further necessary to be added to what I hinted the last Day, which was concerning the Flood of *Noah*, because I find the generality of those who indeavour to give a solution of these *Phænomena*, are inclined to ascribe them to the effects of that Flood, and because what I then said was but in brief, and so possibly what I design might not be so plainly apprehended, or it may be misconstrued, I thought it might be necessary to explain it a little more fully. I said then, that I conceived that those universal *Phænomena* of the remainders of the Sea which are found in all parts almost of the present superficial Parts of the Earth, could not be caused by the general Flood of *Noah*, if the manner of performance and executing thereof were such as is for the generality supposed and explained by Commentators by reason that they make the time of the continuance of the present superficial Parts of the Earth under the Waters to be no longer than the time of the duration of the Flood, as it is recorded in Holy Writ. Supposing that the present Earth and Sea is in the same places with respect to the Body of the Earth as they were before the Flood; nor will the Hypothesis or Explication of the ingenious Author that has lately writ of that Subject, reach it, he supposing there was no apparent Sea before the Flood, but that the Sea was all covered by the Earth, if at least I do rightly comprehend his intentions; for that space of time will not be found of duration long enough to produce *de novo* such multitudes of those Creatures, and to such Magnitudes and Ages of growth as many of them seem to have had, and it will be difficult to be imagined, that such Creatures as do not swim in the Water, should, by the Effects of that Deluge, be taken from their Residences in the bottom of the Sea and carried to the top of the Mountains, or to places so far remote from those Residences. So then, if we will ascribe those Phænomena to that Flood, it will be necessary to consider which way that Catastrophy might be effected that it might be the occasion of such effects. I therefore said, that unless we supposed that there were thereby a change wrought of the superficial Parts of the Globe, and that those Parts which before the Flood were dry Land became Sea, and the Parts which were before covered by the Sea after the said Deluge, became the dry Land, it seems to me, that these appearances cannot be solved by *Noah's* Flood.

Tho' possibly this may seem a little improbable upon the first mentioning of it, yet possibly also upon a little further examination, it may be thought to have somewhat more of liklyhood than is yet imagined, at least I hope the manner will be conceivable.

We have no other means of being informed of the true History of it, but what is to be found recorded in the sacred Writings of *Moses*; and therefore those are to be consulted, and the true meaning of them, as far as can be, must be obtained; for whatever else may be scattered here and there in other Authors that seem to relate thereunto in all probability, were some way or other fetched from his Informations.

[p.413] I conceive then, that considering the Descriptions of *Moses* both of the Generation of the Earth and manner of the Flood, the History of both may be thus explained.

First, For the Fabrick of the Earth;, the Description is but short in the first Chapter of *Genesis* and 2d Verse. *Et terra erat Solitudo & Inanitas, & Caligo Super Facies abyssi, & Spiritus Dei manabat Super facies Aquarum, & dixit Deus sit lux & fuit lux.* This doth seem to represent the order of the four Gradations, Earth in the middle, Water next, then Spiritus, then Light; the Central Earth is described only as a *Vacuum*, and called the *Abysse*, and Darkness inclosing it; that is, the Water follows next above it, which covered it all round: Above this the Air, and lastly the Fire, Æther, or Light in the fourth Verse, according to the Hebrew, *Et divisit Deus inter Lucem & inter tenebras:* The Septuagint renders is, και διεχωρισεν ο θεος ανα μεσον του φωτος και ανα μεσον του σκοτος. That is, God caused a twofold Seperation, one in the middle of the Light or Æther, and another in the middle of the Darkness, which covered the Face of the Abysse or Central Earth; which covering of Darkness was the Water, which is often called the Abysse or great Deep; the former of these is afterwards always called the στερεωυα του ουρανου, the Firmament of Heaven, to distiguish it from the latter, called the στερεωμα εν μεσω του υδατος, the Firmament in the middle of the Waters; for in the sixth Verse it is so exprest, και ειπεν ο θεος γενηθετω στερεωμα εν μεσω του υδατος. Let there be a Firmament in the middle of the Waters, and let it be a division between the Waters and the Waters: And in the seventh Verse; and God made the Firmament, and God made a separation of the Waters that were below the Firmament, from the Waters that were above the Firmament. And in the eighth Verse, και εκαλεσε ο θεος το στερεωμα ουρανον; it is generally rendered, and God called the Firmament Heaven; I conceive it may be rendered, Also God called the Heaven the Firmament, for to shew that there was also a Firmament of the Heavens; for so it is afterward every where called στερεωμα του ουρανου. But this I submit to Divines.

*Hooke's translation follows the Greek words above which, however, do not include the word for "Æther," just "Light."*

*Greek translation = and God called the Firmament Heaven*

*Greek translation = Firmament of the Heavens*

*Greek translation = in the middle of the waters*

This first Firmament then, εν μεσω του υδατος, seems to have been a solid and hard Sphærical Shell, as it were, which incompassed the Ball of the Earth Central, not clear without the Liquid Water, as the hard Shell of the Egg is without the White, and so the Egg-shell doth inclose the whole white of the Egg, as well as the White incloseth the Yolk: But it was, as I conceive, meant, that this Firmament or Sphærical hard Shell was placed, as it were, in the middle of all the White, or of the incompassing Water; the Circumferential half of it being without the Shell, and the Central half of it within the Shell. So then at its first Creation, the order was first the Central Earth or great Abysse; this was in the middle as the Yolk of the Egg round or Sphærical; this was inclosed in Darkness by the Shell of Water underneath the Firmament, being half the whole Body of the Water which was inclosed perfectly within this Firmament, as the White of the Egg by the Egg-shell; and by that hard shell it was perfectly seperated from the other half of the Water which was above the Firmament, and as it were a second White of the Egg without the Shell, and was the Water upon the Face of which the Spirit was said to move; so that the whole Globe, for that time, was all covered with Water. This *Ovid* seems to allude to, when he saith, *Circumfluus humor ultima possedit, Solidumq; coercuit orbem*; which I took notice of in my Lecture about the History of the World, as expressed by that ingenious Mythologic Poet; for he seems to make the Water to be both below the Shell of the Earth and to encompass it.

*What contortions Hooke goes through here in order to interpret the Mosaic account to have water covering the surface of the Earth!*

*He relates this scene to Ovid,*

*Plato* also was of that Opinion in making *Tartarus* the place of the Waters, that is, the middle and Central parts of the Ball or Globe of the Earth; and so the hard part of the Earth to be nothing but this Shell near the Superficies, and it seems also, that *Pythagoras*, yea, and the *Ægyptians* and *Chaldeans* likewise were of the same Sentiment, and divers are of Opinion that *Moses* also understood the same by the great Abysse which he mentions in the Description of *Noah's* Flood. But by this Description of the Creation he seems to be understood otherwise, when he says, *Et terra erat Solitudo & Inanitas & Caligo super facies Abyssi*. For by this he seems to make the form of the Terraqueous [p.414] Globe to be no other than that of a Bubble, such as Children blow into the Air, that is, only a Sphærical Film or Orb of Water, which within it had nothing but *Solitudo & Inanitas* ההו‎בהן‎. *Vacuum & Inane*, αορατος και ακατασκευαστος, and only that this Film of Water was divided in the

*Plato,*

middle by the solid hard Shell of the Firmament which inclosed half the Film; that is, the inner side; and excluded the other half of the Film; that is, the outward superficial Parts: And hence 'tis possible that *Virgil* in the eighth Book of his *Æneids*, says, *Spiritus Intus alit, totamq; infusa per Artus men agitat molem & magno se Corpore miscet. Inde hominum pecudumq; genus vitæq; volantum & quæ marmoreo fert monstra sub æquore pontus,* &c. He seeming there to make the place for the Soul of the World; others there are who would have it to be Fire, and thence to proceed the Causes of the Vulcanos and fiery Eruption; but *Aristotle* there places the pure Element of the Earth; some of the Modern Philosophers would have it to be all one great Load-stone. I could produce various other Opinions, but they are all but Opinions; and it matters not much what the Substance be that fills it, as to the present inquiery; I shall therefore proceed.

*Vergil,*

*and Aristotle.*

This Firmament then in the middle of the Waters, I take to be that which in many places of the Bible is said to be the Foundation of the Earth, as in *Psal.* 24. 2. *The Earth is the Lords and the fullness thereof, for he hath founded it upon the Seas.* (Prov. 3. 19.) *The Lord by wisdom hath founded the Earth, by understanding hath he established the Heavens*, (this seems to refer to the two Firmaments) and in the following Verse, *By his knowledge the depths are broken up, and the Clouds drop down their Dew.* (This seems to refer to the Causes of the general Deluge by opening of those two Firmaments, as I shall by and by shew.) So *Job* 38.6. *Whereupon are the Foundations of the Earth fastned?*

This Sphærical Firmament or Shell then in the middle of the Waters, we may suppose, was in some places raised or forced outwards, and some other parts were pressed downwards or inwards, and sunk lower, when in the ninth Verse, God commanded the Waters under the Heaven to be gathered together to one place, and the dry Land to appear; for by depressing in of some parts of that Sphærical Shell (to make room to receive all the Waters that had before covered the whole) other parts must be thrust out, the Contents within being the same, and so requiring equal Space or Extension; so that what went below the former Sphærical Surface, must be equalled by other parts ascending without that Surface, and so the quantities of the Waters both within it, and those without it, remained each the same, and still distinct and separated by this Firmament in the middle of the Waters, tho' altered from its Sphærical Figure; and the outward Surface of the outward Water, as well as the inward Surface of the inward Water, must remain Sphærical, because of the Power of Gravity from without a Central Earth, or Yolk within, formed of a Sphærical Figure.

In this State the Earth seems to remain till the time of the Flood, which is accounted between sixteen and seventeen Hundred Years according to the Hebrew. When *God looked upon the Earth, and behold it was corrupt, for all flesh had corrupted their way upon the Earth,* as Chap. VI. V. 12. And in the thirteenth told *Noah*, that he would destroy all living Creatures with the Flood. This Destruction began in the six Hundred Year of *Noah's* Life, Chap. VII. V. II. The manner of which was expressed thus in the Septuagint ερραγησαν πασαι αι πηγαι της αβυσσου και οι κατρρακται του ουρανου ανεωχθησαν. *The fountains of the great deep were drawn up, and the windows of Heaven were opened..* This refers again to the twofold Firmament, that εν μεσω τγ υδατο, and that τγ γραvγ. As for that of the Heavens the effects of the opening of them was, that it Rained 40 Days and 40 Nights; but the Consequents of the other are not expressed any otherwise, but that the Flood was upon the Earth forty Days, that is, the Sea continued to flow in upon it. *And the Waters increased, and bare up the Ark and it was lifted up above the Earth,* v. 17. *And the Waters prevailed exceedingly upon the Earth, and all the high Hills, that were under the whole Heaven, were covered,* V. 19. *Fifteen Cubits upwards did the Waters prevail, and the Mountains were covered,* V.20. *Every living Substance was destroyed which was upon the Face of the ground,* Noah *only, and those with him in the Ark, remained a-*[p.415] *live,* V. 23. By which it appears, that not only all Men, Beasts, Cattle, Fowls, Insects, Worms, &c. perished by the Flood, but every living Substance; that is, all Vegetables also; for all Animals were enumerated before: We see therefore that here was a double Cause of the Waters. First, The rain from above. And, Secondly, The pulling up of the Fountains of the great Deep: What I understand by the great Deep, I shewed before; that is, the sinkings inward of the Firmament in the middle of the Waters; and the forcing up of the Fountains of the great Deep, I conceive to signify the raising again of those parts that were before sunk to receive the Sea; and a Consequent of that would necessarily be a sinking of that which was the dry Land, and a Consequent of that, flowing and increasing of the Sea from out of that which was the great Deep, and a prevailing and increasing upon that which was a sinking Earth; and this motion being forty Days in progression before the rising Surface of the Sea, and the sinking Surface of the highest Land met. So long the Waters were said to be flowing and increasing before it was wholly covered; nay, the History goes on with the Journal of its progress, till the Waters were gotten fifteen Cubits above the highest Mountains; but then the account ceaseth, and adds only, that the Waters prevailed on the Earth a hundred and fifty Days, and so long the whole Firmament was covered with Water. So that in probability the progress of the alteration of the Firmament proceeded so far till it recovered its perfect Sphærical Figure truly in the middle of the Waters, as it was at its first Creation placed at about seventy five Days after the forty; but as I conceive it staid not there, but the progression of both the parts went onwards; that is, the sinking parts went as much below the Level, as before they were above, and the rising parts by degrees ascended as much above as they had been below, and that which had been the bottom of the Sea under the Water, became the dry Land, and that which had been before the dry Land, now became the bottom of the Sea, whether the Waters retreated from off these parts which were raised when the Flood was finished; for it is said in the eighth Chap. That God remembred *Noah,* and what was with him in the Ark, to prepare them another Habitation, by making dry Land for them again; and, First, The Heavens were cleared from Raining. And, Secondly, By turning of the Water that had fallen, into Vapours, and by turning all those Vapours, which such a Commotion of the Earth and Sea had caused, into Wind, and by causing the Waters to return from those parts which it had covered into the Deeps that were appointed for their Reception; so that at the end of the hundred and fifty Days the new Earth began to appear. *Ver. 3. And the Waters returned from off the Earth continually, and after the end of the hundred and fifty Days the Waters were abated.* And, *v. 4. And the Ark rested in the seventh Month, on the seventeenth day of the Month, upon the Mountains of* Ararat, which probably was the Name of the Mountain after the Flood in the time of *Moses.* Ver. 5. *And the Waters decreased continually until the tenth month, when the tops of the Mountains were seen. Forty days after this* Noah *opened the Windows of his Ark and sent out a Raven.* So that it

*I believe Hooke is deliberately emphasizing the amount of water described—i.e., 15 cubits above the highest mountains (that's a lot of water considering the heights of known mountain ranges), as he was quite capable of calculating the amount of water necessary to accomplish this task. He obviously knew that 40 days and 40 nights of rain would simply not do it, in spite of the pulling up of the "Fountains of the great Deep," which meant for Hooke the raising again of the parts that had been sunk "to receive the Sea" and the sinking of that which was dry land.*

seems that as the old was forty Days in being covered, so the new was forty Days in being discovered; but *Noah* staid yet many Days longer before the Surface of the Ground was dried. This Explication, I think, doth fully answer to the words of the History of the Flood as they are written by *Moses*, and will likewise shew a probable Cause how those Phænomena of Sea-sand and Shells are become so universal over the Face of the whole Earth, as it is at present, which were the two things which I now indeavoured to make intelligible. I have not, I hope, given any Explication, or made any Supposition, how differing soever it be, from all the Explications I have yet met with, that will any ways distort the plain words of the Text; for I have in this, as near as may be guided my progress by that Direction, and I hope I have hereby shewed a very plain and intelligible way how the Flood became so perfectly universal, and the Earth returned perfectly to its primitive and first Created Figure, without any extravagant supposition of new Created Waters, or bringing them down from above the highest Heavens; nor is here any great need of Calculation to know how great a quantity of Water would need to be new Created and afterwards Annihilated, or first fetch down from the [p.416] Heavens, and then sent back again; nor is there any need of supposing the Earth to be broken to pieces since the Flood, and the Antediluvian World without any visible Sea. And if it were much to the purpose, I could shew how all this, that I have supposed, may be Physically explained, and the Æquilibrium maintained: And, in short, to shew how consonant this Hypothesis may be both to several Expressions in the most antient Authors, and, in a word, with the Rules of Nature itself, of which I have formerly given divers hints to this Society, and may some other time more fully explain, but I fear I have detained you too long at this time.

*Hooke had said plainly several times before that Noah's flood could not have accomplished all that we see, but the prudent Hooke here is attempting to satisfy the prevailing attitude of the time—possibly in answer to criticism.*

*This statement is most likely made tongue-in-cheek, as Hooke, the physicist, could easily make the necessary calculations. For example, even if he did not know about the Himalayas so had to use the Alps for a reference point, say Mt. Blanc at 16,000 feet, simple arithmetic would have told him that the rate of rainfall would have had to be 200 inches per hour to cover Mt. Blanc by "15 cubits" in 40 days and 40 nights.*

**No. 15.** *An extract of a lecture read July 18, 1688.*

[p.428] The Aim of my present Discourse is rather a Progression in the Theory of the Nature of the Air, than of any of the formerly mention'd Effects of Earthquakes, and the rather by the way of Query and Inquisition, than of possitive Theory and Affirmation.

As First, Whether the late Feaverish Distemper that was here so frequent, supposed by some to be inclined to Pestilential, tho' not so Mortal, might not be caused by some Infections or poisonous Vapours cast into it by those late Eruptions in *Italy* or *America*?

*Given the current preoccupation with $CO_2$ increase in the atmosphere and causes for extinctions, it is rather imaginative of Hooke in the 17th century to try to relate earthquakes, volcanic eruptions, etc. to climate changes, poisonous vapors, and human health.*

Secondly, Whether the coldness, unseasonableness of the Spring, the strange Rains, Storms and Tempests, and other such unusual Accidents, that have lately happened in the Weather, may not have been caused by the same Efficients that caused the Eruptions?

Thirdly, Whether it may be reasonable to conceive, that there could be any Communication Subterraneous between these places of those Eruptions in *Naples* and *Lima*; or whether it were Superterraneous through the Air and Æther?

Fourthly, Whether it may be rationally conceived, that Steems raised into the Body of the Air in *Lima* or in *Naples* could be continued so long in it as to be conveyed from either of those places to *England, London,* &c.?

*He was aware that the gases and dust from eruptions can travel around the globe.*

Fifthly, How long time may be judged necessary for such a Conveyance?

Sixthly, Whether such Distempers of the Air may be precedent to the Distempers within the Earth, and so be of the Nature of a Procatarctick cause of the Earthquakes, and if so, whether those Distempers may arise from the Nature of the Air itself, or from some external and influential Cause, either from the Æther, Comets, or some of the more Conspicuous Cælestial Bodies?

Seventhly, Whether there may be not some general, tho' yet unknown, Cause, that may produce both those effects in the Earth and those in the Air, nay, and those in the Æther also, such as Comets and some kinds of Meteors also? because of the usual Concomitance of them; as will in part appear by subsequent Relations.

*As he was wont to do, here he seems to be searching for some general, unifying hypothesis.*

[p.429] These possibly may be looked upon as not very easily solvable, and therefore not so proper to be propounded as Queries, unless they could also be satisfactoraly answered. I must leave every one to Censure as he thinks meet, only this I must add, that the first step towards Knowledge is Inquisition.

And that I may manifest that these Queries are not altogether at random, I shall add some Natural Histories, that may possibly give some hints of their Solution; and those shall be the Accounts of some Accidents or Effects similar to those, which have lately happened at other Times, and in other Places; from whose Congruities one would be apt to conjecture a similitude of Causes, and if not a necessary, yet somewhat more than an accidental Concurrency of Effects, and a kind of Periodick Revolution of them.

In the Year 1672, in the Islands of the *Archipelago*, that is, the Islands of *Greece*, this Winter was so Stormy and Tempestuous, that not only the Trees and Plantations, but the Houses also were destroyed by the Lightning and Hail; so that both the Towns and Villages became almost unknowable, being reduced to Ruins.

In the *Barbadoes* also was a most violent Hurricane, in which many of our Nation Perished.

Near *Ancona, Fauno* and *Rimini*, there were this Year, in *April*, many Houses overturned by an Earthquake; and more especially in *Romania* and St. *Marc*, there were above six hundred People killed, and above quadruple that number hurt: At *Rimini* the Cathedral Church was overthrowed, the Bells shaken out of the Tower, and many People lost their Lives. At *Fauno* twenty eight Persons were killed by the fall of a Bell. The Churches of the *Theatines,* St. *Agnie*, St. *Appollonce*, St. *Mary de la Gomia,* St. *Innocent*, St. *Bernard*, St. *Mary della Colonolla*, and all the others except only those of the *Capuchines*, and of *Maria de Mari*, were endamaged. A great number of Palaces and Houses were ruined: This happened whilst People were at Church; so that above fifteen Hundred were killed, and many more were hurt. At *Pesaro* and *Senegallo* the Walls of the City and many Chimnies were thrown down. *Ancona* and *Rimini* were abandoned by their Inhabitants, who were constrained to lie under the open Canopy of Heaven.

*September* the 30th., of the same Year, there was a Hurricane passed through all *Spain*, but it was most furious about *Madrid*, insomuch that it blew down the Roofs, Chimnies, nay, and the Houses too; as also the Towers and Churches; insomuch that the Damage was exceeding great along the *Prado* and at *Buon Retiro*. But all this was nothing in comparison of what happened the same Day in almost all the Countries of *Spain*; for this furious Tempest caused such Ravages in *Andulasia, Gallicia, Castil, Grenada, Valencia* and *Biscay*, as were truly Amazing: But what was most remarkable was this, that three Days after the Gallions of the Plate-Fleet, which came from the *West-Indies*, being arrived at the Islands of *Terceras*, felt not the least of it.

In this Year were also seen two Comets, ----one in *January* another in *April.*

Eight Years before this, namely, in 1664, were two Comets also; but all the other Natural Histories, or Physical Accidents of that Year I have not yet procured.----

But eight Years after this, *viz.* 1680, which is now also eight Years since; First, For the Comets they are yet in most Mens Memories, and besides there are Histories enough extant; but next for the Earthquakes: First, By a Letter from *Botavia* we have an account of a great Earthquake that happened in *China* about *Peking* the preceding *August*, viz. That the 13/23 of *August* 79, about ten in the Morning, there happened a most terrible Earthquake, which overturned almost all the Houses of that great City and the parts thereabout, whereby a World of People were destroyed in a most dreadful manner, besides multitudes that were hurt, whose number we cannot yet learn to this Hour: Two Heads of certain Beasts, which were Carved and fixed over the Imperial Palace Gate, were beaten off and thrown

*These paragraphs relate occurrences that might yield possible connections between earthquakes/eruptions and other phenomena such as comets and meteors and violent climatic periods.*

*An account of an earthquake in China. 13/23 either is an indication of the uncertainty of the date, either the 13th or the 23rd of August, or it may refer to a date in the Chinese lunar calendar in which a 13th month is sometimes inserted to adjust to the planetary cycle—like our leap year.*

down to the Ground by the force of the shake. All the Palaces of the *Mandarines*, and their Families, and the Courts of Justice round the Palace were tumbled down; the Emperor commanded the principle *Mandarines*, that had command over the five parts of [p.430] the City of *Peking*, to examine themselves in their proper Persons, and to give him an account of all the damages that had happened, that he might the better advise of ways to help the Poor People that had suffered. This they did, and advised, that if his Majesty would distribute to each two or three Crowns of Silver at twenty five Frecks the Crown, it might be a sufficient supply: But he thinking this was too little, commanded ten Thousand Crowns to be taken out of his Treasury and distributed for the present Necessity. The first, second, and following Days that it lasted, the Earth was shaken five, six, or seven times a Day, but not with so much violence as by the first; so that the Inhabitants were in such Consternation as to forsake their Houses; the Soldiers and their Wives were most afflicted having nothing left to subsist; by Day they were exposed to the Sun, and by Night to the cold Heaven, which much incommoded them. The Emperor also was in great pain to know the Damages that had happened in the Neighbouring Parts by this Earthquake, and commanded one of his great *Mandarines*, named *Samolio*, to inquire and inform him of them, who returned this Report, That the 13/23 of *August*, whilst the Heaven was covered all over with dark Clouds, the Earthquake shocked extraordinarily the City of *Tongfu* about a Days Journey from *Peking*, that all the Imperial Magazines there had been overturned, as likewise the old Walls of the City; so that of ten Parts of the City scarce one remained which had not been indamaged, whose pitiful Condition was deplored by every one. The Commissioners of the Magasines Emperial who had escaped, render'd themselves presently to His Imperial Majesty to give him an account of the flying of the under Officers for the Consternation, and of their fear of the Robbing of the remaining Rice and Provisions by Thieves, which caused him to send them sixteen Hundred Soldiers for their Guard. The *Primier Intendant* of the Navy was killed by the overturning of his House. The Emperor had also reported to him, how the Robbers had wasted much of his Treasury in the Magazines that had been overturned; and upon the consideration of the general Calamity, the Emperor makes a most Pious Speech to the principle *Mandarines*, which I shall not trouble you with, only my Author adds, What Christian could have spoke better? Will not such as he rise up in Judgment against many Christians? This is a short Account of what happened to two Cities of *China*: I say Cities; for tho' generally we have only an account of the Damages caused to Cities, Towns and Men, yet we are not to conceive, as if the shaking and disorders of an Earthquake were only aimed at Cities like Marks and Goales to be shot at; no, certainly, there may, in all such Concussions and Devastations have happened much greater and different Effects from those which come to our knowledge; for that the most part of the World have little concern for what may happen in the Mountains, Hills, Plains, Forrests, Seas, &c. which make not any great or publick Calamity to the more considerable sort of Men; wherefore questionless, tho' many strange Effects of this kind may also have happened, and may have been seen and observed by some Men, yet they are but as it were *In transitu*, and quickly forgotten, since there is none to Record them. So that many thousands of such Effects have been swallowed up by the Oblivion of Time, where one has chanced to get by some accidental hint to lie Recorded by chance among the heap of other Histories. Comets indeed, as glaring in every ones Eyes, have found, among the multitude of Observers, some that have Recorded somewhat of

*Samolio is not a Chinese name. Manchu surnames, however, often have 3 or 4 syllables.*

*Hooke attributes these questions concerning Christians to his correspondent, but that Hooke would interject this sentiment in the midst of the earthquake account gives us a clue regarding his attitude toward religion in general.*

them to Posterity, but even among them also, I doubt we shall not find that one of ten has obtained a History. But this Earthquake in *China* was not the only Accident of this time which I would mention; for upon the Coast of *Coromandal*, the Sea so overflowed the Country, that infinities of Men and Cattle were destroyed, many Cities and Villages were drowned. This overflowing was also found at *Jafnapatnam*, where it did much mischief to the Fortification, and to the Country, and the Cattle, but not so much to the Men.

Nor were these kind of Accidents only felt in *India*, but the same Year there happened a considerable Earthquake in *Spain*, and particularly at *Malaga*.

[p.431] All *Spain* was this Year so perished with Drouth, that not only the Pits, Fountains and Rivers were dried, but the Harvest was spoiled, and many perish'd by this means: On the other side in the *Autumn* arose such horrible Tempests and Earthquakes as were felt long after. After the beginning of *September* they had continual Thunder and Lightning, by which divers perished. The Hail fell so on *Pardo*, a Pleasure-house of the Kings of *Spain*, that it rooted up the greatest Trees, and kill'd so many Beast and Foul, that not only the Fields were almost cover'd, but the River *Mancanarez*; it much indamaged the Village *Foncarral*; the old Bridge *de Aranda de Duerro* was born down by the Waters of the River *Tagus*, which run under, and did much damage to *Aranivez*, sweeping away divers People, Cattle, Trees, Bridges, and Houses: The like Ruins were caused almost over all the Kingdom, insomuch that in one Village, only, forty People were lost. The greatest violence was at *Madrid* the twenty sixth of *September*, where the Water overflowed so as to mount into the Garden of the *Augustines* and throw down the Wall; also into the fair Parterres of the Countess *Ognate*, and run into her House, ruin'd the rich Furniture of Pictures, &c. of the lower Story; ruin'd the Stables and razed one House. The River also bore away fifty Foot of a strong Stone-Wall made to stop the Passage into the River *Prado*: This River one of the least in *Spain*, so swelled as to carry away almost all before it, as four Iron Gates, and the Cross of the *Via Sacra*. It beat down the Bridge before *Buon Retiro*, and broke through the middle of the Stone Bank. It rush'd into the Gardens of *Nostre Dame de Arocha* after it had beat down the Wall; it run into the general Hospital carrying with it an Arch of Stones. The twenty seventh the overflowing continued with constant Thunder and Lightning, when the River *Mancanarez* bore down the fair Bridge of *Toledo* of sixty Arches. The twenty eighth the Streams of *Prado* so swelled by the Torrents from the Mountains, that all the Champain near it was drowned, the King and Queen of *Spain* were like to be lost in their return from *Nostre Dame de Arocha*; *Malaga*, a City of the Kingdom of *Granada*, situated on the *Mediterranean*, twenty five Miles from the Streights, a Place Great, Rich and well Peopled, had, the ninth of this Month, such violent Shocks of an Earthquake, that all were frighted, the Sea was so disturbed, that the Fish leaped out of the Water, and the Ships in the Harbour were cast above twenty Foot from their places, which the Mariners believed to be sunk; the Harbours and Walls were sunk, together with the Bulwarks, Towers and Fortifications of four Parishes, of which the City consisted, having 4284 Houses, 1057 were ruined to their Foundation; 1259 so decay'd, that they must be Rebuilt to be Habitable. Divers Churches and Palaces felt the effects also; five Cloysters of Religious with the People were utterly ruined, and above all, that of St. *Francis*, where Stone was not left on

*Another account of an earthquake, this one in Spain.*

Stone, where fourteen Persons Perish'd, four Hospitals, one Colledge, the Bishops Palace, the Palaces of *de Diego de Argote, de Jo. de Torrez, de Diego de Cordua*, and a fair House joining to the Cathedral was thrown down, yet the Church which had been Repairing and Beautifying ever since 1521 scaped, tho' divers times shaken. In the Suburbs *Los Perchelez* two hundred and twenty five Houses were thrown down, so that in all 1282 Houses were destroy'd. Many Houses in the Confines of *Malaga* were overturned; besides the Earth opened in divers places and disgorged Waters in great abundance, which swelled the Rivers and made them overflow. Many Houses in the Villages were destroyed, as at *Pizaria* four miles from *Malaga*, fifteen of twenty four Houses were overturned; some Mountains were displaced, and divers Persons and Cattle lost: The Wall of *Alhaurin de la torte*, two Miles from *Malaga*, opened four Foot, but closed again: The Jasper Columns of the Church were lifted up and setled down again on their Pedestals. At *Competa*, six Leagues from the City, nothing but the Tabernacle and the Cross of the Church remained whole. At *Aloizana* forty Houses were tumbled down, as many at *Cartama*, and thereabout also at *Coin*, and a great number of People perish'd. At great *Alhaurin* two hundred and forty Houses and the Church were destroy'd, of which only fifty three were somewhat Habitable. In tahe City of *Minorz* five Leagues from *Malaga*, thirty seven Houes were tumbl'd down and fif-[p.432]teen Persons crush'd. The Church at the City *Binal-Madera* fell on a heap, and all the Houses render'd unhabitable. The Earth opened at *Veles Malaga*, and so swelled a River, which run some space from thence, that it rose ten Pikes above the tops of the Houses, which it squash'd in running. Many Houses were ruin'd at *Aloro*, others much endamag'd, with the Cloister of St. *Francis*. All the Churches of *Granada* were shaken, and a Chappel in the Church of *Mercy* ruin'd: All which were sad Spectacles.

I have given the Particulars of the whole Relation, most of which concern Buildings, Men and Cattle, those being the Particulars most People are affected with and so observe, and you find only two hints, as it were, of other Effects, the one is of the removing several Mountains, the other of the Earths opening and disgorging a Flood. But 'tis not to be thought but that an inquisitive Naturalist might have found ten times more remarkable Effects in the Country than the shaking down a few Houses in the Towns and Villages, all which, if taken notice of, are soon forgot and lost, and so have been in former Ages, and therefore no wonder if we hear nothing of them in Books: But Nature itself has preserved somewhat of the memory of them by the Medals or indelible Characters of Shells or other Petrify'd, or otherwise preserved Substances, which any, that have Senses and Understanding, may read. But this is not the aim of my present relating these Histories, but to give an example of a Contemporariness of Earthquakes at great distances upon the Earth, and a similitude of Effects with those we have this Year heard of from *Italy* and *America*; nay, and let me add what we have had in *London* and *England, viz.* a kind of Agueish Distemper, yet not Pestilential, which, 'tis well known, has been very general; for I find that in *October* Agues were as frequent this Year in *France* as the late Cold or Distemper was here: It was then that Dr. *Tabour* cured the Prince of *Conde*, and many other Persons of great Quality, among the rest the *Dauphiness* first, and afterwards the *Dauphin* himself, by a Medicine he had invented; tho' *Tabours* demanding five Thousand Crowns for discovering his Receipt, made the *Dauphin* first make use of other means; but without effect. (I will not like an Astrologer name to you the Occurrences that then happened at *Cologne*, nor make comparison

*A naturalist—i.e., scientist—can read the record preserved in nature of the effects of an earthquake and finds it much more interesting than the destruction of some houses.*

*His aim in this discourse is to show that earthquakes can happen almost contemporaneously at places on the Earth far apart from each other. He also tries to relate human illness to such happenings as a result of air or water polluted by poisonous gases from eruptions.*

with the present, but leave those to the Astrologians, &c.) The Plague also this Year 1689, was very much at *Prague*, so that some judged there died in that City thirty Thousand, at *Dresden* above four Thousand, at *Leipsick* about three Thousand; I cannot say there hath been a Comet this Year, but I have been confidently told, that there appear'd one in the Mornings about a Month since, but I could not have the luck to see it, tho' I looked for it divers Mornings after I heard of it, but 'tis more likely it may appear in *October*, or later; but that belongs to another Head, the Affected Earth and Infected Air being those I designed at present to compare; and in these we find the effects in *China* and *Coromandel* eight Years since to answer those of *America* this present Year, and that then of *Spain* to this now of *Italy*; and those then of *France* and *Germany* to the late here in *England*, tho' in all particulars those of the Year eighty seem to exceed those of the present Year. But as the Relations of that are but short and imperfect, so are those of the present as yet much more; but 'twere to be wish'd some more full might be obtained and Recorded before they be forgotten, which a little space of time will otherwise effect, and 'tis not to be doubted but we might hear of much stranger effects of the *Lima* Earthquake, than yet have arrived, if care were taken to procure a fuller account of them. And by the Yesterdays *Brussells* we are informed of a Cleft in a Mountain belonging to the *Marquis de Tarracusa*, of four Spans broad and two Miles long, of which they can find no bottom, and of a Fire shot into the Heavens like a great Beam, of which they lost the sight, not knowing whether it went.

But in the mean time possibly it may not seem altogether unreasonable to suppose, that such an Eruption may emit poisenous Vapours, as well as sometimes poisenous Waters, as appears by that of *France* which I have Printed in one of my Collections. Nor may it seem so strange to suppose its effect may operate at such distances, and not at the very place; when we consider how fiery and volatile such Steams may be, how violently shot into the Air, [p.433] and blow far off the Dust and Ashes of *Hecla, Ætna,* the *Palma,* and many others have been carried in the Air before they have fallen, of which Instances may be produced. And that, in probability, the less active or dead Earthy Materials are those, which fall near the place, whose Qualifications may be of differing Natures. Nor will any very long time be thought requisite for their transport to far distance Countries imbody'd in the Air, when I have proved the velocities of its motions. Nor will it seem strange to one that shall well consider the known Effects of the several Winds, to suppose such kind of transports: But of these Particulars I shall say more upon some other occasion.

*Although Waller states at the beginning of this discourse that this is an extract of a lecture read July 18, 1688, the mention of the plague "this year 1689" must indicate that at least part of this text was taken from lectures in the following year, 1689.*

*Hooke's idea here is that air and water polluted by eruptions could travel around the globe and cause illnesses.*

**No. 16.** *This lecture was read May 29, 1689, to answer two objections.*

[p.433] I Delivered in my last Lecture in this place, the Methods I had made use of for the founding and establishing the Doctrines or Conclusions I had made concerning the Causes and Reasons of the present State and Phænomena of the Surface of the Earth, which was by a methodical Induction from the Phænomena themselves of the most remote, as well as the more approximate and immediate Causes thereof. But notwithstanding all the Arguments I have alledged, and the Proofs I have produced in the delivery of this *Theory*, I still find that there remain upon the Minds of some such Doubts and contrary Persuasions, that they cannot forsake their former Opinions; and therefore (tho' I think I have already fully proved every part, so that the Confutations of such Objections would be but the necessary Corollaries from the said Doctrine, yet since I find they are still insisted on as material Objections that will need a more particular Discussion and Examination) I thought it not impertinent to examine them more strictly, to find the Power and Efficacy, or to discover the Weakness and Insufficiency of them for the purpose they are designed. That thereby the *Idola* (as my Lord *Verulam* says) which pre-possess the Minds of some Men, and molest them in the discovery and imbracing of Sciences may be detected, and, as much as may be, removed and dissolved, thereby to leave the Mind more free to Discourse and Reason aright, without the prejudices of any unsound, unaccountable and unwarrantable Doctrines formerly imbrac'd.

The Objections I shall at present examine are only two, *viz.*

First, That if these large Petrified Bodies, such as the *Ophiomorphite* Stone which I did formerly shew to this Society in the Place, be supposed to have been the Production of this Shell of a certain kind of *Nautilus* of that bigness and shape, which, in preceding Ages of the World, had been produced and perfected to that Magnitude in the Bottom of the Sea, which then was near the place where they are now found, as I have argued for; then it will necessarily follow, say they, that there have been, in former times, certain Species of Animals in Nature, which in succeeding and in the present Age have been and are wholly lost; for neither have we in Authors any mention made of such Creatures, nor are there any such found at present, either near the places of their position (as on the Shores or Sea about this Island) nor in any other part of the World for ought we yet know. Now, to suppose such a Doctrine as doth necessarily infer such a Consequence, is looked upon by such as absurd and extravagant; for that it would argue an imperfection of the first Creation, which should produce any one Species more than what was absolutely necessary to its present and future State, and so would be a great derogation from the Wisdom and Power of the Omnipotent Creator.

*The first objection is that the idea of extinction is not compatible with the idea of perfection of the first Creation and denigrates the wisdom and power of the Creator.*

To this first Objection I Answer, First, That tho' it may possibly be true, that there is at present no such *Nautili* to be found upon the Coast or Shores [p.434] of the Lands where these sorts of figured Stones are found, yet no one is assured that there are not some of the same Species, and as big in some other parts of the World, as possibly at the bottoms of some of the great Oceans. Of such Productions and those Multifarious both Vegetable and Animal, no one can doubt that has found in soundable Depths such variety of testaceous and crustaceous Animals there residing, as in their proper and Natural Regions; which would by no means possibly

*He admits that it may be true that some of the animals represented by fossils might yet be found at remote places.*

be produced or kept alive in parts of the Sea where they should want their natural Accommodations; one of which may possibly be a sufficient degree of Pressure from the incumbent Column of Water, which, if such be necessary to their Life and well Being, we are no more to wonder that they should not be found in shallower Waters, than that Men should not be found inhabiting the tops of the *Andes*, or the *Atlas, Alps,* or *Caucasus*, which from the thinness and coldness of the Air at those heights are no ways fit for Respiration and sustaining Life. Now, that the present Land of *England* may have in former Ages had some such Position with respect to an incumbent Sea, I could produce several Arguments were they now material to the answering the present Objection, but I will not now insist upon it.

*The fossils, however, are found at places which are not their natural habitat, an indication, therefore, that land, like England, was once sea.*

But in the second place I answer, That tho' possibly there may be no such *Nautilus* to be found described in any Natural Historian at this Day, yet 'tis possible there may be many of the same Species, and of as great Magnitudes in divers parts of the World, such as have been either not yet discovered by the *Europeans* or but of late, or but little frequented; and so tho' they may be there frequent and plentiful enough, yet none may have been brought thence into *Europe* as yet, or possibly so much as seen there; 'tis not to be doubted that there really are great multitudes of differing Species of Vegetables, Insects, Beasts and Fishes yet in places less frequented, of which we in *Europe* have hitherto had no knowledge or information; and tho' many strange things have been of late Years brought to our view, yet we may with Reason enough assert, there are many more yet latent, which Time may make manifest: For if we consider the small knowledge of things of this Nature that we yet have acquired, of places remote, even the most frequented, we need not much wonder at the lesser information of such, as are not known or less frequented; for not to insist upon the multitudes of Vegetables that have been newly shewn to us by the Authors of the *Hortus Malabaricus,* and by *Brennius*, and others, we are put in hope, to see the Descriptions of as many more yet by the same Authors, from the same places, which yet are but two small spots in respect of the vast Spaces, and variety of Soils and Climates yet unsurvey'd; and 'tis not to be doubted but that the Earth, and Air, and much more yet, the Seas of several Countries and Climates would afford as great varieties of Birds, Beasts, Insects and Fishes, if there were found knowing and diligent searchers and describers of them: And that this is so, I shall mention only one Instance, because 'tis pertinent to the present Subject, namely, that I have had a peculiar kind of *Nautilus* brought from the *Caribys*, where they are in great plenty, and yet I do not find any Author has taken notice of them, nor could I ever meet with more than one Man that had taken notice or knew any thing of them, tho' the Island has been long inhabited and planted by the *English*; which Shell I have formerly shewn to this Society, who were satisfy'd by the Characteristick that it is a Species of the *Nautili*.

*He grants that Europeans have not seen all the varieties of life forms on Earth.*

*And even when they see a peculiar form they mostly do not take notice of it.*

And as we yet want a *Hortus Sinensis, Japonensis, Tartaricus, Canadensis, Virginianus, Brazilianus, Peruvianus, Americanus,* &c. so we want the Natural Histories of the Animals of most kinds, of those places, and even of the Fishes which are frequently enough met with by Navigators, tho' not further taken notice of than as they may be useful for their present Food, or the like. We are therefore two [*sic*] hasty in our Computations and summing up all we have, and concluding that must be the summe of all that can be had; for that there are yet many particulars behind, that must come into the same account before the inclosure be fully made and the Books be shut, if at least a full Account be expected. We are informed by Mr. -----*Cole*, and divers other late inquisitive Men, how many new things have been discover'd here at home, where [p.435] yet there have not formerly been wanting inquisitive Men; what then may we not expect from other places where none such have ever come, at least, that we know of?

Again, how apt should we have been, if there had been found a Petrify'd *Stella arboresceus Rondeletii*, before we had been certify'd of the existence of such a strange shaped Fish of the Species of the *Stellæ*, to have concluded there had never been such a Fish, because it differs so very much from the Star-fishes or five Fingers, as they term them, commonly taken on our Coasts? The like may be infer'd concerning the strange variety I have seen of the *Echini* brought from several parts; for they differ much more from one another than the Helmet Stones, which I have hitherto seen, do from several sorts of them: The like may be said of the varieties of Sharks Teeth, as to one another, and as to the *Glossopetræ* found upon the Land.

So that upon the whole we may conclude, that it does not necessarily follow, that those *Species* of *Nautili*, must be now wholly lost that produced the moulding Shells of these *Ophiomorphite* Stones, we find here in *England*, because they are not now found upon our Shores, nor because we cannot now certainly affirm where they are to be found, and therefore that the induction or inference is made from too few Particulars, and may, nay, ought to be examined a-new, when we can procure a more full Account of the Productions of the Shores and Oceans, which Time and Industry may possibly effect.

But not further to insist upon this way of Defence, we will, for the present, take this Supposition to be real and true, that there have been in former times of the World, divers *Species* of Creatures, that are now quite lost, and no more of them surviving upon any part of the Earth. Again, That there are now divers *Species* of Creatures which never exceed at present a certain Magnitude, which yet, in former Ages of the World, were usually of a much greater and Gygantick Standard; suppose ten times as big as at present; we will grant also a supposition that several *Species* may really not have been created of the very Shapes they now are of, but that they have changed in great part their Shape, as well as dwindled and degenerated into a dwarfish Progeny; that this may have been so considerable, as that if we could have seen both together, we should not have judged them of the same Species. We will further grant there may have been, by mixture of Creatures, produced a sort differing in Shape, both from the Created Forms of the one and other Compounders, and from the true Created Shapes of both of them. And yet I do not see how this doth in the least derogate from the Power, Wisdom and Providence of God, as is alledged, or that it doth any ways contradict any part of the Scripture, or any Conclusion of the most eminent Philosophers, or any rational Argument that may be drawn from the Phænomena of Nature; nay, I think the quite contrary Inferences may, nay, must, and ought to be made.

*Here Hooke finally gets around to present his true feelings about the issue. His argument runs something like this: Granted, we are not familiar with all the vast varieties of animal and plant life in the world, so that we should not insist that just because we do not know the life forms to be alive, they are necessarily not now existing. But for now, we will assume that extinction has taken place and that the animals represented by their fossils are no more. Some forms used to be big but are now small, and some forms have changed so much that they cannot now be identified as the same species. What he asserts here, he believes, is not against God or scripture.*

For first we do find that all individuals are made of such a Constitution, as that beginning from an Atom, as it were, they are for a certain period of Time increasing and growing, and from thence begin to decay, and at last

Die and Corrupt. And in every part of their Life they are in a continual change or progress, from more perfect to more imperfect, there being a continual growth of Death and Decay to the final Dissolution; yet this is not Argument against the Ominipotence, Providence and Wisdom of the Creator, who thought fit so to Create them. Again, we find that the Powers and Faculties of the animated Bodies do continually exert a succession of differing Effects, and continually change the Figures and Shapes from one degree to another. As we see that there are many changings both within and without the Body, and every state produces a new appearance, why then may there not be the same progression of the Species from its first Creation to its final termination? Or why should the supposition of this be any more a derogation to the Perfection of the Creator, than the other; besides, we find nothing in Holy Writ that seems to argue such a constancy of Nature; but on the contrary many Expressions that denote a continual decay, and a tendency to a final Dissolution; and this not only of Terrestial Beings, but of Celestial, even of the Sun, Moon and Stars and of the Heavens themselves. Nor have [p.436] I hitherto met with any Doctrine among the Philosophers, that is repugnant to this Doctrine, but many that agree with it, and suppose the like States to happen to all the Celestial Bodies, that is, to the Stars and Planets that happen to the Individuals of any Species; and consequently if the Body of the Earth be accounted one of the number of the Planets, then that also is subject to such Changes and final Dissolution, and then at least it must be granted, that all the Species will be lost; and therefore, why not some at one time and some at another? This Objection therefore, I conceive, is of little validity against the Doctrine I have delivered, and therefore I shall proceed to the second Objection, and examine the Validity thereof.

It is Objected then in the Second place, That since it is manifest, that there are many curiously figured Bodies found in the Earth, which cannot be imagin'd to be produc'd by the Causes and Means that I have alledg'd, as the Shapes of Salts, Sparrs, Ores, Chrystals, and divers other kinds of regular mineral Bodies, also, Agates Mochuses, curiously speckled Marbles, and the like. Now, since it must be granted, that they are made by a Plastick Faculty, why may not that Faculty extend also so far as to be the cause of those other Figured Stones, which resemble Shells or other Animal or Vegetable Substances?

To this I answer, That tho' it be manifest, that Salts, Chrystals, Sparrs, &c. do plainly receive their regular Figures from the Texture or Nature of their own Parts, as is evident, most especially in the Chrystallization of Salts, and the Petrifactions of the like Figur'd Substances, yet the Figures, and painted and stained Shapes, as it were of *Agates, Mochus's* and the like, are not to be ascribed to the Designs of Nature, but to the Productions of Chance; for instance, the Pictures that in *Mochus's* seem to represent Trees, Hills, Houses, and other perspective Representations, they are no otherwise caused than by some Clefts, or Flaws in the said Stones, into which some colour'd Juices have insinuated themselves, and by that means formed those Representations which appear in the Body of the Stone, and that this is so, and may be Artificially produc'd by several Bodies and Liquors, which have not affinity, either with *Agate, Mochus,* or Marble, I can make it plainly appear by Experiment, which, if it be thought fit, I shall produce either now, or the next Meeting.

[*The Experiment here mention'd, was by taking two flat Marbles or Glass-plates, and laying upon one of them several drops of a dark Oil-colour,*

---

*Hooke, 170 years before Darwin's* Origin of Species, *is arguing for a progression of changes at the species level from its first appearance ("Creation") to its final extinction ("termination")— "some at one time and some at another."*

*He extends his evolutionary ideas to the celestial domain as well.*

*The first objection, that extinction is not compatible with scripture, therefore, is demolished as being of little validity.*

*The second objection is that the other "figured stones"—minerals and crystals, are formed by some other method, so why cannot the "figured stones" that resemble animals or plants be formed in the same way?*

*Hooke's ideas on the formation of crystals were much ahead of his time and more sophisticated than those of Steno. He believed that the external expression of a crystal is due to the internal arrangement of its particles.*

*Picture agates are caused by impurities deposited along cracks of the stones, and Hooke demonstrates with an experiment.*

*Richard Waller describes the experiment mentioned by Hooke, as he apparently witnessed it.*

*such as Painters use, and pressing the other flat Stone or Glass upon it, by that compressure several curious Representations, like the branchings of Vegetables, and the like, were exhibited; which explain'd the Representations in Agates, &c. a different colour'd mineral Juice insinuating itself into the Clefts or Interstices of the Stone, and afterwards petryfy'd to an equal hardness with the rest of the Stone; tho' many times there is a different hardness in the Veins, or Representations from the other parts of the Stone, as is seen in Marbles and other veined Stones. R.W.]*

*Note by Richard Waller*

**No. 17.** *A Discourse of Earthquakes in the Leeward Islands. Read July 23, 1690.*

[p.416] The greatest Objection that has hitherto been made against a Theory which I have several times discoursed of, to give a rational and probable account of the Reasons of the varieties observable in the present superficial Parts of the Globe of the Earth; which was, that all those inequalities of its Surface had been caused by the Power of Earthquakes, or Eruptions of fiery Conflagrations inkindled in the Subterraneous Regions, which by that means did sometimes raise Mountains, Hills, Islands, &c. and sometimes produce the quite contrary Effects, by levelling of Eminences of sinking of Places, swallowing up Rivers, and making Lakes of Land, or sinking Lands under the Sea, and the like.

*Reiteration of Hooke's "earthquake" theory.*

The greatest Objection, I say, against this, I find hath been, that there were wanting Instances to confirm it from History. For that, all Places, Countries, Seas, Rivers, Islands, &c. have all continued the same for so long time as we can reach backwards with any History: All *Greece* and the *Greecian Islands, Italy, Ægypt,* &c. are all the same as they were above two Thousand Years since, and therefore they were so from the Creation, and will be so to the general Conflagration; and as to the effects of Earthquakes, First, They have happened but seldom: And, Secondly, They have not produced any notable Change, such as I have supposed them to be the Authors and Efficients of; so that it seems but a bare Conjecture and without Ground or Foundation sufficient to found and raise such a Superstructure of Conclusions as I have thereupon raised.

*The greatest objection to his theory is that there is no evidence in written history for more than 2,000 years that the terrestrial surface was ever different from what it is today and therefore it must have always been the way it is since the Creation.*

For Answer to which I shall not now repeat what I have formerly produced here, an alledged to that purpose, such as were the Instances that were to be met with here and there dispersed in antient Writings; since many of those Occurencies having been long since produced, and the relations of them made by such as were not Eye-witnesses, many of the particular matters of Fact have been doubted or disputed; I shall therefore take notice of some particular Instances which have happened within our own Memory, and more particularly of this late instance which hath happened in the *Antilles*, of which we have an account but this last Month in the *Gazet*, namely, in that of *June* the *30th.*, and another in that of *June* the *16th.* preceding; both which Relations, tho' they are but short and imperfect, as to what I could have wished and shall indeavour to obtain, yet as they are, they will be found to contain [p.417] many Particulars,

*For answer, Hooke gives examples of occurrences happening in his and his audience's lifetime.*

which do very much illustrate and confirm my Conjectures. And tho' the particular effects were not so great as to equalize those which I have supposed to have been the productions of former Eruptions, such as the raising of the *Alpes, Pyreneans, Appennines, Andes,* and the like Mountains, or the making of new Lands, Islands, &c. or the sinking of Countries, and drowning of Islands as the *Plantonick Atlantis* and contiguous Islands, yet if they be considered they will be found to be of the same Nature, and to differ only in Magnitude, *Secundum Magis & Minus,* but not in Essence.

*Secundum Magis & Minus = according to a greater or less extent*

The first Account is dated from *Nevis, April* the thirtieth, in these words, ["On *Sunday* the sixth Instant, about five a Clock in the Evening, was, for some few Minutes, heard a strange hollow noise, which was thought to proceed from the great Mountain in the middle of this Island, to the admiration of all People; but immediately after, to their greater Amazement, began a mighty Earthquake, with that violence, that almost all the Houses in *Charles* Town, that were built of Brick or Stone, were in an instance levelled with the Ground, and those built of Timber shook, that every Body made what hast they could to get out of them. In the Streets the Ground in several places clove about two Foot asunder, and hot stinking Water spouted out of the Earth a great height. The Sea left its usual Bounds more than a third of a Mile, so that very large Fish lay bare upon the Shoar, but the Water presently returned again. And afterwards the same strange motion happened several times, but the Water retired not so far as at first. The Earth, in many places, was thrown up in great quantities, and thousands of large Trees went with it, which were buried and no more seen. Tis usual almost at every House to have a large Cistern, to contain the Rainwater of above nine or ten Foot deep, and fifteen or twenty Foot Diameter, several of which, with the violence of the Earthquake, threw out the Water eight or ten Foot high; and the motion of the Earth all over the Island was such, that nothing could be more terrible. In the Island of St. *Christophers* (as some *French* Gentlemen, who are come hither to treat about the exchange of Prisoners, do Report) there has likewise been an Earthquake, the Earth opening in many places nine Foot, and burying solid Timber, Sugar-mills, &c. and throwing down the Jesuits Colledge, and all other Stone Buildings. It was also in a manner as violent at *Antego* and *Montserrat*; and they had some feeling of it at *Barbadoes.* Several small Earthquakes have happened since, three or four in twenty four Hours; some of which made the biggest Rocks have a great motion, but we are now in great hopes there will be no more."]

*A newspaper (the* Gazet) *account of earthquakes in the Leeward Islands*

This is the whole of the Relation from *Nevis.* But the other Account from *Barbadoes,* of the *23d.* of *April,* taketh notice of other particulars than what are mentioned in this Letter; the Printed Account is this that follows. ["About three Weeks since there were felt most violent Earthquakes in the *Leeward Islands* of *Montserrat, Nevis* and *Antego*; in the two first no considerable hurt was done, most of their Buildings being of Timber; but where there were Stone Buildings they were generally thrown down, which fell very hard in *Antego,* most of their Houses, Sugar-mills and Wind-mills being of Stone. This Earthquake was felt in some places of this Island, but did no manner of hurt to Men or Cattle; nor was any lost in the *Leeward Islands,* it happening in the Day-time. It is reported to have been yet more violent in *Martinico,* and other *French* Islands. And several Sloops who came from *Nevis* and *Antego* passing between St. *Lucia* and *Martinico* felt it as Sea: The agitation of the Water being so violent, that

*Another account from the Barbados*

they thought themselves on Rocks and Shelves, the Vessels shaking as if they would break in pieces. And others passing by a Rock and unhabited Island, called *Rodunda*, found the earthquake so violent there, that a great part of that Rocky Island split and tumbled into the Sea, and was there sunk, making a noise as of many Cannon, and a very great Cloud of Dust ascending into the Air at the fall."] ["Two very great Comets have lately appeared in these parts of the World, and in an Hour and a quarters time the Sea Ebbed and Flowed to an unusual degree three times."] In these Relations are many considerable Effects produced, which will much confirm my former Doctrine about the Effects of [p.418] Earthquakes. And First, It is very remarkable, that this Earthquake was not confined to a small spot or place of the Earth, such as the Eruption of *Ætna* or *Vesuvius* out of one Mouth, but it extended above five Degrees, or three hundred and fifty Miles in length; namely, form *Bardadoes* to St. *Christophers*, and possibly, upon inquiry, it may be found to have gone a great deal further, and to have produced Effects in *Statia* St. *Martin, Anguilla, Porco Rico*, or some other of those Islands in the North-west of St. *Christophers*, where, by the Relation, it seems to have been the most violent: And tho' possibly there might not be opportunities of feeling or taking notice of the effects in all places of the Sea where it might have been felt; yet by those few Instances which are related, we may probably conjecture, that its effects might be very considerable, and sensible a great way in breadth under the Sea; for we find that the Strokes or Succusions thereof were felt by the Vessels sayling over some parts of the Sea so affected; and those so violent as if the Vessels had struck upon Rocks, which could be from nothing else but the suddain rising of the bottom of the Sea, which raised the Sea also with it, like Water in a Tub or Dish: And that this was of that Nature does further appear by the unusual Tides at the *Barbadoes* mentioned in the last Relation, *viz*. That in an Hour and 1/4 the Sea Ebbed and Flowed three time in an unusual Degree; which, in probability, were nothing else but Waves propagated from the places where the Ground underneath, and the Sea above, had been by the Concussions of the Earthquake raised upwards. This appears also farther by the recess of the Sea from the Shore at *Nevis* 1/3 of a Mile; for the whole Island being raised by the Swelling or Eruption of the Vapour or Fire underneath, made the Sea run off from the Shores, till it settled down again into its place after the Vapour had broken its way out through the Clefts that were made by those Swellings: From all which Particulars, and several others, 'tis manifest, that the space of Earth raised or struck upwards by the impetuosity of the Subterraneous Powers that caused it, was of great Extent, and might far exceed the length of the *Alps*, or the *Pyreneans*, &c. But there may be other Instances also produced of the great extent of the Powers or Effects of Earthquakes, as those I have formerly mentioned to have happened in *Norway* about thirty Years since; and those which happened in the Northern parts of *America* of a later date.

*Mention of these comets indicates Hooke is still trying to make connections among different effects in search of a unifying hypothesis.*

Another particular notable in this, is the Recess of the Sea from the Shore, and the leaving the Fish upon the so raised bottom; and tho' this part soon after sunk again, so that the Sea returned to its former bounds, yet if some other parts of the Subterraneous Ground had filled up the made Cavity, or that they had so tumbled as to support the so raised parts, Instances of which kinds of Accidents may be produced from other Earthquakes, then it would have left some such kind of Tract as it is now in *Virginia*, where, for many Miles in length, the Low-land is nothing but Sea-sand and Shells, which have been, in probability, so raised into the

*Momentary retreat of the sea, leaving the bottom exposed, shows that such events can happen and sometimes such exposed areas do become lowlands.*

Air, and there supported and continued from sinking again and being covered with the Sea: Of Shells taken up from this Tract, there can be no doubt that they have belonged to Fish of their kind, they remaining hitherto perfect Scallop Shells; of which kind there are some in the Repository.

A Third particular Remarkable, is the overturning and burying of thousands of Trees which were no more seen, being covered by the Earth which was thrown up by the Eruption. This gives us a very plain Instance of the manner how Trees that are now found in divers parts of *England* buried under the Ground, may have come to have been there so disposed and deposited; for tho' possibly in those places there may be no such Trees now growing; and tho' we have no History when there were, or of any such Eruption that might have so overturned and buried them; yet the Records that we have of the antient or former State of those Parts, are not so full and particular, but that we may well enough suppose that such Catastrophies may have happened long since we have begun to have Writings and Records in *England*; that is, since the time that the *Romans* first conquered this Isle; and yet not find any mention thereof there made; since possibly those that might be in or near those Parts might have perish'd with it, and those which were at a [p.419] distance took little notice or regard of what they had little concern for: Besides, in those Days very small were the number that could Write and Read, and fewer were those that minded any thing the effects produced by Nature: What was written was either somewhat relating to Religion or Civil History, very few and rare are the Instances that can be met with of Natural History; and it has not been a Defect peculiar to these parts of the World only, but was taken notice of two Thousand Years since by *Aristotle* upon this very account, as we find in the fourteenth Chapter of his first Book of *Meteors*; "Moreover (says he in the beginning of this Chapter) the same Parts of the Earth are not always dry Land, or always covered with Water, but they suffer a change from the rising of new Rivers, or decay and drying of old; therefore also in places near the Sea there are wont to happen these changes. So that those which are Land, or those which are Sea do not always remain so; but where was Land there is Sea, and where was Sea there is Land; and we are to conlude these changes to happen according to some order. Now (says he) because many of these changes happen but slowly in comparison to the quickness and shortness of the Life of Man, therefore they are hardly taken notice of, a whole Generation having passed away before such changes have come to perfection. Other Catastrophies that have been more quick, have been forgotten, by reason that such as escaped them were removed to some other parts, and there the Memory of them was soon lost; at least a longer tract of Time did quite obliterate the remembrance of them, and the transplanting and transmigration of People from place to place much contributed thereunto." This is made plain enough by the little remembrance was found in *America* of their preceding Estate, when they are first visited by the *Spaniards* and other *Europeans*.

A fourth particular Remarkable in these relations, is the Chapping and cleaving of the Earth and Rocks, and the spouting out of them of stinking Water to a great height; as also of Smoke or Dust, which serves to explain the Reason and Causes of the Flaws and Veins of Marbles and other Stones; for by the Power and Violence of the subterraneous Heavings or Successions the stony Quarries become broken, flawed and cleft, and Subterraneous Mineral Waters impregnated with Saline, Metalline,

*Burial of thousands of trees.*

*Some reasons ancient written accounts are rare: Eyewitnesses perished in the disaster; few could write in olden days; those who could were not interested in the effects of nature.*

*Quote from Aristotle that supports Hooke regarding the exchange of land and sea areas.*

*Another Aristotelian point that Hooke adheres to is that the changes to the terrestrial surface happen slowly, by degrees, in comparison with the lifetime of humankind; and memory is short. In other words, the effect of violent earthquakes is incremental.*

*Earthquakes cause cracks in rocks which can be filled with other minerals.*

Sulphureous, or other Substances are driven into them and fill them up, which having petrified Qualities in them, do, in process of time, petrify in those Clefts, and thereby form a sort of stony Veins of different Colour, Hardness, and other Qualifications, than what the parts of the broken Quarry had before, and oft time inclose divers other Substances by their petrifying Quality, which have happened to fall into those Clefts, and thence sometimes there are found Shells petrified in the middle of the Vein, as I have seen, and other Substances. These Clifts or Chaps happening not only upon the Land, but even under the Sea; so that not only the Sea-water may descend and fill up those Clefts, but it may carry with it Sands, Shells, Mud, and divers other Substances from the bottom of the Sea, that then lay above it, there to be, in process of time, changed into Stone somewhat of the Nature of that which hath been so cleft.

*How the veins in marble are formed.*

Fifthly, 'Tis worth noting, that this Earthquake happened at so great a distance from the main Land and great Continent, and that the noise of the same was first observed to begin at the great Mountain in the middle of the Island of *Nevis*, not but that it might in other parts have begun sooner or at other times; from which I draw these deductions. First, That is seems probable, that this great Mountain may have been formerly visited with Eruptions; and possibly might have been first produced by some such Power, and so have great Cavities within its Bowels produced by such a preceding Eruption, the dislocated Parts not returning each to its own place. And next, that it may hence seem probable, that some such preceding Earthquake (tho' then possibly more violent before the foment of the Fire was by inkindling exhausted) might, not only be the cause of raising this Mountain, but of lifting up from the bottom the whole isle, nay, possibly of all the Islands of the *Antilles*, since one seems as possible as the other, and the Northern of them all seems to hint as much, if considered, in the Map; besides, there seems to [p.420] be many Instances of a like Nature, as in the *Canaries, Tenariff* seems to be a most remarkable Character of such a Supposition; to this may be added *Del Fuogo* among the *Azores* and the Island of *Madera, Sicily, Strombulo,* and *Lipary* in the *Mediterranean, Iseland* in the North Sea, *Mascarenos* near *Madagascar*; to this I may also add the many Islands of the *Archipelago,* which, tho' they have now no great signs of burning Mountains, yet to this Day Earthquakes are very frequent, and antient Traditions do preserve somewhat of the memory of very great alterations that have happened in those Parts by such sorts of Causes; but I will not now meddle with that kind of History, nor of Mythology, having said more concerning it in a more proper place; but I shall rather on this occasion take notice of those Islands that have *Vulcanoes* in or near them, which to me seem to proceed from the same Cause and Principle. And I do not question, but that all those Islands which lie so far in the Sea, if they were thoroughly examined, would plainly manifest whence they have proceeded by Characters of Nature's Writing, which to me seems far beyond any other Record whatsoever. Here I conceive it Lawful and Philosophical to *Jurare in Verba,* when Nature speaks or dictates; however, I shall leave it free to every one to judge, as he thinks most reasonable.

*Reminder: The spelling of names and text is as printed in the 1705 Waller edition.*

*Even though the motto of the Royal Society is not to judge in the words of received opinion, Hooke feels that we can judge in Nature's words when It speaks— i.e., natural records such as those written in rocks and fossils are better documents than those of humans.*

Sixthly, 'Tis very remarkable that the Isle of *Rodunda*, which it seems is all an uninhabited Rock, was split, and a part of it tumbled down and sunk into the Sea; upon which occasion it seems it made a prodigious noise as of many Cannon, and sending up at the same time a great Cloud of Dust, as they term it, which, in probability, was also mingled with Smoak.

Which puts me in mind of the Phænomena I observed lately, when the Pouder-mill and Magazine at *Hackny* blew up; for besides the very great noise of the Blow which I heard, being within a Mile of it, in the Fields, I observed immediately, a great white Cloud of Smoke to rise in a Body to a great height in the Air, and to be carried by the Wind for two Miles and better without dispersing or falling down, but perfectly resembling the white Summer Clouds: But this only by the bye. From these Phænomena of the Earthquake it seems very probable, that it proceeded from such Subterraneous inkindling as resembles Gun-powder, both by the noise it yielded, and in the suddenness of its firing, and its powerful Expansion when fired; for the noise was as of many Cannon; this alone proves it to be very suddain. Next the splitting of the Rocky Island proves its Power to be very great; this is proved yet farther by the Blow and Strokes it communicated to the Sea, and so to the Ships that failed upon it; for no slow motion whatever could have communicated such a Concussion through the Water to the Vessels upon it; but it must be as suddain as that of Pouder, otherwise the stroke of the Earth upon the incumbent Seas, would never have had the like success; for it had been a gradual rising of the bottom, the Sea would gradually have run off from it, and upon its sinking again have gradually returned, and the Vessels on it would only have been sensible at most but of a Current or Running of the Water to or from the place sinking or rising, somewhat like the effect that happened at *Nevis*; which doth plainly shew, that, besides the suddain Strokes or Concussions; there was also a considerable rising and sinking of the whole Island as to the level of the Sea. But that which I principally note under this Head is, that a good part of the said Island tumbled down and was sunk into the Sea, which gives an account how many parts of the Earth come to be buried under Ground and displaced from their former Situations, and thence how Ships, Ankers, Bones, Teeth, &c. that have sometimes been digged up from great depths, may have come to be there buried.

*The power of splitting (or faulting) of rock is likened to that of a great explosion of gunpowder.*

Seventhly, 'Tis remarkable also, that this Eruption sent up into the Air great Clouds of Dust and Smoke, which for the most part must soon fall down again into the Sea, or contiguous parts of the Island. This will give a probable account how the Layers of the Superficial Parts of the Earth may come to be made; for the bigger part of this Dust must come down to the bottom first and settle to a certain thickness and make a Bed of Gravel, then will follow Beds of coarse Sand, then Beds of finer and finer Sand, and last of Clays or Moulds of several sorts; again much of that which fell upon the [p.421] higher parts of the Island, will, by the Rivers, be washed down into the Vales, and there produce the like Beds or Layers of several kinds, and so bury many of the parts that were before on the Surface. Thus Plants and Vegetable Substances may come to be buried, and the Bones and Teeth of the Carcasses of Dead Animals: These may also sometimes be buried under Beds or Crusts of Stone, when the parts that thus make the Layers chance to be mixed with such Subterraneous Substances as carry with them a petrifying Quality. But I shall not detain you any longer with farther Deductions from these few Remarks we find in these two casual Relations of this Earthquake; I shall only add, that I could heartily wish that some care were taken that a more particular account might be procured of it whilst the effects thereof are fresh in Memory, that they might be Recorded and added to the Collections of Natural History. And for the same end it were desireable to know what former Earthquakes have been taken notice of in these, or any other of these Islands, as *Jamaica, Cuba,*

*While Steno's concept of strata is explicit in his Law of Superposition, Hooke's is implicit. Typically, however, Hooke extends his idea of layers further to explain also the gradation of grain size in æolian (after an eruption), fluviatile, and marine deposits. See Drake and Komar (1981).*

*Hispaniola, Porto Rico,* &c. for that the Memory of such Accidents, if they be not Collected and Recorded whilst the Spectators are in being, are soon forgotten and lost or not regarded by the succeding Generations, as *Aristotle* has taken notice of also in the Chapter I before quoted.

What is most remarkable in these Earthquakes in the *Leeward* Islands, is, that they have all happened to places not far distant from the Sea, or even under the Sea itself, though the Eruptions have been, for the most part on the Land. So that there doth seem to be somewhat of Reason to Conjecture as *Signor Bottoni* in his *Pyrologia Topographica*, that the saline Quality of the Sea-Water may conduce to the producing of the Subterraneous Fermentation with the Sulphureous Minerals there placed, which the Experiment lately here exhibited at a Meeting of this Society, does yet make more probable; for by that it was evident, that the mixing of Spirit of Salt with Iron, did produce such a Fermentation as did produce a Vapour or Steem, which by an actual Flame was immediately fired like Gun-pouder, and if inclosed, would, in all probability, have had a like effect of raising and dispersing of those parts that bounded and imprisoned it. Now, 'tis evident that the melted Matter which was vomited out of *Ætna* in the Year sixty nine (of which we have a part now in the Repository) was very much like to melted or cast Iron, and I doubt not but that there may be much of that mineral in it; besides, the Foot of that Mountain does extend even to the very Sea, and in al probability may have Caverns under the Sea itself, which is argued also from the Concurrency of the Conflagration of *Strombolo* and *Lipary*, Islands considerably distant from it by Sea, at the same time, where it is generally believed that there may be Subterraneous Cavernous Passages between them, by which they communicate to one another; so that sometimes it begins in *Ætna,* and is communicated to *Strombolo*, and reciprocally communicated to *Mongibel*.

This possibly may afford a probable Reason why Islands are now more subject to Earthquakes than Continents and inland Parts; and indeed how so many Islands came to be dispersed up and down in the Sea, namely, for that these Fermentations may have been caused in the parts of the Earth subjacent to the Sea, which being brought to a Head of Ripeness, may have taken Fire, and so have had force enough to raise a sufficient quantity of the Earth above it, to make its way through the Sea, and there make itself a vent, as that of the *Canaries* did in the Year 39, which, if sufficiently copious, may produce an Island as that did also for a time, though it hath since that time again sunk under the Surface of the Sea. But the Island of *Ascension*, which, by all appearances, doth seem to have been the same way produced, doth still remain as a witness to prove this Hypothesis. A like Testimony to this, of the Cause and Manner of their Production, I take the Island and Pike of *Tenariff* to be, so *Hecla* of *Iceland*, so *Bearenberg* of *John Mayens* or *Trinity* Island, so *Del Fuego* of the Icelnds [sic] of *Cape Verd*, so *Ternate* of the *Moluccas*, and the Island of *Mascarenas*, of the Islands about *Madagascar* among the *Antillas* or *Caribes*, all which do seem to me to be remaining Testimonies how, and in what manner, and by what means those other Islands which have now worn out the marks of their first Origination, were at first [p.422] produced. And tho' the Fires be extinct in many of the other Islands, yet 'tis observable, that the prodigious high Mountains or Sugar-lofe Pikes or Hills do yet remain as marks of what they had been heretofore; so the Pike of *Fayal* among the *Terceras*, and the whole Island of St. *Helena*, and several others of those about *Madagascar* and of the *East-Indies*, and of those of the *Antilles*, and that of St. *Martha* mentioned by *Dampire*, do seem to me to be plain evidences of the former and Original Causes of them all.

Nor do I in the least doubt but that an inquisitive Person who should purposely survey all other Islands that wanted these Marks or Tokens of such Eruptions, might find enough of other Indications to manifest by what means they so came to be placed in the Sea, so far from any part of the Continents they are opposite to. Nor do I conceive they were all thus formed at once, but rather successively, some in one, some in other Ages of the World, which may probably be in some measure collected from the quantity or thickness of the Soil or Mould upon them fit for Vegetation; whence the Island of *Ascension* may be rationally concluded to have been a Production of not many Ages, and the *Bermoodas* [*sic*] also of not very many more, because of the thinness of such a Soil. So also the Island of *Barbadoes*, and some others, whose Mould is yet but thin in respect of what it is in some others, and especially in those of greater Magnitude and in the greater Continents.

*Hooke hypothesizes on the ages of these islands—that some of them are very young by the thinness of the soil that formed from the erupted material—another astute observation.*

Hereupon possibly it may be inquired why those greater Islands and Continents should be of great Antiquity than the smaller Islands. To which I answer, that in the first Ages of the World there were much greater Magazines, or Stores of the Materials fitted for this purpose, which being first kindled threw up from under the Sea, with which they were covered, vast quantities of it all at once, and thereupon those Magazines became in a manner exhausted, yet not so totally as not to leave some smaller parcels of those Substances so disposed, as not to be ready for inkindling together with those greater; besides there remained other smaller parcels of it disposed and placed in other parts of the Globe sufficiently distant from them, not to be affected or inkindled at the same time, as those I have mentioned to have been the causes of the Islands far distant from the Continents. Nor do I conceive that all those Clusters were all thrown up at once, as the *Greecian* Islands in the *Archipelago* the *East-Indian* in that part called the *South-Sea*, The *Maldivia Islands* near the Coast of *Malabar*, the Islands scattered at the North of *Madagascar*, the Islands to the Southwest of St. *Helena* in the *Atlantick* Ocean, *Finiduda dos picos*, the Isles of *Cape Verd, Canaries, Terceras, Orcades,* &c. also the *Gallopegas* and others in the *Pacific* Sea or *Mar del Zur*; but rather that some were made in one Age, some in other Ages of the World. And this was timed as the several Magazines came to be ripened and then fired; they only indicating, as I conceive, that in those places of the Terrestrial Globe, there were placed the proper mineral Foments or Seeds as it were of them, which, when the convenient times were come and accomplished, then they were put into Act, and then they produced their Effects, which are the Islands that now remain the lasting Monuments of them. Nor can I suppose that all the Magazines of the Earth of this kind are blown up and spent, but that there may be many other yet remaining for future Ages to be made sensible of their Effects. Nor can I be fully satisfied that all the main Continents were thrown up or made Land all at once. Nor have we any proof that the Continent of *America* was in the time of *Noah*'s Flood, nor indeed how large the habitable World then was, but certain we are, that what was then in being was all overflowed and drowned by it, and all living Creatures, except those preserved in the Ark with *Noah,* perished by it. But whether the dry Land that appeared after the Flood, were the same with that before the Flood, is a question not easily determinable; to me it seems that the

*Although the term "magma chambers" is unknown to Hooke, he seems to be describing something similar in his word "magazines."*

*The "stores of the materials" in these "magazines" can come to the surface at any time to form volcanic islands.*

preceding Earth was wholly changed and destroyed, and that there was produced a new Earth which before that had not appeared; and this Doctrine seems to be indicated by that Text in *Genesis* vi. 13. *And God said unto* Noah, *the end of all Flesh is come before me; for the Earth is filled with violence through them, and behold I will destroy them with the Earth.* And Again *Chap.* viii. 21. *I will not again curse the ground any more for Man's sake.* And 2 *Pet.* iii 5. *By* [p.423] *the word of God the Heavens were of old, and the Earth standing in the Water and out of the Water.* (ver. 6.) *Whereby the world that then was, being overlowed with Water, perished.* But the clearing this Doctrine by the Expressions in Scripture I shall leave to the Divines; nor shall I in the least interfere with them: However, it seems to me, that the Expression of *Breaking up the Fountains of the great Deep,* might signify the raising up of the bottom of the Sea; and the *Water prevailing so as to cover the top of the highest Mountains,* might denote to us the sinking or subsiding of the former part of the dry Gound: So as *the former was wholly drowned and destroyed,* which was *Cursed* for Man's sake, so a new one was raised, which God promised should not be *Cursed for Man's sake,* as the former had been ; but this only by the bye. Certain I am that I have never yet met with my self, or heard of any other that hath any Records of the Age of *America*, which, for any thing appears, may have been much younger than the Flood of *Noah*: Nay, I believe it will be pretty difficult to prove even these Islands of *England, Scotland* and *Ireland* to have been in being ever since that Flood, and much more that there were such before it. And tho' some may Conjecture that they have been so (which is the most that any one can do) yet others may Conjecture that they have not (which is every deal as valid). The same may be said of a very great part of the Earth, without any trespassing upon our Faith or Religion; nay, it was we know, not long since, that a Bishop was condemned of Heresy because he asserted Antipodes. So skillful were some of our Fore-fathers in the Geography of the Habitable Parts of the World, or of the Figure of the Earth; and I do very much question, whether any Inhabitant of *Europe, Asia,* or *Africa* had ever any knowledge of *America* till within these last three hundred Years. But my present subject is not so properly to search and inquire into the History, as to find out what have been the Natural or Physical Causes of their Productions, Situations and Forms, and that, I think, I have shewn to have been in probability some preceding Earthquakes, which Earthquakes may have been caused by Subterraneous Fermentations and Accensions.

But some perhaps may except against this Doctrine as supposing it Derogatory to Divine Providence to assert any other Cause but the immediate Hand of God. To which I Answer, That 'tis not denying of Providence to inquire into, or to assign the Proximate Causes of Phænomena in Physical Subjects. For that we have Instances in the sacred Scriptures of such Explications, as in the case of the *Israelites* through the Red Sea; where 'tis said, *The Lord caused an East Wind to blow, which made the Sea to go back and to leave the bottom dry Ground.* And at the Waters of *Marah* God shewed *Moses* a Tree, which, when he had cast into the Waters, the Waters were made sweet. So in the Description of the Deluge, we find that God caused it to Rain forty Days and forty Nights, and the Foundations of the great Deep to be broken up, and the Windows of Heaven to be opened; which denote by what Natural Means God was pleased to effect and Collect the great quantity of Water that was to drown and overflow the then Habitable Earth; and many other such Instances there are to be met with in Holy Writ, where the Physical Causes are explained, for it is the same Omnipotent Power which does influence the remote

*Given the religious attitudes of the time, Hooke was compelled to bring God into the picture, but mostly to argue that his own inquiries into the physical causes do not in any way deny the role of God.*

*Hooke says here, that one can debate whether any island or continent was there before or after the flood without trespassing on one's religious beliefs.*

*It is apparent in this passage that Hooke would have deplored the condemnation of this bishop.*

*Hooke's argument: The ultimate causes are the domain of God. Humans should be concerned only with "proximate" causes, and in discovering these physical causes, they do not deny God's role.*

Causes as well as the proximate and the universal Providence that ordereth all the effects, doth also determine and appoint all the Causes and Means conducing thereunto; nor is there a necessity of supposing new created Causes for all the effects that we are ignorant how they are brought to pass, or to belive every thing effected supernaturally, of which we cannot find out the Natural Cause; the Divine Providence is not less Conspicuous in every Production that we call Natural, and think we know the Causes of it, than in those we are less skillful and knowing in: 'Tis the Contemplation of the wonderful Order, Law and Power of that, we call Nature, that does most magnify the Beauty and Excellency of the Divine Providence, which has so disposed, ordered, adapted and impowered each part so to operate, as to produce the wonderful Effects which we see; I say wonderful, because every natural Production may be truly said to be a Wonder or Miracle, if duly considered; for who can tell the Cause of the Growth, Form, Figure, and all the Qualifications and peculiar Proprieties [424] of each, or any one Vegetable or Animal Species or individual? An observing Naturalist may perhaps tell the Steps or Degrees he has taken notice of in its Progress from the Seed to the Seed: Again, how he has observed the Seed to sprout, how that Sprout increaseth and forms itself of this or that Magnitude, Shape, Colour, &c. and how it produceth such a Flower, and after that Flower such other Seeds as that from which it sprung; He may also tell the Times and Seasons in which these Progresses have been or will be performed; but if it be inquired how the Progresses come to be acted, what is the moving Power, or what is the inlivening Principle that orders, disposes, governs and performs all these wonderful Effects, there he finds the *Ne plus ultra*, there is the Miracle that he may truly admire but cannot understand; however, *Est aliquid prodire tenus si non dature ultra*, let us first find the proximate Causes, and then proceed to the more remote; I think no one ought to be blamed or discouraged from searching after these Causes and Reasons of Natural Productions so far as the Powers he is endowed with will enable him; for this will more powerfully convince him of a Divine Providence that Rules and Regulates the things of this World, than all the other methods of Contemplation or Argumentation whatsoever.

*Hooke's eloquent argument for the study of the "Order, Law and Power" of the natural world—a process that magnifies the "Beauty and Excellence of the Divine Providence" rather than detract from it.*

**No. 18.** *The Fable of the Rape of Proserpina. Read April 12, 1693 although Waller dates it March 8.*

[p.402] I have some Years since propounded in this place my Conjectures for the explication of the Mythology delivered by *Ovid* in his *Metamorphosis*, namely, That he thereby designed to comprise a History of the Changes which had happened to the World from the beginning thereof to the times wherein he lived, which he signifies Ænigmatically by the first four Verses. The method of which is by personating Things and Powers: The one by Mortals, namely, material Things, the other by Immortals, namely, Powers, or Energies and comprising therein a three fold *Cabala* or Tradition, namely, A *Physical*, comprehending the Causes, Effects and Reasons; an *Historical*, comprehending the Times, Ages, Persons and Places, And a *Moral*, to make them Instructive and Useful for the Regiment, and moralising the more vulgar part of Mankind. In which he has indeavoured to follow the method of the Greek Poets, who, as I have formerly exemplified in *Homer, Hesiod*, and the rest of the Mythologers, did prosecute the same design. By the interpretation of which Mythology, if we could discover and find out the true Key, I conceive it would open and make manifest much of the History of the Catastrophies that have happened in the World, and of the places and Ages wherein they were produced.

That which makes me repeat this Notion at this time, is the dreadful Effects of the late Earthquake in *Sicily*, which put me in mind of what I had here formerly instanced, on the Story as it is delivered by *Ovid* in his fifth Book concerning the Rape of *Proserpina* by *Pluto*, which I then conceived and am still of the same Opinion (however others were of a contrary) was designed by the Poet to represent some dreadful Earthquake that had formerly happened in *Sicily*, not far, nor much differing from the Place and Effects of this late dreadful Catastrophy: Save only that we do not yet hear of the swallowing up, or the sinking down into, or under the Water of any considerable Country or Town, which I conceive is plainly specified by the Poet to have been effected by that which he designs to delineate: For *Proserpina* or *Abrepta*, as the Name signifies, is plainly described to have been seated in that place of the Island where the Lake *Pergusa* was in *Ovid's* Time, and where it remains, I suppose, to this Day: It is described to be the Daughter or Off-spring of *Ceres*, that is, a City or Place flourishing in a much civilised and well cultivated Country; for that *Ceres* doth plainly denote, and is so signified by the Poet.

*Prima Ceres Unco Glebam Dimovit aratro,*
*Prima dedit Fruges alimentaq; mitia terris;*
*Prima dedit Leges: Cereris sunt Omnia Munus.*

Ceres *with crooked Plough Gleabes first did turn,*
*And first taught Men to feed on Fruits and Corn:*
*She first gave Laws* Ceres *did all adorn.*

This Place or Country was also very pleasant and flourishing with Fruits, Flowers, Fountains, especially Woods, in which *Proserpina* is said to be disporting and innocently gathering Flowers, when *Pluto* suddainly seised her and carried her into the Earth; that is, whatever the place were, whether Town, Village or City before the Rape, it seems to have the same fate with St. *Euphæmia* mentioned by *Kircher* in his *Mundus Subterraneus*;

*A late (recent) earthquake in Sicily prompts Hooke to return to his idea that many ancient writings depict real natural histories couched in mythology, e.g., the fable of the rape of Proserpina by Pluto is interpreted to represent a dreadful earthquake that happened in Sicily in earlier times.*

that is, to have been by an Earthquake swallowed up into the Earth, and to have sunk so low, as to have left a Bason for a Lake for after Ages to this Day. Now, that this is plainly the Physical meaning of the Poet will appear plain if the whole Story be taken in. First, The Poet tells you the place to be *Trinacria* or *Sicily*. This Country, tho' most Delicious, Pleasant, rich in Soil, well Cultivated and Tilled, and [403] very much Civilised and Governed by good Laws, as is figured by *Ceres*, as I have mentioned, was yet seated on a Sulphureous, and Fiery, Cavernous Foundation, subject to Heavings, Tremblings and Earthquakes, expressed by the Giant *Typhæus*, who had endeavour'd to invade Heaven, that is, to belch up Flames, and throw Stones and Rocks against it, but at length came to be covered with this Island. Verse 346, to 356. Hereupon *Dis* or *Pluto*, that is, the Spirit of the Earth is fained by the Poet to be roused up to see least the Vault of the Earth over his place of Residence, should be broken by the fury of the Giant *Typhæus*; that is, by the working of the Sulphureous Vapours and Fulminations, Ver. 356----*Et Rex pavet ipse silentum*, Ver. 362. That is, the Earth rose and swelled, and there were Eruptions of Smoke signified by the Black Horses his Chariot was drawn by, and there was a general Earthquake over all the Island, but the great Eruption was at the place where the Lake *Pargusa* is. Here the Poet fains that *Pluto*, or the Subterraneous Powers were in Love with the Beauty of this Goodly Place personated by *Proserpina*. *Venus*, or Youthful Beauty or Thoughts doth excite *Cupid* or Love and Desire in this Terrestrial or Subterraneal Power to take away this pleasant Place, *Proserpina* from its curious Situation, and swallow it up, or hurry it away on a sudden into his own, or the Subterraneous Regions. Ver. 363, to 385. All Poetical to express how *Dis* was thus enamored of *Proserpina*, from 385, to 395. The pleasant Situation of *Proserpina* is described from 395, to 408. The Catastrophy is described, which manifests, that this terrible Earthquake extended from the place where the Lake *Pergusa* now is to the Place where the City of *Syracuse* stood, now called *Saragosa*, where the River *Anapis* runs into the Sea: By this the Lake *Anapis* was broken open and made a Bay to the Sea, and no remainder of *Cyane*, but some small Brooks that run into that Bay. This I take to be the meaning of the Poet from 409, to the end of 437. The remaining part of the Story seems to be a description of the Devastations made in the Country, being made unfit for Tillage, and therefore *Ceres* is said to seek her Daughter *Proserpina* all over the World, the remaining Husbandman seeking for other places fit for Tillage, but returns to *Sicily*, at Verse 463.

It would be too long to interpret all which I think I could easily do, and shew plainly the meaning of the Physical *Cabala*. But I designed to mention this only at present as an example pertinent to this present time, when we yet have the noise of the *Sicilian* Earthquake, and some others, yet sounding in our Ears: As to the Moral *Cabala* many have handled it; and for the Historical, I shall take some other time to Discourse of it, and to give my Conjectures.

**No. 19.** *Read May 31, 1693.*

[p.389] I have formerly discoursed concerning the great and strange Effects that have been produced on the superficial Parts of the Earth by means of Earthquakes, the raising of Hills, the sinking of Vallies and Lakes, the swallowing and new producing of Rivers, the raising and sinking of Islands, the cleaving of Hills and Rocks, and the tumbling and disordering of the superficial Parts of the Earth, by which means have been produced the Veins and various mixtures in Marbles and other kinds of Stone, and most of the petrifactive Productions, besides the Production of Mines and Metalline Bodies, as well as of other Saline, Sulphureous and divers other mineral Substances. And in short I conceive that the whole Surface of the Earth, as it is at present, has been some ways or other influenced and shaped by them: I have on several occasions alledged several Arguments and Observations to make these Conceptions probable, and have produced several Histories that seem to be that way conducing. But most of the greatest Mutations having in probability been performed in the Αδηλον or μηθιηον the uncertain or fabulous Times, as they are termed by *Varro*, there is not to be found in the Historical time very many that do make much for it; the greatest Instance I conceive to be had of it, is the History of *Phaeton*, which, tho' among the Greeks it be included within the fabulous times, yet it seems by that Passage of *Plato* which he relates concerning what *Solon* had learned from the *Ægyptian Preist* [sic] that the *Ægyptians* had Records thereof in their History, as in probability they had of many others, of which the *Grecians* were wholly ignorant, as may in part appear by the Relation of the *Atlantis*; for the *Greeks* had nothing of History elder than the Flood of *Oygges*, which, as *Eusebius* says, happened about the times of *Jacob*, which was long after that of *Noah,* and long before that of *Deucalion,* which was about the latter end of *Moses*'s Life. All which time according to *Varro*, and many hundred Years after even to the beginning of the *Olympiads* (which was but 776 Years before Christ) was included in the Fabulous Age, which was likewise 776 Years after *Moses* his Death, he dying in the 1552 Year before Christ's Nativity; within which space of time the Catastrophy Mythologised by the Story of *Phaeton* seems to have happened; for *Orosius* relates it to have been much about the time of the *Israelites* departure out of *Ægypt*; as he doth also assert that of *Deucalion*'s Flood, in which the greatest part of the People of *Thessaly* were lost, only some few escaping who fled to the Mountains, especially *Parnassus*, near the Foot of which *Deucalion* then reigned. Now if we consider the Story as it is related by *Ovid* in the Second Book of his *Metamorphosis*, making allowance for what is Poetically spoken, one may plainly enough from the whole drift of the Fable conjecture at the History or Tradition that is couched under it, as well as somewhat also of the Philosophy; as for the morality thereof enough have taken notice of and writ concerning it. As for the time of it, *Ovid* places soon after the Fable of *Deucalion* which is the seventh Fable of his first Book, and the eighth, ninth, tenth, eleventh, twelvth, are of Matters consequential of that Flood which must have followed it in a very short time (as I may on some other occasion make more probable) or rather prævious to this, as being indeed part of it. But to let that pass for the present, I shall only take notice now of the Physical or Philosophical part thereof, which to me seems to contain a Description of some very great Earthquake or fiery Eruption which affected a great part of the World then known.

First then we find *Phaeton* to be termed a Son or production of the Sun, which is the biggest and most powerful Fire of the World, that we who

*The story of Phæton, though deemed to have happened in fabulous or uncertain times, was recorded by the Egyptians.*

live upon the Earth do know, but by the Mothers Side, to be the Son of *Clymene* which is an epithite of *Pluto* and denotes *Phaeton*, or this aspiring Fire to be generated by the Sun in the Bowels of the Earth; all the proeme of the [p.390] Story is Poetical and of a moral Signification to denote a Genius aspiring and undertaking more than what it was able or fit to perform and manage, yet it is so ordered as to comprise the main Design and Physical meaning of the Poet, *viz.* that by some extraordinary or universal influence of the Suns Beams the Subterraneous Vapours had been kindled, and that a fore-runner of this was Lightning and Thunderings in the Air, which seems to be expressed by the description of the Horses that drew the Chariot of the Sun.

*The fable of Phæton interpreted.*

> *Interea volucres Pyroeis, Eous & Æthon*
> *Solis equi, quartusq; Phlegon, hinnitibus auras*
> *Flammiseris implent, pedibusq; repagula pulsant.*

> Mean while the Suns swift Horses, hot Pyroeis,
> Light Æthon, *fiery* Phlegon, *bright* Eous,
> Neighing aloud inflame the Air with heat,
> And with their Thundring Hoofs the Barriers beat.
> Metam. Lib. 2. v. 153, &c.

The Managery and Course of the Horses and Chariots through the Heavens is all poetical, accommodated to shew the Constellations of *Aratus*, and to the Cosmography of the Poets, to signify the concurrence of the other Celestial Bodies and Powers: But the effects it produced on the Earth as the flaming and burning of Mountains, the cleaving and chopping of the Earth, the swallowing up of Rivers, the rising of Lands out of the Sea, as especially that about *Ægypt*, and the Sandy Deserts on the West side of it, seem to be Historical as well as Poetical.

But I confess the whole is so Poetical that much certainty of History cannot be fetched out of it; yet for the present let me add thus much that I conceive may be deduced therefrom, and that is this, That there was an ancient Tradition among the *Greeks*, and that there was an ancient History among the *Ægyptians* of some very great and almost general Conflagration or Eruption of fiery Streams which made very great Devastations on the Earth, especially of those parts mentioned by the Poet in this Relation; such as *Athos, Ida, Oete, Tmolus, Taurus, Helicon, Æmus, Ætna, Parnassus, Othrys, Cynthus, Erix, Mimas, Rhodope, Dindyma, Caucasus, Mycale, Cytheron, Pindus* and *Ossa, Olympus,* the *Alpes* and *Appenine,* all which Mountains are said to have been on Fire, and to have cast up Smoak, Ashes, and burning Coles, and to have thickned and darkned the Air.

*This fable, Hooke admits, is so "poetical" that nothing of certainty can be derived from it. But both the ancient Greeks and Egyptians seem to have a traditional story of some very great conflagration or eruption that covered a huge area.*

> *Tum facta est Libye raptis humoribus æstu*
> *Arida, tum Nymphæ passis, fontesq; lacusque*
> *Deflevere comis, &c. v. 237.*

> Then a dry Desert Libya became,
> Her full Veins empty'd by the thirsty Flame;
> With their scorcht Hair the Nymphs the dry'd up Streams
> And Lakes, their ancient seats, bewail.

There were cast up the *Libyan* Desarts and many Lakes and Rivers swallowed up and perverted, the names of which the Poet mentions, which are too many now to repeat; then the other parts of the Earth were cleft and tumbled to and fro.

> *Dissilit omne Solum penetratq; in Tartara Rimis*
> *Lumen & Infernum terret cum Conjuge Regem.*
> *Et mare Contrahitur Siccæq; est Campus arenæ,*
> *Quod modo pontus erat; quosq; altum texerat æquor*
> *Existunt montes, & Sparsas Cycladas augent.* V. 260.

> [p.391]*Earth cracks, to Hell the hated Light descends*
> *And frighted* Pluto *with his Queen offends;*
> *The Ocean shrinks and leaves a Field of Sand,*
> *Where new discovered Rocks and Mountains stand,*
> *Which multiply the scatter'd Cyclades.*

Then was the Sea contracted into a narrower but deeper Cestern, the Hills and Lands on each side of it raised from under the former Sea and made dry Lands and Mountains, the Islands that are now dispersed in it were thrust up out of its bottom, and stand in that Position to this time: In short not to detain you at present too long upon this Mythologick Story, I conceive it to contain the History or *Cabala* of the Production or Birth of the present *Mediterranean, Ægean* and *Euxine* Seas, and of all the bordering Shores and Countries near adjacent to them, together with all the Islands, Peninsula's, Cliffs, Promontories, Mountains, Hills, Lakes, Rivers and Countries which had been before that time all covered with the Sea, but by a prodigious Catastrophy which Divine Providence then caused to be effected, the former Face of those Parts was transformed and metamorphosed into much what it is now found, in General, tho' not in all Particulars; for that there may have since been by the same Divine Providence produced other particular Catastrophies and Mutations, of which there are many Instances mythologically Recorded in this our Author, some of which I have already mentioned, and divers others which I may have occasion to mention some other time, besides divers others of which we have plain and not hitherto doubted or disputed Histories. Now, tho' I confess what I have here asserted to be seemingly very Extravagant and Heterodox from the general Conceptions of most that have had occasion to mention this Fable; and tho' it had been less improbable, I should not have expected any Concurrence of Opinion: Yet possibly when the Matter has been more sedately and without prejudice thought of and examined, it may, as well as some of my former Extravagancies, receive at least a more mild Censure, tho' it should not be wholly accommodated to the Gusto of every such Examinant. In these Matters Geometrical Cogency has not yet been applied, and where that is wanting, Opinion, which is always various and unstable, prevails. However, I may on some other occasion shew that there is to be found in Physick, as well as Geometry, unanswerable Probation.

And when the Extravagancy and Novelty of the Doctrine has run the Gauntlet of Censures, I shall indeavour to add somewhat to cover and cure its Scars.

---

*The myth is interpreted as narrating the birth of the Mediterranean, Ægean, and Euxine Seas and their bordering countries.*

*Hooke admits he has been "extravagant" in these interpretations, but in the absence of "geometrical cogency" (i.e., this is not an exact science), opinion, though various and unstable, prevails. But even in the exact sciences, there are matters that are unanswerable.*

**No. 20.** *More on the same topic as previous lecture. Read June 7, 1693.*

[p.391] I did the last day indeavour to shew what I conceived was veiled by the Poet under the Story of *Phaeton*, and that was this, That by this Mythology the *Grecian* and *Latin* Poets did preserve the memory of some extraordinary great Catastrophy, which all the parts of the Earth or Countries not far removed from the *Mediterranean, Ægean, Euxine,* and *Caspian* Seas had suffered by fiery Eruptions or Meteors, effecting Earthquakes.

This to me seems probable from the Order and from the Manner of the whole Relation.

For the Order of it; we find it placed by *Ovid* soon after the Flood of *Deucalion*, and so we find it is related by *Paulus Orosius* (which I hinted the last day) for in the ninth and tenth Chap. of his first Book of Historys he makes the Flood of *Deucalion* to have happened much about the time of the Plagues of *Ægypt*, and the Passage of the *Israelites* through the Red Sea, by which Flood the greatest part of the People of *Thessaly* were destroyed. *Quo* (says *Orosius*, speaking of that Flood) *Major pars populorum Thessaliæ absumpta est, paucis perfugio Montium Liberatis. Maxime in monte Parnasso, in Cujus Circuitu Deucalion tune Regnabat; qui ad se confugientes Ratibus Suscepit & per gemina* [p.392] *Parnassi Juga fovit aluitq; ob idq; locum fecit Fabulæ ut ab eo Reparatum Genus humanum diceretur. His etiam temporibus adeo jugis & gravis æstus incanduit ut Sol per Devia transvectus, universum orbem non calore affecisse, sed igne torruisse Dicatur. Impressnmq; Fervorem & Æthiops plus Solito, & insolitum Scytha non tulerit. Ex quo etiam quidam, dum non concedunt Deo ineffabilem potentiam suam, Inanes Ratiunculas conquirentes Ridiculam Phaetontis Fabulam texuerunt.* Thus far he, by which it seems that *Orosius* did, in the Stories of *Deucalion* and *Phaeton* for the main, believe the Matters of Fact to be true, but he was not for giving a Philosophical Conjecture at the Causes of it, or the ascribing them to the Pagan Deities, but for ascribing it immediately to the ineffable Power of God.

Now I do not conceive it doth any ways detract from the Omnipotency and Power of God, to explain the Causes that he was pleased to make prævious to those Effects: For the Power of God is not less wonderful, in producing and disposing the Causes of things, than in producing the things more immediately. But such a Story as this Fable of *Phaeton* is, and to give such an account of its Causes, as the Poets have there given, if understood literally, seems sufficiently ridiculous, and impious. But it is easy enough to be seen that those who made this Fable knew better things, and only made use of Mythology to conceale their knowledge from the Vulgar, and yet communicate it to such as had the Key to unfold the Mystery contained therein.

And this appears plain enough from the whole series also of the History; for as I noted before, *Phaeton* is said to be produced or generated by the Sun in the Womb of *Clymene*, an Epithite of *Pluto*, that is, in the Subterraneous Regions; and that it is so understood, appears plainly by the behaviour of *Clymene*, who is said, after the Death of her Son, to have been *Lugubris & amens, & Laniata Sinus totum percensuit orbem,*

*Hooke's defence of inquiries into the causes of things: It doesn't detract from God's power to do so; i.e., God is not "less wonderful" because He created the causes of things rather the things themselves. The fact that in this period of his writing Hooke very often needs to make such defences shows that there was a return to the more literal interpretations of scripture in society following a period of free-thinking spirits at the inception of the Royal Society.*

*exanimesq; Artus primo, mox ossa requirens.* Which seems to denote the murmuring and tumbling in the Earth that continued after the Conflagration was over, and the Story of the Sisters of *Phaeton* seems very consonant also thereunto if I had time now to consider them.

*Phaeton* being grown to maturity, is said to have a great desire to know his Father, whom *Clymene* directs to go to the Palace of the Sun; that is, the Vapours being copiously generated in the Earth are expelled into the Air ascending towards the Sun. *Phaeton* is said to have come at length to the Palace of the Sun, and there to have been much pleased with the glorious work thereof, and more especially with the Workmanship of *Vulcan* in the Gates. *Nam Mulciber illic Æquora cœlarat medias cingentia terras, terrarumq; orbem Cœlumq; quod imminet orbi. Cœruleos habet unda Deos, Tritona Canorum, Proteaq; ambiguum Balenarumq; prementem, Ægeona Suis immania terga lacertis, &c. Terra viros, Urbesq; Gerit, Sylvasq; Ferasq; Fluminaq; & Numphas & Cœtera numina Ruris. Hæc Super imposita est Cœli fulgentis Imago Signaq; sex foribus dextris totidemq; sinistris, &c.* Then approaching the Sun -----. *Sedebat, in Solio Phœbus claris Lucente Smaragdis, a Dextra Lævaq; Dies & Mensis & Annus, Sæculaq; & positæ Spatiis æqualibus Horæ. Verq; Novum Stabat Cinctum florente Corona: Stabat nuda Æstas & Spicea certa gerebat. Stabat & Autumnus calcatis sordidus uvis, Et Glacialis Hyems Canos hirsuta Capillos.* The meaning of all which seems to be this, That the state of the World before this Catastrophy was much the same (*facies non omnibus una, nec diversa tamen*) with the State of it afterwards; that is, the Course of the Sun was through the twelve Signs; there was a Spring, Summer, Autumn, and Winter, as there has been since; no alteration of the Axis or obliquity of the Ecliptick; But there were Ages, and Years, and Months, and Days, and Hours as now; and *Phœbus* describing the way to drive his Chariot through, doth name the same Constellations: So that the Philosophers who made the Theory, or the Poets that made the Fable, did not understand or suppose the obliquity of the Ecliptick to be made by that Deviation of the Chariot, or that this Catastrophy had altered the Axis of the Earth, with respect to the Heavens. But neither did they design to signify, even by this Story, the Deviation of the Sun it self at that time, as if that had descended and fired the Earth: For *Phœbus* did not accompany the Chariot, *Occupat ille levem juvenili corpore currum,* v. 150. But [p.393] they rather seem to make *Phaeton* a fiery Meteor proceeding from the East, and moving Westward by another way and course than the Sun usually took and differing from the Direction that *Phœbus* had given to *Phaeton* to observe: But his Horses now mounted upwards towards the fixt Stars, now downwards towards the Earth, now far to the North, then as much to the South, and last of all he was broak all to peices by Lightning, and fell down like a Meteor upon the Earth, and like some such Meteors as have of late Years been observed, but much greater. *At Phaeton, Rutilos flamma populante capillos, volvitur in præceps, Longoq; per aera tractu Fertur, ut interdum de cælo Stella Sereno, quæ si non cecidit potuit cecidisse videri.* Whether there might ever have been any such Comet as in its Course might come so near the Earth as to set the superficial part on Fire, and to kindle or excite the Subterraneous, Sulphureous and Nitrous Minerals, or whether it were some Exhalation collected into a great Body in the upper Regions of the Air, and being kindled might seem to pass near those Constellations, through which *Phaeton* is said to be hurried and to come so near the Mediterranean parts as to burn the superficial Parts, and to inkindle the Subterraneous Mines of combustible and inflammable

*Phæton is interpreted as representing a meteor or comet that broke up before reaching the Earth but set the surface on fire as well as kindled the subterraneous parts. Considering the present-day controversies regarding the causes for the extinctions at the K-T boundary, this hypothesis does not now seem to be so "extravagant."*

Substances; or whether it were some prodigious quantity of inflammable Steams collected in the Air, and so burnt off by continual Lightning, it is hard positively to determine, because that part of the Story I conceive to be Hypothetical, and Conjectural, or Philosophical, and not meerly Historical. But the Effects produced, those I conceive to be Historical; that is, that there were divers parts, which were before covered by the Sea, that by this Eruption, were raised from under it and left dry. *Tum facta est Libye, raptis humoribus, æstu arida: Tum Numphæ passis fontesq; lacusq; Deflevere comis.* 237, 238. *Et Mare contrahitur, Siccæq; est campus arenæ, quod modo pontus erat.* 262, 263. Then also were raised from under the Sea both Islands and Mountains. *Quosq; altum texerat æquor Existunt montes, & Sparsas Cycladas augent.* 263, 264. Then also did other parts sink under the Water. *Ipsum quoq; Nerea fama est, Doridaq; & Natas, tepedis latuisse sub undis.* Other parts were overflowed by the Sea and again deserted. *Ter Neptunus aquis cum torvo brachia Vultu, Exerere ausus erat, ter non tulit aeris ignes.* 272. Then also caused great Earthquakes, and overturning and tumblings of the Earth. *Alma tamen Tellus, ut erat circumdata ponto, inter aquas Pelagi, contractosq; undiq; fontes, Qui se condiderant in opacæ viscera Matris, Sustulit Omniferos Collo tenus arida Vultus: Opposuitq; manum fronti,* Magnoq; *tremore omnia Concutiens Paulum Subsedit; & infra, quam Solet esse fuit.* Then also was the Air filled with Fumes and Smokes, and the Surface of the Earth covered with Ashes and Cinders. 231, 232. *Et neque jam Cineres ejectamq; favillam ferre potest, calido involvitur undiq; fumo.* And again, 283, 284, speaking of the Earth, *(Presserat ora vapor) tostos en aspice crines, inq; oculis fumum; volitant Super ora favillæ.* The superficial parts of the Earth, Vegetable and Animal, were destroyed. 210, &c. *Corripitur flammis quæq; altissima tellus, fissaq; agit Rimas, & Succis aret ademptis, pabula canescunt; tum frondibus uritur arbor, Materiamq; Suo præbet Seges arida Damno: Flumineæ volucres medio caluere Caystro.* The Earth was rent and cleft, and all the high Hills on Fire like *Ætna* or *Vesuvius:* Those I named the last day. By this means many Rivers were swallowed up into the Earth; others dried up by evaporation and boyling Heat. *Mediis Tanais fumavit in undis.* 243, &c. *Nili Ostia septem Pulverulenta vacant, Septem sine flumine Valles.* And, to be short, all the effects that have ever been observed in Earthquakes, are here eminently expressed. So that there can be no manner of doubt of the design of the Story, *viz.* That it was designed to denote or describe a Catastrophy of the Mediterranean parts of the Earth by Earthquakes; since all things are so properly delineated and represented for that end, as if the Poet or Maker thereof had been spectator or Eye-witness of it, or at least a Contemporary with it. And we may here find the whole Progress or Phænomena of an Earthquake from its very first beginning to its very last end, and the effects also that precede it, and those that are subsequent to it, as I could plainly shew if it were not too much for this present Discourse, by explaining the Mythologick Histories immediately prefixt and following it. Nay, there has not been in this late Earthquake in *Sicily* [p.394] (which seems to be the greatest mentioned in History) any one Phænomenon which cannot be shewn in this of *Phaeton*, and indeed most of the Phænomena mentioned in this of *Phæton* have been exhibited or exemplified in this last of *Sicily*; which I could easily manifest by comparing *Ovid*'s Description with that of the *Italian* Frier; but I shall pass it by for the present.

*Even though Hooke feels that this part of the story is hypothetical or conjectural, the effects, he feels, are historical: i.e., Land was formed from the bottom of the sea.*

*The story of Phæton, therefore, represents the description of an earthquake catastrophe that happened in the Mediterranean region.*

*Every phenomenon described by Ovid in this story of Phæton occurred in the recent earthquake in Sicily.*

**No. 21.** *Richard Waller found the following loose leaf among Hooke's papers entitled "A Copy of Dr. Thomas Gale's Papers concerning Giants." This undated letter must have followed a lecture Hooke gave on July 13, 1693.*

[p.384] SIR,

In Answer to your Question about the word Rephaim *and* Gigantes, *I make this short return.*

*1. There is no rudical word in the Hebrew Language whose signification doth at all lead us to understand* Gigantes *by the word* Rephaim, *so that the Radix of* Rephaim *is either lost as to the present Hebrew Language (as many others are) or else that word* Rephaim *is a foreign word to that Language, as many more such are now found in the Bible.*

*2. The Septuagint Translators do often render Hebrew words not according to their Natural Sense, but with respect to some History or Tradition, or general belief prevailing at that time: The reason was because those Translators lived among Greeks at* Alexandria*: And they were desirous to shew that the Bible was not unaquainted with the Greek-Stories, where the thing could be done without injury to their Books.*

*3. In their Rendring of the word* Rephaim *by* Gigantes *and* Mortui, *and the Verb_____they plainly point at the Story of the* Titanes, *who in the Greek Mythologies are said* ταρτηρωθηναι. St. Jude *uses the same word when he speaketh of the Hellish Angels,* εταρτηρωθησαι. *Another Greek Translator rendereth the same word* Rephaim *by* Titanes.

*4. As to the suspicion that the* Gigantomachia *was an Earthquake, or perhaps several Earthquakes, but by the Poets put altogether, the true notation of the word* Gigas *seems to make for you. In Hebrew the Radix* Gagash *is* terra commota fuit. *And the Substantive* Gigas, *tho' commonly taken for a Greek word, is indeed of Hebrew or Phœnician Original. In that place of* Isaiah *where the 70 use* Gigantes, Symmachus *uses* θεομαχοι, *both alluding to the Poetical Fable, but the 70 do it more warily,* Symmachus *more plainly.*

[p.385] But to me I confess it seems rather to allude to the fourth Verse of the sixth Chapter of *Genesis*, where it is said, that there were Giants in the Earth in those Days, because the word γιγαντες is made use of by the Septuagint οι δε γιγαντες ησαν επι της γης εν ταις ημεραις εκειναις, and it seems to be a full Period, besides we find that God immediately after this Passage, is said to be very highly displeased with the wickedness of Mankind at that time upon the Earth, and to resolve their Destruction and Extirpation, which shews that there is a great agreement of the Poets Mythology with this History of *Moses:* For *Ovid* makes this *Gigantomachia* to precede the Flood of *Deucalion*, as the Scripture doth make this to precede that of *Noah*. And besides joins the Fable of *Lycaon* to that of his Giants, which seems plainly to allude to the wickedness of Men mentioned by *Moses* upon this occasion. Further, I do not know whether the word may not sometime have been used to denominate Earthquakes, or subterraneous Powers; for in the ninth Verse of the fourteenth of *Isaiah* where the same word is used by the seventy. It seems plainly to signify some such thing; but this is besides my Province, and I shall rather leave it to the Divines to determine: For *Gigas* is the same

---

*Hooke's July 13th lecture on mythologic giants is not included in the Waller edition.*

*Thomas Gale (1635–1702) was also educated at Westminster under Dr. Busby; he was the same age as Hooke but was educated at Cambridge where he was professor of Greek. A member of the Royal Society, he became the dean of York in 1697. He is now chiefly known as the author of the inflammatory inscription on the base of the London Monument designed by Hooke.*

*The Septuagint = the oldest Greek text of the Old Testament. There were 70 Septuagint translators so that they are further referred to, both by Gale and by Hooke, as "the 70."*

*Greek translation = thrown into hell*

*Greek translation = fighting against the gods*

*The next paragraph seems to be either some notes of Hooke to himself or a part of a lecture or letter.*

*Greek translation = The giants were on the Earth in those days*

*An attempt to bring Ovid's mythology in concordance with the Old Testament.*

word with the Greek word γιγαζ, which *Eustachius* derives from γη and γαω that is an Off-spring or Progeny of the Earth, *i.e.* somewhat generated in the Bowels or Womb of the Earth and thence Born, brought forth or protruded, which is a very proper Appellation and Description of that production of Nature, wherewith the Earth seems to be first impregnated and made tame, then to be in great Agony and Pangs, and to have many pangs and throws before it is delivered of it; and last of all to produce Islands, Mountains, or the like Monsters, which seem to threaten or aspire at the Celestial Mansions.

**No. 22.** *Read August 2, 1693.*

[p.385] When I gave an account the preceding Meeting, *July* the thirteenth, of what I conceived the Poets meant by the Mythology of the Giants warring with the Gods, some of the Society then present were very Inquisitive to be informed what should be meant by the History of *Python* which was destroy'd by *Apollo*, of which though I had made some mention in a former Discourse concerning the Mythology mention'd by *Ovid*, yet being then only mention'd *in transitu*, I have now somewhat more particularly drawn up my Sentiment concerning it. I mention'd before then *Ovid b*y this Mythology (as I conceived) did design to describe the state of the Earth from its first beginning and formation out of a *Chaos*, through all the various Alterations, Changes and Metamorphoses it had undergon even to that time in which he lived. And therein to comprise the Traditions and opinions of the Antients, and possibly also some of the Moderns of his Times, and some also of his own, thereby to give some Account and some Reasons of the then present Phænomena of the World. I need not repeat what I have formerly instanced in, about the *Chaos* and the Ages succeeding, nor what I said concerning the Fable of the *Giants*: But to make the probability of my Conjectures the more manifest, I would observe to you the Co-hærence and Connexion of the Mythologies, as they are ranged in this first Book. After the War of the *Giants* which had raised up Mountains that seemed to threaten the very Heavens by their height, and the disturbances that had thereby been caused in the Air by Lightning and Storms which he makes to be the means by which the Gods destroy'd their fury, he comes to consider the Face of the Earth as it was left, which he Mythologizes by the Story of *Lycaon*, whereby he describes the confusion there was left by the subversion, sinking, overwhelming and destructions that had been made, the *Rustica Numina* as the *Fauri, Nymphæ, Satyri,* and the *Sylvani* of the Mountains, were all likely to be destroyed for the future; that is, the fine Plains, the Woods, the Rivers and Rivulets, the Woods on the Hills were all deformed, confounded, and put into confusion, and not only so but the Air itself was from the Clefts and Chasms poisoned and continually filled with noxious Expirations out of the Earth, the People remaining were distracted and grown barbarous, preying upon and destroying one another; it was thought therefore by *Jupiter*, i.e. Divine Power, necessary, that all must be set to rights again by a general Deluge, whereupon the Poet brings in *Jupiter* Swearing,

*The story of Python interpreted.*

[p.386] *Nunc mihi qua totum Nereus circumsonat orbem*
*Perdendum est mortale Genus: Per flumina juro*
*Infera, subterras Stygio labentia Luco.*
*Cuncta prius tentanda, sed immedicabile Vulnus*
*Ense recidendum est, ne pars sincera trahatur.*

*Now whereso'ere resounding Waves are spread,*
*All mortal Beings must die; by Streams that run*
*Beneath, I swear, Streams that ne'er see the Sun.*
*All ways first try; But th' incurable Wound*
*Must be cut off, lest it infect the Sound.*

The Flood then follows that was to reduce this torn and confounded Face of things into some better Form and Order, by which the Caverns left should be filled, the ruggednesses plain'd, the superficial Parts, now Rocks and Stones, and the Recrements of the Eruptions should be cover'd by a more soft, and fine, and fatter Skin of Earth, which should be fit to produce and nourish Vegetables and Animals as before. The Poet then describes the Flood, and thereby makes all Men and other Creatures to perish by it, except only *Deucalion* and *Pyrrha*, who were to be the restorers of Mankind, whom he supposed to have somewhat more Divine than all the rest of the Creatures, which he conceived to be generable out of Corruption, as you will see by and by; but Man only by propagation, yet his method of Propagation looks at first glance but very extravagant, namely, from Stones cast behind them by *Deucalion*, and *Pyrrha*, *Deucalion*'s being generated into Men, and *Pyrrha*'s transformed to Women (*quis hoc credat nisi sit proteste Vetustas*) says *Ovid*; and I am very apt to think that *Ovid* himself was one of the Unbelievers, notwithstanding the Testimony of the old Traditions, that is, that he did not take it to be a truth in the plain Sense of the Words, tho' he seems to draw a Consequence from them. [*Inde genus durum sumus experiensq; Laborum.*] But that he understood what was meant or intended to be signified by this Mythologick Description [*Et Documenta damus qua sumus origine nati.*] But to proceed. After he has told us how Mankind was preserved and propagated after the Deluge, he next comes to the other Creatures.

*Cætera diversis, Tellus animalia, formis*
*Sponte sua peperit; postquam vetus humor ab igne*
*Percaluit Solis, Cœnumque udæq, Paludes*
*Intumuere œstu, fœcundaq semina rerum*
*Vivaci nutrita solo, seu Matris in alvo*
*Creverunt, faciem aliquam cepere morando.*

*All other Creatures took their numerous Birth*
*And Figures voluntary, from the Earth,*
*When slimy Marshes from the Suns vast heat;*
*And with his Power impregnated grow great*
*With Child, and Seeds, as from the Mothers Womb,*
*By Steps and Time both Growth and Shape assume.*

And here he is for *Æquivocal* Generation to the height, if you understand him literally, or according to the words, *Quippe ubi temperiem Sumpsere Humorq; calorq; concipiunt & ab his oriuntur cuncta duobus.* All came from two Principles; for he seems to make all things to arise or be generated out of a temperature of Heat and Moisture, and by that means the

Earth, when left by the Deluge, abounding with muddy and boggy Places the heat of the Sun working thereupon produced, according to him, not only all the several Creatures anew which had been lost and destroyed by the Deluge, but divers others of strange, and before unknown, and monstrous Forms, which were terrible and destructive to Mankind, and amongst the rest he mentions a strange, venomous and prodigious *Serpent*, which he calls *Python,* which he [p.387] relates to be killed or destroyed by the Darts of *Apollo*. By which I conceive no more is meant, but that those boggy Places after a time corrupted and produced pestilential, dark, Clouds and Vapours, which frighted and was noxious both to Men and Beasts.

But that in some time after the Rays of the Sun and Lightning having prevailed, did thereby burn off and discharge the poisonous Exhalations, and put an end to that monstrous off-spring, nor need we be much concerned for what the *Dæmonologers* had thereupon superstructed for the promoting and carrying on of their *Theourgy*. After this drying of the boggy places of the Earth by the Sun; we have the account of the production of Woods and Trees by the Power of the Sun in the Story of *Daphne*: And then the description of the Rain, Dew and the Foggs that moistened the Air, and made Rivulets and Streams producing Grass in the Fields, and greenness on Trees and Plants by *Io* then *Juno*, the Air finding these Vapours to be drawn up into her Bed or Residence by the *Sun* or *Jupiter*, is said out of jealousy to set *Argus*, that is, the Stars to watch it by Night and cause it to fall: But *Mercury*, or the light of the morning cuts off the head of *Argus*, that is, makes the Stars disappear and the Sun return to raise them, and *Io* is then restored to her former Shape, or the Dew or Moisture on the Ground is raised into Vapours. By the bye he inserts the Generation of Water, and River-plants by *Syrinx*, and the production of the *Rain-bow* by the Head of *Argus*, placed by *Juno* or the power of the Air in the Feathers of *Juno's* Bird, which are the Clouds of the Air. By these Mythologies having described the postdiluvian state of the Waters, and the Air and watery Meteors, he ends the Book with the Pedigree of *Phaeton* which he compleats in the beginning of the next, of which hereafter.

But as to *Python*, which gave the occasion of my present Discourse, 'tis plain that its Name signifies Corruption, and by the manner of its Generation, 'tis evident that he supposes this Corruption to be caused by the Bogginess or Floods that remained in the Plains, Lakes, or Holes, lower Grounds or Vales incompassed with higher Grounds that the Water could not run off: From the fermentation of the softned Earth he supposes the Animals to be formed that were of the same form with the *Antediluvian*; but from a longer stay of the Waters this fermentation turned to Corruption, and then produced not only Monstrous Creatures, but noxious and dreadful Exhalations, whence proceeded Distempers and Diseases, because these Waters by several Streams moved (as most commonly they do) to lower Places and Cavities and there made a great Body which possessed a considerable part of the incompassing Hills or Mountains: *Apollo* or *Jupiter*, that is, the Sun by many Days and Years irradiating with its Darts, Rays or Beams, doth partly dry by Exhalations, partly by flashes of Lightning, dissipate, and dispel, and last of all it causeth Clefts and openings of the Earth which swallow it up, and leave those Cavities like the black Wounds which the Poet affirms to remain for a witness to Posterity.

*Python represents corruption, or decay.*

> --------*Sed te quoq; maxime Python*
> *Tum genuit: Populisq; novis, incognite serpens*
> *Terror eras; tantum spatii de monte tenebas.*
> *Hunc Deus arcitenens, & nunquam talibus armis*
> *Ante, nisi in damis Caprisq; fugacibus usus,*
> *Mille gravem telis, exhausta pæne Pharetra*
> *Perditit effuso per vulnera nigra veneno.*
> *Neve operis famam posset delere vetustas,*
> *Instituit sacros celebri certamine Ludos,*
> *Pythia perdomiti serpentis nomine dictos.*

> [p.388] *Huge* Python *th' Earth against her will then bred,*
> *A serpent whom the new-born People dread:*
> *Whole bulk o're so much of the Mountain spread.*
> *The dazling God that bears the silver Bow,*
> *(inured before to strike the flying Doe)*
> *That Terror with a thousand Arrows slew,*
> *His Quiver empty'd, and the Poison drew*
> *Thro' the black Wounds: Then least the Memory*
> *Of such a work in after times should die,*
> *He instituted celebrated Games*
> *Which from this Serpent he the* Pythian *names.*

The Earth produced various Creatures some monstrously shaped, these were *invita terra* contrary to its proper teeming Vertue brought forth: Of these one was more corrupt than the rest, and more contrary to Nature; this possessing so much room of the Mountains, wrigling on all sides by the Rills that ran into its vast Body or Lake, by its Poison became dreadful to the new produced Creatures: This Celestial Power that kept the Tower of Heaven (so I English *Arcitenens*) that is, the Sun, Fire, or Heat, by its Rays and by thousands of flashes of Lightnings (insomuch that one would have thought they had been all spent and the whole stock fired off and whereas those Rays before had been only used to dispel and scatter small Clouds or Foggs) did hereby at last destroy or disperse this stagnant and corrupted Body of Water, by causing it to rise into Thunder Clouds discharging by Lightning its poisonous Vapours with which it swelled; besides the heat of the Sun and the Lightning also kindling the Subterraneous Spirits, caused Clefts and Chasms in the Earth, which swallowed up most of the remaining stagnant Waters, and so destroy'd the Cause or Original of those Evils, leaving in several places divers of those Chasms or black Wounds which the Poet describes.

To this purpose there is a notable Passage in *Lucian*, which, among others, to another intent, is quoted by Dr *Burnet, Theor. Sacr.* Part 2. Chap. 4. "These are the Matters (says *Lucian*) which the Greeks have related concerning the Flood of *Deucalion*. But among the things that have happened soon after it, there is a certain relation of the Inhabitants of *Hierapolis*, which is justly looked upon with great admiration, namely, that in their Country there had happened to be made a great Chasme in the Earth, which had swallowed up all the remaining Waters; whereupon *Deucalion* had built Altars and a Temple dedicated to *Juno* over the same. Now for a sign that this Relation is so, they do thus twice every Year, Water is brought from the Sea to this Temple, and not only the Priests bring it, but all *Syria* and *Arabia*, and many which dwell beyond the *Euphrates*, go to the Sea and fetching the Water from thence bring it to this place: And first indeed they pour it out into the Temple, and then it runs into the Chasm, and tho' this Chasm be but small, yet it swallows an immense quantity of Water. When they perform this Ceremony, they say that *Deucalion* instituted this Rite and Law of this Temple, that it might be a Memorial as well of the Destruction by as of the Deliverance and Safety procured against the Flood. This (says *Lucian*) is the old Story concerning this Temple." This Tradition, 'tis very probable, *Ovid* was not ignorant of and might therefore add to his Relation *Fuso per vulnera nigra veneno. Neve operis famam posset delere vetustas.* And 'tis very probable also that the Mythology of *Argus* has a respect to the Generation of the Rainbow soon after the Flood as it is mentioned by *Moses*. For 'tis plain that their Signs or Hieroglyphical Representations and Notions, were many of them abundantly more incongruous with the things signified than this is; for Clouds may by an easy Figure be fancied the Fowls or Birds of the Air, as we usually say when great flakes of Snow fall, the Winter is plucking its Geese or Fowls: And which among all Fowls, or indeed Creatures, does better represent the Rainbow then the Peacock when it spreads its Tail, whereby it represents such a glorius Arching of a most stupendious Va-[p.389]riety of Colours as numerous and as resplendent as the very Rainbow. And to make the coherence the greater those Rings being made up of a Circular Order of beautiful Spots, what could he better Metamorphose it from than from a Head adorned with abundance of Eyes, which he makes to be of one *Argus?* I suppose for want of Microscopes he knew not that the Eyes of Flies were planted in so curious an order, otherwise possibly that might have served for a Hieroglyphick for the Star-light-Night as well as *Argus*.

## No. 23. Read July 25, 1694.

[p.446] The Memoires of the *Parisian Academy* have furnish'd many curious Discoveries both Mathematical and Physical, yet divers of them or of the same kind have been first discover'd in this Society, tho' not entertain'd with that approbation, which they have there met withal; nor are the *English* so nimble in Publishing what they discover themselves, nor so sharping to arrogate to themselves what they know to have been first discover'd by others; (as I do find divers to be) who will leave no means unattempted to make all their own, tho' there be never so evident Arguments against their Cause. But tho' this be a Practice to be abhor'd by every ingenious Man, and the Bashfulness of the other be blameable, yet there is somewhat to be said both for the one and the other Party, that may seem to countenance these proceedings of them. As first, 'tis a discouragement to any one to Publish that which he finds by Discourse is generally disapproved. A Man may rationally enough distrust his own Thoughts and Reasons, nay, and even his Senses too, if he finds those he converses with to be of another Opinion, tho' acquainted with the Arguments that prevail'd with him, at least 'till he finds, that it was done for some Sinister Designs to defraud him of his Discovery. Next, when by publishing, more Opponents or Emulators (which are both Enemies) are produced, than approvers or indifferent Persons, who at best will do him no

*Hooke has had bitter personal experience in this respect so writes here with much feeling. In this tirade Hooke, the patriot, defends the honor of his countrymen against the pirating French.*

good; 'tis thought better to abstain with quietness, than with Labour and Industry, to create new Troubles. But on the other side 'tis certain, however, that ambitious Minds will try all means to obtain their Designs; they find that such Practises often prevail, and therefore *Quid tentare Nocebit*; they find that the generallity of Men are not much concern'd for the first Discoverer, and that they usually take him for such, who first acquainted them with it; and for one Reader that can disprove them, or detect them of Plagiary, there are a thousand that can not, and for those that can, they find ways to evade and by Confidence carry the point, and even with a general Approbation and Advantage: 'Tis, I confess, a general Observation, that seldom the first Inventer reaps either Honour or Advantage by his Invention, but on the contrary, those that come in at a second Hand acquire them both. But be it as it will, certain it is, that many Discoveries pretended to in the Works, of the *French Academy*, were first made here and elsewhere, nay, and many of them publish'd too in Print, and some of them also in the *French* Language, which yet they will not own, or mention to have seen. I shall instance but in two or three things: The first is that of *Torricellius* a-[p.447]bout his Invention and Demonstration of the *Solidum Acutum Hyperbolicum*, which was Publish'd by him, together with his other Works at *Florence* in the Year 1644, and that without Contradiction by *Roberval* ever since; yet now a Letter is trumpt up, and some Papers found that must needs persuade us that *Torricellius* stole it from *Roberval*. The like slur is cast upon the Works of Mr. *James Gregory*; both which Persons have given sufficient Proofs by their other Works, that they had very little need of stealing from *Roberval*, who has not yet made it evident, nor any other for him, that he was Master of either of their Problems, 'till since the publication of them by the said Authors.

The Second is the discovery of the Glade of Light observable in the Evenings in *Febr.* and *March* each Year, which was first made by our Dr. *Childrey*, and an Advertisement of it Publish'd in his *Britannia Baconica*, in the Year 1660; which Book was Translated into *French*, and Publish'd at *Paris* soon after, which was long enough before it is pretended to be discover'd there. However, the second Person has the Title of the discovery, and the first is defrauded of his due Praise. I could add a hundred other Instances to prove this Assertion; but I shall not at present spend time thereon, tho' it may possibly not pass without some Reflections on another Occasion, that every one, as near as may be, may have his due Praise. For my own part, I think it ingenuous to mention any thing of theirs, which I have occasion to make use of, and to own all such things as theirs, as I find to be new or ingenious; and that First, Because I would give every one that which is due to him. But, Secondly, Because I find it necessary to back a Doctrine with a *French Approbation*. I know there are many things will not be regarded, 'till they have that Stamp to make them current, and then they will readily pass with the present Age and Humor.

In the Memoir of the 31st. of *June* 1692, (so 'tis marked) I find an Observation concerning a Petrify'd Substance produc'd and examin'd by the *Royal Academy*, with some Reflections on it made by Mr. *De la Hire*, which because consonant to some Discourses I have formerly made in this place, I thought might countenance somewhat the Doctrine I then deliver'd, I have also render'd the same in *English* before I make Reflections upon the same.

"The Cabinets (says he) of the curious are fill'd with all sorts of Bodies Petrify'd, as of Plants, Fruits, Woods, and of divers parts of Animals, but

---

*Quid tentare Nocebit = What harm in trying?*

*The examples of piracy Hooke gives here have nothing to do with his own experience in being often plagiarized by others—an indication of his distaste for such priority battles in his own behalf, an assessment that contradicts the opinion of some that he was always ready to cry plagiary of his works and ideas.*

*Hooke was a most fair and honest man.*
*Giving everyone his due is a motto that Hooke seemed to have lived by. These lectures are proof that he did what he preached. Unlike some of his contemporaries (e.g., Newton) and later writers (e.g., Hutton), Hooke was meticulous in giving credit where due.*

*Hooke now even mentions some observations on fossils made by a Frenchman to give him his due, as they also happen to support his own ideas.*

Naturalists are not yet agreed about the cause of their Production; some supposing them to be Stones so shaped by accident, but others suppose them produced by a Water that has a power of converting those several Substances into Stone, after it has long pickled them; probable Reasons are alledg'd for each Opinion.

"Mr. *L'Abbe de Louvoys* sent to the *Academy*; a Petrifaction, which may serve to decide this Controversy, namely, two peices of the Trunk of a Palm converted into Stone, they were brought from *Africa*, with two other pieces of a Palm just like them, but not Petrify'd, the better to compare them together; the Petrifactions are *True Flints*, as appears by their hardness, by their Colour, and somewhat of Transparency, by their Sound, which is clear and sonorous, and by their Gravity, which is more than ten times that of the unpetrify'd; yet these two Flints are so like to the two pieces of Wood, that there is no shew of Reason to conceive, they should be so formed by chance.

"One of these Flints which is two Foot long, and about four or five Inches Diameter, is a piece of the Trunk of a Palm Barked of its Rind; in this may be seen all the Fibres of the Wood of the bigness of 2/3 of a Line, some of which are forked; they run the length of the Trunk and are hollow like Pipes. The Pulp, which is between the Fibres, which serves to join them together, is chang'd into a kind of *Gluten*, but very hard.

"Mr. *De la Hire* gives a Reason of the hollowness of the Pipe, *i.e.* that the outward Parts being dry'd before the middle, when they are dry, they are by the outward Parts kept from shrinking, and so the Pipes become stretched from the Center outward (*which is the same Reason with that I have given for the blebbs that appear in the Glass drops.*)

[p.448] "Now, tho' some might fancy (yet without the least probability) that this with straight Fibres might thus be formed by chance, yet 'tis impossible to conceive so of the other piece, which is a part of the bottom of the Trunk; for this is not only compos'd of streight Fibres as the other, but its Bark is all garnish'd with small Roots as big as one's little Finger, and about three Inches long, which is cover'd with a thin Skin, which contains an infinite of small Fibres like Hairs; in the middle of each of these Fibres is a ligneous Chord, that one may call its *Nuel* or Pith, about 1/3 of the bigness of one's Finger, whose hollow was fill'd with a Pithy extended Substance. All which Parts are also exactly shaped in the Flint, where are visible not only the long streight Fibers, but the Roots and all the small Fibres of a blackish transparent Substance, but the Pith in the middle is of a whitish opaque Substance, and in the most of the small Roots it is hollow; which Mr. *De la Hire* conceives to proceed from the same Cause that he before assigned.

"It is evident therefore (says the Author) that this was no *Lusus Naturæ*, but that these two Flints were originally two pieces of the Trunk of a Palm afterwards chang'd into the Substance of a Flint; and what *Father Duchatz* reports in his Physical and Mathematical Observations, doth decide the Controversy, and leaves it without doubt.

"This Father there says, that the River that passes by *Bakan* in the Kingdom of *Ava*, has, for the space of ten Leagues, or twenty eight Miles, the vertue of Petrifying Wood, and that he had seen great Trees Petrify'd

**346**  DISCOURSE OF EARTHQUAKES

thereby so high as the Surface of the Water reached, but that the other parts of them remained still dry Wood. He adds, that those Petrify'd Woods were as hard as the Flints of a Fire-lock; and such indeed was the hardness of the two pieces of which we have been speaking.

"This Account of *Duchatz* is to be found in the Second Volume of Observations made in the *Indies* by the *Jesuits*, sent thither by the King of *France*, but Corrected and Printed by the care of *P. Gouye*; I have not yet seen the Book, but by the Account of it I find in these Memoires, I conceive it will be well worth the procuring, as containing many other curious Observations, and Histories of Matter of Fact."

This Mémoire of Monsieur *De La Hire* is much the same with what I *have formerly presented to this Honourable Society*, and have Printed among some other Observations made with *Microscopes*; wherein I examine'd the Shape, the Colour, the Hardness, the Weight, the Brittleness, the Incumbustibleness, the Solidity, &c. of it; for I found it to be for its appearance to the naked Eye, perfectly like a piece of Wood, and to have the visible Grain of Wood, and farther by a Microscope, I found it to have all the Microscopical Pores like Wood; I found it of the colour of Wood, but of the hardness of a Flint, and that it would cut Glass: I found its Weight to be to Water as $3\frac{1}{4}$ which seems to be much the same with this of Mr. *De la Hire;* only he compares its weight to that of the Palm Wood, which, by his description, must be much lighter than Water, and mine was only comparative to Water. I found it incumbustible in the Fire, tho' dissolvable by corrosive Liquors. I found it Brittle and Friable like a Flint, and to feel cold to the touch, as a Stone, or Mineral Body usually doth; from all which I concluded it to have, at first, been a piece of Wood, and afterwards, by some Petrifying Water or Vapour, converted into the Substance of a Stone or Flint. And I find that from the very same Arguments, the *French Academy* draw the same Conclusions as to this Substance, and they confirm it by the Observation of *P. Duchatz*; this therefore passing there for a good Argument, I see no reason why it may not also be a good Argument here, and why the same will not also pass for the Petrifactions of other Bodies both Vegetable, as Leaves, Fruits, Roots, and also Animal, as Shells, Bones, Teeth, Scales, *&c.* which are found to have the same Qualifications, that is, the Shapes, Colours, Textures, *&c.* of those animate Substances, nay, and often times the very Bodies themselves not Petrify'd, tho' included in Petrify'd Bodies, as Stones or Minerals; must these be questioned or rejected, only because such Substances are found in places where we cannot give particular Histories of their pristine Estate, and how they come to be there placed and transformed, or so inclosed; or because [p.449] possibly we are not able to produce patterns of Creatures now at hand, and in being, which are exactly of the same Shape and Magnitude as the Academy did produce, to Authorize, or at least incline them to be of that Sentiment; certainly the same Argument that is cogent for the one, ought not to be less valid for the other; for if the finding of Coines, Medals, Urnes, and other Monuments of famous Persons, or Towns, or Utensils, be admitted for unquestionable Proofs, that such Persons or things have, in former Times, had a being, certainly those Petrifactions may be allowed to be of equal Validity and Evidence, that there have been formerly such Vegetables or Animals. These are truly Authentick Antiquity not to be counterfeited, the Stamps, and Impressions, and Characters of Nature that are beyond the Reach and Power of Humane Wit and Invention, and are true universal Characters legible to all rational Men.

*Hooke's* Micrographia *which was published in 1665, indeed predates de la Hire's publication in the mémoires of the French Academy by some 30 years.*

*What a great source of irritation and frustration it must have been for Hooke to have to keep insisting on the organic origin of fossils to his audience in the Royal Society for over a period of some 30 years, when even the French Academy seems to have accepted this idea as valid by publishing de la Hire's Mémoire! Hooke was often a very unlucky man, though talented as he was in so many fields.*

Now, if these are such (as to me they seem to be, notwhithstanding I cannot tell the time when, or certain History how, they came to be there disposed and ordered as they are now found) then certainly it cannot be irrational to conclude at least, that there have been some precedent means that have produced these Effects; and that those means have been such, as we have from Histories and Relations within the times of our own Memory, Experience and Information, that they have produced much the like, which tho' they are not exactly the same, nor possibly by much so great and powerful as they must necessarily be granted, that did effect those we now discover; yet I think it not unreasonable to conceive, that there may have been much greater and more powerful Agents than those we now have had, yet still of the same kind, and acted by the same Powers; for if there are now newly such as have raised, removed, cleft and torn Mountains, have made Lakes, fill'd and levelled Plains, stopped and turned Rivers, spouted out Sea-water at a great distance from the Sea; raised the Sea-shore above the Surface of the Sea and left it dry, with the Fish, and the remainders of them to cover the Surface of it; at other places to raise the bottom of the Sea, which was many Fathoms under Water, and place it above the Surface, and many such other wonderful Effects; then certainly it cannot be unreasonable to suppose, that there may have been much greater in former Times, whilst the matter was yet unconsumed and dispers'd up and down in more places, and more Copiously, and that more Powerful and Effective.

But it is Objected by some, That for such Persons, Places, or Things, of which we find now the Relicks; we have Histories that tell us what, who, and when they were; whereas for the other we have no such Histories in being, nor during the times whereof we have any Histories, can we find any parallel Instances that can countenance such Mutations, Changes, and Catastrophies, as are, and must be supposed to solve the Phænomena. *Greece, Ægypt, Italy, Spain* and *France* have continued the same; no new Lands have been raised out of the Sea, much less Hills or Mountains. Besides, there are many of those Bodies that we now find, both Animal and Vegetable Substances, that are as perfectly like the Species of those supposed Creatures now in being; and therefore we are not to suppose, that any Species could be utterly destroyed, which yet that Supposition seems to make necessary, if well consider'd, and the Consequences thereof produced.

*The same objection against his theory, to which he answered on July 23, 1690, four years before, is still being uttered: no history exists that document the changes he claims.*

To which I Answer, First, That tho' we have no true History, when, or by whom, or by what means the *Pyramids* of *Ægypt* were built; yet all that have seen them do conclude that they were built by Men, and that those Men were good Masons and Architects and Engineers; and that they were not produced of that Shape or Magnitude, by a *Vegetative Power*, or by a *Plastick Faculty, or by meer chance, or the accidental concurrence of Petrifactive Atoms.* Nor can I see any reason to conclude, that the vast *Obelisks* that have been transported from place to place, and erected, were so ordered by Conjureing or Diabolical Magick, tho' I may not be able to tell by what means they become so ordered; I should rather be inclin'd to believe that they were so made and placed by the Industry, and Invention, of some knowing and ingenious Mechanick, who had some Contrivances to perform his undertaking that I am ignorant of. Nor do we make it an Argument that these [p.450] *Pyramids* were never made by Men, because no History does tell us when the like have been made since. Besides, I conceive it would have been a very absurd Conclusion, if any one should

*Hooke's argument: Just because there is no document detailing the building of the Pyramids does not mean we cannot conclude that they were built by men.*

have asserted that those Horns, I lately mention'd here, were a *Lusus Naturæ*, and not the parts of any living Animal, because he could not tell of what Creature they were; or if he should have concluded that the Species of the Creature that produced them were lost, because he knew not where to find it. Certainly there are many *Species* of Nature that we have never seen, and there may have been also many such *Species* in former Ages of the World that may not be in being at present, and many variations of those *Species* now, which may not have had a Being in former Times: We see what variety of *Species*, variety of Soils and Climates, and other Circumstantial Accidents do produce; and a *Species* transplanted and habituated to a new Soil, doth seem to be of another kind, tho' possibly it might return again to its first Constitution, if restored to its first former Soil.

> *By the same token, just because we do not find a living animal like those "Horns" (i.e., ammonites) does not mean that the fossils should be considered a* Lusus Naturæ, *a trick of nature.*

> *Species change to adapt to the environment.*

But I say again, that we have, since the times wherein Histories have been Written, many Instances of the like Changes and Catastrophies, as I have suppos'd to be the necessary Consequences of this Theory of Petrifaction, and several so lately, that the sound of them is hardly out of our Ears; so that we need not be beholding to antient Historians, to tell us when and where they have actually been produced; for first there is no place in the Earth that we do know, nor can we indeed know any such, that is now and ever has been exempt and free from such Mutations, as I have supposed; who can tell what part of it hath ever been and ever will be exempt and free from Earthquakes? And Tho' Histories should inform us that during the times of which they writ, there had been no such Crisis of Nature (which yet would be a very improbable Assertion as being a Negative) yet it were impossible to be assured by them, that there had never been any before that time, nor never would be for the future.

> *This question—Who can tell what part of the Earth will be exempt and free from earthquakes?—is still one of the major unanswered questions in geology today.*

And, Secondly, There is no impossibility in the Supposition that every part hath, at some time or other, been shaken, overturned, or some way or other subject to Earthquakes, and transformed by them; and when we consider how great a part of the preceding Time has been *adelon*, or unknown, and unrecorded, one may easily believe that many Changes may have happened to the Earth, of which we can have no written History or Accounts. And to me it seems very absurd to conclude, that from the beginning things have continued in the same state that we now find them, since we find every thing to change and vary in our own remembrance; certainly 'tis a vain thing to make Experiments and collect Observations, if when we have them, we may not make use of them; if we must not believe our Senses, if we may not judge of things by Trials and sensible Proofs, if we may not be allowed to take notice of and to make necessary Consectaries and Corollaries, but must remain tied up to the Opinions we have received from others, and disbelieve every thing, tho' never so rational, if our received Histories doth not confirm them; this will be truly *Jurare in verba Magistri*, and we should have no more to do but to learn what they have thought fit to leave us: But this is contrary to the *Nullius in verba* of this Society, and I hope that sensible Evidence and Reason may at length prevail against Prejudice, and that *Libertas Philosophandi* may at last produce a true and real Philosophy.

> Jurare in verba Magistri = *To judge in the words of the master*
> Nullius in verba = *Nothing in words; i.e., not bound by received opinion*

**No. 24.** *Read May 26, 1697.* Richard Waller's note: *This Lecture treats of Animal Substances found buried in the Ground in several parts of the World, and of a Ship found in* Switzerland *with the Bodies of forty Men in it at a considerable depth under ground* [p.438].

We have lately had several Accounts of Animal Substances of various kinds, that have been found buried in the superficial Parts of the Earth, that is not very far below the present Surface; as particularly the parts of the Head of an *Hippopotamus* at *Chartham* in *Kent*, that of the Bones of the *Mammatoroykost*, or of a strange Subterraneous Animal, as the *Siberians* fancy, which is commonly dug up in *Siberia,* which Mr. *Ludolphus* judges to be the Teeth and Bones of Elephants; and indeed that peice which I saw of it was much like Ivory in its Texture, only the out side of it seem'd to have been cover'd by a kind of Skin, which I never heard of or saw any Elephants Tooth so cover'd with; then the Bones and Teeth of a large Elephant lately dug up in *Pomerania*, of which I some while since transcrib'd the Relation out of one of the late Monthly Mercuries, and read it at one of the Meetings of [p.439] this Society; also the great Bone in the Repository presented to the Society by Sir *Tho' Brown*, which was found upon the foundering or calving of some Cliff in *Norfolk*, which seems to have been the Leg-bone of some Elephant, if it be not some Bone of the fore Fin of some Whale; 'tis equally admirable which soever it may be found to be by one skill'd in the Osteology of those Creatures; and lastly the great Hornes that have been often found and dug up in *Ireland*, of which the account is printed in the last Transaction; all which, and divers others which I could mention, do shew that the present superficial Parts of the Earth have suffer'd very great Alterations, which I in my Lectures in 1664, indeavour'd to prove to have been the effects of some preceding Earthquakes, without which Supposition I cannot conceive any probable Cause can be assigned, much less can there be any such rational Cause assigned, for the Position of many other Phænomena which have been observ'd of such like Substances found and dug up at much greater depths, that is, of more than two or three Fathoms below the present Surface, at which depth those I have mention'd are said to be found. I conceive it will be very improbable to assign the Cause to the universal Deluge of *Noah*, and much more so to ascribe it to any particular Deluge, as to that of *Deucalion,* &c. for how could the Flood bury the Shells of Fishes in the middle of some of the highest *Alps*, and cover them with a prodigious height or thickness of Rocky Mountains? Or how should the bottom of the Sea come to be raised to such a prodigious height above the present bottom of the Sea at the Shore next such places? To me, I confess, it seems a most improbable, and groundless supposition: Improbable, for that 'tis hardly conceivable how the Water should heap up these Substancess, such prodigious masses of Stony of Earthy Concretions; and groundless, for that we have no mention in Sacred or Prophane History of any such effects produced by a Flood. However, tho' we should grant that Elephants might be carry'd by the Waters of the universal Deluge from the more Southern or Æquinoctial Parts to those Northern of *Siberia* or *Pomerania*, yet how shall we conceive by what means the universal Deluge should bury a Ship and forty Men at a hundred Fathom under Ground, and that at so great a distance from the Sea, as *Switzerland* now is, of which nevertheless we have an undoubted History? I say undoubted, because I have not found any Author that has question'd the truth of this Relation. Now, tho' I confess I did not know 'till lately (upon perusing Dr. *Wagners* curious Natural History of *Switzerland*) who inform'd me who was the first

*In this lecture Hooke uses recent reports of discoveries of mammalian fossils and even of a ship buried deep in mud to support his ideas which he has espoused over a period of over 30 years.*

Historian that had acquainted the World with this discovery; tho' I had met with the account in several other Historians, yet none of them speaking of it with any doubting Expression I conceiv'd it must be related by some Historian of good Repute. This Enquiry then Dr. *Wagner* answer'd by telling me the first relater of it, which was *Baptista Fulgosi* Duke of *Genoua*, which Author's Book I have since procur'd, and have read his Account of it, which I will presently give you as I find it express'd by *Camillus Gilnus* in Elegant Latine, being by himself, and his Father translated from the Original, Publish'd by the Author in the Year 1483, but the Book translated into Latin was Printed 1565. In this Book I find an account of the Author, and the Esteem he had, and the occasion of the writing of it, which was partly to drive away melancholy Reflections of his past Misfortunes, having lost his Dukedome, and partly for Instruction to his own son. In which Relation 'tis remarkable, that this Ship and Men should be buried so deep in the Earth as a hundred Fathom or six hundred Foot. Next, that the Bodies of forty Men should be found in the Ship itself. *3dly.* That this should be a Ship of the Ocean, and not of some River, because of the great distance of it from the Sea. *4thly.* That the Anchors and Sails, tho' torn, should yet remain and be plainly discoverable. *5thly.* That he did not take this Story from uncertain Report, but from divers grave Men, who had been Eye-witnesses of it, who had inform'd him themselves. *6thly.* That it was so remarkable in that time, that the Learned Men had meditated and reason'd on it to assign the Cause of it; that is, to give a rational Hypothesis, by which to shew how it might come to pass, they having it seems pitch'd upon two especially, which do both of them to me seem very insufficient, not to say very absurd. So that upon the whole Matter, there seems to me no Reason or Cause to doubt the matter of [p.440] Fact or the οτι, but all the difficulty lies in the διοτι that then shall be the next thing to be examin'd, and that the rather, because this seems to be a true *Experimentum Crucis* to distinguish between my Hypothesis and those of some other Authors. As first, concerning the two Solutions specify'd by the Author, not as his own, but as of some other Philosophical Men, who then lived, and who were satisfy'd, it seems, of the truth of the discovery, and 'tis not unlikely it might be some of those. *Plurimi Graves viri qui rem perspexerunt & qui in Re presenti fuere a quibus ipse accepit.* For as for himself he ventures not at any Solution, but says only *Cæterum utcunq; res fuerit admirationis non Mediocres relinquit Causas.*

First then, for the Hypothesis of *Noah*'s Flood, 'tis not said in any History, that Navigation, especially on the Ocean, was grown to such a perfection in *Noah*'s time as to make Ships of that Bigness and perfection of Anchors, Sail and Rigging, as this by this short Description seems to have been; and 'tis very likely if any such Navigation had been, it would have been taken notice of in the History of the Bible; for it cannot be suppos'd that *Noah* should not be inform'd of it, if any such Art had been then practiced in any part of the World how remote soever from the place of his Abode. Next, if such should have been, it might have happen'd that some other Men or Creatures might have escap'd with Life besides those in the Ark. Next, supposing that there had been such a perfection of Navigation at the time of the Flood, I cannot conceive how a Ship of that bigness, as this seems to have been, should be carry'd down so deep under the Surface of the Earth as 600 Foot: Certainly a twelve Month soaking of the Earth, much less forty Days, could not reduce the superficial Parts to such a hasty pudding Consistence as this Phænomenon does seem to

οτι = *matter*
διοτι = *cause*

*The following paragraph, I think, reveals the extent of Hooke's skepticism regarding the effects of Noah's flood. "Ships of that Bigness and perfection" refers to the buried ship found; i.e., if a ship with such advanced technology could be built at the time of Noah, the latter would have heard of it. Further, others besides those in the Ark could have survived the flood.*

*Also, how can such a big ship be carried down 600 feet below the Earth's surface? Surely the 40 days of Noah's rain could not have made the Earth so soft.*

require, since I doubt whether there can be found in the World any part of the bottom of the Sea, that has been soaked for some thousands of Years, that is so softned.

Next for the second Hypothesis of a subterraneous Navigation, to me, I confess it seems a ridiculous Supposition, tho' I know a late Author has imbrac'd such an Hypothesis to solve the Phænomena of Sea-shells, and the like Substances found in Mountains and Mines; tho' Mr. *Purchas* has Publish'd a like Story of *Andrew Knivet*, but I am apt to think that most Readers will look upon it as told by a Seaman and a Traveller.

But the Matter of Fact being so well attested, it must at least be suppos'd to be there plac'd by some Natural Cause, as must also all those other Phænomena I have mention'd.

Now for assigning a Cause sufficient, I conceive there cannot be a more probable one, that the effect of Earthquakes, which have, and do still produce as considerable Effects as any of these; the late Relations we have had of the effects wrought by them in *Lima, Jamaica,* among the *Cariby* Islands, among the *East-India* Islands, about *Vesuvius,* in *Norway,* and in the Island of *Sicily,* will furnish us with Phænomena almost as strange; besides it seems rational to believe, that Earthquakes in former Ages before we had History, were not only more frequent, but much greater and more powerful.

These, I conceive, have not only produc'd wonderful Effects in this or that part of the Earth at one time, but at many times successively, possibly at the distance of many Ages; so that at one time they may have raised the bottom of the Sea to make a dry Land, and sunk other parts so as to be overflow'd by the Sea, which were before far above the Surface of the Water, or to make Inland Seas or Lakes, as that of *Geneva* and divers others thereabouts: But by succeeding Earthquakes those effects may have been quite differing, so as to sink again those parts it had raised, and raise again and fill up with other Earthy or Stony Matter, those it had formerly sunk, and so also by various Efforts at various Times it may have overturn'd and turn'd upside down, or otherwise tumbl'd and confounded the parts of the Earth, which seems plainly to be hinted to us by the Mythologick Story of the Giants fighting with the Cælestial Powers, and heaping Mountains upon Mountains; and (I do confess) I conceive there can be nothing more reasonable and conformable to the proceeding of Nature in these Times, than to suppose there have been the like and much greater, in former Ages of the World.

[p.441] I conceive then, that whenever that part of *Switzerland* was the bottom of the Sea, this Vessel (which the Author calls *Navis* or a Ship) was upon that Sea over this very place, when there happen'd an Earthquake just underneath it, which did raise the same above the level of the Water, as much as it now is; and that by this there having happen'd to be an Opening, Cleaving, or Chasm in the Ground under it which swallow'd up some of the Sea, and with it this Vessel, and afterwards closed again, and inclosed what it had swallowed; or else that this part had been some very deep Inland Lake, as that of *Geneva*, and divers others there about, that this vessel was Navigating in this place when some Earthquake happen'd, which overthrew some Neighbouring Mountain, Hill or Lands, which falling into this Water, did not only sink the Ship, but fill'd up and levell'd

*Even the bottom of the sea, having been soaked for thousands of years, is not so soft.*

*Hooke always returns to a rational explanation—a natural, rather than supernatural cause.*

*Reiteration of earthquakes as the cause of terrestrial surface changes.*

*Successive earthquakes cause the exchange of land and sea areas.*

*This passage is significant in that Hooke conceived the possibility of rocks being "overturn'd and turn'd upside down." His examination of the rocks along the shore of his own birthplace, the Isle of Wight, must have also given him reinforcement in this idea. Some of the strata along the south shore show uplift and overturning. For a summary of the geology of the Isle of Wight, see Chapter 2.*

*Speculations on how this ship came to be where it was found.*

the Lake with the Contiguous Lands or Shoars of it; neither of which ways of explicating it do need any other effect, but such as we are by antient and much more later Observations ascertain'd, and the usual effects of Earthquakes.

**No. 25.** *Read February 22, 1699.* Richard Waller's note: *An account of a Ship found in the bottom of a Lake in Italy, supposed to be ever since* Tiberius's *Time with several Deductions and Queries thereupon.*

[p.441] I have consider'd the Passage mention'd by *Leo. Bapt. Alberti* in the fifth Book and Twelfth Chapter, concerning *Trajan*'s Ship found in his time in *Italy*. Now, I find that this *Alberti* was a *Florentine* Gentleman, who flourish'd about the Year 1483, and was accounted the *Vitruvius* of his Time: He being a Scholar, an excellent Painter, Sculptor and Mechanist, and an excllent Architect, he was the first that indeavour'd the Explication of *Vitruvius*, in which he made great progress, much to the improvement of that Age; in order to which he survey'd and measur'd the remainders of Antiquity; he understood Perspective also, and writ a Book on that Subject, which was not well understood by the Antients, nor much by the Moderns in this Time. But my present Inquiry is chiefly about this Passage mention'd in his Book *De Re Ædificatoria*, produc'd the last Day by Mr. *Bridgman* concerning *Trajans* Ship discover'd in *Alberti's* Time, which had lain sunk in a Lake of *Italy*, which he calls *Nemorensis*, ever since the time of *Trajan*, which was near one hundred Years after Christ, for he died in the Year ninety eight, which is now full sixteen Hundred Years since, and so was more than thirteen Hundred in the time of *Alberti*. The Passage is as follows. *Leo Baptista Albertus De Re Ædificatoria. Parisiis, 1512 8°. Libro V. Capite XII. Materiam omnemreprobant quæ fissilis, fragilis, sindens, putricosaque sit, clavosq; & ligulas æneas præferunt ferreis, ex Navi Trajani, per hos dies dum quæ scripsimus commentarer, ex lacu nemorensi eruta, quo loci annos plus milte trecentos demersa & destituta jacuerat, adverti pinum, materiam, & cupressum egregie durasse, in ea tabulis extrinsecus duplicem superextensam & pice atra perfusam, tela ex lino adglutinarant, supraque id chartam plumbeam claviculis æneis coadfirmarant. (Lacus Nemorensis) a dix huit Milles de Rome vers l'Orient, il s'appelle aujourd'huy Lago di Nemi.* What this Ship was, and the History of it, I have not met with, nor can I find any such Lake as is call'd *Lacus Nemorensis*, or *Nenorensi Lage*, as *Petrus Laurus*, in his Translation of this Book into the Vulgar *Italian* renders it. *Bartoli*, who Translated this Book into *Italian* after *Caius* renders it *Lago della Riccia*. *Pliny* indeed mentions a Ship of *Iayus*, which was purposely sunk at *Ostia* to found the Mole upon; but he could not say any thing of this, he dying almost 20 Years before *Trajans* time; nor do I treat upon what occasion it was that caus'd them to dig it out, nor at what depth it was found, nor whether it were buried in the Ground, or were only sunk into the Mud: If any have met with any further information concerning it in their Reading, I should be glad to be inform'd concerning it. *Fulgosus* having writ his Book much about the same time that *Alberti* writ this, I thought I might have met with some account of it in him, it being somewhat Analogous with his Relation of the Ship found in *Switzerland*, about the same time; but I do not find he hath any mention of it. It seems pretty strange how either of these Ships should come to be transported into the places where they are said to be found; but 'till we know the History we can at best but conjecture concerning them. There are many other particulars I should have desir'd information of besides those which he has mention'd, and 'tis very likely some of them may have been taken notice of in the Relation of its discovery, which I am inclin'd to believe must be somewhat more at large and more fully related than we find it here, which only hints two Remarkables [p.442] proper to the purpose, for which it is mention'd, *viz.* About the durableness of Timber fit for building of Ships in its own Nature: And, Secondly, Of the way of securing it against the Corrosion of Worms, which it seems was so long since taken notice of and provided against by the Shipwhrites of *Trajan*'s Time; which they perform'd by a double Sheathing; the first, next the double Planking, (*Tabulis extrinsecus duplicem Superextensam & pice atra perfusam tela ex lino adglutinarant*) was a kind of Tarpollin, they covering the Planks with Pitch, and that Pitch with Linnen-cloth sticking to it; the second was a thin sheet of Lead fastn'd by Brass Nails to the Plank; that it was very thin, I think is denoted by *Charta Plumbea*, that is, such kind of thin Lead as they formerly us'd for Writing on, much like the thinnest sort of Mill'd-Lead now made by the new Engine; which how they made is not known, nor do we certainly know how they make the like Sheets of Lead in *China*, of which kind I have seen a great variety, and all of it very even and regular: The Plumber will tell you 'tis done by Casting the Lead on Ticking, but that I conceive will not make it so thin and even as I have seen it; we have a way of beating it after the manner of Gold-beating, which doth foliate it very thin and even, 'tis commonly call'd *Tin Foile*, and 'tis us'd for foiling Looking-glasses; 'tis a mixture of Lead and Tin, as is also the *Tootenag* of *China*, and possibly theirs may be done the same way; but the Rowlers in the Mill I take to be much the better way; 'tis by somesuch Engine they foliate Brass and Copper in *Germany*, tho' they do some sorts also with the Hammer, as Kettles, and the thin Iron Plates for Latton by beating many of them together at once, as they do also Leaf-gold, Silver and Brass; but Asidue somewhat thicker, is done by an Engine with Rowlers, as they flatten Wire for Threads; and so also is a sort of Sheet Brass somewhat thicker: Possibly both ways may be known and made use of in *China*, where they have many other curious Inventions which we have not yet attain'd, and 'tis not unlikely but that the Antient *Romans* might for this foliating of Lead, have somewhat the same.

*Hooke, the walking encyclopedia of his day, digresses onto the subject of making thin metal foil for shipbuilding.*

Now as to the use of it for Sheathing of Ships, I find the *Spaniards* make use of it at this time, and have done so for a long time. This I find Sir *Rich. Hawkins* takes notice of in the account of his Voyage to the South Sea, Page 87, which see.

Here we have an account of all the ways of Sheathing of Ships he knew, and his Judgment or Censure of them, which how just they are must be left to Experienc'd Men; however, I have been lately inform'd that the *Spaniards* make use of the same way still for their Gallions, which 'tis not likely they would if they knew any way better; they had indeed another help to keep out the Water in case of any failure in the outward Plank, and that is the filling all the Space between the Ribbs and Planks with a certain sort of Plaister which may be a security to the innermost Plank, but not at all to the outermost against the Worm or Springing of them;

*He has a fascination for the Chinese culture, its language, its customs, its technological advances, and its history.*

however, 'tis of good use to keep off a suddain overflow or entrance of the Water in case of either Defect. But the best way of all seems to be the *Chinese*, by the Varnish, which neither Worms nor Water, nor Heat will damnify; nor in their way of building their Junks, do they leave any vacuity in the thickness of the sides to need Plaister, but what is fill'd with Damar, which is in itself lighter than Water, and will swim on it. But that way is not practicable here in *Europe* where we want the Varnish, whereas the others are, especially that of Sheet Lead, of which Metal this Nation affords us great plenty, and the late invented Mill doth certainly out do all other for giving it a proper Form; besides, if Plaister were necessary, we have as good as the World affords, or which possibly may be better, we can have Pitch enough (much of the same Nature and Use with Damar) to prevent any suddain gushing in of the Water: But this only by the bye.

The strangeness of the Relation or History of the Ship found sunk in a Lake, some where in *Italy*, mention'd by *Leo. Bapt. Alberti*, and the shortness and imperfection thereof as deliver'd by him, made me very desirous to get a more full and perfect Relation thereof. I thought *Baysius* in his Treatise *De Re Navali* might have taken notice of it, he having Written since that time; but he has never a word concerning it as I can find, nor do I find any [p.443] mention of it in *Dassie*'s Book *de L'Architecture Navale*; but *Pere Fournier* in his Hydrography (Book the Fourth, Chap. the First) treating of the Navigation which was before the universal Deluge of *Noah*, says, it seems rational to think that (considering the long Life of Men before the Flood, and the populousness of those Times) there was no part of the World uninhabited, tho' we have no History of them but the Bible, and tho' that has not one word concerning it; and that not only the great Continents of Land, but there being Islands both in the Seas and Rivers, those also were inhabited which could not be suppos'd without the use of some kind of Navigation. Add to this in the third place, that 'tis reasonable to think that the *Antediluvians* were as ingenious, if not much more, than the *Postdiluvians*, for the inventing of Ships, and for the use of them, for the transplanting of Colonies, for Trading and for War. Morever (says he) in the Year 1462, as is Recorded by *Fulgosus,* at *Bern* in *Switzerland*, as they were working in the Mines, at above a hundred Fathom deep in the Earth, there was found an old Wooden Ship built as ours are, whose Anchors were of Iron and the Sails of Linnen, with the Carkasses of forty Men. *Peirre Naxis* Relates a like History of another, such a one as found under a very high Mountain. In like manner the Jesuite *Eusebius Neurembergius*, in the Second Chapter of the Fifth Book of his Natural History, says, "That near the Port of *Lima* in *Peru*, as they were working a Mine for Gold, those which follow'd the Vein in the Mountain found an old Ship, which had many old Characters very differing from ours, which all People believed to have been there buried by the universal Deluge.--------------*Namq; Juxta portum Limæ in Peru cum eviscerarat avaritia terram, insecuta auri venam, Navigium inventum est sub ipso monte, quod a nostris, & hactenus fama & Scriptis antiquorum notis plurimum dissidebat. Creditumq; ab universis illuvie fuisse humatum.*" There was found also in a very high Mountain of *Mexico* a prodigiously large Elephants Tooth, tho' in all *America* there was never yet found any Elephant. "Without doubt (says he) all these things have been thus buried by the tumbling and overturnings of a universal Deluge, as well as the Wrecks of other Vessels which have been found at three Thousand Stadiums or Furlongs from the Sea, as *Strabo* relates in his First Book." Thus far Father *Fournier* to this purpose. Nor do I find that he

*This passage intimating that the antediluvians were much more ingenious than the postdiluvians not only shows his humor but also his attitude toward religion— the antediluvians, after all, were supposed to be so* bad *in the eyes of God that they had to be destroyed!*

hath taken any notice or made the least mention of this Vessel, mention'd by *Alberti*, which, methinks, he should not have been ignorant of, especially considering the great Pains he has taken, and great Learning he hath shewn concerning the Subject of Shipping.

The *Heer Witsen* in his Book intitled, *Ael Oude en Heden dueysche Scheeps Bouven Bastier*, in the Fourth Chapter of his First Book, hath given us a somewhat larger account and more particular than *Alberti*, but quotes not the Authors from whom he receiv'd it; so that we must rely on his Reputation 'till we can be better inform'd. His Relation in *Dutch* is to this effect. "In the time that the Pope, *Pius* the Second, possest the Chair (which I find was from *August 1458,* to August 1464) Men found in the *Numidische* Lake twelve Fathom under Water, in the Mud, a Ship, in length thirty Foot, and in breadth proportionable; built of *Cypress* and *Larix* Wood (which is a Species of Pine-Tree Wood) which was become of such an hardness, that it could neither be burnt nor broken, if it were needful. This Ship had lain under the Water for fourteen Hundred Years without the least perceivable Rotting to decay it: It was on the Deck done over with Pitch, and that cover'd with a Coat or Crust of a certain Pap or Morter made of Clay and Iron well temper'd or beaten together, which art of mixture is now conceil'd; tho' others are of Opinion that this mixture was not made of Clay and Iron, but of Clay and Pitch well kneaded together. The Deck was cover'd with Paper, Linnen Cloth, and Plates of Lead, which were nailed to the Planks with Copper Nails guilded. This Ship (a wonder) was found so stanch, that not the least drop of Water was found to have soaked into its Hold; it had the length of an old *Trireme* Vessel, and the breadth of a Hulk. In the Hold was found the Hangings of fine Velvet of an Orange Green, and in the middle of the Floor a Copper Coffer fastned by four black Strings, which being open'd there appear'd an Earthen Urne or Vessel, which was ornamented [p.444] with a Gold Plate, and fill'd with Ashes; and because Men saw the Name *Tiberius* several times engraven upon some Leaden Plates about the Border of it, they conceiv'd this might be the place of his Sepulture."

*A ship found in Italy under the bottom of a lake during the time of Pope Pius II (August 1458 to August 1464) dating back to the Roman Emperor Tiberius. The details of the preservation quoted by Hooke from Witsen's book are rather fantastic.*

This Account, tho' in divers Particulars different from that of *Alberti*, yet seems to be translated from the same Original History, which neither for them having mention'd by what Author it was written, we are yet to seek of the true account, which probably may be much more particular than either of these, or both of them put together; for that it is usual in second Hand Relations, to take notice of such Passages of the Original, as concern the present Subject they are treating of, and to omit many other Particulars, tho' in themselves much more remarkable; this therefore I further sought for in divers other Authors; and in *Riccioli's Hydrography*, I found a further account of it, which also gave me a hint of the true Author: *Riccioli*'s Account is this, Chapter the thirty ninth of the Tenth Book, which whole Chapter treats of Ships that have been much celebrated for their Magnitude, Splendor, Voyages, or other very remarkable Conditions; among which, Page 340, he brings in the Ship of *Tiberius* as one very remarkable instance, whose History he thus describes, *Narrat Æneas Sylvius, suo tempore repertum in lacu Numicio Cubitis 12 Sub aquam, navem ex Larice Cubitorum 20 Bitumine & mixtura ferri terræque; nescio cujus incrustatam, quæ per annos 1400. non computruerat. Siquidem in multis canalibus, ac fistulis incisum erat Tiberii Nomen; existimatumq; in ea Cineres illius Tyranni inclusos fuisse.*

*Hooke, therefore, tries to confirm the truth and accuracy of this account by researching into other writings and finds another account of this ship of Tiberius in Riccioli's book on hydrography.*

Thus we have found at length the Bush where this Game is seated, and whence it is to be started if we will have it, and I have follow'd it by its scent and Foot-steps to its Seat; but in what part of the Volume of the Works of *Æneas Sylvius* it is to be found I cannot yet discover, for his Tracts are many and make a bulky Volume together, which, whether it contain all that he writ I am not yet well inform'd; for he wrote very many particular Tracts, and left some imperfect and not ready for the Press, as *Conrad Gesner* informs us. This was the Man that, in *August* 1458, was made Pope, and who died in *August* 1464, so that he possest the Chair six Year; within which time it seems both these discoveries were made, (if at least they were two differing Discoveries, for possibly they may be only two differing Relations of the same Discovery) the one noting one sort of Circumstances, and the other, another. I cannot so well judge of the matter, 'till I find this Relation of *Sylvius*; however, 'tis obvious that what *Riccioli* makes to be only twelve Cubits, Mr. *Witsen* makes twelve Fathom, which is four times as much; and possibly this twelve Fathom or seventy two Foot *Fulgosus* might make one hundred Cubits, and yet all of them innocently without a design of imposing on their Readers, they writing from the Relations of others, and possibly from the failing of their own Memory to boot; for we find how rare a thing it is to find out the truth of a Fact, tho' 'twas done but Yesterday and almost at next Door, if allowances are not made for the Circumstances of the Relators, and the defects of every one's Memory and Comprehension; upon which account it is that I could wish that Relation concerning the Elephant lately found in *Germany* and made by the Colledge of *Gothan* might be inserted into a Transaction as well as that of *Tentzelius*, that Men might see how much the Humour and Inclinations of the Relators will diversify the Relation, and confound the Apprehension and Judgment of the Reader; and therefore I conceive it would not be amiss also to add to this last account the Sentiments of this Society, or at least of some of their Members, concerning the Substances sent by *Tentzelius* to be perus'd and examin'd by them; for there is no better way, I conceive, in the World to give a satisfactory account to Posterity of this Fact than this Course; for there cannot be made a good History, either of things Natural or Artificial, without curious judicious and accurate Observations, and Pertinent and Critical Experiments, that may be as thoroughly examin'd and verify'd, as a Geometrical Proposition by Persons sufficiently accomplish'd for such a Task. 'Tis not one possibly of a hundred is fit for such a Business, and yet such are necessary, and hence I conceive it is, that we have such a multitude of medicinal Observations made or pretended to be made by young Physicians, and possibly not one of five Hundred of any manner of real [p.445] Use or Benefit; for that the most of such Writers are two much biassed by precarious Hypotheses, and many likewise compose and Publish them only for Interest, that is, as Advertisements to make themselves the more known, and so to get Practice; and tho' this or that Symptom may be true and matter of Fact, yet the true Cause of the Distemper, and the reason of the Cure or Miscarriage of the Patient possibly was really quite differing from those assign'd by them; and tho' some of them may have been truly describ'd, yet those that know how small and inconsiderable Circumstances in themselves will yet make great and most considerable alterations in the Effects, will be more cautious than to take them all for true which are in reality quite otherwise; those therefore that relate an Experiment or Observation, should be both very understanding in the Subject, and very diligent in taking notice of, and relating the Circumstances of it; for that all that can be done in this way will be little enough of information to him, that is to make use of it for making Deductions and Inferrences therefrom, and indeed it will be hazardous to build any thing upon Foundations so uncertain; for even in the most perfect Accounts of this Nature, a Writer or Applier of it for the founding or examining a Theory thereby will find a necessity of ocular inspection and examination proper and fitted to his present Subject, either to obviate some Objection, or to give some further Light; for oft times the most considerable part of the whole Experiment may lie in some one trivial Circumstance, which not one of a thousand would otherwise have thought worth taking notice of, yet to him that knows what that Circumstance is that makes for or against his Theory which he is inquiring into, will judge it very considerable, and be sure not to omit the Scrutiny and Test thereof; and 'tis preposterous for any one to write an Experimental Natural History without making and examining the Experiments needful to the perfecting thereof, without making the Experiments himself, nay, and without the repeating of them, as Doubts may arise after the first Trial, or as he may need further information upon them; nay, without making them whilst he is writing, that he may trust, as little as may be, to his own Memory and Judgment. Thus in Anatomical Experiments and Observations, how many considerable Discoveries do we owe to such repeated Trials omitted wholly, or scarce hinted at in many preceding; for every discovery gives a new set of Doubts and Inquiries, as well as a new Light, not only δις και τρις *Sed etiam decies repetita placebunt*, as I have very often experimented myself; nay, I have found it absolutely necessary, and even that not enough to make some Spectator to apprehend the Consequences thereof: But this only by the bye. Before I leave this Subject I cannot but take notice of a Doubt that arises from the variety of these Relations, and that is, whether the Sheet Lead were used for the Sheathing of the outsides of the Ship under Water, or only for the Covering and Housing of the Deck, as the *Heer Witsen* makes it; nor know I how to solve it without seeing the Original Relation, only I must not omit one Passage of *Riccioli*, which seems to hint the use of Sheet Lead somewhat Analogous to Sheathing, and that is this, describing the Ship of *Hieron*, whose Architect was *Archimedes*.----*Dimedia Pars navis per 300 operarios sex mensibus absoluta, rimæ asserum laminis plumbeis tectæ,* &c. My doubt on this Passage is, whether the Vessel were Caulked and Pitched in the Joints of the Planks under the Sheet Lead, or whether the Plates of Lead were only made use of instead of Caulking and Pitching, the description is at large in *Athenæus* which I have not by me, and he, it seems, had it from --------, who writ a whole Book of the Description of it: It was in this Ship where *Archimedes* made use of his admirable Invention of his helical Pump, which he himself hath no where describ'd.

*Greek translation = two and three times; in other words and including the meaning of the Latin: not only two and three times but tens of times repeated*

*After rambling on about the importance of being skeptical about observations and narrations and of making experiments by oneself and replicating such experimentation, Hooke finally returns to the original topic of this lecture, which was the nature of the sheathing of the ship that was found preserved at the bottom of a lake.*

The small number of Authors that have recorded so remarkable a Phænomenon as this, informs us how little curious the World have been in the matter of Philosophical History, and thence how vain a thing it is to expect to find every such accident as this to be Recorded, tho' very remarkable in its self; for if these Ships were differing, then they have each but one Original Historian; for all the other Authors that have since mention'd them, seem to have borrow'd the Accounts from these two; but if the Relations were [p.446] only of one and the same Vessel (as methinks the Circumstances of the time and the being sunk deep into the Earth seem to intimate) then we have but two Historians that take notice of so remarkable a Fact; and those so discordant in their Stories, that one knows not which of them to give Credit to; the one making it to be found in the

*Lago de Nemi*, about twenty Miles from *Rome* towards the West; the other making it to be found near *Berne* in *Switzerland*, when 'tis not known that ever there was any Lake there, as *Fulgosus* mentions and Objects. It is therefore unreasonable to reject all Hypotheses that suppose other Accidents to have been the occasions of producing Petrify'd Substances, than those Recorded in History, especially if they happen'd before Printing was in use, or possibly Writing commonly known; for even since that time many considerable Phænomenoa have been very slightly hinted only, and scarce taken any notice of; as for instance, the Comet that appear'd in 1580, which produc'd but one diligent Observer and Historian, which was *Mich. Mastlin*, and the great Earthquakes and Catastrophies in *China*, which are Recorded in the *Mercurie Hollandois*, and no where else that I know: So 'tis probable this newly happening Earthquake at *Constantinople* would have been quickly forgotten, and probably never recorded to Posterity, if the *Gazett* and News Papers had not taken notice of it: But this only by the way.

*A major objection to Hooke's theories of earthquakes and petrifaction is that they lack written documentation. This long discourse has been to show that one should not rely too much on recorded history. The two accounts of the same buried ship are so different that one cannot be certain of the facts. Hooke's conclusion then is that one should not discard [his] hypotheses such as that concerning petrified substances (i.e., fossils) just because they were not documented in written history.*

As to this accident of the Ship, I conceive it to have afforded so many particular Informations worthy to have been Recorded, that I could wish it had happen'd in a more curious Age; at least I conceive it very desirable, that the Original History of it, such as it is, might be sifted out and inserted in a more proper place to be found, than where it is said to be at present.

**No. 26.** *A Discourse of the Causes of Earthquakes. Read July 30, 1699.*

[p.424] I mentioned in some Lectures that the Earth did seem to grow old and to have lost many of those Parts, which, in the younger times of the World, it seemed to me to have more abounded with; that which I instanced in, was the Foment or Materials that serve to produce and effect Conflagrations, Eruptions, or Earthquakes. These Materials I conceive to be somewhat analogous to the Materials of Gun-pouder, not that they must be necessarily the very same, either as to the Parts or as to the Manner and Order of Composition, or as to the way of Inkindling and Accension; for that as much the same Effect may be produced by differing Agents, so the Methods and Order of proceeding may be altogether as differing: A clear Instance of this we may find in the Phænomena of Lightning, wherein we may observe, that the Effects are very like to the Effects of Gun-pouder.

For we have first the flash of Light, which is very suddain, very bright, and of very short continuance, being almost momentaneous, at least every single flash is so, tho' the kindling of several parts at some distance from one another does sometimes continue a succession or longer duration of the Light.

Next we may observe the violence of the Crack or Noise which is likewise as momentaneous as the Fire, if it be single, but if there be many particular flashes that contribute to this effect, and those made at several distances, then the duration of the Thunder heard is longer than the duration of the flashes of Lightning, which proceeds, as I conceive, from two Causes; First, For that those flashes that are farther distant, have their Thunder a longer time in passing to the Ear, than those which are nearer, by reason, that tho' the Passage or Motion of Light be almost instantaneous, yet the progression or motion of sound is temporaneous, and requires a certain sensible time to pass a sensible space, and the times are proportionably longer as the spaces passed are greater. But a second Cause of the duration of the Thunder, I conceive, proceeds from Echoes that are rebounded both from parts of the Earth, and likewise parts of the Air, as from charged Clouds; of both which I am sensibly assured both by natural Reasoning and from sensible Observations, and I have observed much the same Effects produced by the Echoeing and Rebounding of the sound of a peice of Ordnance, from places at several distances adapted for the production of such Repercussions.

*It has been said that Hooke would never admit to a finite speed of light. The use of the words "almost instantaneous" here shows that he realizes that light velocity is finite, but he was a practical man; for all practical purposes, and given the precision of instruments at the time, light might as well be considered instantaneous, while the speed of sound can be sensible—i.e., detected by the senses.*

But, Thirdly, We have also the Power and Violence of the force of the Fire and Expansion, in fireing several things that are combustible, in suddenly melting of Metals and other Materials, which are difficult and slow enough otherwise to be made to flow, in rending, taring [sic], throwing down and destroying whatever stands in its way, and the like; and yet after all, that [p.425] which causeth these and many other strange Effects resembling those of Gun-pouder, seems to be nothing but a Vapour or Steem mixed with the Body of the Air, which is inkindled, not by any actual Fire, but by a kind of Fermentation or inward working of the said Vapour. Again, we find that the *Pulvis Fulminans* as 'tis called, which hath some of its materials differing from that of common Pouder; as also *Aurum Fulminans*, which is yet more differing both as to its materials and as to its way of kindling, have yet most of the same effects with Gun-pouder, both as to the flashing and thundring Noise, and as to the Force or Violence. So that as these are differing in many particulars, and yet produce much the same effects; so 'tis probable, that what is the cause of Earthquakes and Subterraneous Thundring, Lightning and violent Expansion, as I may so call those Phænomena observable in those *Crises* of Nature, may be in divers particulars differing from every one of these, both as to the materials, and as to the form and manner of Accension, and yet as to the Effects they may be very Analogous and Similar. So that tho' I cannot possibly prove what the materials are, yet the Effects speak them to be somewhat Analogous to those of Gun-pouder, or *Pulvis Fulminans, Aurum Fulminans* or Lightning, which, tho' they seem very differing in many particulars, yet when I come to shew the Causes and Reasons of those Effects, I shall manifest, that 'tis but one Operation in Nature, and that which causes the effect in one causes the effect in all the rest, and the outward appearances of the differing materials, and the differing way of Operating, are nothing but the Habits, and Dresses, and Vizards of the Actors, and the differing Modes and Dances by which they Act their several Parts, which, when they have done, they are at an end, and have exerted their whole Power, and there must be a new set of Actors to do the same thing again; the Oil of the Lamp will be turned all into Flame, but you must have fresh Oil, if you will have the Flame continued. So the Materials that make the Subterraneous Flame or Fire, or Expansion, call it by which name you please, is consumed and converted to another Substance, not fit to produce any more the same Effect; and if the Conflagration be so great as to consume all the present Store, you might safely conclude that place would no more be troubled with such Effects; but if there be remainders left, either already fit and prepared, but sheltered

*Not knowing about radioactivity, which was not discovered until centuries later, Hooke's remarkable mind is groping toward what could be a source for subterraneous heat that becomes "consumed and converted to another substance."*

from Accension by other interposing incombustible Materials; Or that there be other parts not thoroughly Ripe, and sufficiently prepared for such Accension, then a concurrence of after Causes may repeat the same Effects, and that *toties quoties* 'till all the Mine be exhausted, which I look upon as a thing not only possible, but probable, nay, necessary, for that I find it to be the general method of Nature, which is always going forward, and continually making a progress of changing all things from the State in which it finds them in at the present; all things as they proceed to their Perfection, so they proceed also to their Dissolution and Corruption, as to their preceding Estate; and where Nature repeats the process, 'tis always on a new Individual.

Now, tho' it may be Objected, of the material that produceth Lightning, tho' it seemeth to be all kindled and so burnt off by the flash, yet we find that after some time the same is again renewed, and so from time to time, and therefore as one Operation doth destroy and consume it, so another doth generate and produce it anew, and therefore it doth seem probable that the same may be done in the Subterraneous Regions, and thence, tho' there were many Accensions and Consumings of the foment of Earthquakes in former Ages, yet if Nature did thus again repair it, there would be little reason to suppose, that former Earthquakes should be greater than those which have in later, or in this present Age, been observed; to which I Answer, That tho' it seems plain, that the foment of Lightning is renewed, yet I conceive that to be only by new Emanations from the proper Minerals in the Body of the Earth, and not for that the same Substance which is burnt off in the Lightning, is again restored into its former State and made fit for a second Accension; for tho' there may be necessary a prævious Digestion of the Steams, which is performed by the Air and heat of the Weather, yet that does only prepare it with a proper fitness, but it must be some proper Mine-[p.426]ral that must furnish the Materials: And the same thing is more evident in *Vulcanoes* and burning Mountains, which are there only observed to break forth and burn where there is plenty of Brimstone and other proper Substances for such Conflagrations; for if the same were only a continual new Generation of Combustible Materials for the Fire, then I see no Reason why those *Incendiums* should not be equally frequent and equally great in all places, as well as in those where they are now frequently observed; for why should it not as frequently happen in our Hills and Mountains, as it does in *Sicily,* or *Island,* in *Ætna,* or *Hecla,* the one being as much colder then we, as we are then the other? It follows therefore, that it must be caused, not by the Renovation of the Foment, but from the Duration of the Mines or Minerals that supply fit Materials, and consequently, that when those shall be quite consumed, then, and not 'till then will the Fire go quite out. Nay that there are some such Instances of preceding *Vulcanoes*, which have heretofore burned and are now almost quite spent, may be concluded from the *Pike* of *Tenariff*, which, by all Circumstances, seems to have been formerly a burning Mountain, but is now quite extinct, and the Island of *Ascension* seems to be another such an Instance. All which Conflagrations are the several Symptoms of the progress of Nature in the determined Course and Method, which tho' it be differing from that of Life or Vegetation in lesser Bodies, yet it may be possibly as Natural and Necessary in the greater.

*He feels that the replenishing energy is some "proper mineral."*

I cannot therefore see any Absurdity in thinking or asserting that this Globe of the Earth on which we inhabit is in a state of Progression from one degree of Perfection to that of another degree, which may be termed of Perfection, for as much as it is the Progress and Operation of Nature; and at the same time it may be conceived in a progress to Corruption and Dissolution in as much as it continually changed from its preceding State, and acquires a new and differing one from what it had before, which new Estate may be upon some accounts considered as more perfect, tho' upon other accounts it may be accounted corrupting and tending to its final Dissolution; and as 'tis certain that it is continually older in respect of Time and Duration, so I conceive also that it grows older as to its Constitution and Powers, and that there have been many more Effects produced by it in its more Juvenile Estate, than it doth or it can now produce in its more Senile, as more particularly to Earthquakes and Eruptions; for to me it seems most evident and past doubting, that there have been in some preceding Ages of the World Eruptions and Conflagraions which have infinitely surpassed any that have happened of later Years, or indeed any that we have any certian account of in History. Some kind of Memory of some antient Traditions concerning a very great one that sometimes happened, seems to be preserved by the Poetical or Mythological History of *Phaeton*, of which *Plato* also tells us, that the *Ægyptians* had a more perfect knowledge and account, than ever the *Greeks* were Masters of, who, at best, as to Histories of preceding Ages, were, by the *Ægyptian* Priests, accounted but Boys and Children; however, *Ovid*, by his wording of that Fable, does seem plainly to have had some knowledge of what was meant or understood thereby; and tho' he seems to ascribe the Cause thereof to some extraordinary heat of the Sun, yet that might be nothing else but the relating the Opinion of the Antients preserved by the same Tradition, by which the Memory of the prodigious effects that had been wrought had been retained.

*The Earth grows old. Much more volcanic seismic activity must have occurred when the Earth was younger.*

In which case we are to distinguish between Histories of Matters of Fact, and those of Opinion; and *Plato* takes notice of as much when he mentions the Relation. The Matters of Fact seem to have been the Conflagraion of many parts of the Earth at once, and those the most eminent, such as the Mountains, which, whether they were in being before the Conflagration, or made by that Eruption, does not appear by the Story, but it seems most probable, that that was the time of their Production; and the calling of them by several Names, yet retained, does signify no more, but that those Mountains, which are now called so or so, were then on Fire and burning.

*Where the energy is spent there are extinct volcanoes.*

[p.427] But having before explain'd this Fable of *Phaeton*, and several others of that ingenious Mythologick Poet *Ovid*, I shall forbear the repetition of them here, and for the present would only infer, that in former and younger Ages of the World those kind of effects, produced by Eruptions and Earthquakes, have been much more considerable than those which are now produced, or which have been produc'd since we have had any Records kept of such Events; and therefore we are not to conclude that such huge Mountains, as the *Alpes,* the *Andes, Caucasus, Atlas,* or the like, could never be produced by means of Earthquakes or Eruptions, because we do not now find Instances of Effects of the same Grandure produced in our present Age, or in the Ages of which we have some more perfect account; for that in the former Ages there have been a much greater plenty of those kinds of Minerals which have been consumed, and for that the Relicts which are now left are but very small, and in probability not so apt for Conflagration, nor so strong and efficacious in their Operations; besides many of their Substances that were left may have since been

*Not only was the activity more frequent, but it was also more considerable—i.e., larger and more violent.*

petrify'd and converted into Substances, wholly unfit for the Foment or Fuell of such kinds of Fire; for that such Mutations have been effected by length of Time, I think no one that has observ'd and consider'd the Nature of Petrifaction can at all doubt, any more than he can whether there be any such Substance as Stone; for that all Places and quarries especially will furnish him with Evidences enough to convince any that will not be wilfully ignorant.

This effect of Petrifaction is a Symptom of old Age; for as plenty of *Spirituous, Unctuous* and Combustible or Inflammable Juices and Moisture is a sign of Youth: So the want of them, and of the Effects Produced by them, is a sign of old Age, in which those unctuous Juices are consumed and the Spirituous Fluids wasted, and the Parts become dry, and hard, and Stiff, and unactive; neither fit to inkindle the active Flame or to maintain it; neither fit to make other Substances fluid, nor to be made fluid themselves; which Fluidity is an inseparable Concomitant of that we call Spirituous Substances: And 'tis the plenty of those kind of Substances that maketh the Youthful Ages both of Plants and Animals to Flourish, and the Consumption and want of them, that makes both Plants and Animals to decay and grow old, as we call them, to grow stiff, and dry, and rough, and shrivelled; all which Marks or Symptoms may plainly be discovered also in the Body of the Earth, and I am apt to believe would be very much more if we could be truly inform'd of the former and younger Condition thereof; for I have very good Reason to believe, that there has been times of the Earth wherein it hath had a much smoother and softer, and more succous Skin than now it hath, when it more abounded with Spirituous Substances, when all its Powers were more strong and vegete, and when those Scars, Roughness and Stiffness were not in being; and tho' possibly some may think all these Conceptions to be groundless and meerly Conjectural, yet I may in good time manifest, that there are other ways of coming to the discovery of many Truths than what have been to this purpose hitherto made use of, which yet are not less capable of Proof and Confirmation, than Histories or Records are by Coins, Inscriptions or Monuments. And tho' it may seem difficult to understand or be informed of the State of the subterraneous and inaccessible Regions, and of the Ages before History, yet I do not look upon either as an impossibility, no, nor as insuperable by the Industry of a few, nay, of a single Person. And possibly I may some other time shew divers other ways of Inquiry, and other Methods of Demonstration of Causes than what have been yet applied to those purposes.

Nor is this Assertion of the growing old of the Earth to be looked on as so great a Paradox, or as Heterodoxical, or Scismatical, for we find in Scripture that the Kingly Prophet *David* in the 102 *Psalm* has an Expression that doth plainly assert it, not only of the Earth but even of the Heaven. *Of old hast thou laid the foundations of the Earth, and the Heavens are the works of thy hands, they shall perish, but thou shalt indure; yea all of them shall wax old like a garment, as a vesture shalt thou change them, and they shall be changed.* Which Expression is almost verbatim repeated by the Prophet *Isaiah*, Chap. 51.v.6. *Lift up your eyes to the heavens and look upon the earth beneath, for the heavens shall vanish*[p.428]*away like smoke, and the Earth shall wax old like a Garment.* Nay, this Expression of the *Psalmist* is verbatim repeated by St. *Paul* in the 10.11. and 12. Verses of the Epistle to the *Hebrews.* By all which it is evident at least, that *David, Isaiah* and St. *Paul* were all of that belief. I

---

*Because the fuel of the Earth is consumed and becomes spent, so the world grows old. While Hooke had in other lectures described the cyclical nature of some terrestrial processes, the Earth as a whole progresses and advances toward an end. In this concept with its implicit idea of limits and the second law of thermodynamics, he may have been more sophisticated than later geologists who were steeped in a world without end.*

*Hooke's conviction that it is possible to acquire knowledge about former ages before history or about the interior of the arth, is a fundamental tenet necessary for the development of geology as a science.*

*Having been a curate's son, Hooke could always quote scripture to back up any statement he wished—for the sake of silencing the "Objectors."*

could produce many Expressions to the like purpose both in Sacred and Prophane Histories, both of Christian and Heathen Writers, but those I have quoted I suppose may be sufficient to answer Objectors of that kind.

As for any other Objections that may be brought against this Doctrine, such as the equal Stature and Ages of Men for so long time as we have had any History; from the want of Histories of such Juvenil Estates, from the Permanency and Duration of all the Species of Plants and Animals in the same Estate, from the Incorruptibility of the Heavens and Cælestial Bodies, and so of their Influences, Causations, and many other of the like Nature; I doubt not to be able to give a satisfactory Answer if any of them shall be pressed or insisted upon, tho' at the same time I cannot hope that all will be convinced, much less, that all will confess themselves to be so, tho' really they are. All I can say, is *Valeat quantum valere potest*, let every one enjoy his own freedom.

*At age 65 Hooke begins to show some resignation and cynicism toward his fellow men. Unlike in his younger days, he doesn't much care now whether or not he convinces anyone.*

**No. 27.** *Read January 10, 1700.*

[p.436]    We have lately had an Account from Mr. *Tentzelius* Historiographer of the Duke of *Saxony*, of the Skeleton of an Elephant found buried in *Germany*, at the Foot of a Hill or Mountain at fourteen Foot deep, and cover'd with several Layers of Earth, but buried in a Sand, which the whole adjacent Mountain is found to consist of, being at a place call'd *Tonna* near *Erfond* in *Germany*. [I take notice of these Particulars, because they may be found to give some light as to the explication of an other Phænomenon which I shall by and by relate.] Now, tho' *Tentzelius* really judged and pronounced it to be the Skeleton of an Elephant, yet it was not without the Contradiction of many others of divers differing Opinions; the greatest number of which were for asserting it to be a *Lusus Naturæ*, as it seems the whole Colledge of *Gottha*, and divers other Learned Professors; but their Arguments are prov'd insignificant, and his own Doctrine sufficiently Establish'd in the Epistles which he wrote to Snr. *Magliabechi* and Publish'd in Print, and this Honourable So-[p.437]ciety were yet farther convinc'd of the certainty of it, by the Fragments and Specimens of the trials he had made of several of those Bones: But, after all, great Difficulties arose concerning the Means and Cause of the burying of it at such a place, and at such a depth and the covering of it, to be the natural Layer of the Earth, and not the Artificial filling up of a Grave or Pit dug by Art to bury it: Some attributed it to the effect of *Noah's* Flood, as 'tis usual for most to do in the like Cases, where they can think of no other Cause; to me, I confess, it seem'd rather to be the effect of some preceding Earthquakes, as I formerly here deliver'd in a Discourse on that Subject, when I first met with a Relation of it, as I have in other Discourses also about *Lignum Fossile* or Subterraneous Trees, and other Substances found buried, and now dug out from uner the Ground, not only in *Italy, Germany* and *France*, &c. but even in *England, Scotland* and *Ireland*. Now, because by our forreign Gazets, and also by our own from them of a late Phænomenon in the *East-Indies*, we have the History of a late Earthquake that happened there this present Year, whose effects do give an

*The skeleton of an elephant found in Germany where the "greatest number" still look upon fossils as tricks of nature.*

evident Proof of the Doctrine which I supposed, and indeavour'd to maintain, I thought it would not be improper to mention it here, and to add it as a further addition to the History of Nature. I shall indeavour to get the full Account of it Printed at *Batavia* in *Java*, an Abstract of which was Printed in the *Harlem* Currant in *October* last, and an Epitomy of that in our Weekly News-papers, which was this. Transcrib'd out of the *London* Post for *Sep.* 30. 1699. Printed for *Ben. Harris*.

"*Amsterdam October* 2. Our Letters from *Batavia* in the *East-Indies* of the 8th of *February*, say, That on the fifth of *January*, about two in the Morning, a most terrible Earthquake happen'd, which was so violent, that one and twenty Brick Houses, and twenty others were overturn'd, so that if it had lasted a little longer they must have been all thrown down. About 40 or 50 Persons were Buried alive under the Ruins of the Houses that fell, and near the same number were Lamed. Some small time before the Earthquake, the Blew Mountain, otherwise call'd Mount *Sales*, burst with such a terrible Flame and Noise, that it was both seen and heard there, tho' six Days Journey distant. Next Morning the River which falls into the Sea here, and has its Rise from that Mountain, became very high and Muddy, and brought down abundance of Bushes and Trees half Burnt; and the Passage being stopt, the Water overflow'd the Country round, all the Gardens about the Town, and some of our Streets; so that the Fishes lay dead in them: It was a whole Month before the River could be clear'd, altho' 3000 *Indians* were daily imploy'd to clear the same, during which time we were oblig'd to fetch fresh Water from *Bantam*, which is forty Miles. All the Fish in the River, except the Carps, were kill'd by the Mud and dirty Water: A great number of drown'd Buffaloes, Tigers, Rhinocero's, Deer, Apes, and other Wild Beasts, were brought down by the Current; and nothwithstanding a Crocodile is Amphibious, several of them were found Dead among the rest."

The Phænomena of this Earthquake, tho' they afford a probable solution of the more common Phænomena of fossile Trees, Wood, Nutts, Leaves, &c. of Vegetables, and of the fossile parts of Animals, &c. such as Teeth, Hornes, Bones; yet there are some other strange Phænomena, which I conjecture to have been effected by the same efficient Cause. If it be inquir'd what those strange Phænomena are, that I may give you an Instance, I shall acquaint you with one I late met with and receiv'd from a curious Person, who made the Observations himself, of which I have since been confirm'd by another curious Person who had seen and observ'd all the same Particulars. The Relation, in short, is this.

*A Description of the Ridge of* Mary Burrow *in the Queens County in* Ireland.

"This Ridge runs North and South, from *Tymohoe* to *Mary-Burrow*, about seven Miles, from thence towards *Montmelick*, four Miles further, and as this Author was inform'd through the King's County of *Westmeath*, towards [p.438] *Athlone,* but in these last mention'd Countries is much lower than in the Queens County.

"From the said *Tymohoe* to *Montmelick*, being both in the Queens County, it is about fourteen or fifteen Foot high, where highest, as near as this Author can conjecture, being laid as irregular as the Sands are usually laid by the Waves on the Sea-shoar, with several bendings in and out, high

*The north-south trending, elongated and sinuous hill described by Hooke's correspondent is probably a glacial deposit—possibly an esker.*

and low; the Sides so steep, that in most places not easy to ride up, and in many places Trees growing on the Sides, and a little thin Skin of Grass, apt to be burnt or scorch'd with the least dry Weather.

"It is so broad on the top as to afford room enough for four Horse-men to ride a Breast, the Road, in many places, lying on the top thereof.

"It is compos'd altogether of small rough Pebble grayish Stones about the bigness of a Mans Fist, and other smaller ones mix'd with Sand or Gravel, but no mixture of Clay or Loam, as this Author ever observ'd, which several times he sought after as he travell'd that way.

"None of the Lands adjacent to this Ridge have any of the materials whereof it is compos'd, mix'd with their Soils; in most places there are Boggs to within a very few Yards of its Foot, and where any Arrable lies near it, there is no mixture of the above Pebble or Sand therewith.

"So that it should seem probable that this Ridge of Pebble and Sand was brought from some remote places by some violent motion of Waters, and dispos'd into the form it now remains in, which induc'd the author several times to say, he believ'd it to be the effects of *Noah's* Flood, the Consideration whereof he refers to better Judgments.

"If any farther Particulars relating to this Ridge are desir'd, and a few Lines sent by the Post directed to the Author at *Rathdowney* near *Burris*, in *Ossory, Ireland*, they shall be carefully inquir'd into and answer'd by

*Nov. the* 14th                                        *Your most humble Servant*
1699.                                                       Ric. Prior.
    This Ridge is distant from the Sea about thirty Miles."

The same curious Person who is now return'd to his Estate, which lies in the Queens County not far from it, has promised me to make many other Observations about it, which I desir'd, and has promis'd to send me an account of his success, by which I hope I shall be better enabl'd to explain the Cause and Reasons thereof; 'till when I shall forbear for the present to make any further Reflections on it.

FINIS

*Hooke died three years after he wrote this last paragraph included in his* Discourse of Earthquakes, *curious to the end, searching for "Causes and Reasons."*

# *APPENDIX A*

The following are Waller's notes inserted in the Discourse [p. 281] explaining that the first five tables were drawn by Hooke himself with full description for only the first three tables:

> *As to the figured Stones or Petrifactions here mention'd, I found only one Sheet of the Descriptions of several of the* Cornu-Ammonis *sort, with some of the* Echini, *or Helmet-stones, which Descriptions follow: As for the Designs of them, they were, I know not by what means, not to be found amongst his Manuscripts; but by the Favour of Dr.* Sloane, *into whose Hands they happily fell, I procured them for the Graver, to whom the World and my self are obliged for this, as well as for other more valuable Communications. The Five first Tables were design'd by Dr.* Hook *himself; and tho' he has not perfected the Descriptions of them all, yet I have procured them all to be graved, supplying in some measure my self those Figures which were left undescribed by him. The Two last I drew my self from some figured Stones I happen'd to meet with, not far from* Bristol, *some Years since; about which time I gave the Designs to Dr.* Hook, *together with a Particular Explication of the Figures; but by Misfortune did not keep an exact copy of what I then gave him, which, amongst others of his Papers, is lost. I have endeavoured to supply this, as well as I now can, as the Reader may see by a short Account of a Letter I then sent him, with these Draughts; which I have so far presumed upon the Readers Acceptance, as to insert after the Explications of the Author's own Draughts.* R. W.

The following is the letter Waller mentioned in the note above and his descriptions for Tables VI and VII [p. 286-287]:

S I R, *Bristol, Aug. 17. 1687.*

In answer to some of your Enquiries, as to the *Cornua Ammonis,* and other Shell-like Stones found about *Keinsham*, and other Places, I shall give you this short Information of my Discoveries, and present you with the Draughts of some I happen'd to meet with there, and in other Places not very far distant, that is, in Part of *Gloucester* and *Somersetshire*.

The *Cornua Ammonis*, near *Keinsham*, lie most of them upon a little Hill, or rising Ground, above *Keinsham*-Bridge; the Place, as I take it, is about 18 Foot above the River: The River there runs half round the Foot of the Hill, where they lye very thick almost to touch each other, and are all of the large sort bedded in an hard Rock or Stone; some also I found near a Mile from thence in the Stone-walls of their Fields, and on the way in the Lanes; and at *Stowey*, four or five Miles from *Keinsham*, I saw some Snake-stones, Oyster, and Cockle-shells petrified and bedded in hard Stone, where is also a petrifying Spring incrustating the Moss and Grass, and all the wooden Troughs, by which it is conveyed with a stony Substance. Where they are not found fastned to a Rock, I found them encompassed in a pretty large irregular Mass of Stone, not sandy, but rather like a whitish Clay harden'd, and these stony Masses bedded in a loamy kind of Earth; in which soft Earth are also found Star-stones, and a sort of petrified Cockle-shells, such as Fig. 7. Tab. VI. I found none of the Snake-stones to have above 6 Turns, except one Mr. *Beaumont* shewed me of seven, and another amongst Mr. *Cole*'s Rareties. But that which I esteemed the greatest Curiosity, was a large Stone of the true Shape and Figure of the common *Nautilus*, or Mother of Pearl-shell, which tho' but a part of the whole Shell, weighs about 30 pound: This I found in one of the dry Walls, near *Keinsham*, and not far from it another Piece of the same; these are figured in the VIIth Table, Fig. 1,2,3,4; and in these not only the Diaphragms are very visible, but the Holes also in the middle of them, by which the Gut or String passes from one to another, in all respects answering to that of the *Nautilus*-shell; which, I think, will evince this at least to be a petrify'd Shell, tho' much larger than any of that kind that have been yet mention'd. Going down from *Mendip*-hills to *Okey*-hole, I found a small Muscle-shell petrify'd, on this the Shell was yet discoverable in some Places; this is figured, Tab. VI. Fig. 1.

On the Face of the Rocks that are on the sides of the *Avon*, not far from St. *Vincent*'s Rock, I found several wood-like Pieces of Stone standing out a little from the Rock it self; some of which I have broke off, and have represented them in the Figures, Tab. VII. Fig. 5,6,7, and 8, and some such like Bits of Wood in the Earth between the Layers of Stone unpetrify'd. But I shall at present detain you no longer, but for a more exact Information refer to the Figures themselves; which I can assure you are truly design'd, and are all Stones or Petrifactions, except the 6th Figure.

## TABLE VI.

Fig. 1. represents two Views of a sort of Muscle-shell found near *Okey*-hole.
Fig. 2. a sort of Cockle, on this likewise part of the Shell was visible.
Fig. 3. a Piece of a *Belemnites*, these are of the true bigness.
Fig. 4. a large *Cornu-Ammonis*, about 18 Inches Diameter; there are much larger to above 2 Foot, but this was one of the most perfect and neat I could find; on this the curious Foliage (as I may call it) of the Diaphragms was very visible, as is represented in the Figure, and near the Center several small Shells petrify'd: This I had from *Keinsham*.
Fig. 5. I know not what to make of, except it be one joint of the Spine of the Back of some Fish; something like it I have seen in the Backs of Salmons.
Fig. 6. a sort of *Nautilus*-shell not petrify'd, being still visibly a Shell broken, squeezed and flatted to the thickness of an Half-crown; the several Fragments were each roundish, and stuck together with a kind of blewish Clay. This was given me by Mr. *Cole*, who told me there had been several of them found amongst the Rocks and Stones in Quarries.
Fig. 7. another sort of Muscle or Cockle; these are frequently found about *Bristol*, particularly on the top of St. *Michael*'s Hill on the Road, and near the Gallows bedded in the Earth, not Stone, but are themselves an hard Stone, and very thick and strong.
Fig. 8. a Piece of a broken *Cornu-Ammonis*, in which several of the Diaphragms are very visible, the hollows not being fill'd up, but shot on the sides with a sort of flinty, hard, and transparent Spar.
Fig. 9. a small *Cornu-Ammonis* of but three Turns, yet seems perfect and unbroken; I know not whether it has any Diaphragms.

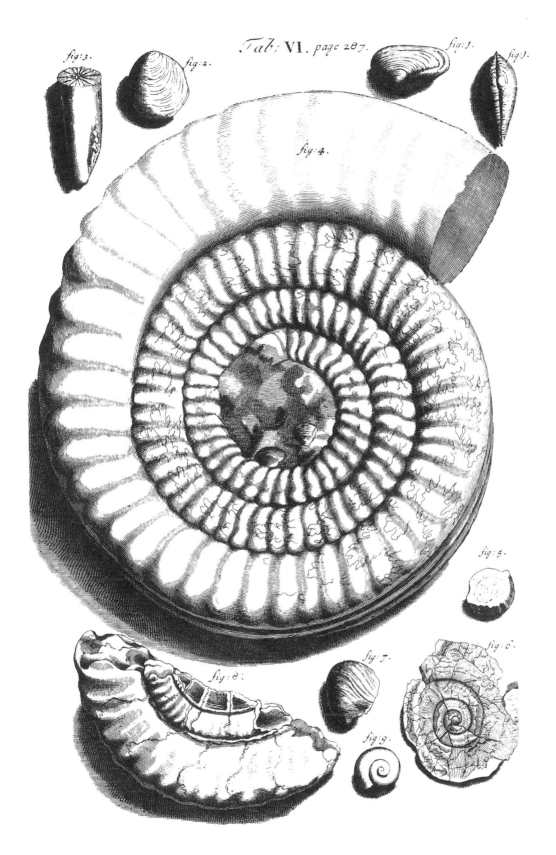

*Table VI. Waller's plate in Hooke's* Discourse of Earthquakes *(1705).*

**TABLE VII.**

Fig. 1. shews the large Stone of the Common Nautilus-shape, in which *d d d d* shews the Diaphragms to be seen on the outside of the Stone; as far as *f* is the larger Piece, which weigh'd near 30 Pound; from *f* to *g* is a lesser Piece, in which also the Diaphragms are visible on the outside *d d d*, and likewise on the inside, where the Piece was broken cross, some of the Diaphragms appear; as at *a a a, e e*, shews part of the Shell still sticking to the outside: *i i i* is a small Piece of the Center of the Stone, being only as much as makes 3 Diaphragms; the prick'd Lines *b* and *c* shew where the Stone should have been to have made it perfect: This is drawn not a quarter so large as the Stone it self.

Fig. 2. another Piece of the same sort found at another Place in *d d d*, are the Diaphragms, *e* a part of the Shell remaining: These shelly Parts are of a different Substance from the rest.

Fig. 3. a Piece of the same Stone near the Center, with the lesser part *c*; these have at *a* and *c* a protuberant part, being the hole of the Diaphragm.

Fig. 4. a Piece taken off from the former, in which at *a* and *c* are two small Cavities, answering the Protuberances *a* and *c* in the third Figure; *e* the edge of the thin Shell which covers all the part from *e* to *a*.

Fig. 5,6,7,8, Pieces of Stone resembling Wood petrify'd, of which the 6th is of several small Bits sticking to the hard Stone; this I broke off from the Rock; as also the 7th, wherein the cross Lines shew the ends of the long Fibres cut aslope; this exactly resembled a small Stick cut slanting. In the middles of the 5th and 7th was a Cavity in the place of the Pith. The middle of the 8th was filled with a stony Concretion very hard, as were all these Pieces, but something different from the rest of the Stone.

*Yours,* R. W.

Table VII. *Waller's plate in Hooke's* Discourse of Earthquakes *(1705)*.

# APPENDIX B

In addition to the London Monument and other city buildings and structures, Hooke designed and built the following (Batten, 1936-37 and Colvin, 1987):

The Royal College of Physicians, 1672-78 (moved in 1825; theatre demolished in 1866; the rest burnt down in a fire in 1879), Fig. 1-20.

Merchant Taylors' Hall, Threadneedle Street, London, carved wooden screen, 1673, destroyed during the Blitzkrieg, 1940-41.

Merchant Taylors' School, Suffolk Lane, 1674-75. Demolished 1875.

Bethlehem Hospital, known as Bedlam, was built in fifteen months between April 1675 and July 1676. Evelyn describes it as "magnificently built, and most sweetly placed in Moorfields." Building was pulled down in 1814 when it was found that the ground was "very unfit for the erection of so large a building." (Wheatley, 1891), Fig. 1-21.

Montague House, Bloomsbury, London. Hooke designed and built this palatial house including the gardens for Lord Montague in 1676, created Earl 1689 and Duke 1705. Unfortunately, on January 19th, 1686, the palace was gutted by fire. Upon the site of the Montague house was built the present British Museum where Hooke's original iron works of the gates survived until the 19th century, Fig.1-22.

The Parish Church, Willen, Buckinghamshire. Designed and built by Hooke for Dr. Busby of Westminster School. This fine church still stands and services are held there, Fig. 1-2.

Ragley Hall, Warwickshire. Designed by Hooke for Lord Conway. This house, now owned by the Marquess of Hertford, is open to tourism today. If one visits the Great Hall at Ragley and notes the black and white checkerboard floor pattern and filligreed ironwork for heating and notes the same style at the little Willen church, it is not difficult to conclude that the same person designed both of these buildings, Fig. 1-23.

Alderman Aske's Almshouses, Hoxton. The Haberdashers Company engaged Hooke for the design and construction of this building as a bequest by the alderman. The building has since been destroyed.

Canterbury Cathedral panelling in choir 1676. Removed c. 1826.

Magdalene College, Cambridge, the Pepysian Building. Designed 1677.

Escot House, Devonshire. Designed for Sir Walter Yonge, 1677-8.

Londesborough House, Yorkshire. Designed for the 1st Earl of Burlington, including garden layout, 1676-8. Demolished in 1811.

Houses on St. James's Square, London, now demolished.

Houses in the Strand, London, for John Hervey, 1678.

Somerset House, the Strand, London, The Stables for Queen Catherine of Braganza, c. 1669-70. Demolished c. 1780.

Spring Gardens, London. Designed house for Sir Robert Southwell, 1684-85. Demolished.

Whitehall, London, house in Privy Gardens for the 20th Earl of Oxford, 1676. Demolished 1691 or 1698.

Lowther Church, Westmorland, for Sir John Lowther, 1686.

Royal Dockyard, Hamoaze, the Officers' Dwelling-Houses, Great Storehouse, Ropehouse, etc. c. 1690-1700. Hooke drawings are preserved in British Museum (B.M.Add. MS. 5238°).

Ramsbury Manor, Wiltshire, for Sir William Jones, Attorney General, c. 1680-83. Ramsbury Manor is now owned by Harry Hyams, but the house is described by John Julius Norwich in his book, *The Architecture of Southern England*.

Shenfield Place, Essex, for Richard Vaughan, 1689.

In addition, many other houses and structures are also associated with Hooke. For example, it seems clear that Hooke worked on various parts of Westminster Abbey for his old master and friend Dr. Busby. Hooke was responsible for re-paving the choir for which he received only three guineas from Dr. Busby in 1676. He also was consulted about Solomon's Porch and repairs to Henry VII's Chapel, the North Window and various other parts of the Abbey.

# INDEX

Entries in bold font signify major topics with subentries.

*A New Theory of the Earth*, 74
Abrepta; *see* Proserpina
Académie des Sciences, 31, 122
Accademia del Cimento, 113
Acipio Africanus, 186
Acosta, Josephus, 195
Adams, Frank Dawson, 75
Ægean Sea, 103
Ægean, myths on origin of, 334, 335
Æneas Sylvius; *see* Pope Pius II
Æneis, 212
Ætna, 188, 212, 213, 278, 290
age of the Earth, 96–103, 207, 271, 278, 288–290
Agricola, 113, 301
air circulation, 51, 268
air pump, 14, 19
Albertus, Leo Baptista, 352, 354, 355
Albritton, Claude, 117
Aldrovandus, 236, 301
All Saints Church, 10, 129n.
All Souls College, Oxford, ix
Alps, 174, 198, 207, 213, 216, 232, 241, 317, 361
Alum Bay, 62
*American Journal of Science*, ix
**ammonites,** 87, 116, 161–165, 176, 208, 217, 235, 236, 238, 239, 241, 242, 244, 299, 348
  descriptions of, 162–167
  drawings of, 164
  fimbriæ or suture of, 162–163
  origin of, 300
*An Inquiry into the Original State and Formation of the Earth*, 121
Anapis River, 331
anchor escapement, 40
Andes, 194, 207, 216, 317, 361
Andrade, E. N. da C., 5, 9, 45, 47, 129n.
Andromeda, 280, 284
Anglesy, trees buried in, 207
antediluvians, 97
Antilles, 326
Apollo or Jupiter, 339
Appenines, 174, 198, 207, 213
archbishop of Canterbury, 54
arches, principle of, 35
*Architecture Navale, de L'*, 354

Arequipa, 195
Aristarchus Samius, 246
Aristotelian sphere, 30
Aristotle, 70, 238, 288, 297, 307, 323
Aristotle on land/sea exchanges, 303
Aristotle's *Meteorologica*, 85
Arno Valley, 120
Ascension Island, 360
Ashmolean Museum, Oxford, 70, 75, 130n.
assistants, training of, 50–51
Astroites, 168
Athens, 271, 272, 256
Atlantis, legend of 101–102, 103, 194, 273, 274–275, 280, 294, 295, 332
Atlas, 280, 281, 282, 283, 317, 361
atmosphere, shape of, 267–269
atmospheric pressure, 219–225
Aubertin-Potter, Norma, ix
**Aubrey, John,** 4, 10, 16, 50, 76, 105, 106–111, 129n., 133n.
  and Ray's plagiarism, 106–111
  *Memoires of Naturall Remarques in the County of Wilts*, 106, 109
Augustus, 244
*Aurum Fulminans*, 205
Auvergne, 122
axial displacement, 85, 87–95, 210, 245; *also see* polar wandering
axis of magnetic pole variation, 218
Azores, 187, 188, 189, 195, 208, 324

Bacon, Francis, 80, 123, 237, 253
Bacon, Roger, 243, 244
Bailey, E. B., 121
balance wheels for watches, 47
Ball, Dr. Peter, 232
Ball, William, 235
Bancroft's School, 134n.
Bannat of Temeswar, 122
barometer, 38
basalt, origin of, 122
Bauer, Georg, 113
Bauhine, John, 237, 301
Baysius, 354
*Beagle*, 98
Beale, Mary, 11, 130n.
Beaumont, John, 76, 134n.

Bedlam, see Bethlehem Hospital
belemnites, 168–169
Bell, Walter, G., 34
Bellonius, 238, 255, 301
Berkeley, Lord, 22, 86
Berkelius, Abrahamus, 273
Bermudas, 327
Berry, Joyce, ix
Bethlehem Hospital, 37
Bevis, John, 122
Bible, 75, 96; insufficient time in, 102
biblical account, 232
biblical chronology, 69, 217–218, 271, 288, 295
biogeography, 95
biological evolution, 85, 97–100
Bishop of Chester, 86
Bishopsgate, 34
Black Notley, 108
Black, Joseph, 124
Blackbourne, Dr. Richard, 106
blood transfusion on dogs, 23
Bochart, 273
Bodleian Library, ix, 20, 50, 129n., 135n
Boethius, Hector, 194
Born, Baron I. von, 122, 136n.
Bosphorus, 195
Bottoni, Signor, 326
Bow Curch, 33
Boyle's Law, 14
Boyle's, laboratory, 14–16
Boyle, Robert, 5, 14, 20, 22, 40, 48, 55, 82, 104, 130n., 133n., 219, 233
Bradley, James, 42
*Britannia Baconica*, 184, 188, 191, 193, 196, 344
British Isles raised from sea, 209
British Isles under water, 174
British Museum (Natural History), ix
Britton, John, 110
Brouncker, Lord, 22, 40. 132n., 219
Brown, Thomas, 349
bubble level, 38
Buddington family, 134n.
Burford-stone, 208
Burnet, Bishop Thomas, 71–73, 74–75, 76, 85, 86, 90, 134n., 136n., 210, 269, 277, 286, 305, 342
Busby, Dr. Richard, 11
Butler, Samuel, 20 130n.
Button-stones, see echinoderms
Byzantinus, Stephanus, 273

Cæsar, Sir Julius, 134n.
Calceolarius, 301
calculator, 50
calculus, the, 33
Cambden, 201, 202
Canary Islands,174, 186, 187, 188, 195, 208 275, 324, 326
Canterbury, archbishop of, 103, 108–109, 111
capillary attraction, 16
Cardan, 238, 301
Caribbeans, 326
Cartesian universe, 77
cartography, 51
Caspian Sea, origin of, 335
Cassini, Giovanni, 22, 48, 253
catastrophism, 135n., 248
catenary line, 131n.
Caucasus, 174, 182, 207, 216, 317, 361
Cayenne, 90, 249
celestial mechanics, Hooke's, 42–43, 77–78
cellular structure, 23, 25
center of gravity changes, 247
centripetal force, 32, 78, 159, 200, 257
chalk cliffs, 60–67, 234
Charles II, 5, 6, 20, 21, 104, 130n.
Cheapside, 33
Cheshire, trees buried in, 202, 207
Chester, bishop of, 103
Chichester Cathedral, 264
Childrey, 188, 191,193, 196, 201, 202, 344
Chinese age of the Earth, 86
Chinese chronology, 101, 103, 217, 218
Chinese language and culture, 243, 244, 353
Christ Church College, ix, 11
chronology from fossils, 233
chronometer, 40
Church of All Saints, 60
circular pendulum, 38
circulation of air, 51
circumference of the Earth, 38
Cisalpinus, 301
City Surveyor, 34–35
Clark, Alan J., ix, x, xiv
Clarke, Trevor, x, 129–130n., 134n.
Cleeveley, Ron, ix
climate zones, 87
cloth-making, 39
Clymene, 335, 336
Cohen, I. B., 14
Columba, Fabius, 301
combustion experiments, 21
combustion theory, 27
Comet of 1677, 48
*Cometa*, 48
comets, 22, 47, 322
Conduitt, 130n.
congruity, 220, 225, 252
consolidation, 125–126, 180; see also petrifaction
constancy of interfacial angles, 115
continental drift, 79, 94
continents, origin of, 174
Cooper, John, ix
Copernican system, 41, 42

Copernicus, 88, 217, 246, 249
Coriolis effect, 78
Cornhill Street, 34
*Cornua Ammonis*; see ammonites
Cosimo III, 114
Courtillot and Besse, 95
crab, fossillized, 169–170
cratering experiments, 216
craters of the Moon, 29, 30–31, 216
Creation and extinction, 318, 319
Creation, 99, 316
crinoid stems, 168–169
Critias, 271
Cromwell, Oliver, 13
Cromwell, Richard, 13
Croone, Dr., 23
Crosby, Sir John and Lady, 133n.
crystal forms; see crystallography
crystallization, 178
crystallography, 29, 118–119, 161, 178, 319
Cumming, Sandra, ix
Cupid, 331
Curator of Experiments, 17
Cutler, Sir John, 18, 35, 39, 55, 133n.
Cutlerian lectures, 18, 35, 41–49
cyclic theory, 71, 84–85, 86, 100–101, 155, 176, 190, 200, 201

Darwin, Charles, 98
Dassie, 354
Davenant, Sir Will, 107
David, King, 362
Davies, Gordon, 114, 121, 128
*De Mundo Subterreano*, 83
*De potentia Restitutiva*, 48
*De Re Ædificatoria*, 352
*De Re Metallica*, 113, 301
*De Re Navali*, 354
de Milt, Clara, 27
Dead Sea, 193
Dean of York, 338
Derbyshire, 121
Descartes, 219, 286, 294
Desmarest, Nicholas, 122
Deucalion's flood, 332, 335
Deucalion, 340, 342, 343, 349
Deutsch, Hans, 113
Dick, Oliver Lawson, 129n.
*Dictionary of National Bography*, 105
Dillon, Joseph, 57
Disbrowe, General John, 13
*Discourse of Earthquakes*, 83, 123, 158–365
Dorset, 61, 66
Dott, R. H., Jr., vii, 126
Draco, 42
Drake, Charles Whitney, Jr., x, 134n.
Drake, Judith Ellen, x
Drake, Robert Charles, x

Drake, Sir Francis, 195
Duchatz, Father, 345, 346
Duke Humphrey Library, ix
Duncan, Robert, ix, 95
Durdans, 22, 86
dust deposits, 189
dyes, experiments with, 39
Dymond, Jack, ix

**Earth**
  age of, 96–103, 207, 271, 178, 288
  an oblate spheroid, 87–95, 242, 245
  center of gravity changes, 174, 210–211
  core of, 215
  evolution of, 361, 362
  fuel spent, 361, 362
  gravitation of, 258
  Hooke's system of the, 42–43, 77–86
  shape of, 87–95, 135n. 242, 245, 246, 249, 251–256, 294, 302
  teleology, 362
**earthquakes**, 174, 181, 182–195, 197, 270
  as cause of Earth's surface, 174, 182–210, 240, 248, 249, 303, 323, 351
  causes of, 247, 358–363
  four effects of, 157, 183
  Hooke's meaning of the word, 174, 181, 194, 216
  in Barbados, 321–322
  in Batavia, 364
  in Leeward Islands, 320–329
  in Mediterranean region, 337
  in Sicily, 330
  of 1672 in the Greek Islands, 311
  of 1680 in China, 311–312
  of 1680 in India, 313
  of 1680 in Spain, 313–314
  raising of the land, four "species" of, 183
  sinking of the land, four "species" of, 183
**echinoderms**, 166–168, 172, 235
  descriptions of, 166–168
  drawings of, 167
Egyptian chronology, 101, 217
Egyptian pyramids, 254–255, 278
elephant, fossil of, 363
England in Torrid zone, 87, 242
England under sea, 174, 317
English Rose, 265
environmental adaptation, 98, 99, 217 247
Epsom, 22, 86
Eratosthenes, 295
eruptive gases, 314–315
*Essay toward a Natural History of the Earth*, 70–71, 75
'Espinasse, Margaret, 5, 9, 11, 18,
Eusebius, 354
Euxine Sea, myths on origin of, 334, 335
Evelyn, John, 22, 34, 189

evolutionary theory, 96–103, 116, 174, 239, 318, 319
extinct volcanoes, 361
extinction, concept of, 30, 70, 71, 98–99, 174, 239, 316, 318, 319
extinction, objection to, 316, 319

Fabianus, 85, 201
fable as history, 269–284, 285–293, 297, 330–334, 339–343
fables, interpreted by Hooke, 280–284
Farber, Paul, ix
Fauconerius, 238
Fayale, 209, 213, 326
Fellow of the Royal Society, 18
felt-making, 22, 47
Ferber, J. J. 122, 136n.
Ferdinand II, 113
Fernie, 135n.
figure of the Earth, *see* Earth, shape of
figured stones; *see* fossils
Firman, R. J., 82
Fish Street Hill, 35, 50
Fisher, John, ix
Flamsteed, John, 50, 51
flying experiment, 15
foraminifera, 29, 61, 63
fossilization, 215
**fossils**
  chronology from, 233, 304
  occurrence of, 71, 172–173, 182, 213, 232, 235, 317
  origin of, 28, 69, 83, 112, 114, 116–117, 161–177, 182, 206, 228–231, 346
  seven "propositions" of, 156
  study of, 229
  substance of, eleven "propositions," 156
  usefulness of, 97–99, 209, 233, 255, 279, 304
Fournier's *Hydrography*, 354
Fracastorius, 301
Freind, Dr. John, 106
French Academy, 343–346
French Guiana, 90
Freshwater Bay, 63, 66, 67, 233
Freshwater, 10, 58, 60, 134n.
Friedman, Gerald, vii

Galapagos Islands, 327
Gale, Thomas, 338
Galileo, 30, 219, 220, 253, 257
Garver, Cynthia, ix
gastropods, 168–169
generation of new species, 98–99, 174
*Genesis*, 70, 86, 96, 120, 213, 288, 300, 328
  Hooke interpretation of, 305
Geologia, 76
Geological Society of America (GSA), vii
geological timescale, 103

Gesner, Conrad, 112, 236, 273, 275, 301, 356
giants in myths, 338, 339
Gilbert, William, 78
Gilbert, G. K., 30, 32
Giles, Ann, 130n.
Giles, Cicely, 10, 61
Gimcrack, Sir Nicholas, 20
Glanvill, Joseph, 20
Glen Tilt, 61
glossopetræ; *see* sharks' teeth
Goethe, 122
Goldreich and Toomre, 95
Goldsmiths' Row, 33
Gorgons, 280, 283
Grace Church Street, 34
Grace Hooke, 54, 133n.
Graham, George, 42
Gratorix, 14
gravitation, 32–33, 42–43, 77–79, 159, 200, 227, 268–269
gravity deficiency in mountains, 199, 208
Great Fire of London, 32, 33–39
Greenwich Observatory, 50
Gregory, James, 344
Gresham College, 13, 17, 22, 41, 42, 43, 106, 107, 117, 235
Gresham lectures, 18
Gresham Professor of Astronomy, 34
Gresham Professor of Geometry, 17, 34
Gresham, Sir Thomas, 134n.
Grew, Nehemiah, 41, 50
Group, the 1645, 13
Guildhall Library, ix
Guildhall, 33, 34
Gunther, R. T., 14, 16, 18, 30, 110, 129n., 30n.
Habichtswald, 122
Hadley cells, 268
Hadley, John, 51
Hall, A. R., 40, 132n.–133n.
Hall, M. B., 132n.–133n.
Halley, Edmund, 22, 48, 54, 56, 90, 91, 92, 249
Hambly, E. C., x, 35
Hampshire Hookes, 10, 34
Hampshire-Dieppe Basin, 61
Hanno's *Periplus*, 85
Haswell, J. E., 16
Hearne, Thomas, 134n.–135n.
Heath, G. Ross, ix
Hecla, Iceland, 326
helioscopes, 35, 38, 44
Helmet-stones, *see* echinoderms
Herodotus, 190, 297, 298, 299
Hesiod, 271, 276, 285, 330
Hevelius, Johannes, 29, 30, 43, 254, 260
Hipparchus, 29
Hire, de la, 344, 345, 346
*History of Earth Sciences Journal*, ix
History of Earth Sciences Society (HESS), vii

History of Geology Division of GSA, vii
Hobbes, Thomas, 105, 106, 130n., 294
Hodges, William G., ix
Holder, William, 105
*Holologium Oscillatorium*, 43
Homer, 271, 276, 285, 330
Hook Hill, 58, 60
Hooke *Diary* entries, xiii, 129n., 130n., 133n., 134n.

**HOOKE, ROBERT**
Hooke's life (in chronological order)
  physical description, 4
  birthplace, viii, 10
  *Diary*, ix, 129n., 130n., 133n., 134n.
  early years and family, 9–11
  inheritance, 129n.–130n.
  Westminster School/Dr. Busby, 11
  gastrointestinal problems, 54, 129n.
  Oxford days, 11–16
  Royal Society, 17–59
  Curator of Experiments, 17–59
  salary, 17, 18, 35, 39, 52–53
  Gresham Professor of Geometry, 17
  Fellow of Royal Society, 18
  *Micrographia*, 23–32
  correspondence with Newton, 32
  rebuilding of London, 33–39; see also Appendix B
  refusal to sign watch agreement, 40
  secretary of the Society, 50
  member of the Council, 53
  coffeehouses, 54
  Cutlerian lectureship, 18, 35, 41–49
  training of assistants, 50–51
  lawsuit against Cutler settled, 55
  honorary doctoral degree, 54
  death of his niece Grace, 54, 56
  last appeal for justice against Newton, 54
  last days, 56–57
  congestive heart failure, 56
  Hooke's grave, 56
  fortune found in trunk, 57
  Hooke's library, 58
  "Worthies" window, St. Helen's Church, 56, 57
  commemorative stone block, 58

**Hooke's discoveries, inventions, and experiments (not an exhaustive list)**
  air pump, 14, 19
  arches, principle of, 35
  balance wheels for watches, 47
  barometer, 38, 39
  biological evolution, 96–103, 116, 174
  blood transfusion, 23
  bubble level, 38
  capillary attraction, 16
  cartography, 51
  celestial mechanics, 32–33, 42–43
  cellular structure, 23, 25
  centripetal force, 32, 159
  circulation of air, 51
  circumference of the Earth, 38
  clock-driven telescope, 43
  cloth-making, 39
  combustion theory, 27
  congruity and incongruity, 220, 225
  conical pendulum, 38, 43
  crystallography, 118–119, 161
  dyes, 39
  felt-making, 22
  frequency of musical sounds, 22
  gravitation, principles of, 32–33, 42–43, 77–79, 159, 200, 227, 268–269
    lecture of May 23, 1666, 77–78
  helioscope, 35, 38
  Hooke's Law, 48–49
  Hooke's Spot, 22
  hygrometer, 27
  hygroscope, 38, 39
  iris diaphragm, 51
  Jupiter, rotation of, 249, 253
  Jupiter's giant spot, 22, 249
  light, wave theory of, 27
  lunar craters, 29, 30–31, 216
  methodology, 159–160, 226–229
  meteorology, 27
  microorganisms, 23, 25
  optical telegraph, 52
  optics, 27
  parallax determination, 42
  pendulum clock, 16
  periodicity of comets, 22
  petrifaction, 174–182
  picture agates, 319–320
  polar wandering, 79–80, 84, 85, 87–95, 126–127, 242, 245–247, 249–250, 251–256, 259, 260, 269
  pressure/volume apparatus, 15
  quadrant, 131n.
  respiration experiments, 27, 39
  seawater sampler, 21, 22, 38
  sextant, 22
  shape of the Earth, 246, 249, 251–256
  sounding gear, 21
  spring-controlled watch, 16
  stellar aberration, 42
  telescope, reflecting, 38
  telescopic sights, 260–262, 264
  terrestrial magnetism, 51
  thermometer, 27, 38
  transportation modes, 53
  universal joint, 44
  wheel barometer, 27
  wind guage, 27, 38, 39

**Hooke's attitude toward**
  credit where due, 344
  fables, legends, and myths, 280–284, 332–334,

336–337
God, Bible, and religion, 96–103, 201, 213, 312, 328–329, 335, 354
Henry Oldenburg, 117
Noah's flood and ark, 119–120, 201, 350–351
plagiarism by the French, 343–344
power of technology, 243
scriptural chronology, 86, 97, 102, 119–120
the environment, 159, 348
the medical profession, 356
time and age of the Earth, 100–103, 207
**Hooke's geological contributions**
chronology from fossils, 233
comparison with Hutton, 120–128
   Hooke's influence on Hutton, 120–128
comparison with Steno, 116–120
concept of strata, 117–118, 325
crystallography, 118–119, 161
cyclicity of erosion, deposition, uplift, 71, 84–85, 86, 100–101, 118, 176, 190, 200, 201
*Discourse of Earthquakes*, 120, 122, 123, 159–365
Earth in space, 79
"earthquake" theory, 124–125, 302, 320
**evolution theory**, 96–103, 116, 174, 239, 318, 319
   extinction/generation of species, 97–100, 174
formation of islands, 326
formation of mountains, 118
formation of rocks, 81–82
Hooke's importance in history of geology, 128
isostatic compensation, 80
land-sea exchanges, 124–125, 183–195, 208
"magazines" for subterraneous heat, 327
organic origin of fossils, 113, 116, 228–231
origin of ores and minerals, 81–82, 191
petrifaction, 174–182
polar wandering, 79–80, 84, 85, 87–95, 126–127, 242, 245–247, 249–250, 251–256, 259, 260, 269
present is key to past, 240
shape of the Earth, 246, 249, 251–256
system of the Earth, 42–43, 77–86
unconformity implicit in cyclicitiy, 100, 128, 190
uniformitarianism and catastrophism, 135n., 240, 248
usefulness of fossils, 233, 279
Hooke *Diary* entries, ix, 129n., 130n., 133n., 134n.
Hooke-Newton controversy, 32–33, 158
Hooke Museum, 129n.
Hooke's bad press, 58, 105, 110
Hooke's methodology, 80–81, 159–160, 226–229
   *Method of improving Natural Philosophy*, 123
Hooke's salary, 17, 18, 35, 39, 52–53

Hooke's Spot, 22
Hooke, John (Robert's father), 10, 61, 131n.
Hooke, John (Robert's brother), 10
Hooke, Katherine, 10
Hooke-Wallis letter, 135n.
Hoskins, John, 11
*Hudibras*, 130n.
Hungary, 122
Hunter, Michael, 9, 110
Hurst Castle, 232
**Hutton, James,** 82, 86, 112, 120–128, 184, 188
   *Abstract*, 122
   argues against aqueous solution 125–126
   argues against axial shift, 126–127
   debate with Hooke, 124–127
   similarity with Hooke, 120–121, 124–127
   *Theory of the Earth*, 121
Huttonian theory, 182
Huttonian world-without-end, 103
Huttonians vs. Wernerians, 125
Huygens, Christian, 16, 33, 43, 44, 45, 219, 132n.
hydrologic cycle, 85
hydrothermal vents, 326
hygroscope, 38

impact theory, 30–31
Imperatus, Ferranti, 301
incongruity, 220
interbreeding, 100
interference colors, 27, 28
International Geological Congress, Bologna, 114
iris diaphragm, 51
Isaiah, 362
islands, origin of, 155, 326
Isle of Portland, 67
**Isle of Wight,** viii, 58, 60–68, 80, 84, 86, 96, 100, 120, 129n., 134n., 176, 181, 232, 233, 241, 351
   **geology of Isle of Wight**
   Atlantic opening, 68
   Bright stone anticline, 61
   Brixton anticline, 61
   calcareous rock, 65
   conglomerate rock, 64
   Corallian beds, 68
   *Cornu Ammonis*, 67
   Cretaceous transgression, 65, 68
   Cretaceous, 61, 68
   dissolution of chalk, 65
   Eocene, 62, 66
   erosion of cliffs, 65, 66, 67
   fossiliferous rock, 64
   *Galba*, 66
   invertebrate fossils, 66, 67
   Jurassic, 68
   Laurasia, 68
   Mesozoic Era, 68

Middle Chalk, 65
Oligocene, 61, 66
Paleocene, 61, 66
Permo-Triassic time, 68
*Planorbis*, 66
polar wandering, 67–68
Sandown anticline, 61
Tertiary, 68
Torrid Zone, 67
*Unio*, 66
Upper Chalk, 65
vertebrate fossils, 66, 67
*Vivaparus*, 66
isostasy, 208
isostatic compensation, 80
Israelites' passge through Red Sea, 335
Italy, buried ship found in, 352–358
Ito, Yushi, 70, 76, 85

Jagger, Rev. Ian, 12
James II, 104
Jerman, Edward, 34
Johnston, 236, 238
JOIDES *Resolution*, 95
*Journal of Geological Education*, ix
Juno, 342
Jupiter Ammon, 280, 284, 299
Jupiter or Apollo, the god, 280, 284, 299, 339
**Jupiter, the planet**, 253
   its giant spot, 22, 249
   rotation of, 249
*Jurare in verba Magistri*, 80

Kassel, 122
Kentish stone, 237
Kepler, Johannes, 249
Kerr Library, OSU, ix
Keynes, Geoffrey, 109
Kircher Athanasius, the Jesuit, 83, 109, 179, 180, 186, 187, 192, 208, 296, 330
Kite, Margaret, 104
Komar, P. D., viii

Lake Pergusa, 330, 331
*Lampas*, 45, 47, 117
land displacements, 196
**land/sea exchange**, 174, 182–210, 240, 248, 249, 303, 323
   raising of the land, 183
   sinking of the land, 183
Landgrave Collections, 135n.–136n.
*lapides sui generis*, 168
latitude determination of, 265–266
latitude of London, 250
latitude, measuring a degree of, 254, 256
Laudan, Rachel, 82, 112, 122
law of gravitation, 77–79, 159, 200, 227, 268–269

lecture of May 23, 1666, 77–78
Lawrence, Robert, ix
Leeuwenhoek, Anton van, 48, 51
Leeward Islands, 320–329
legends as history, 285–293, 297, 330–334, 339–343
Leibniz, Gottfried, 33, 51
Lely, Sir Peter, 11, 61, 130n.
Lely, U. P., 93
Leonardo da Vinci, 112
Leopold, 113
Letzmann, J., 93
Levi, Shaul, ix
Lhwyd, Edward, 75, 76, 80
light, theory of, 27
light, velocity of, 359
*lignum fossile*, 203
Ligurian Sea, 113
Linschoten, 184, 189, 194
Linus, Franciscus, 130n.
Lipary, 325, 326
Lisbon earthquake, 103, 123
Lister, Martin, 50
lit-par-lit molecular replacement, 82, 234
Lithgow, 193
**London**
   Bishopsgate, 34
   Bow Church, 33
   Cheapside, 33
   Cornhill Street, 34
   Fish Street Hill, 35
   Goldsmiths' Row, 33
   Grace Church Street, 34
   Great Fire of 1666, 32, 33–39
   London Bridge, 33
   Mermaid Tavern, 33
   Pudding Lane, 33, 35
   rebuilding of, 33–39
   Royal Exchange, 33, 34
   St. Paul's Cathedral, 33, 35
   Thames, 33, 36
   The Mitre, 33
   The Monument, 35–36, 50, 338
longitude determination, 40, 41, 256, 264, 265–266
Lucasian Professor, Cambridge, 134n.
Lucian, 343
Ludolphus, 349
Lunar Society, Birmingham, 124
lunar craters, 29, 30–31, 216
*lusus naturæ*, 83
Lycaon, legend of, 291, 339
Lyellian uniformitarianism, 84

Macdougall, Alexander, 134n.
*Machina Cælestis*, 260
Magdalen College School, Oxford, 130n.
magma chambers, 327

magnetic force, 78
magnetic poles, variation of, 210–211, 218, 247, 250
magnetism, terrestrial, 268–269
Maldive Islands, 327
Malta, 114
mammal teeth, 169–170
mammalian fossils discovery, 349
marble, veins in, 324
Marcus, Clifford Stephen, ix
Mary II, 96
Mascarenes, 325, 326
mass deficiency beneath mountains, 80
matter within the Earth, 80
May, Hugh, 34
Mayow, John, 27, 104
McInnis, Helen, xiii
McVicar, Barbara, ix
Mead, Dr. Richard, 106
Medicis Chapel, San Lorenzo, 114
Medicis, Court of, Florence, 113
Mediterranean region earthquakes, 337
Mediterranean, myths on origin of, 334, 335
Medusa, 280, 282, 283
Memphis, 298
Mercator, 51
mercury ingestion, 54
Mercury, 219–224, 253
Mermaid Tavern, 33
Merton, Robert, 112
*Metamorphoses*, 85, 86, 102, 109, 211, 278
   Hooke interpretation of, 285–293
Meteor Crater, 131n.
meteorites, 31
*Meteorologica*, 85
meteorological instruments, 51
*Meteors*, 323
*Method of improving Natural Philosophy*, 123
methodology, Hooke's, 80–81, 159–160, 226–229
Meyer and Co., Messrs., 57
Michell, John, 123–124
*Micrographia*, 14, 15, 20, 23–32, 48, 83, 86, 98, 113, 220, 346
   cells in cork, 23, 25
   dedication to Charles II, 23, 24
   eye of a fly, 23, 26
   fossil Foraminifera, 23
   Hooke's microscope, 23, 25
   house fly, 23, 26
   louse, 23, 26
   microfungi, 23, 25
   title-page, 24
Mills, A. A., 32
Mills, Peter, 34
Mills, Reginald A., ix
mineralization, 208
minerals, origin of, 191

Moluccas, 326
Molyneux, Samuel, 42
Mons Olympus, 29
Montagu, Edward, 104
Montagu, Ralph, 58
Montagu, Sir Sidney, 104
Montague House, 38, 131n.
Monument, The, London, ix, 35, 36, 50, 338
Moon, 29, 30–31, 216, 253
Moray, Sir Robert, 17, 40, 132n.
More, Henry, 45, 94, 109
More, L. T., 130n.
Mosaic account, 74, 75, 155
Moses, 213
motion, first law of, 257
mountains, origin of, 183–184
*Mundus Subterraneus*, 179, 180, 186, 187, 192, 330
*Musæum*, 189
musical sounds, 22
mythology, 269–284
myths as history, 269–284, 285–293, 297, 330–334, 339–343

Naples, 184, 188
*Natural History of Switzerland*, 349–350
Nauenberg, Michael, 33
**nautilus**, 165–167, 236, 238, 239, 241, 242, 244, 316
   descriptions of, 165–167
   drawing of, 167
Needles, The, 60, 61, 62, 63, 66 232, 233, 241
Neile, William, 13
Nell Young, 133n., 134n.
neoplatonist, 94
neptunism vs. plutonism, 125–126
Neshyba, Linda Drake, viii, x, 134n.
Neshyba, Steven, x
*New Experiments*, 130n.
Newton reference in *Earthquakes*, 285
**Newton, Isaac**, 10, 12, 32–33, 37, 45, 50, 54, 74–76, 88, 90, 91, 92, 120, 128, 131n., 134n., 135n., 158, 184, 200, 210, 227, 235, 242
   and Creation, 75
   and *Micrographia*, 23
   and theory of combustion, 27
   *annus mirabilis*, 32
   Cartesian influence on, 75
   corpuscular theory of light, 28
   correspondence with Hooke, 32
   Hooke-Newton controversy, 32–33, 158
   law of gravitation, 32, 54
   letter to Oldenburg, 33
   "Newton's rings," 27
   path of a falling body, 50
   *Principia*, 32, 78, 88, 90, 91, 226
   principle of unsociableness, 33
   support of Burnet's theory, 76

untruthfulness, example of, 33
*vis centrifuga*, 32
Nile delta and deposition, 84, 190, 200, 298
Noah's Ark, 71, 327
Noah's flood, 28, 69, 70, 71, 72, 82, 91, 96, 97, 108, 113, 190, 198, 201, 202, 208, 217–218, 240, 300, 304–309, 327–328, 349, 350, 363, 365
Norfolk, Duke of, 236
Northamptonshire, 237
Northhamptonshire-stone, 208
*Novum Organum*, 123
*Nullius in Verba*, 69, 80

"oblate," meaning of, 88
oblate spheroid Earth; *see* Earth
Oldenburg, Henry, 33, 41, 45, 46, 47, 50, 107, 116, 117, 132n.–133n., 296
Oldroyd, D. R., 80, 85, 121, 280
Olearius's travels, 182, 241
Olmstead, J. W., 30, 131n.
Oregon State University (OSU), viii, ix
ores, origin of, 191
original horizontality, law of, 115
Orosius, Paulus, 332, 335
Orpheus, 276, 288
**Ovid**, 70, 85, 86, 102, 109. 212, 228, 277, 278, 279, 306, 330, 332, 335, 338, 340, 343, 361
 and Earth's history, 285–293, 297
 *Metamorphoses*, 85, 86, 102, 109, 211, 278
Oxford English Dictionary, 89
Oxford Group, 13
Oxford Philosophical Society, 85, 91, 94
oxygen, discovery of, 27

Palissy, Bernard, 112
Pangæa. 94
 lecture of May 23, 1666, 77–78
Papin, M., 50
parallax, determination of, 42
*Parentalia*, 34
parliamentary "Visitors," 13, 130n.
Patterson, Louise, 33
pendulum clock, 16
pendulum, conical, 43
Penrose Conference of GSA, vii–viii
**Pepys, Samuel,** 11, 21, 22, 23, 47, 104, 111, 132n.
 family connection, 111
 secretary of the Admiralty, 111
 taking bribes, 111
periodicity of comets, 22, 47
Periplus of Hanno, 85, 273, 275, 295, 297
Perseus, 280, 282, 283
petrifaction, 86, 125–126, 174, 177–182, 203–205, 214, 234
Pettus, John, 83
Petty, Dr. William, 13, 22, 86

Phæton, fable of, 332–334, 335, 336, 361
*Philosophical Collections*, 50, 52, 76, 296
*Philosophical Transactions*, 45, 46, 47, 50, 52, 56, 83, 123, 296
Phœnicians, 297
Pickering, Sir Wm., 134n.
picture agates, origin of, 319
*Pilgrims*, 184
Pisias, Nick, xiii
Piso, Gulielmus, 238
plagiarism in 17th century, 104–111, 248
"plastick Vertue," 173
plate tectonics, 125
Plato, 70, 194, 271, 272, 273, 279, 288, 306
Plato's *Republic*, 271
Plato's *Timæus*, 85, 86, 103
Playfair, John, 122
Pliny, 61, 109, 181, 185, 187, 193, 194, 196, 197, 238, 244, 295
Plot, Robert, 52, 70
Pluto, 330, 333, 335
plutonism vs. neptunism, 125–126
Po River, 84, 200
pocket-watch, 40
**polar wandering,** 79–80, 84, 85, 87–95, 126–127, 242, 245, 246, 247, 249, 250, 254, 251–256, 259, 260, 269
 distinction from continental drift, 93–94
 relationship with biogeography, 95
"pole-flight" force, 93
Pope Pius II, 355, 356
Popham, Colonel Edward, 105
pornography, insinuation of interest in, 58
Porter, Roy, 75, 123
Portland stone, 116, 163, 181, 208, 235, 236, 237, 241
postdiluvians, 97
*Posthumous Works*, 40, 120–121, 155
Powell, Anthony, 104, 106, 110
Power and Towneley hypothesis, 15
Pratt, Roger, 34
precession of equinoxes, 88, 210
pressure/volume apparatus, 15
Prince of Wales tavern, Oxford, ix
*Principia*, 32, 78, 88, 90, 91, 135n., 226
*Prodromus*, 108, 114–115, 117, 119
Proserpina, rape of, 330–334
Ptolemaic system, 41, 42
Ptolemy, 244, 249, 250, 260
Pudding Lane, 33, 35
Pugliese, Patri J., 33
Pullen Sue, ix
Purbeck Peninsula, 61
Purbeck stone, 235, 237, 181, 208
Purchas, 184, 185, 189, 243
Pyramids of Egypt, 252–255
Pyrenees, 207, 213
pyrites, 237

*Pyrologia Topographica*, 326
Pyrrha, 340
Pythagoras, 228, 246, 249, 288, 297
Python fable interpreted, 339–343

quadrant, Hooke's, 131n.
Queen's College, Cambridge, 123
Queens County, Ireland, 365
Quicksilver, *see* mercury

radioactivity, 359
Ragley Hall, 38, 131n.
Ranelagh, Lady, 133n.
rape of Proserpina, fable of, 330–334
Rappaport, Rhoda, 129n., 155
**Raspe, Rudolf Erich,** 122–124, 135n.–136n.
   *Travels of Baron Münchausen*, 122
   *Specimen*, 122
   translations of travels, 122
   volcanic origin of basalt, 122
Raven, C. E., 110
**Ray, John,** 71, 86, 94, 99, 106–111, 239
   *Chaos and Creation*, 110
   *Miscellaneous Discourses*, 108
   objection to axial shift, 109–110
   plagiarizing Hooke, 106–111
   *The Wisdom of God Manifested in the Works of the Creation*, 108–109
   *Three Physico-Theological Discourses*, 110
Red Sea, 335
reflection, principle of, 46, 51
refraction, principle of, 46, 51
regression of the sea, 322
Reinert, David, ix
respiration experiment with dog, 39
Réunion hotspot, 95
Ricaut, Paul, 195, 199
Riccardus, Franciscus, 187
Riccioli's *Hydrography*, 355
Ricciolus, 29, 41
Richer, Jean, 90, 135n.
Roberval, 344
Robinson and Adams, 41
Robinson, Tancred, 76
Robinson, Thomas, 134n.
Robison, John, 27, 130n.
rock fracturing, 325
rocks and minerals, formation of, 81–82
Roemer, Olaus, 51
Rome, Dom Remacle, 116
Rooke, Lawrence, 13
Rossiter, A. P., 98
rotational axis shifts; *see* polar wandering
Royal College of Physicians, 37
Royal Exchange, 33, 34
Royal Observatory, 50
**Royal Society, London,** ix, 13, 16–23, 32, 34, 39, 69, 76, 83, 112, 116, 122, 135n., 205, 343
   attending Hooke's funeral, 56
   censure of Hooke, 39
   charter, 16
   criticized, 226
   failure to support Hooke in watch dispute, 40
   meetings in Gresham College, 41
   motto, 324, 348
   Newton letter to Huygens, 33
   payment to Hooke "by results," 52–53; *see also* Hooke's salary
   recess during plague years, 22
   repository, 235, 236, 349
Royal Society, Edinburgh, 120, 122
Rozet, M. le Captaine, 32
Rubriques, Gulielmus de, 243
Rudbeck, Dr. 254
Russell, Sir Francis, 13

*Sacred Theory of the Earth*, 71–73, 74–75, 76, 86, 269, 286, 305, 342
Salisbury Cathedral, 264
Sand Horns; *see* ammonites
Sandys, 184, 188
Santerinum, Island of, 187
Santorini, 102
Saragosa, 331
Saturn, 249, 253
Saxony, 113
*Sceleta Serpentum*; *see* ammonites
Schaffer, Simon, 9
Scheiner, 253
Schneer, Cecil, 98, 119
Screw-stones; *see* gastropods
scriptural chronology, 86, 97, 102, 119–120
sea bucket for sampling, 38
seawater sampler at any depth, 21
Seneca, 70, 85
Septuagint, 338
serpentine-stones; *see* ammonites
sextant, 22
Shadwell, Thomas, 20, 226
shape of the Earth, 87–95
Shapin, Steven, 5, 12, 18, 58
sharks' teeth, 112, 113, 114, 168–169, 318
sheathing of ships, 357
Sheldonian Theatre, Oxford, 36
Sherley, Thomas, 70, 82
ship-building, 353
Short, T., 82
Siberian mammalian fossils, 349
Sicily, earthquake in, 331
Small, Larry, ix
Smith, Barnabas, 130n.
Smyrna, 199
snail-stones. *See* ammonites
snake-stones. *See* ammonites
Socrates, 271, 272

Sodom and Gomorrha, 193
solar wind, 78
Solon, 103, 271, 272, 294, 332
sounding gear, 21
species and environmental change, 98, 99, 217, 247
species, extinction of, 97
species, generation of, 97–99
*Specimen*, 122
Spencer, Sir John, 134n.
Sprat, Bishop Thomas, 19
spring-controlled watch, 16, 40
St. Clair Humphreys, A., 32
St. Euphæmia, 192 330
St. Helen's Church, ix, 34, 134n.
St. Helena, 90, 195, 249, 326, 327
St. James's Park, 232, 233
St. Mary's Church, Cambridge, 108
St. Paul's Cathedral, 33, 35, 177, 131n.
Star-stones, 168
*State of Physic* (1718), 106
stellar aberration, 42
Stelluti, Francesco, 203
Steno's Law, 115, 119
**Steno, Nicholas,** 47, 83, 107, 112, 113
   *Canis carchariæ dissectum caput*, 113
   comparison with Hooke, 116–120
   constancy of interfacial angles, 115
   conversion to Catholicism, 114, 119
   dissection of a shark's head, 113
   geological cross sections, 115
   history of Tuscany, 115
   on crystallography, 118–119
   on formation of mountains, 118
   on fossils, 116
   on scriptural chronology, 119–120
   on strata, 117–118
   on the deluge, 119–120
   ordained priest, 114
   original horizontality, law of, 115
   *Prodromus*, 114
   self-deprivation, 114
   Steno's Law, 115, 119
   superposition, law of, 115
Stenonis, Nicolaus; *see* Steno
Stensen, Niels; *see* Steno
Stephens, Elizabeth, 57
Stevinus, 301
Stow, Mr. 196
Strabo, 109, 260, 295, 354
Straits of Magellane, 195
Stromboli, 325, 326
Strongylus, 194
Stubbe, Dr. Henry, 20
subduction, 125
substances, transformation of, 203–204
subterraneous eruptions; *see* volcanic eruptions
**subterranean fires or heat**
   cause of, 216
   decay of, 214–215
   source of, 359–360
   storage in "magazines," 327
superposition, law of, 115
Switzerland, buried ship found in, 349
Syracuse, 331

tectonics and cyclicity, 84–85
telescope set-up in Gresham College, 43
telescope, clock-driven, 43
telescope, reflecting, 38
telescopic sights, 260–262
Tenerife, 174, 188, 194, 207, 209, 275, 295, 296, 324, 326, 360
Terceras, 174, 184, 189, 195, 326, 327
Thales, 297, 299
Thames, 33, 200
*The History and Philosophy of Earthquakes*, 122–123, 124
The Mitre, 33
*The Origin of Continents and Oceans*, 93
*The Sceptical Chemist*, 82
*Theoria Sacra*; see *Sacred Theory of the Earth*
Theories of the Earth, 69–76
*Theory of the Earth*, Hutton's, 121
Thera, 187
thermal contraction, 125
thermometer, 38
Thessaly, 332
Thevenot, 254
*Three Physico-Theological Discourses*, 86
Tiberius, Emperor, 97, 355
Tillotson, John, 54, 103, 108–109, 111
*Timæus*, 85, 86, 194, 271, 275, 279
time, duration of, 100–103, 207; *also see* scriptural chronology
Tompion, Thomas, 16
tonguestones; *see* sharks' teeth
Torricelli, 219, 220, 344
Torrid Zone, 87, 242
Totland Bay, 63
Transylvania, 122
*Travels of Baron Münchausen*, 122
trees buried by earthquakes, 323
Trinacria, 331
Trinity College, Cambridge, 33, 40
Trinity Island, 326
Tuscany, geology of, 113, 115
Tycho Brahe, 249
Tychonic system, 41, 42
Typhæus, 331

uniformitarianism, 135n., 248
Universal Deluge; *see* Noah's flood
universal joint, 44
universal law of gravitation, 32, 54
University College, 14

Ussher, Archbishop, 100, 207

Valle, Peter de la, 190, 255
Varro, 332
velocity of light, 28, 51
Venice, Gulf of, 84 200
Venus, 253, 331
verge escapement, 40
Vergil, 307
Vernatti, Sir Philiberto, 205
Verulam, Lord, 123, 237, 253, 316
Vesuvius, 188, 189, 194, 213
*vis centrifuga*, 32, 78, 90
"Visitors," parliamentary, 13, 130n.
volcanic eruptions, 155, 186–190, 212, 310, 326, 359–361
volcanic islands, origin of, 326–327
volcanic theory, 30–31

Wadham College, 13, 14, 86
Wagners, Dr. 349–350
Waller, Richard, 4, 9, 15, 40, 54, 56, 129n., 155, 319–320
Wallis, John, 13, 85, 90, 91, 94, 105, 127, 135n., 260, 302
   dispute with Thomas Hobbes, 105
   dispute with William Holder, 105
   dispute with Hooke, 91–92
Ward, John, 129n.
Ward, Seth, 13, 16
Warren, Erasmus, 75–76
Warren, Sir W., 104
watch dispute, 40–41, 45, 132n.
water erosion, 200, 202, 204
weather-wiser, 50
Wegener, Alfred, 73, 93, 94, 252
Weinstock, Robert, ix, 78
Weir, H. E., xiii
Wendy, Lady Letice, 109
Werner, Abraham Gottlob, 126
Wernerian School, Freiberg, 112

Westfall, Richard, 6, 78, 130n., 133n., 220
Westminster Abby, 19, 58; *also see* Appendix B
Westminster School, 11
Wheeler, Sir George, 256
Whistler, Dr. 23
Whiston, William, 33, 74, 134n.
Whitehurst, John, 82, 121, 124
Wight, see Isle of Wight
Wilcock, Bruce, ix
Wilkins, John, 12, 13, 14, 15, 17, 22, 42, 55, 86, 103
Willen Church, 11, 12
Willen, 58
William III of Orange, 96
Willis, Dr. Thomas, 14
Willtshire, 76
Wilson, Craig, ix
Wilson, June, ix
Wiltshire Topographical Society, 106, 110
Wiltshire, 106–111
Winchester Cathedral, 264
wind deposition, 205
wind erosion, 204
wind guage, 38
Wing, John, ix
Witsen, Heer, 355, 356
Wood, Anthony, 109
Woodward, John, 70–71, 106
Woodwardian Museum, Cambridge, 70
Worlington, Suffolk, 75
Wormius, Olaus, 189, 301
"Worthies" window, 57, 58
**Wren, Sir Christopher**, ix, 6, 13, 14, 33, 34, 35, 36, 37, 42, 54, 78, 265
   dislike of Oldenburg, 132n.
   Monument design, 131n.
Wright, Michael, 40

York Cathedral, 264
Yorkshire stone, 237
Young, Nell, 133n., 134n.